T0192087

UNITEXT for Physics

Series editors

Michele Cini, Roma, Italy
Attilio Ferrari, Torino, Italy
Stefano Forte, Milano, Italy
Guido Montagna, Pavia, Italy
Oreste Nicrosini, Pavia, Italy
Luca Peliti, Napoli, Italy
Alberto Rotondi, Pavia, Italy
Paolo Biscari, Milano, Italy
Nicola Manini, Milano, Italy
Morten Hjorth-Jensen, Oslo, Norway

More information about this series at http://www.springer.com/series/13351

Emilio d'Emilio · Luigi E. Picasso

Problems in Quantum Mechanics

with Solutions

Second Edition

 Springer

Emilio d'Emilio
Dipartimento di Fisica
Università di Pisa
Pisa
Italy

Luigi E. Picasso
Dipartimento di Fisica
Università di Pisa
Pisa
Italy

ISSN 2198-7882
UNITEXT for Physics
ISBN 978-3-319-85109-9
DOI 10.1007/978-3-319-53267-7

ISSN 2198-7890   (electronic)

ISBN 978-3-319-53267-7   (eBook)

1st edition: © Springer-Verlag Italia 2011
2nd edition: © Springer International Publishing AG 2017
Softcover reprint of the hardcover 2nd edition 2017
This work is subject to copyright. All rights are reserved by the Publisher, whether the whole or part of the material is concerned, specifically the rights of translation, reprinting, reuse of illustrations, recitation, broadcasting, reproduction on microfilms or in any other physical way, and transmission or information storage and retrieval, electronic adaptation, computer software, or by similar or dissimilar methodology now known or hereafter developed.
The use of general descriptive names, registered names, trademarks, service marks, etc. in this publication does not imply, even in the absence of a specific statement, that such names are exempt from the relevant protective laws and regulations and therefore free for general use.
The publisher, the authors and the editors are safe to assume that the advice and information in this book are believed to be true and accurate at the date of publication. Neither the publisher nor the authors or the editors give a warranty, express or implied, with respect to the material contained herein or for any errors or omissions that may have been made. The publisher remains neutral with regard to jurisdictional claims in published maps and institutional affiliations.

Printed on acid-free paper

This Springer imprint is published by Springer Nature
The registered company is Springer International Publishing AG
The registered company address is: Gewerbestrasse 11, 6330 Cham, Switzerland

Preface to the Second Edition

The present edition of the book follows the first one by about five years.

Apart from the corrections of a few minor misprints and the updating of the precise values of the fundamental constants (and related problems 2.2, 2.3), the main reason for presenting this second edition lies in the fact that in 2015 also the companion textbook "Lectures in Quantum Mechanics" by one of us (LEP) was published by Springer: such textbook contains an extra chapter, 'Elementary Theory of Scattering', a subject not dealt with by the first edition of "Problems". As a consequence the authors felt the necessity to fill the gap.

Indeed the present edition has one more chapter, the final one, with 21 further problems. Some of them are simple. Others instead, much in the spirit of the rest of the book, offer – in pills – important subjects that often are just hinted at in textbooks. In the end the chapter proposes a number of concrete examples that provide a sort of guided tour through the main topics in potential scattering theory.

Concerning the last chapter, we find appropriate to repeat here the recommendation given in the preface to the first edition: the student should try all the problems hard and should not feel discomforted if he or she will have to resort to the solutions – he or she will, in any event, learn something more.

Pisa, December 2016

Emilio d'Emilio
Luigi E. Picasso

V

Preface to the First Edition

This book stems from the experience the authors acquired by teaching Quantum Mechanics over more than two decades.

The necessity of providing students with abundant and understandable didactic material – i.e. exercises and problems good for testing "in real time" and day by day their comprehension and mastery of the subject – confronted the authors with the necessity of adapting and reformulating the vast number of problems available from the final examinations given in previous years. Indeed those problems, precisely because they were formulated as final exam problems, were written in a language appropriate for the student who is already a good step ahead in his preparation, not for the student that, instead, is still in the "middle of the thing".

Imagining that the above necessity might be common to colleagues from other Departments and prompted also by the definite shortage, in the literature, of books written with this intent, we initially selected and ordered the 242 problems presented here by sticking to the presentation of Quantum Mechanics given in the textbook "Lezioni di Meccanica Quantistica" (ETS, Pisa, 2000) by one of us (LEP).

Over time, however, our objective drifted to become making the present collection of problems more and more autonomous and independent of any textbook. It is for this reason that certain technical subjects – as e.g. the variational method, the virial theorem, selection rules etc. – are exposed in the form of problems and subsequently taken advantage of in more standard problems devoted to applications.

The present edition – the first in English – has the advantage over the Italian one ["Problemi di Meccanica Quantistica" (ETS, Pisa 2003, 2009)] that all the material has by now been exhaustively checked by many of our students, which has enabled us to improve the presentation in several aspects.

A comment about the number of proposed problems: it may seem huge to the average student: almost certainly not all of them are necessary to have a satisfactory insight into Quantum Mechanics. However it may happen –

particularly to the student who will take further steps towards becoming a professional physicist – that he or she will have to come back, look at, and even learn again certain things ... well, we do not hide our intent: this book should not be just for passing exams but, possibly, for life.

Here are a few further comments addressed to students who decide to go through the book. Firstly, some of the problems (also according to our students) are easy, standard, and just recall basic notions learned during the lectures. Others are not so. Some of them are definitely difficult and complex, mainly for their conceptual structure. However, we had to put them there, because they usually face (and we hope clarify) questions that are either of outstanding importance or rarely treated in primers. The student should nonetheless try them using all his or her skill, and not feel frustrated if he or she cannot completely solve them. In the latter case the solution can be studied as a part of a textbook: the student will anyhow learn something new. Second, despite our effort, it may happen (seldom, we hope) that a symbol used in the text has not been defined in the immediately previous lines: it can be found in the Appendices. Our claim also is that all the problems can be solved by simple elementary algebra: the more complicated, analytic part of the calculation – when present – should take advantage of the proposed suggestions (e.g. any awkward, or even elementary, integral supposed to appear in the solution is given in the text) and should be performed in such a way as to reduce all the formulae to those given in the Appendices.

A last comment concerns the way numerical calculations are organized, particularly in the first chapters. We have written dimensionless numbers as the ratio of known quantities, e.g. two energies, two masses ... (so that a better dimensional control of what is being written is possible at a glance and at any step of the calculation – a habit the student should try hard to develop) and we have used the numerical values of these known quantities given in Appendix A: this is quicker and safer than resorting to the values of the fundamental constants.

Among the many persons – students, colleagues, families – who helped us over years in this work, three plaied a distinguished role. We are thankful to Pietro Menotti, maybe the only one of our colleagues with a more long-lasting didactic experience of the subject, for the very many comments and suggestions and for having been for one of us (EdE) a solid reference point along the twenty years of our didactic collaboration. Stephen Huggett helped us with our poor English. Bartolome Alles Salom, in addition to having gone through the whole book with an admirable painstaking patience, has a major responsibility for the appearance of the present English edition, having driven and convinced us with his enthusiasm to undertake this job.

Of course all that could have (and has not yet) been improved is the authors' entire responsibility.

Pisa, May 2011

Emilio d'Emilio
Luigi E. Picasso

Contents

7   Time Evolution

Time evolution in the Schrödinger and Heisenberg pictures; classical limit; time reversal; interaction picture; sudden and adiabatic approximations.

8   Angular Momentum

Orbital angular momentum: states with $l = 1$ and representations; rotation operators; spherical harmonics; tensors and states with definite angular momentum ($l = 1$, $l = 2$).

9   Changes of Frame

Wigner's theorem; active and passive point of view; reference frame: translated, rotated; in uniform motion; in free fall, rotating.

10   Two and Three-Dimensional Systems

Separation of variables; degeneracy theorem; group of invariance of the two-dimensional isotropic oscillator.

11   Particle in Central Field

Schrödinger equation with radial potentials in two and three dimensions; vibrational and rotational energy levels of diatomic molecules.

12   Perturbations to Energy Levels

Perturbations in one-dimensional systems; Bender–Wu method for the anharmonic oscillator; Feynman–Hellmann and virial theorems; "no-crossing theorem"; external and internal perturbations in hydrogen-like ions.

13   Spin and Magnetic Field

Spin $\frac{1}{2}$; Stern and Gerlach apparatus; spin rotations; minimal interaction; Landau levels; Aharonov–Bohm effect.

14   Electromagnetic Transitions

Coherent and incoherent radiation; photoelectric effect; transitions in dipole approximation; angular distribution and polarization of the emitted radiation; life times.

15   Composite Systems and Identical Particles

Rotational energy levels of polyatomic molecules; entangled states and density matrices; singlet and triplet states; composition of angular momenta; quantum fluctuations; EPR paradox; quantum teleportation.

16   Applications to Atomic Physics

Perturbations on the fine structure energy levels of the hydrogen atom; electronic configurations and spectral terms; fine structure; Stark and Zeeman effects; intercombination lines.

17   Elementary Potential Scattering

One-dimensional systems; time-delay; optical theorem; hard sphere; spherical barrier; spherical potential well; spherical Dirac delta–shell; resonances at low energies; bound states and virtual levels as poles of the S-matrix; Breit-Wigner formula; Jost functions; Levinson theorem; Ramsauer-Townsend effect; Yukawa potential; Bragg reflection; identical particles.

1

Classical Systems

Atomic models; radiation; Rutherford scattering; specific heats; normal modes of vibration.

1.1 According to the model proposed by J.J. Thomson at the beginning of the 20th century, the atom consists of a positive charge Ze (Z is the atomic number) uniformly distributed inside a sphere of radius R, within which Z pointlike electrons can move.

a) Calculate R for the hydrogen atom ($Z = 1$) from the ionization energy $E_I = 13.6\,\text{eV}$ (that is, the minimum work necessary to take the electron from its equilibrium position to infinity).

b) If the electron is not in its equilibrium position, it performs harmonic oscillations within the sphere. Find the value of the period. Assuming it emits radiation with the same frequency, find the wavelength λ of the emitted radiation and say in which region of the electromagnetic spectrum it falls. (For visible radiation $3900\,\text{Å} \le \lambda \le 7500\,\text{Å}$, $1\,\text{Å} = 10^{-8}\,\text{cm}$.)

c) Determine the polarization of the radiation observed in the direction of the unit vector $\hat{n}$ if: *i*) the electron oscillates in the direction of the z axis; *ii*) the electron moves in a circular orbit in the plane $z = 0$.

1.2 In Thomson's model for the hydrogen atom (see Problem 1.1) and neglecting radiation, the electron moves inside a distribution of positive charge and performs a harmonic motion that we shall assume rectilinear and with amplitude $A_0 \le R$ (R is the radius of the distribution).

a) Take radiation into account and assume $A_0 = R$. Calculate the power that should be supplied to the electron from the outside so that the amplitude of its oscillations stays constant in time. Take for R the value found in the solution of the previous problem.

If no power is supplied to the electron, the amplitude $A(t)$ of its oscillations is a decreasing function of time. We want to estimate the *lifetime* of the atom, i.e. the time τ necessary for the energy of the oscillator to be reduced by a

© Springer International Publishing AG 2017
E. d'Emilio and L.E. Picasso, *Problems in Quantum Mechanics*,
UNITEXT for Physics, DOI 10.1007/978-3-319-53267-7_1

factor e = $2.71828\cdots$, assuming $A(t)$ is a slowly varying function, namely that over a period $\Delta A \ll A$ (underdamped oscillator).

b) Write the total (kinetic + potential) energy E of the oscillator as a function of the amplitude $A(t)$; put $dE/dt = -W$, where W is the radiated power as function of E, and determine τ.

c) Compute the quality factor $Q \equiv \omega\tau$ of the oscillator ($Q = 2\pi \times$ the number of oscillations in the time interval τ).

1.3   Consider Thomson's model for the helium atom (He: $Z = 2$, $R \simeq 1\,\text{Å}$).

a) Find the equilibrium positions for the two electrons.

b) Compute the first ionization energy (the minimum work required to take just one electron to infinity) and the energy necessary to completely ionize the atom.

c) Determine the normal modes of vibration for the two electrons (it may be convenient to use the center-of-mass and relative coordinates of the two electrons).

In the dipole approximation the observed radiation is associated with the normal modes of vibration in which the electric dipole moment $\vec{d}$ is nonvanishing.

d) Say which are the frequencies (or the frequency) of the dipole radiation emitted by the atom.

1.4   Consider the scattering of α particles off gold (Au: $Z = 79$) nuclei (Rutherford scattering).

a) Assume Thomson's model for gold nuclei with $R = 1\,\text{Å}$ and neglect the presence of the electrons. Say what is the maximum value allowed for the energy of the α particles such that deflections of $180°$ are possible in a single collision.

b) In the scattering of α particles of energy $E = 10\,\text{MeV}$ the nucleus of gold behaves as if it were a pointlike charge. What conclusion can be drawn about its dimension?

1.5   In the experiments by Geiger and Marsden (1909) α particles with velocity $v_\alpha = 2 \times 10^9\,\text{cm/s}$ were scattered off a golden (atomic weight $A = 197$) foil of thickness $s = 4 \times 10^{-5}\,\text{cm}$: one particle in 2×10^4, on the average, was back-scattered (i.e. the deflection angle was greater than $90°$). We want to show that this result is not compatible with Thomson's model.

a) Knowing the mass density of gold is $19.3\,\text{g/cm}^3$, estimate the radius R of the atoms.

b) For α particles $m_\alpha c^2 \simeq 4 \times 940\,\text{MeV}$. Express their energy in MeV.

If for gold nuclei Thomson's model with the value of R determined above is assumed, the maximum deflection that α particles (with the given $v_\alpha = 2 \times 10^9$ cm/s) may undergo is $\delta \simeq 2 \times 10^{-4}$ radians.

c) Estimate the number of collisions an α particle undergoes when crossing a golden foil with the given thickness. Show that, even in the most favourable conditions, a particle cannot be deflected by an angle greater than $90°$.

1.6   A mole of a monoatomic gas contained in a box of height h is subject to the action of gravity.

a) Find how the single-particle partition function $Z(\beta)$ ($\beta \equiv 1/k_BT$) depends on β and compute the internal energy $\mathcal{U}$ of the gas.

b) Compute the molar heat C_V and its limits for $T \to 0$ and $T \to \infty$.

1.7   A mole of a gas of polar molecules, whose intrinsic dipole moment has magnitude d, is subject to a constant uniform electric field $\vec{E}$.

a) Compute the internal energy $\mathcal{U}$ of the gas and the polarizability α_e of the single molecule:

$$\alpha_e \equiv \frac{1}{N_A}\left(-\frac{1}{2}\frac{\partial^2 \mathcal{U}}{\partial E^2}\right)_{E=0} \qquad (N_A \text{ stands for Avogadro's number}).$$

b) Compute the molar heat C_V and its limits for $T \to 0$ and $T \to \infty$.

1.8   A one-dimensional model for a crystal consists of $N \simeq 10^8$ identical atoms of mass $m \simeq 30 \times 10^{-24}$ g whose equilib-

rium positions are $x_i^0 = i \times a$, $i = 1, \cdots, N$, where a is the lattice spacing of the crystal. It is assumed that each atom interacts only with its nearest neighbours, i.e. the two atoms adjacent to it; for small displacements from the equilibrium positions the interaction between any pair of atoms is approximated by an elastic force whose constant is k. It is also assumed that the two ends of the crystal are held fixed (see figure).

a) Write the Hamiltonian of the system as a function of the Lagrangian coordinates $\xi_i = x_i - x_i^0$, $i = 1,\cdots,N$, and of the respective canonically conjugate momenta $p_i = m\dot{\xi}_i$.

b) Show that the problem of finding the frequencies ω_n relative to the normal modes of vibration of the crystal (that will be explicitly found in Problem 1.9) may be traced back to that of determining the eigenvalues of the $N \times N$ real matrix:

$$\mathcal{V} = 2I_N - B; \quad B = \begin{pmatrix} 0 & 1 & \cdots & 0 & 0 \\ 1 & 0 & \cdots & 0 & 0 \\ & \vdots & \ddots & & \\ 0 & 0 & \cdots & 0 & 1 \\ 0 & 0 & \cdots & 1 & 0 \end{pmatrix}, \quad I_N = N \times N \text{ identity matrix}.$$

c) The matrix $D_{ij} = (-1)^i \delta_{ij}$ satisfies $BD = -DB$. Deduce from this that if the matrix B has the eigenvalue μ, also $-\mu$ is an eigenvalue. Show that from this and the positivity of the potential energy it follows that $0 < \omega_n < 2\omega$, where $\omega = \sqrt{k/m}$.

d) Let $\mu_{\min}$, $\mu_{\max}$ be the minimum and maximum eigenvalue of some Hermitian matrix A; let v be any unit vector ($\sum_i v_i^* v_i = 1$). Show that:

$$\mu_{\min} \le (v, Av) \le \mu_{\max}, \qquad (v, Av) \equiv \sum_{ij} v_i^* A_{ij} v_j .$$

e) Let $v_1 = \dfrac{1}{\sqrt{N}}(1, 1, \cdots, 1)$, $\quad v_2 = \dfrac{1}{\sqrt{N}}(1, -1, 1, -1, \cdots, (-1)^{N-1})$.

Exploit the preceding result with $A = \mathcal{V}$, at first with $v = v_1$, then with $v = v_2$, to show that the distances of $\omega_{\min}$ and $\omega_{\max}$ respectively from 0 and 2ω are decreasing functions of N.

f) Explain why v_1 and v_2 have been chosen to approximate respectively the minimum and the maximum eigenvalue of the matrix $\mathcal{V}$. (Hint: think of the analogous, but simpler case of two coupled pendulums.)

1.9   Consider the one-dimensional model for a crystal described in the previous problem.

a) Verify that the vectors $v^{(n)}$, $\quad 1 \le n \le N$, $\quad$ with components:

$$v_j^{(n)} = \sin\left(j \, \frac{n}{N+1} \, \pi\right), \qquad\qquad 1 \le j \le N$$

are the eigenvectors of the matrix B defined in the previous problem. Find the corresponding eigenvalues μ_n .

b) Find the characteristic frequencies ω_n of the crystal and numerically estimate the ratio $\omega_{\max}/\omega_{\min}$ for $N \simeq 10^8$.

The velocity of sound in the crystal is determined by the low frequencies ω_n, $n \simeq 1$: $v_s = \lambda_{\max} \omega_{\min}/2\pi$, where $\lambda_{\max}$ is twice the length $N \times a$ of the crystal.

c) $v_s \simeq 10^3 \div 10^4$ m/s and $a \simeq 1$ Å being known, estimate ω, $\omega_{\min}$, $\omega_{\max}$.

1.10   The one-dimensional crystal described in Problem 1.8 may be considered as the discretization of an elastic string (or of a spring endowed with mass): one obtains the continuous system when the limits $N \to \infty$, $a \to 0$, $m \to 0$, $k \to \infty$ are taken under the conditions $Na = l$, $m/a = \mu$, $ka = \tau$, l being the length of the string, μ its mass density and τ its tension. When such limits are taken $\xi_i(t) \to \xi(x,t)$, $0 \le x \le l$.

a) Write the equations of motion for the discrete variables ξ_i, $1 < i < N$, and obtain the equation for the elastic string as the limit of such equations.

b) Show that the frequencies relative to the normal modes of the discrete system tend to the frequencies of the stationary waves of the string.

Solutions

1.1

a) Inside a uniformly charged sphere, whose total charge is $Z\,e$, the electric field and the potential ($\varphi(\infty) = 0$) are:

$$\vec{E} = \frac{Ze}{R^3}\,\vec{r}, \qquad \varphi = -\frac{Ze\,r^2}{2R^3} + \frac{3Ze}{2R}, \qquad r \le R.$$

The equilibrium position for the electron is the center of the sphere, which is a position of stable equilibrium for negative charges; the minimum work to take the electron at infinity is $-(-e)\,\varphi(0)$, therefore:

$$\frac{3}{2}\frac{e^2}{R} = 13.6\text{ eV} = 2.2 \times 10^{-11}\text{ erg} \quad \Rightarrow \quad R = 1.6 \times 10^{-8}\text{ cm} = 1.6\,\text{Å}.$$

b) The restoring force is harmonic, its angular frequency is $\omega = \sqrt{e^2/m_e R^3}$. Then, rewriting ω as $(c/R) \times \sqrt{r_e/R}$, ($r_e \equiv e^2/m_e c^2$ is the classical electron radius) one has:

$$T = \frac{2\pi}{\omega} = 2\pi \times \sqrt{\frac{R}{r_e}} \times \left(\frac{R}{c}\right) = 8 \times 10^{-16}\text{ s}$$

and the wavelength of the emitted radiation is $\lambda = cT = 2.4 \times 10^{-5}\text{ cm} \simeq 2400\,\text{Å}$, in the ultraviolet region.

c) In the dipole approximation, if $\vec{d}(t)$ stands for the dipole moment of the sources and $\ddot{\vec{d}}(t) = -\omega^2 \vec{d}(t)$ (harmonic oscillator), at large distances in the direction of the unit vector $\hat{n}$ one has:

$$\vec{E}(\vec{r}, t) = \frac{\omega^2}{rc}\left(\vec{d} - (\vec{d} \cdot \hat{n})\,\hat{n}\right), \qquad \vec{d} \equiv \vec{d}(t - r/c)$$

and the polarization is given by the trajectory of the vector

$$\vec{e}(t) = \vec{d} - (\vec{d} \cdot \hat{n})\,\hat{n}$$

which is the projection of the vector $\vec{d}(t)$ onto the plane orthogonal to the direction of observation $\hat{n}$. So, if $\vec{d} \parallel \hat{z}$, in every direction $\hat{n}$ different from the direction of the z axis (where the electric field is vanishing), the radiation is linearly polarized in the plane containing $\hat{n}$ and the z axis and is orthogonal to $\hat{n}$; if the electron follows a circular trajectory in the $z = 0$ plane, the projection of the orbit onto the plane orthogonal to $\hat{n}$ is an ellipse; the latter may degenerate into a segment, if the orbit is projected onto a plane orthogonal to the orbit itself, or may be a circumference, if the orbit is projected onto a plane parallel to it. In summary, the polarization is linear in all directions orthogonal to the z axis, circular in the z direction, elliptic in the remaining cases.

1.2

a) The radiated power is given by the Larmor's formula:

$$W = \frac{2}{3} \frac{e^2 \overline{\vec{a}^2}}{c^3}$$

where $\vec{a}$ is the acceleration, $\overline{\vec{a}^2}$ is the average of $\vec{a}^2$ over one period; in the case of a harmonic oscillator of angular frequency $\omega = \sqrt{e^2/m_e R^3}$:

$$W = \frac{2e^2}{3c^3} \times \frac{1}{2}\omega^4 A_0^2 = \frac{e^2 A_0^2 \omega^4}{3c^3} = \frac{e^6 A_0^2}{3m_e^2 R^6 c^3}$$

and since $A_0 = R$ and from Problem 1.1 $e^2/R = (2/3) \times 13.6$ eV,

$$W = \frac{1}{3}\left(\frac{r_e}{R}\right)^2 \times \frac{e^2}{R} \times \frac{c}{R} = 1.7 \times 10^9 \text{ eV/s}.$$

An equal power should be supplied from the outside.

b) If A is the amplitude of the oscillations, the (kinetic+potential) energy is

$$E = \frac{1}{2}k A^2 = \frac{1}{2}m_e \omega^2 A^2 = \frac{1}{2}\frac{e^2}{R^3} A^2.$$

The power W as a function of the amplitude has been determined above:

$$W = \frac{e^6 A^2}{3m_e^2 R^6 c^3} \quad \Rightarrow \quad W = \frac{2e^4}{3m_e^2 R^3 c^3} E = \frac{2e^2\omega^2}{3m_e c^3} E.$$

Energy balance:

$$\frac{dE}{dt} = -W = -\frac{2e^2\omega^2}{3m_e c^3} E \quad \Rightarrow \quad E(t) = E_0 e^{-t/\tau}$$

$$\tau = \frac{3m_e c^3}{2e^2\omega^2} = \frac{3m_e^2 R^3 c^3}{2e^4} = \frac{3}{2}\left(\frac{R}{r_e}\right)^2 \times \frac{R}{c} = 2.6 \times 10^{-9} \text{ s}.$$

c) $Q = \omega\,\tau \simeq 2 \times 10^7$: in spite of the radiation loss and of the short lifetime, the atom is a very good oscillator.

1.3

a) The two electrons must be on the same diameter at the same distance d from the center of the spherical distribution. One must have:

$$\frac{2e^2d}{R^3} = \frac{e^2}{(2d)^2} \quad \Rightarrow \quad d = R/2 \; .$$

b) While one electron is taken away, the second electron gets closer to the center of the distribution, the center being its equilibrium position when the first is at infinity. The required work is, for example, the sum of the work made to remove one electron – the other being kept fixed – plus the (negative) work made to take the remaining electron at the center of the distribution:

$$E_{\rm I}^{(1)} = \left(-\frac{e^2}{2d} + \frac{11}{4}\frac{e^2}{R}\right) - \frac{1}{4}\frac{e^2}{R} = \frac{3}{2}\frac{e^2}{R} = 21.8 \text{ eV} \; .$$

The full ionization energy obtains by adding the work necessary to take the second electron at infinity to the work calculated above:

$$E_{\rm tot} = E_{\rm I}^{(1)} + 3\frac{e^2}{R} = \frac{9}{2}\frac{e^2}{R} = 65.2 \text{ eV} \; .$$

c) The potential energy is, up to the constant $-3Ze^2/R$ $(Z=2)$:

$$U = \frac{1}{2}\frac{Ze^2}{R^3}(r_1^2 + r_2^2) + \frac{e^2}{|\vec{r}_1 - \vec{r}_2|} \; .$$

Putting $\vec{\xi} = \vec{r}_1 - \vec{r}_2$ and $\vec{\eta} = \frac{1}{2}(\vec{r}_1 + \vec{r}_2)$ one has:

$$U = U_1(\xi_1, \xi_2, \xi_3) + U_2(\eta_1, \eta_2, \eta_3)$$

$$U_1 = \frac{e^2}{2R^3}(\xi_1^2 + \xi_2^2 + \xi_3^2) + \frac{e^2}{\sqrt{\xi_1^2 + \xi_2^2 + \xi_3^2}} \; ; \quad U_2 = \frac{2e^2}{R^3}(\eta_1^2 + \eta_2^2 + \eta_3^2) \; .$$

Correspondingly the kinetic energy is

$$E_k = \frac{1}{2}\mu\left(\dot{\xi}_1^2 + \dot{\xi}_2^2 + \dot{\xi}_3^2\right) + \frac{1}{2}M\left(\dot{\eta}_1^2 + \dot{\eta}_2^2 + \dot{\eta}_3^2\right), \quad \mu = \frac{m_e}{2}, \quad M = 2m_e \; .$$

In order to find the normal modes and their frequencies it is necessary to diagonalize the matrix of the second derivatives of U evaluated at the equilibrium position $\bar{\xi}_i$: $\bar{\xi}_1^2 + \bar{\xi}_2^2 + \bar{\xi}_3^2 = R^2$, $\bar{\eta}_i = 0$, that consists of two 3×3 blocks:

$$\frac{\partial^2 U_1}{\partial \xi_i \partial \xi_j} \Rightarrow \frac{3e^2}{R^5}\begin{pmatrix} \bar{\xi}_1^2 & \bar{\xi}_1\bar{\xi}_2 & \bar{\xi}_1\bar{\xi}_3 \\ \bar{\xi}_2\bar{\xi}_1 & \bar{\xi}_2^2 & \bar{\xi}_2\bar{\xi}_3 \\ \bar{\xi}_3\bar{\xi}_1 & \bar{\xi}_3\bar{\xi}_2 & \bar{\xi}_3^2 \end{pmatrix}; \quad \frac{\partial^2 U_2}{\partial \eta_i \partial \eta_j} \Rightarrow \frac{4e^2}{R^3}\begin{pmatrix} 1 & 0 & 0 \\ 0 & 1 & 0 \\ 0 & 0 & 1 \end{pmatrix}.$$

One way to find the eigenvalues and eigenvectors of the first matrix consists in observing that, when applied to any vector $(\alpha_1, \alpha_2, \alpha_3)$, it gives the vector $(\bar{\xi}_1, \bar{\xi}_2, \bar{\xi}_3)$ multiplied by $(3e^2/R^5)\,(\bar{\xi}_1\alpha_1 + \bar{\xi}_2\alpha_2 + \bar{\xi}_3\alpha_3)$, therefore $(\bar{\xi}_1, \bar{\xi}_2, \bar{\xi}_3)$ is an eigenvector corresponding to the eigenvalue $3e^2/R^3$ and all the vectors orthogonal to it correspond to the eigenvalue zero. Another way: performing a rotation of the axes that brings the x axis in the direction of the line joining the two charges at the equilibrium position, one has $\bar{\xi}_2 = \bar{\xi}_3 = 0$, $\bar{\xi}_1^2 = R^2$ and the matrix becomes diagonal. The normal mode belonging to the nonvanishing eigenvalue corresponds to the oscillations where only the distance between the electrons varies and its angular frequency is

$$\omega_1^2 = \frac{3e^2}{\mu R^3} = \frac{6e^2}{m_e R^3} .$$

The two vanishing eigenvalues correspond to displacements of $\vec{\xi}$ orthogonal to $\vec{\xi}$, i.e. to free rotations of the system.

The second matrix says that the center of mass of the two electrons is a three-dimensional isotropic harmonic oscillator with angular frequency

$$\omega_2^2 = \frac{4e^2}{MR^3} = \frac{2e^2}{m_e R^3} .$$

d) The dipole moment of the system is $\vec{d} = -e\,(\vec{r}_1 + \vec{r}_2) = -2e\,\vec{\eta}$, then the emitted radiation is due only to the oscillations of the center of mass: ignoring quadrupole radiation, the spectrum of He should consist of only one spectral line with frequency

$$\nu_2 = \frac{\omega_2}{2\pi} = \frac{\sqrt{2}}{2\pi} \times \sqrt{\frac{r_e}{R}} \times \frac{c}{R} = 3.6 \times 10^{15} \text{ s}^{-1}.$$

1.4

a) The potential at the center of a uniformly charged sphere of total charge $Z\,e$ and radius R is

$$\varphi(0) = \frac{3}{2}\frac{Z\,e}{R} .$$

As a consequence the α particles may be deflected by 180° only if they have a vanishing impact parameter and energy less than $2e\,\varphi(0)$:

$$E_\alpha < 3 \times 79 \times \frac{e^2}{R} = 237 \times 14.5 \text{ eV} = 3.4 \text{ keV} .$$

b) Let R_{Au} the nuclear radius, i.e. the largest between the dimension of the charge distribution and the distance within which the non-Coulombic forces (nuclear forces) are different from zero. If the nucleus behaves as if it were a pointlike charge, then R_{Au} is smaller than the least distance r_{min} reached by the α particles:

$$r_{\text{min}} = \frac{2Ze^2}{E} \quad \Rightarrow \quad R_{\text{Au}} < \frac{2Ze^2}{E} = 2Z \times \frac{e^2/R}{E} \times R = 2.3 \times 10^{-11} \text{ cm} .$$

1.5

a) A mole of atoms of gold occupies the volume $197/19.3 = 10.2 \text{ cm}^3$, then the volume per atom is $10.2/N_A = 1.7 \times 10^{-23} \text{ cm}^3$, whence:
$R = \frac{1}{2}(17 \times 10^{-24})^{1/3} \simeq 1.3 \text{ Å}$.

b) The α particle consists of two protons and two neutrons, all having a mass of about $940 \text{ MeV}/c^2$, then $(v^2/c^2 \simeq 4.4 \times 10^{-3} \ll 1)$:

$$E_\alpha = \frac{1}{2}m_\alpha v^2 = \frac{1}{2}m_\alpha c^2\left(\frac{v^2}{c^2}\right) = \frac{1}{2}\,4\times940\cdot4.4\times10^{-3} = 8.35\text{ MeV}\,.$$

c) In crossing the golden foil, each α particle interacts with about $s/2R = 4\times10^{-5}/(2.6\times10^{-8}) = 1540$ nuclei and $1540\,\delta = 0.31$ radians $\simeq 17° < 90°$.

1.6

a) The single-particle partition function is

$$Z(\beta) = \int \exp\left[-\beta\left(\frac{p^2}{2m}+mgz\right)\right] d^3p\,d^3q$$

$$= \left(\int \exp\left[-\beta\frac{p^2}{2m}\right] d^3p\right) \times \left(\frac{V}{h}\int_0^h \exp\left[-\beta\,mgz\right] dz\right).$$

Up to factors not depending on β (then irrelevant for the calculation of the internal energy), the first integral equals $(\beta)^{-3/2}$, the second equals $V\times(1-e^{-\beta\,mg\,h})/\beta\,mg\,h$, whence for the mole:

$$U = -N_A\frac{\partial\log Z(\beta)}{\partial\beta} = \frac{3N_A}{2\beta} + \frac{N_A}{\beta}\left(1-\frac{\beta\,mg\,h}{e^{\beta\,mg\,h}-1}\right).$$

b) Putting $R = N_A k_B$ and $M = N_A m$ one has:

$$C_V = \frac{\partial U}{\partial T} = \frac{3}{2}R + R\left(1-\frac{(Mgh/2RT)^2}{\sinh^2(Mgh/2RT)}\right) \rightarrow \begin{cases} \frac{5}{2}R & (T\to0) \\ \frac{3}{2}R & (T\to\infty)\,. \end{cases}$$

1.7

a) In addition to the contribution to the internal energy due to the translational and rotational degrees of freedom and given by $(\nu/2)\,RT$ (where ν is the number of degrees of freedom), there is the energy of interaction with the electric field, then:

$$U = \frac{\nu}{2}RT - N_A\frac{\partial}{\partial\beta}\left(\log\int \exp\left(\beta\,\vec{d}\cdot\vec{E}\right) d\Omega\right).$$

One obtains ($x \equiv \cos\theta$):

$$\int \exp\left(\beta\,\vec{d}\cdot\vec{E}\right) d\Omega = 2\pi\int_{-1}^{+1} \exp\left(\beta\,dE\,x\right) dx = 4\pi\frac{\sinh\beta\,Ed}{\beta\,Ed} \quad\Rightarrow$$

$$U = \frac{\nu}{2}RT + RT\left(1-\frac{ED}{RT}\coth\frac{ED}{RT}\right), \qquad D \equiv N_A d\,.$$

The polarizability is

$$\alpha_e = \frac{1}{N_A}\left(-\frac{1}{2}\frac{\partial^2 U}{\partial E^2}\right)_{E=0} = \frac{d^2}{3k_B T}\,.$$

b) $C_V = \frac{\nu}{2}R + R\left(1-\frac{(ED/RT)^2}{\sinh^2(ED/RT)}\right) \rightarrow \begin{cases} \left(\frac{\nu}{2}+1\right)R & (T\to0) \\ \frac{\nu}{2}R & (T\to\infty)\,. \end{cases}$

1.8

a) The kinetic and potential energies respectively are:

$$K = \frac{1}{2m} \sum_{i=1}^{N} p_i^2, \quad V = \frac{k}{2} \left(\xi_1^2 + (\xi_1 - \xi_2)^2 + \cdots + (\xi_{N-1} - \xi_N)^2 + \xi_N^2 \right)$$

$$= \frac{m}{2} \omega^2 \sum_{ij} \xi_i \, V_{ij} \, \xi_j \geq 0 \; .$$

b) Expanding the squares one realizes that

$$V = \begin{pmatrix} 2 & -1 & \cdots & 0 & 0 \\ -1 & 2 & \cdots & 0 & 0 \\ & \vdots & \ddots & & \\ 0 & 0 & \cdots & 2 & -1 \\ 0 & 0 & \cdots & -1 & 2 \end{pmatrix} = 2I - B \; .$$

In order to identify N uncoupled harmonic oscillators (the so called "normal modes") the matrix V has to be diagonalized (see Problem 1.3): let R be the orthogonal matrix such that $R \, V \, R^{-1} = \Delta$ with $\omega^2 \Delta_{nm} = \omega_n^2 \, \delta_{nm}$ (ω_n^2 are the eigenvalues of $\omega^2 V$). One obtains

$$H = \frac{1}{2m} \sum_n (R \, p)_n \, (R \, p)_n + \frac{m}{2} \sum_n (R \, \xi)_n \, \omega_n^2 \, (R \, \xi)_n$$

and then, putting $\eta^{(n)} = \sum_i R_{n\,i} \, \xi_i$, $\pi^{(n)} = \sum_i R_{n\,i} \, p_i = m \, \dot{\eta}^{(n)}$ ($\eta^{(n)}$ are the "normal coordinates" and $v_i^{(n)} \equiv R_{n\,i}$ the n-th eigenvector of V), one arrives at:

$$H = \sum_{n=1}^{N} \left(\frac{1}{2m} \left(\pi^{(n)} \right)^2 + \frac{1}{2} m \omega_n^2 \left(\eta^{(n)} \right)^2 \right) \; .$$

c) If $v^{(n)}$ stands for the eigenvector of B corresponding to the eigenvalue μ_n, then $D \, v^{(n)}$ is an eigenvector of B corresponding to the eigenvalue $-\mu_n$: $B \, D \, v^{(n)} = -D \, B \, v^{(n)} = -\mu_n \, D \, v^{(n)}$. Then the $\omega_n^2 = \omega^2 \, (2 - \mu_n)$ are symmetrically distributed around the point $2 \omega^2$. In addition, from the positivity of the potential energy one has $\omega_n^2 > 0$, whence $0 < \omega_n^2 < 4\omega^2$.

d) If μ_i are the eigenvalues of A and $v^{(i)}$ the corresponding normalized eigenvectors ($(v^{(i)}, v^{(j)}) = \delta_{ij}$) one has:

$$v = \sum_i c_i \, v^{(i)}, \quad \sum_i |c_i|^2 = 1; \quad A \, v = \sum_i c_i \, \mu_i \, v^{(i)}$$

$$\Rightarrow (v, A \, v) = \sum_i |c_i|^2 \, \mu_i \Rightarrow (v, A \, v) \geq \mu_{\min} \sum_i |c_i|^2 = \mu_{\min} \; .$$

Similarly $(v, A \, v) \leq \mu_{\max}$.

This result is known as the "minimax principle" and will be used in the sequel to find an upper bound to the lowest eigenvalue of Hermitian operators (variational method).

e) One has:

$$B\,v_1 = \frac{1}{\sqrt{N}}\begin{pmatrix}\frac{1}{2}\\[2pt]\vdots\\[2pt]2\\[2pt]\frac{1}{2}\\1\end{pmatrix} \quad\Rightarrow\quad (v_1,\,B\,v_1) = \frac{1}{N}\bigl[\,2(N-2)+2\,\bigr] = 2\Bigl(1-\frac{1}{N}\Bigr)$$

and likewise, or even from $v_2 = D\,v_1$, $\ (v_2,\,B\,v_2) = -2\Bigl(1-\dfrac{1}{N}\Bigr)$, then:

$$\omega_{min}^2 \le \frac{2\,\omega^2}{N}, \qquad \omega_{max}^2 \ge \omega^2\Bigl(4-\frac{2}{N}\Bigr) \quad\Rightarrow$$

$$\omega_{min} \le \omega\,\sqrt{2/N}, \qquad \omega_{max} \ge \omega\Bigl(2-\frac{1}{2N}\Bigr), \qquad N \gg 1\,.$$

f) The vector v_1, which is not is an exact eigenvector of B (and therefore of V) but it 'almost' is such, is the analogue of the oscillation mode in which two coupled pendulums keep their relative distance unchanged ("symmetrical mode"): indeed, since all the masses undergo the same displacement in the same direction, only the first and the last spring change their length ($\xi_i - \xi_j = 0$) and as a consequence the motion of the system is slow, i.e. mainly low frequencies are involved; v_2 is instead the analogue of the "antisymmetric mode" of oscillation of the two pendulums, all the springs change their lengths and high frequencies intervene in the motion of the system.

1.9

a) Once the eigenvalue equation is written, one takes advantage of the identity:

$$\sin\Bigl((j-1)\frac{n\,\pi}{N+1}\Bigr) + \sin\Bigl((j+1)\frac{n\,\pi}{N+1}\Bigr) = 2\cos\Bigl(\frac{n}{N+1}\,\pi\Bigr)\sin\Bigl(j\,\frac{n\,\pi}{N+1}\Bigr)$$

and the eigenvalues read:

$$\mu_n = 2\cos\Bigl(\frac{n}{N+1}\,\pi\Bigr), \qquad 1 \le n \le N\,.$$

b) The characteristic frequencies are obtained from $\omega_n^2 = \omega^2(2-\mu_n)$:

$$\omega_n = 2\,\omega\,\sin\Bigl(\frac{n}{N+1}\frac{\pi}{2}\Bigr)\,.$$

For $N \simeq 10^8$ one has:

$$\omega_{min} \simeq \frac{\pi}{N}\,\omega, \qquad \omega_{max} \simeq 2\,\omega \quad\Rightarrow\quad \frac{\omega_{max}}{\omega_{min}} \simeq N \simeq 10^8.$$

c) $\omega_{min} \simeq \dfrac{\pi\,\omega}{N} = \dfrac{2\pi\,v_s}{\lambda_{max}} = \dfrac{2\pi\,v_s}{2N\,a} \quad\Rightarrow\quad \omega \simeq \omega_{max} \simeq \dfrac{v_s}{a} = 10^{13} \div 10^{14}\,\mathrm{s}^{-1}$

$\omega_{min} \simeq 10^5 \div 10^6\,\mathrm{s}^{-1}.$

1.10

a) The equations of motion for the ξ_i, that can be derived either directly or from the Hamiltonian found in Problem 1.8, are:

$$\ddot{\xi}_i(t) = -\frac{k}{m}(2\,\xi_i - \xi_{i-1} - \xi_{i+1}) = \frac{\tau}{\mu}\,\frac{\xi_{i+1} - 2\,\xi_i + \xi_{i-1}}{a^2}\;.$$

In the right hand side one can recognize the discretization of the second derivative with respect to x; putting $v = \sqrt{\tau/\mu}$, one obtains, in the continuum limit,

$$\frac{\partial^2\xi(x,t)}{\partial x^2} - \frac{1}{v^2}\frac{\partial^2\xi(x,t)}{\partial t^2} = 0$$

which is the equation of the elastic string, where v is the velocity of propagation of (longitudinal) waves.

b) The frequencies of the normal modes, found in Problem 1.9, are:

$$\omega_n = 2\sqrt{\frac{k}{m}}\,\sin\Big(\frac{n}{N+1}\frac{\pi}{2}\Big) = \frac{2v}{a}\,\sin\Big(\frac{n}{N+1}\frac{\pi}{2}\Big)$$

and for $N \to \infty$

$$\omega_n \to \frac{2v}{a}\frac{n\,\pi}{2N} = 2\pi\,n\,\frac{v}{2\,l} = 2\pi\,\frac{v}{\lambda_n}\;, \qquad n = 1,2,\cdots$$

where $\lambda_n = 2\,l/n$ are the wavelengths of the stationary waves in the string.

2

Old Quantum Theory

Spectroscopy and fundamental constants; Compton effect; Bohr–Sommerfeld quantization; specific heats; de Broglie waves.

Note. *The problems in this chapter are based on what is known as Old Quantum Theory: Bohr and de Broglie quantization rules. Those situations are treated in which the results will substantially be confirmed by quantum mechanics and some problems of statistical mechanics are proposed where the effects of quantization are emphasized.*

2.1 The visible part of the electromagnetic spectrum is conventionally thus divided:

4000 Å	4680 4860	5390	5900 6200			7500
violet	blue green		yellow orange		red	

wavelengths being given in Å.

a) Convert the above wavelengths into the energies of the associated photons, expressed in eV.

2.2 The dimensionless *fine structure constant* is defined as $\alpha \equiv e^2/\hbar c$.

a) Show that the Rydberg constant $R_\infty \equiv m_e e^4/4\pi\hbar^3 c$ may be written as $R_\infty = \alpha^2/2\lambda_c$ ($\lambda_c \equiv h/m_e c$ is the Compton electron wavelength) and the ionization energy of the hydrogen atom (in the approximation of infinite proton mass) as $E_I = \frac{1}{2}\alpha^2 m_e c^2$.

According to the present day (2016) available data in the field of spectroscopy one has:

$$R_\infty = 109\,737.315\,685\,08(65)\,\text{cm}^{-1}; \qquad \alpha = 7.297\,352\,5664(17) \times 10^{-3}$$
$$m_e = 0.910\,938\,356(11) \times 10^{-27}\,\text{g}; \qquad \frac{m_e}{m_p} = 5.446\,170\,213\,52(52) \times 10^{-4}$$

and in addition, by definition, $c = 299\,792\,458\,\text{m/s}$.

© Springer International Publishing AG 2017
E. d'Emilio and L.E. Picasso, *Problems in Quantum Mechanics*,
UNITEXT for Physics, DOI 10.1007/978-3-319-53267-7_2

b) Calculate the relative standard uncertainties for the values of R_∞, α, m_e.

The Rydberg constant R_H for the hydrogen differs from R_∞ because of the finite proton mass.

c) Calculate R_H and the Planck constant h with the correct number of significant figures; also give the relative standard uncertainties of the results.

2.3 The frequency of an absorption transition from the $n = 2$ level of hydrogen was measured in a high precision spectroscopy experiments. The measured frequency was $\nu_H = 799\,191\,727\,409\,\text{kHz}$.
Owing to relativistic corrections and other minor effects, the energy levels of hydrogen are not exactly those given by the Bohr theory. Nonetheless:

a) Find the value of n for the final level.

In deuterium (the isotope of hydrogen with $A = 2$) the same transition gives rise to an absorption line whose frequency is $\nu_D = 799\,409\,184\,973\,\text{kHz}$.

b) Assuming the difference between ν_D and ν_H is mainly due to the different masses of the nuclei, calculate (with no more than three or four significant figures) the value of the ratio between the deuterium nuclear mass and the electron mass. (Use the numerical data given in Problem 2.2.)

2.4 Positronium is a system consisting of an electron and a positron (equal masses, opposite charges) bound together by the Coulomb force.

a) Calculate the value of positronium binding energy E_b (i.e. the opposite of the energy of the ground state).

One of the decay channels of positronium is the annihilation into two photons: $e^+ + e^- \to 2\gamma$ (the lifetime for this channel being $\tau_{2\gamma} \simeq 1.25 \times 10^{-10}\,\text{s}$).

b) Compute the energy and wavelength of each of the two photons in the center-of-mass reference frame of positronium.

The decay photons are revealed by means of the Compton effect on electrons.

c) Calculate the maximum energy a photon can give to an electron at rest.

d) Assume the electrons are in a uniform magnetic field $B = 10^3\,\text{G}$ with the energy found in the previous question. Calculate the radius of curvature of the trajectories described by the electrons.

2.5 Muonium is an atom consisting of a proton and a μ^- meson. It is formed via radiative capture: the proton (at rest) captures a meson (at rest) and this reaches the ground state by emitting one or more photons while effecting transitions to levels with lower energy (radiative cascade).

a) Calculate the mass of the μ^- meson, given that the maximum energy of the photons emitted in the radiative cascade is $2.5\,\text{keV}$.

b) Calculate the characteristic dimension of muonium in its ground state.

c) Say what is the resolving power $\Delta\nu/\nu$ necessary to distinguish – by measuring the frequency of the photons emitted during the radiative cascade – whether the μ^- has been captured by a proton or by a deuteron (the latter being the nucleus of deuterium: the bound state of a proton and a neutron).

2.6 The purpose of this problem is to show that any quantum state (i.e. in the present case: any energy level), relative to a one-dimensional system quantized according to the Bohr rule, occupies a (two-dimensional) volume h in phase space.

Consider a one-dimensional harmonic oscillator quantized according to the Bohr rule.

a) Compute the volume of phase space bounded by the surface of energy $E_n = n\,\hbar\,\omega$ and that of energy E_{n-1}.

Consider now a particle constrained to move on a segment of length a; its energy levels E_n are obtained by means of the Bohr quantization rule.

b) Compute the volume of phase space bounded by the two surfaces of energy E_n and E_{n-1}.

c) Show that the same result obtains for any one-dimensional system with energy levels E_n obtained through the Bohr rule. (Hint: use Stokes' theorem.)

Consider now an isotropic three-dimensional harmonic oscillator.

d) Use the Bohr quantization rule in the form $\sum_i \oint p_i\,dq_i = n\,h$ to show that the energy levels still read $E_n = n\,\hbar\,\omega$ and that the (six-dimensional) volume of phase space bounded by the surface of energy E_n has magnitude $n^3 h^3/6$.

2.7 When a system with several degrees of freedom enjoys the possibility of the *separation of variables* – i.e. there exists a choice of q's and p's such that the Hamiltonian takes the form $H = H_1(q_1, p_1) + H_2(q_2, p_2) \cdots$ – it is possible to use the Bohr–Sommerfeld quantization rules $\oint p_i\,dq_i = n_i h$ for all $i = 1, \cdots$ relative to the individual degrees of freedom.

a) Find the energy levels $E(n_1, n_2, n_3)$ of an *anisotropic* three-dimensional harmonic oscillator. Exploit the fact that its Hamiltonian can be written in the form:

$$H = \frac{p_1^2}{2m} + \frac{1}{2}m\omega_1^2 q_1^2 + \frac{p_2^2}{2m} + \frac{1}{2}m\omega_2^2 q_2^2 + \frac{p_3^2}{2m} + \frac{1}{2}m\omega_3^2 q_3^2 \,.$$

Consider now an *isotropic* three-dimensional harmonic oscillator. The number of states corresponding to a given energy level $E_n = n\,\hbar\omega$ (the "degeneracy" of the level) is the number of ways the three quantum numbers n_1, n_2, n_3 can be chosen such that $E(n_1, n_2, n_3) = E_n$.

b) Compute the degeneracy of the energy levels for an isotropic three-dimensional harmonic oscillator and the number of states with energy $E \le E_n$.

c) Find the energy levels of a particle confined in a rectangular box with edges of lengths a, b, c.

d) Still referring to the particle in the rectangular box (of volume $V = abc$), compute the number of states enclosed in the phase space volume:

$$V \times \left[(|p_1| \le p_{n_1}) \times (|p_2| \le p_{n_2}) \times (|p_3| \le p_{n_3}) \right]; \qquad p_{n_1} = \frac{n_1 h}{2a}, \qquad \text{etc.}$$

and show that, just as in Problem 2.6, the volume-per-state is h^3.

2.8 A particle of mass m in one dimension is subject to the potential $V(x) = \lambda\,(x/a)^{2k}$ with $\lambda > 0$ and k a positive integer.

a) Show that the energy levels obtained through the Bohr quantization rule are:

$$E_n = n^{2k/(1+k)} \left(\frac{h\,\lambda^{1/2k}}{\sqrt{8m}\,a\,C_k} \right)^{2k/(1+k)}, \qquad C_k = \int_{-1}^{+1} \sqrt{1 - x^{2k}}\ \mathrm{d}x .$$

b) Explicitly write the energy levels for $k = 1$ and $k = \infty$. Which well known potential does the case $k = \infty$ correspond to?

2.9 Consider a nonrelativistic electron in a uniform magnetic field $\vec{B}$, moving in a plane orthogonal to $\vec{B}$.

a) Find the energy levels (*Landau levels*) by means of the Bohr quantization rule $\oint \vec{p} \cdot \mathrm{d}\vec{q} = n\,h$, paying attention to the fact that, in presence of a magnetic field, $\vec{p} \ne m\vec{v}$.

b) Calculate the distance between energy levels for $B = 1\,\mathrm{T} = 10^4\,\mathrm{G}$.

2.10 A particle of mass m in one dimension is constrained in the segment $|x| \le \frac{1}{2}a$ and is subject to the potential:

$$V(x) = \begin{cases} 0 & |x| > \frac{1}{2}b \\ -V_0 & |x| \le \frac{1}{2}b \end{cases} \qquad b < a, \qquad V_0 > 0 .$$

a) By use of the Bohr quantization rule determine the energy levels with $E_n < 0$, the condition for the existence of at least one level with negative energy, and the number of levels with negative energy.

b) Determine the energy levels with $E_n \gg V_0$ (neglecting terms of order V_0^2/E^2).

c) Show that the corrections to the 'unperturbed' levels (i.e. those with $V_0 = 0$) found in the previous question, coincide with $-V_0 \times$ (probability of finding the particle with $|x| \leq \frac{1}{2}b$), where such a probability is the ratio between the time spent in the segment $|x| \leq \frac{1}{2}b$ and that spent in the segment $|x| \leq \frac{1}{2}a$.

2.11   Consider a gas of atoms (or molecules) with a ground state $E_0 = 0$, an excited state E_1, a third level E_x with $0 \leq E_x \leq E_1$, as well as other energy levels $E_n \gg E_1$ (a three-level system). Let us consider the contribution to internal energy and heat capacity exclusively due to the three energy levels E_0, E_x and E_1.

a) Calculate the contribution of the three levels to the internal energy as a function of the temperature T and of E_x. For what range of T is it legitimate to ignore the levels with $E_n \gg E_1$?

The three curves (a, b, c) in the figure represent (not necessarily in the same order) $C_V(T)$ for three different values of E_x: $E_x = 0$, $E_x = E_1$, $E_x = \frac{1}{10}E_1$.

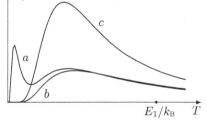

b) Identify the value of E_x for each curve and explain qualitatively their different features: more precisely, why is the maximum in c higher than in b and why are there two maxima in a?

2.12   Consider a particle of mass m constrained in a segment of size a.

a) Show that, for high values of the temperature T, the quantum partition function $Z(\beta) = \sum_n \exp\left[-\beta E_n\right]$ ($\beta \equiv 1/k_B T$) is well approximated by the classical partition function divided by the Planck constant h. Explain what 'high values of T' means.

2.13   Consider the gas consisting of the conduction electrons of a conductor with given volume V. The conductor being neutral, the ions of the crystal lattice partially screen the charge of the electrons, nearly making their repulsion vanish. In a first approximation the conduction electrons may therefore be considered as a gas of free particles.

a) In Problem 2.7 it has been shown that the phase space volume taken by each quantum state is h^3. Calculate the number of (quantum) electron states with energy $\vec{p}^2/2m_e$ less than E_F.

Due to the *Pauli principle*, at most two electrons are allowed to occupy the same quantum state; furthermore, at temperature $T = 0\,\mathrm{K}$, the gas has, compatibly with the Pauli principle, the lowest possible energy.

b) Let N be the number of conduction electrons in the volume V. Calculate the maximum energy E_F a conduction electron may have at $T = 0\,\mathrm{K}$. (E_F is known as the *Fermi energy* of the system.)

c) Under the same conditions specified above, calculate the value E of the total energy of the gas (approximate sums with integrals) and its pressure p. (For $T = 0\,\mathrm{K}$ the pressure is $p = -\partial E/\partial V$). Verify that $pV = \frac{2}{3}E$. (Actually this relation holds also for $T > 0$.)

d) Knowing that for silver the density is $10.5\,\mathrm{g/cm^3}$, the atomic weight is $A = 108$ and that one conduction electron is available for each atom, calculate the value (in atmospheres) of the pressure p at $T = 0\,\mathrm{K}$ and the value of the *Fermi temperature* $T_\mathrm{F} \equiv E_\mathrm{F}/k_\mathrm{B}$ for the electron gas.

2.14 Neutrons produced in a nuclear reactor and then slowed down ('cold' neutrons) are used in an interferometry experiment. Their de Broglie wavelength is $\lambda = 1.4\,\mathrm{\AA}$.

a) Calculate the energy of such neutrons and the energy of photons with the same wavelength (neutron mass $m_\mathrm{n} \simeq 1.7 \times 10^{-24}\,\mathrm{g}$).

The neutrons are fired at a silicon crystal and the smallest angle θ (see the figure), for which Bragg reflection is observed, is $\theta = 22°$.

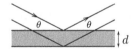

b) Calculate the distance d between the lattice plains of the crystal responsible for Bragg reflection.

c) Say for how many angles Bragg reflection can be observed.

2.15 'Ultracold' neutrons are free neutrons whose de Broglie wavelength is some hundred $\mathrm{\AA}$.

a) Calculate the speed and energy of neutrons with $\lambda = 900\,\mathrm{\AA}$ and their 'temperature' ($T \equiv E/k_\mathrm{B}$).

A way to obtain ultracold neutrons is to inject cold neutrons vertically into a tower of height $D \simeq 35\,\mathrm{m}$.

b) Say what the initial wavelength λ_i of the cold neutrons must be in order that, at the top of the tower, the final wavelength is $\lambda_\mathrm{f} = 900\,\mathrm{\AA}$.

A nonabsorbing material behaves for neutrons as a region where the potential is about $V_0 \simeq 10^{-7}\,\mathrm{eV}$ (a *repulsive* potential). For aluminium $V_0 = 0.55 \times 10^{-7}\,\mathrm{eV}$.

c) Calculate the refractive index of aluminium (i.e. the ratio between the wavelengths in vacuum and in the medium) for the neutrons with $\lambda = 900\,\text{Å}$.

Neutrons with $\lambda = 900\,\text{Å}$ impinge on the surface of a plate of aluminium.

d) Say for what range of angles (see the figure) does total reflection occur.

Solutions

2.1

a) Since for photons $\lambda[\text{Å}] \times E[\text{eV}] = 12400$ eV Å one has:

7500 Å		6200	5900		5390		4860 4680		4000
I		I	I		I		I I		I
1.65 eV		2	2.1		2.3		2.55 2.65		3.1
	red		orange	yellow		green	blue	violet	

2.2

a) One has:

$$R_\infty = \frac{m_e e^4}{4\pi\hbar^3 c} = \frac{e^4}{\hbar^2 c^2} \times \frac{m_e c}{4\pi\hbar} = \frac{\alpha^2}{2\lambda_c}; \quad E_I = R_\infty hc = \frac{\alpha^2 hc}{2\lambda_c} = \frac{1}{2}\alpha^2 m_e c^2.$$

b) $\dfrac{\Delta R_\infty}{R_\infty} = 5.9 \times 10^{-12}; \quad \dfrac{\Delta\alpha}{\alpha} = 2.3 \times 10^{-10}; \quad \dfrac{\Delta m_e}{m_e} = 1.2 \times 10^{-8}.$

c) With μ_e the reduced mass of the (e, p) system,

$$R_H = R_\infty \times \frac{\mu_e}{m_e} = \frac{R_\infty}{1 + m_e/m_p}$$

$$\frac{\Delta R_H}{R_H} = \frac{\Delta R_\infty}{R_\infty} + \frac{\Delta(m_e/m_p)}{1 + m_e/m_p} \simeq \frac{\Delta R_\infty}{R_\infty} + \Delta(m_e/m_p)$$

$$= 5.9 \times 10^{-12} + 0.05 \times 10^{-12} = 6 \times 10^{-12}$$

then R_H has 12 significant digits as R_∞: $R_H = 109\,677.583\,4063(7)$ cm^{-1}.

$$h = \frac{\alpha^2 m_e c}{2R_\infty} \quad \Rightarrow \quad \frac{\Delta h}{h} = 2\frac{\Delta\alpha}{\alpha} + \frac{\Delta m_e}{m_e} + \frac{\Delta R_\infty}{R_\infty} = 1.2 \times 10^{-8}$$

$(\Delta R_\infty/R_\infty, \Delta\alpha/\alpha \ll \Delta m_e/m_e)$, then $h = 6.626\,070\,040(80) \times 10^{-27}$ erg s.

2.3

a) $\nu_{\rm H} = R_{\rm H} c \left(\dfrac{1}{4} - \dfrac{1}{n^2} \right) \quad \Rightarrow \quad \dfrac{1}{n^2} = \dfrac{1}{4} - \dfrac{\nu_{\rm H}}{R_{\rm H} c}$.

If $n^2 \lesssim 10^4$ it is sufficient to make calculations with 6 significant digits ($R_{\rm H}$ is given in the solution of Problem 2.2):

$$\dfrac{1}{n^2} \simeq \dfrac{1}{4} - \dfrac{0.799192 \times 10^{15}}{109677 \cdot 299792 \times 10^5} = 0.007 \quad \Rightarrow \quad n^2 \simeq 143 \Rightarrow n = 12 .$$

b) As the frequencies are proportional to the reduced masses, one has:

$$\dfrac{\nu_{\rm H}}{\nu_{\rm D}} = \dfrac{1 + m_{\rm e}/m_{\rm d}}{1 + m_{\rm e}/m_{\rm p}} \quad \Rightarrow \quad \dfrac{m_{\rm e}}{m_{\rm d}} = \dfrac{\nu_{\rm H}}{\nu_{\rm D}} \left(1 + m_{\rm e}/m_{\rm p} \right) - 1$$

and, with $m_{\rm e}/m_{\rm p}$ given in the text of Problem 2.2,

$$\dfrac{m_{\rm e}}{m_{\rm d}} = \dfrac{799\,192}{799\,409} \times 1.000544 - 1 \simeq 2.724 \times 10^{-4} \quad \Rightarrow \quad \dfrac{m_{\rm d}}{m_{\rm e}} \simeq 3670 .$$

2.4

a) Positronium differs from the hydrogen atom only for the value of the reduced mass, which is a half of the mass common to electron and positron. Then:

$$E_{\rm B} = \dfrac{e^2}{4 a_{\rm B}} = \dfrac{1}{2}\, 13.6\,{\rm eV} = 6.8\,{\rm eV} .$$

b) In the center-of-mass reference frame the two photons have the same energy $m_{\rm e} c^2$ (binding energy neglected):

$$E_\gamma = m_{\rm e} c^2 = 0.51\,{\rm MeV} , \qquad \lambda = \dfrac{hc}{E_\gamma} = \dfrac{12400}{0.51 \times 10^6} = 0.024\,{\rm \AA}$$

which is the Compton electron wavelength $\lambda_c = h/m_{\rm e} c$.

c) The maximum release of energy from the photon to an electron takes place when the photon is scattered backwards ($\theta = 180°$). In this case the wavelength of the scattered photon is

$$\lambda(\pi) = \lambda(0) + 2\lambda_c = 3\lambda_c \quad \Rightarrow \quad E_\gamma^{\rm f} = \dfrac{1}{3} E_\gamma^{\rm i}$$

and as a consequence the energy released to the electron is

$$E_{\rm e} = \dfrac{2}{3} E_\gamma^{\rm i} = \dfrac{2}{3} m_{\rm e} c^2 = 0.34\,{\rm MeV} .$$

d) The momentum of the electron is

$$p = \dfrac{E_\gamma^{\rm i}}{c} - \dfrac{-E_\gamma^{\rm f}}{c} = \dfrac{4}{3} m_{\rm e} c$$

so the radius of curvature of the electron trajectory is

$$\rho = \dfrac{pc}{eB} = \dfrac{4}{3} \dfrac{m_{\rm e} c^2}{eB} = \dfrac{4}{3} \dfrac{hc}{4\pi\,\mu_{\rm B} B} = 2.3\,{\rm cm}$$

where $\mu_{\rm B} \equiv e\,\hbar/2 m_{\rm e} c = 5.8 \times 10^{-9}\,{\rm eV/G}$ is the Bohr magneton.

2.5

a) The energy levels of muonium differ from those of the hydrogen atom only because of the different value of the reduced mass. The highest energy of the emitted photons is equal to the ionization energy of muonium and is $2.5 \times 10^3/13.6 = 184$ times that of the hydrogen atom, therefore the reduced mass μ of the system $(\mu^- p)$ is 184 times the electron mass:

$$m_\mu = \frac{m_p \, \mu}{m_p - \mu} = \frac{1840 \, m_e \times 184 \, m_e}{1840 \, m_e - 184 \, m_e} = 204 \, m_e \, .$$

b) Also the dimensions of the orbits of the μ^- meson are reduced by a factor 184 with respect to those of the electron. As a consequence the size of muonium in its ground state is $a_B/184 = 0.53 \, \text{Å}/184 = 2.9 \times 10^{-3} \, \text{Å}$.

c) The reduced mass of the system $(\mu^- d)$ is $193 \, m_e$, whence:

$$\frac{\Delta \nu}{\nu} = \frac{\Delta \mu}{\mu} = \frac{193 - 184}{184} \simeq 5\% \, .$$

2.6

a) The curve described by the equation $p^2 + m^2 \omega^2 q^2 = 2m E_n$ is an ellipse whose semiaxes are $\sqrt{2m E_n}$ and $\sqrt{2E_n/m\omega^2}$, so the enclosed area is

$$A_n = \frac{2\pi E_n}{\omega} = n \, h \quad \Rightarrow \quad A_n - A_{n-1} = h \, .$$

b) In the case of a particle in a segment, the Bohr quantization rule gives $p_n = \pm n \, h/2a$, so the volume of the phase space where $E \leq E_n$ is the area of the rectangle whose base and height respectively are a and $2|p_n|$, therefore:

$$A_n = 2|p_n| a = n \, h \quad \Rightarrow \quad A_n - A_{n-1} = h \, .$$

Equivalently:

$$A_n = \int_{E \leq E_n} dq \, dp = a \int_{-\sqrt{2m E_n}}^{+\sqrt{2m E_n}} dp = 2a \, n \sqrt{\frac{h^2}{4a^2}} = n \, h \, .$$

c) The volume of the phase space where $E \leq E_n$ is

$$A_n = \left| \int_{E \leq E_n} dq \, dp \right|$$

and by Stokes theorem (the surfaces are oriented):

$$\int_{E \leq E_n} dq \, dp = -\oint_{E = E_n} p \, dq$$

(indeed, the flux of the curl of the two-dimensional vector $\vec{B}$ with components $B_q = -p$, $B_p = 0$, $\text{curl} \, \vec{B} = \partial B_p/\partial q - \partial B_q/\partial p = 1$, equals the circulation of the vector $\vec{B}$) therefore, owing to Bohr quantization rule, $A_n = n \, h$.

d) One has:

$$\oint \sum_i p_i \, dq_i = \sum_i \int_{\text{period}} p_i \, \dot{q}_i \, dt = 2 \, \frac{2\pi}{\omega} \, \overline{E_c} = \frac{2\pi}{\omega} \, E_n = n \, h \;\Rightarrow\; E_n = n \, \hbar \, \omega \,.$$

Making the change of variables $p_i = \sqrt{m\omega} \, p_i'$, $q_i = q_i'/\sqrt{m\omega}$ (the Jacobian is 1), the surface of energy E_n becomes the surface of the sphere of radius $\sqrt{2E_n/\omega}$. The volume of the sphere of radius R in d dimensions is

$$V_d = \frac{\pi^{d/2} R^d}{\Gamma\!\left(d/2 + 1\right)} \quad\Rightarrow\quad V_{E \le E_n} = \frac{n^3}{6} h^3$$

where use has been made of the fact that $d = 6$ and that, for integer k, the Euler Γ function has the value $\Gamma(k) = (k-1)!$.

The meaning of the obtained result is that the number of states of the oscillator with energy $E \le E_n$ is of the order of $n^3/6$ (approximately one state for each cell of the phase space with volume h^3).

2.7

a) As the Hamiltonian H is a separate variables one: $H = H_1 + H_2 + H_3$, its energy levels are:

$$E(n_1, n_2, n_3) = n_1 \hbar \omega_1 + n_2 \hbar \omega_2 + n_3 \hbar \omega_3 \,.$$

b) In the case of an isotropic oscillator $\omega_1 = \omega_2 = \omega_3 \equiv \omega$ and

$$E(n_1, n_2, n_3) = (n_1 + n_2 + n_3)\, \hbar\omega \equiv n\, \hbar\omega \,, \qquad n = n_1 + n_2 + n_3 \,.$$

Choosing $n_1 = n - k$, $(k = 0, \cdots n)$, n_2 and n_3 may be chosen in $k+1$ ways: $n_2 = k$, $n_3 = 0$; $n_2 = k-1$, $n_3 = 1$; $\cdots$ $n_2 = 0$, $n_3 = k$. So the degeneracy of the level E_n is

$$g_n = \sum_0^n (k+1) = \frac{(n+1)(n+2)}{2}$$

and the number of states with energy $E \le E_n$ is

$$\sum_0^n g_k = \frac{1}{2} \sum_0^n (k^2 + 3k + 2) =$$

$$= \frac{1}{2}\left(\frac{n(n+1)(2n+1)}{6} + 3\frac{n(n+1)}{2} + 2(n+1) \right) = \frac{(n+1)(n+2)(n+3)}{6} \,.$$

Compare this result – that will be confirmed by quantum mechanics – with what has been found in question d) of Problem 2.6.

c) Also in the case of a particle in a box the Hamiltonian is a separate variables one: $H = p_1^2/2m + p_2^2/2m + p_3^2/2m$, therefore:

$$E(n_1, n_2, n_3) = \frac{n_1^2\, h^2}{8ma^2} + \frac{n_2^2\, h^2}{8mb^2} + \frac{n_3^2\, h^2}{8mc^2} = \frac{h^2}{8m}\left(\frac{n_1^2}{a^2} + \frac{n_2^2}{b^2} + \frac{n_3^2}{c^2} \right) \,.$$

d) Due to $p_{n_1} = n_1 h/2a$, $p_{n_2} = n_2 h/2b$, etc. the required volume is given by $V \times 2^3 \, p_{n_1} p_{n_2} p_{n_3} = n_1 n_2 n_3 \, h^3$ and, since the number of states with quantum numbers less or equal to n_1, n_2, n_3 is $n_1 n_2 n_3$, the result follows.

2.8

a) $\oint p \, dx = \oint \sqrt{2m\left(E - V(x)\right)} \, dx = \sqrt{2mE} \oint \left(1 - \dfrac{\lambda}{E} \dfrac{x^{2k}}{a^{2k}}\right)^{1/2} dx$.

Putting $y = (\lambda/E)^{1/2k} \, x/a$,

$$\oint p \, dx = 2\sqrt{2m\,E}\; a \left(\dfrac{E}{\lambda}\right)^{1/2k} \int_{-1}^{+1} \sqrt{1 - y^{2k}} \; dy = n\,h \quad \Rightarrow$$

$$E_n^{(1+k)/2k} = n \, \dfrac{h\,\lambda^{1/2k}}{\sqrt{8m}\; a\, C_k} \quad \Rightarrow \quad E_n = n^{2k/(1+k)} \left(\dfrac{h\,\lambda^{1/2k}}{\sqrt{8m}\; a\, C_k}\right)^{2k/(1+k)} .$$

b) For $k = 1$ $C_1 = \pi/2$ and $E_n = n\,\hbar\,\sqrt{2\lambda/ma^2}$: these are the energy levels of a harmonic oscillator with $\frac{1}{2}m\omega^2 = \lambda/a^2$.
For $k = \infty$ the potential is that of an infinite potential well of width $2a$ ($x^{2k} \to 0$ for $|x| < 1$, $x^{2k} \to \infty$ for $|x| > 1$, $C_\infty = 2$ and the energy levels are $E_n = n^2 h^2/8m(2a)^2$.

Only in the two cases $k = 1$ and $k = \infty$ (up to the additive constant $\frac{1}{2}\hbar\omega$ in the case of the oscillator) the energy levels found by means of the Bohr quantization rule will turn out to be identical to those predicted by quantum mechanics: in general, the energy levels calculated using the Bohr–Sommerfeld quantization rule agree with those predicted by quantum mechanics only for large values of the quantum number n.

2.9

a) The electron follows a circular trajectory with cyclotron angular frequency $\omega_c = eB/m_e c$ (twice the Larmor frequency) and velocity $v = \omega_c r = eB\,r/m_e c$. One has:

$$\vec{p} = m_e\,\vec{v} - \dfrac{e}{c}\vec{A}; \qquad \oint \vec{p}\cdot d\vec{q} = \dfrac{2\pi}{\omega_c} m_e\, v^2 - \dfrac{e}{c}\oint \vec{A}\cdot d\vec{q}$$

and, thanks to Stokes' theorem,

$$\dfrac{e}{c}\oint \vec{A}\cdot d\vec{q} = \dfrac{e}{c}\pi\,r^2 B = \dfrac{2\pi}{\omega_c} \times \dfrac{1}{2}m_e \dfrac{e^2 B^2 r^2}{m_e^2 c^2} = \dfrac{2\pi}{\omega_c} \times \dfrac{1}{2}m_e\, v^2$$

then (the energy is only kinetic):

$$\oint \vec{p}\cdot d\vec{q} = \dfrac{2\pi}{\omega_c} \times E_n = n\,h \quad \Rightarrow \quad E_n = n\,\hbar\,\omega_c = n\,\hbar\,\dfrac{eB}{m_e c} .$$

b) The Bohr magneton is defined by (see also Problem 2.4):

$$\mu_B \equiv \dfrac{e\,\hbar}{2m_e c} = 0.93 \times 10^{-20}\,\text{erg/G} = 5.8 \times 10^{-9}\text{eV/G}$$

so the distance between Landau levels is

$$\Delta E_n = 2\mu_B B = 1.16 \times 10^{-4}\,\text{eV}\;.$$

2.10

a) For negative energies the particle is confined in the region $|x| \le \frac{1}{2}b$, whence:

$$p_n = \sqrt{2m(E_n + V_0)} = \frac{n\,h}{2b} \quad \Rightarrow \quad E_n = \frac{n^2 h^2}{8mb^2} - V_0$$

$$E_1 < 0 \quad \Rightarrow \quad V_0 > \frac{h^2}{8mb^2}\,; \qquad E_n < 0 \quad \Rightarrow \quad n < \frac{2b\sqrt{2mV_0}}{h}$$

and the number of levels is given by the integer part of $2b\sqrt{2mV_0}/h$.

b) For $E \ge 0$ the Bohr condition reads:

$$\left[(a-b)\sqrt{2mE_n} + b\sqrt{2m(E_n+V_0)}\right] = \frac{n\,h}{2}$$

that, for $E_n \gg V_0$ and up to the first order in V_0/E_n, takes the form:

$$\sqrt{2mE_n}\left[(a-b) + b\left(1 + \frac{1}{2}\frac{V_0}{E_n}\right)\right] = a\sqrt{2mE_n} + \frac{mb\,V_0}{\sqrt{2mE_n}} = \frac{n\,h}{2}$$

that gives, upon solving and neglecting the terms of order V_0^2/E_n^2,

$$E_n = \frac{n^2 h^2}{8ma^2} - \frac{b}{a}V_0\,, \qquad n \gg \frac{2a\sqrt{2mV_0}}{h}\;.$$

c) In one period, the time spent by the particle in a given segment, is twice the ratio between the length of the segment and the velocity of the particle: in order to find the result to the first order in E/V_0 we must take the velocity of the unperturbed motion (that with $V_0 = 0$), then:

$$t_b = 2\frac{b}{v}\,; \quad t_a = 2\frac{a}{v} \quad \Rightarrow \quad -V_0\frac{t_b}{t_a} = -\frac{b}{a}V_0\;.$$

2.11

a) Putting $E_0 = 0$ one has:

$$\mathcal{U} = \frac{E_x e^{-\beta E_x} + E_1 e^{-\beta E_1}}{1 + e^{-\beta E_x} + e^{-\beta E_1}}\;.$$

It is legitimate to neglect the levels with $E_n \gg E_1$ when their population is negligible with respect to that of the level E_1, namely when $e^{-\beta(E_n-E_1)} \ll 1$, i.e when $T \ll (E_n - E_1)/k_B$.

b) Note that, when $E_x = 0 = E_0$, the degeneracy of the level E_0 is 2, when $E_x = E_1$ the degeneracy of E_1 is 2, while for $E_x = \frac{1}{10}E_1$ the lowest energy level is "quasi degenerate" with E_x. So, for high temperatures $(k_B T \gg E_1)$, i.e. in the limit of equi-population, if $E_x = E_1$, the internal energy tends to a value that is twice that of the case $E_x = E_0$ ($2E_1/3$ in the first case, $E_1/3$ in the second) and almost twice ($2/1.1$) that of the case $E_x = \frac{1}{10}E_1$, and then grows more than in the other cases. For this reason the specific heat of the case $E_x = E_1$ (the curve labeled by c) is greater than in the other cases.

If $E_x = \frac{1}{10}E_1$, the level E_x becomes immediately populated (i.e. for temperatures $T \simeq E_x/k_B$) and the heat capacity grows accordingly; then, as long as $k_B T \ll E_1$, the system behaves as a two-level system, therefore C_V decreases towards zero to start a new growth when the level E_1 starts populating: in conclusion the curve labeled by a corresponds to the case when the lowest energy level is quasi degenerate: $E_x = \frac{1}{10}E_1$.

2.12

a) The classical partition function is

$$Z_{\rm cl} = \int \exp\left[-\beta p^2/2m\right] dq\, dp = 2a \int_0^\infty \exp\left[-\beta p^2/2m\right] dp$$

$$\simeq 2a \sum_n \exp\left[-\beta p_n^2/2m\right] \times \Delta p_n$$

and, if we take $p_n = n\,h/2a$, $\Delta p_n = h/2a$, the thesis follows. Let us now examine the conditions under which approximating the integral by the series is legitimate. One has:

$$Z_{\rm cl}/h = \frac{2a}{h} \int_0^\infty \exp\left[-\beta p^2/2m\right] dp = \frac{2a}{h} \sum_{n=0}^\infty \int_{p_n}^{p_{n+1}} \exp\left[-\beta p^2/2m\right] dp$$

$$= \frac{2a}{h} \sum_{n=0}^\infty \exp\left[-\beta \bar p_n^2/2m\right] \times \Delta p_n = \sum_{n=0}^\infty \exp\left[-\beta \bar p_n^2/2m\right]$$

where $p_n < \bar p_n < p_{n+1}$. The maximum of the difference with respect to the sum with p_n instead of $\bar p_n$ is obtained if one replaces $\bar p_n$ with p_{n+1}: in this case the two sums differ by the first term that equals 1. The quantum partition function and $Z_{\rm cl}/h$ differ by a function of β (the $\bar p_n$ do depend on β) bounded by 0 and 1; since $\int e^{-ax^2}\, dx = \sqrt{\pi/a}$, one has:

$$Z_{\rm cl}/h = \frac{a}{h}\sqrt{\frac{2\pi m}{\beta}}$$

and in conclusion, if $Z_{\rm cl}/h \gg 1$ namely for $\beta \ll ma^2/h^2$ ($k_B T \gg h^2/ma^2$), the difference is negligible.

2.13

a) Since the energy of the electrons is $\vec p^2/2m_e$, putting $p_F = \sqrt{2m_e E_F}$ one has:

$$\int_{E \leq E_F} d^3q\, d^3p = V \times 4\pi \int_{p \leq p_F} p^2 dp = \frac{4\pi}{3}V\left(2m_e E_F\right)^{3/2}.$$

The number of states is $n = \dfrac{4\pi V}{3h^3}(2m_e E_F)^{3/2}$.

b) The energy is a minimum if all the states with energy less than E_F are occupied and there are two electrons per state, so:

$$N = 2\,n = 2 \times \frac{4\pi V}{3h^3}(2m_e E_F)^{3/2} \quad \Rightarrow \quad E_F = \frac{h^2}{8m_e}\left(\frac{3N}{\pi V}\right)^{2/3} .$$

c) The total energy is

$$E = 2 \times \sum_{n_1,n_2,n_3} E_{n_1,n_2,n_3} = 2 \times \sum_{n_1,n_2,n_3} \frac{1}{2m_e}(p_{n_1}^2 + p_{n_2}^2 + p_{n_3}^2)$$

where $p_{n_1} = n_1 h/2a$, etc. and the sum is performed on all the quantum numbers such that $E_{n_1,n_2,n_3} \le E_F$. The points $\vec{p} = (p_{n_1}, p_{n_2}, p_{n_3})$ in the octant $p_i > 0$ ($i = 1,\ 2,\ 3$) of momentum space give rise to a lattice with unit steps $h/2a$, $h/2b$, $h/2c$. So, replacing the sum with the integral:

$$E = 2 \times \frac{1}{8}\frac{8V}{h^3}\, 4\pi \int_0^{p_F} \frac{p^2}{2m_e}\, p^2\, dp = \frac{4\pi V}{5m_e h^3}\, p_F^5 = \frac{3\,h^2 N}{40\,m_e}\left(\frac{3N}{\pi V}\right)^{2/3}$$

and since E is a homogeneous function of V of order $-2/3$:

$$pV = -V\frac{\partial E}{\partial V} = \frac{2}{3}E \quad \Rightarrow \quad p = \frac{2}{3}\frac{E}{V} = \frac{\pi h^2}{60\,m_e}\left(\frac{3N}{\pi V}\right)^{5/3} .$$

d) A mole of atoms of silver occupies the volume $108/10.5 \simeq 10\,\text{cm}^3$, so:

$$N/V \simeq 6 \times 10^{22}\,\text{cm}^{-3} \quad \Rightarrow \quad p \simeq 2 \times 10^{11}\,\text{dyn/cm}^2 = 2 \times 10^5\,\text{atm} .$$

$$E_F = 9 \times 10^{-12}\,\text{erg} = 5.6\,\text{eV} \quad \Rightarrow \quad T_F = 6.5 \times 10^4\,\text{K} .$$

2.14

a) While for a photon:

$$E_\gamma = h\nu = \frac{hc}{\lambda} \simeq \frac{12400\ \text{eV Å}}{1.4\ \text{Å}} = 8.9 \times 10^3\,\text{eV} ,$$

for a particle of mass $m \neq 0$, if m_e stands for the electron mass:

$$\lambda = \frac{h}{p} = \frac{h}{\sqrt{2mE}} \quad \Rightarrow \quad \lambda\sqrt{E} = \frac{hc}{\sqrt{2m_e c^2}}\sqrt{\frac{m_e}{m}} = 12.4\sqrt{\frac{m_e}{m}}\ \text{Å (eV)}^{1/2}$$

and, as a consequence, for neutrons of mass $m_n \simeq 1.7 \times 10^{-24}\,\text{g} = 1840\,m_e$ one has:

$$E_n = \left(\frac{12.4}{1.4}\right)^2 \times \frac{1}{1840} \simeq 4.3 \times 10^{-2}\,\text{eV} .$$

b) From the Bragg relation $2\,d\sin\theta = n\lambda$ with $n = 1$ one obtains:

$$d = \frac{\lambda}{2\sin\theta} \simeq 1.9\,\text{Å} .$$

c) The number of angles for which there occurs Bragg reflection is the integer part of $2d/\lambda$, namely 2.

2.15

a) $v = \dfrac{p}{m_n} = \dfrac{h}{m_n \lambda} = 4.3\,\text{m/s}$; $E_f = \dfrac{h^2}{2m_n \lambda^2} = 10^{-7}\,\text{eV}$; $T = 1.1 \times 10^{-3}\,\text{K}$.

b) The difference between the initial and final kinetic energy is $3.7 \times 10^{-6}\,\text{eV}$, that practically is the same as the initial energy; so, if the energy is expressed in eV and the wavelength in Å (see Problem 2.14), one has:

$$\lambda_i = \frac{12.4}{\sqrt{E_i}}\sqrt{\frac{m_e}{m_n}} \simeq 150\,\text{Å}$$

or, since λ is inversely proportional to the square root of the energy,

$$\lambda_i = \lambda_f \sqrt{E_f/E_i} = 900\sqrt{10^{-7}/3.8 \times 10^{-6}} \simeq 150\,\text{Å}\;.$$

c) In vacuum $\lambda_0 = h/p_0 = h/\sqrt{2m_n E}$; in the medium $\lambda = h/p = h/\sqrt{2m_n(E - V_0)}$, therefore $n \equiv \lambda_0/\lambda = \sqrt{1 - V_0/E} = 0.67$ (note that $n < 1$).

d) Note that, contrary to the convention used in optics, here the incidence angle is measured from the reflection plane. So total reflection occurs for angles $\theta < \theta_r$ where $\cos\theta_r = n$, namely $\theta < 48°$. Equivalently, if $\vec{p}_0$ is the momentum of the neutron in vacuum and $\vec{p}$ is the momentum in the medium, taking the y axis normal to the surface and the x axis in the plane containing the incident beam, one has:

$$E = \frac{p_{0x}^2}{2m_n} + \frac{p_{0y}^2}{2m_n} = \frac{p_x^2}{2m_n} + \frac{p_y^2}{2m_n} + V_0\;.$$

Since $p_x = p_{0x}$ and $p_{0y} = p_0 \sin\theta$, there occur both reflection and refraction when $p_y^2 > 0$, i.e. $E\sin^2\theta > V_0$, therefore $\sin^2\theta_r = V_0/E$, namely $\cos\theta_r = \sqrt{1 - V_0/E}$.

3

Waves and Corpuscles

Interference and diffraction with single particles; polarization of photons; Malus' law; uncertainty relations.

Note. *The exercises in this chapter regard the fundamental concepts at the basis of quantum mechanics: Problem 3.3 exposes all the "interpretative drama" of quantum mechanics, which is why its somewhat paradoxical aspects are discussed in detail in the solution.*

3.1 A beam of monochromatic light, with wavelength $\lambda_0 = 6 \times 10^3$ Å (sodium yellow light), enters from the left the Mach–Zehnder interferometer represented in the figure. The mirrors s_1 and s_4 are semi-transparent: s_1 transmits the fraction a^2 of the intensity of the incoming light and reflects the fraction

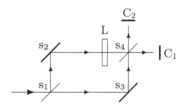

b^2 (a, b positive, $a^2 + b^2 = 1$), whereas s_4 transmits and reflects the 50% of the incident intensity. By varying the inclination of the thin glass plate L, whose width is d (and therefore the length of the optical path), it is possible to vary the phase of the wave that follows the path $s_1 \rightarrow s_2 \rightarrow s_4$, with respect to that of the wave that follows $s_1 \rightarrow s_3 \rightarrow s_4$.

a) If n is the refractive index of glass, calculate the phase difference φ_L due to the plate. If $n = 1.2$ and $d = 1$ mm, by what angle α should the plate be rotated in order to have a variation of the phase difference $\delta\varphi = 2\pi$?

b) Given the intensity I_0 of the incoming light, calculate the intensity $I_1(\varphi)$ ($\varphi \equiv \varphi_{s_1 \rightarrow s_2 \rightarrow s_4} - \varphi_{s_1 \rightarrow s_3 \rightarrow s_4} \equiv \varphi_1 - \varphi_2$) of the light seen by the counter C_1 and the 'visibility' (or contrast) V of the interference fringes:

$$V \equiv (I_1^{\max} - I_1^{\min})/(I_1^{\max} + I_1^{\min}) .$$

c) Calculate the intensity $I_2(\varphi)$ of the light seen by C_2.

Assume that instead of L there is a plate of absorbing material that completely absorbs the light in the path $s_2 \rightarrow s_4$.

d) Calculate the intensities of the light seen by C_1 and C_2.

© Springer International Publishing AG 2017
E. d'Emilio and L.E. Picasso, *Problems in Quantum Mechanics*,
UNITEXT for Physics, DOI 10.1007/978-3-319-53267-7_3

3.2   The intensity of the light ($\lambda = 6000\,\text{Å}$) entering the interferometer of Problem 3.1 (without the glass plate L) is $I_0 = 5 \times 10^{-3}\,\text{erg s}^{-1}\text{cm}^{-2}$ and the section of the beam is $25\,\text{mm}^2$.

a) Calculate the average number per second N of photons entering the interferometer. Assuming that the distance between s_1 and the counters is 60 cm, how many photons are, on the average, inside the interferometer?

b) With the notations of Problem 3.1, let $\varphi = 0$ and $a^2 = 0.7$, $b^2 = 0.3$ the values of the transmission and reflection coefficients of the semi-transparent mirror s_1. What are the average counting rates N_1, N_2 of the counters C_1 and C_2 and the respective standard deviations ΔN_1, ΔN_2?

3.3   Consider the experiment with the Mach–Zehnder interferometer described in Problems 3.1 and 3.2, with the following variation: in the path $s_2 \to s_4$ an optical fiber is inserted so that its length is increased by some tens of meters. The photons entering the interferometer are taken from a sodium lamp and are emitted in the transition from the first excited level to the ground state. The lifetime of the excited level, i.e. the average decay time, is $\tau = 1.6 \times 10^{-8}\,\text{s}$.

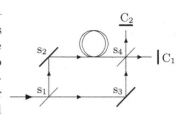

a) Calculate the average 'length' of the emitted photons.

b) Always assuming that the transmission and reflection coefficients of the semi-transparent mirror s_1 are a^2 and b^2, and that $N \gg 1$ is the rate of photons entering the apparatus, calculate the average counting rates at C_1 and C_2.

3.4   The Bonse–Hart interferometer for neutrons is similar to the Mach–Zehnder interferometer for light (the mirrors are silicon crystals by which neutrons are reflected à la Bragg).

Neutrons (mass $m_n = 1.7 \times 10^{-24}\,\text{g}$), whose de Broglie wavelength is $\lambda = 1.4\,\text{Å}$, are sent horizontally in a Bonse–Hart interferometer positioned in such a way that their paths are in a vertical plane. The difference in height between the paths $s_2 \to s_4$ and $s_1 \to s_3$ is d (see figure). Assume the propagation of the neutrons between the mirrors is rectilinear.

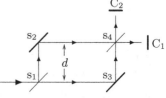

a) Let $k \equiv 2\pi/\lambda$ be the neutron wavenumber and g the gravitational acceleration. Neglecting terms of order g^2, calculate the difference $\Delta k \equiv k - k'$ between the wavenumbers in the paths $s_1 \to s_3$ and $s_2 \to s_4$ due to the difference in potential energy.

b) Assume the paths $s_1 \to s_3$ and $s_2 \to s_4$ both have length L and that also the paths $s_1 \to s_2$ and $s_3 \to s_4$ are identical. Calculate the phase difference φ between the de Broglie waves that arrive at s_4 via the 'low' path ($s_1 \to s_3 \to s_4$) and via the 'high' path ($s_1 \to s_2 \to s_4$). Calculate φ when $d = 3$ cm, $L = 7$ cm .

The interferometer is rotated around the direction of the incident beam (the direction $s_1 \to s_3$) in such a way that the difference in height between the paths $s_2 \to s_4$ and $s_1 \to s_3$ is varied.

c) Calculate the number of maxima in the countings at C_1 ('number of fringes') for a rotation from $-30°$ to $+30°$ with respect to the vertical plane.

3.5 A Bonse–Hart interferometer for neutrons (see Problem 3.4) with the neutron paths in a horizontal plane, is at rest in a frame rotating with constant angular velocity ω around a vertical axis (z axis). The Hamiltonian of a free particle of mass m in the rotating frame is

$$H = \frac{\vec{p}^2}{2m} - \omega\, L_z\,, \qquad L_z = q_1 p_2 - q_2 p_1 \,.$$

a) Write the Hamilton equations (note that $\vec{p} \neq m\,\vec{v}$) and verify that they give rise to the equation of motion $\vec{F} = m\,\vec{a}$, $\vec{F}$ being the sum of centrifugal and Coriolis forces.

Let l_x be the length of the paths $s_1 \to s_3$, $s_2 \to s_4$ and l_y that of the paths $s_1 \to s_2$, $s_3 \to s_4$ (see Problem 3.4). The rotation of the interferometer causes a phase difference φ between the de Broglie waves that arrive at the semi-transparent mirror s_4 via the different paths $\gamma_1 = s_1 \to s_2 \to s_4$ and $\gamma_2 = s_1 \to s_3 \to s_4$. (This is the Sagnac effect for neutrons.)

b) Let E be the energy of the neutrons in the rotating frame. Calculate the wavenumber $k = p/\hbar = 2\pi/\lambda$ in the different paths of the interferometer to first order in ω: to this end assume that the propagation of the neutrons between the mirrors is rectilinear.

c) Calculate the phase difference φ to first order in ω. What is the numerical value of φ if ω is the angular velocity of the Earth and $l_x \times l_y = 9$ cm^2 ?

[The Sagnac effect was first measured for light in 1925 (Michelson et al. using an interferometer with $l_x = 613$ m and $l_y = 340$ m) and for neutrons in 1979 (Werner et al. with the above given area).]

3.6 A pointlike light source S is located at a distance D from a screen in which a small circular pin hole of diameter a has been made. The emitted light, whose wavelength is λ, is recorded on a photographic plate parallel to the screen a distance L away from it.

a) Making use of diffraction theory and not paying attention to factors of the order of unity, estimate the dimension of the spot produced on the photographic plate.

Assume now that only one photon arrives on the photographic plate.

b) What does one observe on the plate: a faint spot whose dimension is that calculated in a), or a more intense and (practically) pointlike spot?

c) Assume that only the source – screen distance D is known, whereas the position of the source along the dashed line in the figure (in any event not too far from the central position) is not known. Assume also that only one photon arrives on the photographic plate. With what uncertainty is the position of the source known? What if N photons arrive?

3.7 Consider Young's double-slit interference experiment carried out by sending one photon at a time. A system has been suggested to establish which slit does any photon, arriving in the central interference fringe, come from: a hole is made in the screen S (see the figure), in correspondence with the central fringe and of 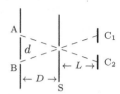 the same dimension, so that the photons coming from slit A trigger counter C_2, while those coming from slit B trigger C_1. In this way the interference pattern is not destroyed, for all the photons that trigger the counters belong to the central fringe. (The same system may be used for other fringes.) Both the distance D between S and the slits, and the distance L from S to the counters are much greater than the distance d between the slits and the width a of the central fringe. Assume finally that the dimension of the slits is small compared to d.

a) Calculate the width a of the central fringe (i.e. the distance between two adjacent points where the intensity vanishes).

b) Making use of diffraction theory, show that the proposed device does not allow one to establish which slit does any single photon come from.

3.8 The electric field of a plane wave propagating in the direction of the z axis is described as:

$$\vec{E}(z,t) = \frac{E_0}{2} \left[(\cos\vartheta\, e^{i\,\varphi_1}\, \hat{e}_1 + \sin\vartheta\, e^{i\,\varphi_2}\, \hat{e}_2)\, e^{i\,(kz-\omega t)} + c.c. \right]$$

where $\hat{e}_1$ and $\hat{e}_2$ are the unit vectors along the x and y axes, and c.c. stands for complex conjugate. The polarization state, either of the wave or of the single photon, is described by the complex unit vector:

$$\hat{e}_{\vartheta\varphi} \equiv \cos\vartheta\, \hat{e}_1 + \sin\vartheta\, e^{i\,\varphi}\, \hat{e}_2, \qquad \varphi = \varphi_2 - \varphi_1; \qquad \hat{e}_{\vartheta\varphi}^* \cdot \hat{e}_{\vartheta\varphi} = 1.$$

a) Write the vectors that describe the states of circular polarization and show that – up to an overall inessential phase factor – they do not depend on how the x and y axes, orthogonal to the direction of propagation, are chosen.

A polaroid is a plastic sheet capable of transmitting the component of the light–wave polarized parallel to a given direction (transmission axis) and of absorbing the orthogonal component. We assume that the both the transmission and the absorbtion coefficients are 100%.

b) Calculate what fraction of the intensity of a wave, whose electric field is the one given above, is transmitted by a polaroid sheet whose transmission axis is in the direction of the x axis (Malus' law). What is the probability that a single photon is transmitted by the polaroid? What is the probability that a single photon in a circular polarization state is transmitted by the polaroid? What is its polarization state after it has been transmitted?

Photons in the polarization state $\hat{e}_1$ impinge on a succession of N parallel polaroid sheets oriented in such a way that the transmission axis of the first makes an angle $\alpha = \pi/2N$ with the direction $\hat{e}_1$, that of the second 2α and so on (the axis of the last polaroid having the direction of $\hat{e}_2$).

c) Calculate the probability for a single incident photon to be transmitted by all the polaroids in the following cases: $N = 2$ ($\alpha = \pi/4$), $N = 90$, $N \to \infty$.

d) Calculate the same probability as in c) for a circularly polarized incident photon.

3.9   Consider the Young's interference experiment with two slits A and B, performed with monochromatic light produced by a source S. The light is polarized, but its polarization state is not known. Let $\hat{z}$ be the direction of propagation of the light (see figure). A polaroid sheet $\mathrm{P_S}$, whose transmission axis is parallel to the plane containing the slits, is interposed between the source and the screen with the slits.

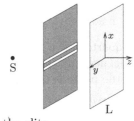

a) It is observed that as the polaroid is rotated around the z axis, neither the position nor the intensity of the interference fringes changes. What can be said about the polarization of the light emitted by S?

The polaroid sheet $\mathrm{P_S}$ is removed.

b) How does the interference pattern change?

Two polaroid sheets $\mathrm{P_A}$ and $\mathrm{P_B}$ are now placed just in front of the two slits, with the transmission axes both parallel to the x-y plane. The axes of $\mathrm{P_A}$

and P_B make respectively angles α and β with the x axis. (Remember, P_S is no longer there.) The electric field on the plate L near $y = 0$ is

$$\vec{E}(x,t) = E_0\,\hat{e}_\alpha \cos\left(\omega t + \varphi_A(x)\right) + E_0\,\hat{e}_\beta \cos\left(\omega t + \varphi_B(x)\right)$$

where $\hat{e}_\alpha$ and $\hat{e}_\beta$ respectively are the unit vectors describing the rectilinear polarization of the light that is transmitted by P_A and P_B, and $|\alpha - \beta| \le \pi/2$. The phases $\varphi_A(x)$ and $\varphi_B(x)$ are proportional to the length of the optical paths from the two slits.

c) Calculate the intensity $I(x)$ on the plate L and the visibility V – or contrast – of the interference fringes (see Problem 3.1). For what values of α and β is V a maximum? For what values is it a minimum?

Keeping P_A and P_B in place, a further polaroid sheet P_L is interposed some-where between the slits and the plate L: its transmission axis makes an angle γ with the x axis and an acute angle with respect to both $\hat{e}_\alpha$ and $\hat{e}_\beta$: $|\alpha - \gamma| \le \pi/2$, $|\beta - \gamma| \le \pi/2$.

d) Calculate V as a function of γ and find for what value of γ V is a maximum. Consider the particular case when $\hat{e}_\alpha = \hat{e}_1$, $\hat{e}_\beta = \hat{e}_2$ ($\hat{e}_1$ and $\hat{e}_2$ being the unit vectors of the x and y axes). Say what difference is observed on the plate L in the two cases: i) P_L is absent, ii) P_L is present and its axis makes $45°$ with the x axis.

3.10 In an experiment a beam of light is available about which it is known that all the photons are in the same state of circular polarization, but their chirality (i.e. whether they are either left or right) is not known.

a) Can chirality be determined by making measurements with an apparatus made only of polaroid sheets and counters?

A quarter-wave plate is a birefringent crystal (e.g. calcite or quartz) of suitable thickness that induces the following transformation on the polarization state of the photons that go through it:

$$\cos\vartheta\,\hat{e}_1 + \sin\vartheta\,e^{i\varphi}\,\hat{e}_2 \;\to\; \cos\vartheta\,\hat{e}_1 + \sin\vartheta\,e^{i(\varphi+2\pi/4)}\,\hat{e}_2$$

where $\hat{e}_1$ and $\hat{e}_2$ stand for the two states of rectilinear polarization orthogonal to each other and parallel to the so called fast and slow axes of the crystal.

b) Show that, by using a quarter-wave plate and then making *only one mea-surement* with a polaroid sheet on *one single photon*, the chirality of the beam can be established.

An arbitrarily large number of photons, all in the same unknown state of polarization $\hat{e}_{\vartheta\varphi} \equiv \cos\vartheta\,\hat{e}_1 + \sin\vartheta\,e^{i\varphi}\,\hat{e}_2$, is available.

c) By exploiting Malus' law, which information can one obtain about the state of polarization by means of measurements that make use only of polaroid sheets (and counters)? What if also a quarter-wave plate is used?

3.11   A plate of an optically active substance induces a rotation of the direction of the (rectilinear) polarization of the photons crossing it:

$$\cos\vartheta\,\hat{e}_1 + \sin\vartheta\,\hat{e}_2 \;\to\; \cos(\vartheta+\alpha)\,\hat{e}_1 + \sin(\vartheta+\alpha)\,\hat{e}_2$$

and has a linear behaviour. (As usual, $\hat{e}_1$ and $\hat{e}_2$ stand for mutually orthogonal rectilinear polarization states.)

a) Given the basis $(\hat{e}_1,\hat{e}_2)$, write the matrix U that implements the linear transformation induced by the optically active plate on a generic polarization state.

b) Which polarization states are left unaltered by the plate?

c) Which polarization states are left unaltered by a quarter-wave plate (see Problem 3.10)?

d) When photons in an arbitrary state of polarization are sent through either a quarter-wave plate or an optically active plate, is one making a measurement on the photons?

3.12   Uncertainty relations allow one to re-establish for particles some results of interference and diffraction theory in optics. Particles (e.g. neutrons, electrons, photons ...) orthogonally cross a screen in which a pin hole of width a has been made.

a) By using the uncertainty relation in the form $\Delta y\,\Delta p_y \simeq h$, show that the particles that go through the hole come out 'diffracted' with an angular width of the order of λ/a, where $\lambda = h/p \lesssim a$ is the de Broglie wavelength (and the y axis is orthogonal to the direction of the incoming particles).

Consider now a double-slit interference experiment (Young's experiment), with a the width of the slits and $d \gg a$ the distance between them.

b) State whether the uncertainty on the y component of the momentum of the photons that have crossed the screen with the slits is $\Delta p_y \simeq h/d$ or $\Delta p_y \simeq h/a$ and accordingly write down the product $\Delta y\,\Delta p_y$.

3.13   Consider Young's double-slit interference experiment with the following variation: just after one slit there are two parallel mirrors with an inclination of $45°$ with respect to the incoming photons (see the figure). The upper one (thinner in the figure) can move in the direction orthogonal to itself, so that when it is hit by a photon it recoils. The photons are then revealed on a photographic plate. For each photon arriving at the photographic plate it is possible to decide which slit has been crossed by revealing whether the mirror has recoiled or not ("which way" experiment).

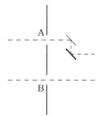

a) If $\Delta q^{(m)}$ is the uncertainty in the position of the mirror (see the enlarged view in the figure), say which is the uncertainty ΔL in the length of the optical path from the upper slit to the screen where the arrival of the photons is recorded.

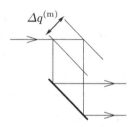

b) Which condition must $\Delta q^{(m)}$ satisfy in order that the interference pattern is not destroyed?

c) If $\Delta p^{(m)}$ is the uncertainty in the momentum of the mirror and p the momentum of the incoming photons, which condition must $\Delta p^{(m)}$ satisfy in order that the recoil of the mirror can be detected?

d) Are the conditions on $\Delta q^{(m)}$ and $\Delta p^{(m)}$ determined above consistent with the Heisenberg uncertainty relations?

3.14   The photons emitted in the transitions between the energy levels of a given atom give rise to the atomic spectrum. The spectral lines, i.e. the observed frequencies ν_i, have a nonvanishing width $\Delta\nu_i$ called "natural width of the line", due to the lifetime of the transition, namely the average decay time τ (see Problem 1.2).

a) If τ is the lifetime of a given transition, what is the average length of the emitted photons, i.e. the uncertainty Δx in their position?

The yellow light of a sodium lamp has a wavelength $\lambda \simeq 5890\,\text{Å}$ and is emitted in the transition from the first excited level to the lowest energy level of sodium atoms. The lifetime of the transition is $\tau = 1.6 \times 10^{-8}\,\text{s}$ (see Problem 3.3).

b) Calculate the uncertainty Δp in the momentum of the emitted photons and, as a consequence, the uncertainty $\Delta\nu$ in their frequency. What is the value of the frequency ν?

The value of $\Delta\nu$ calculated above is the natural width of the yellow line of sodium.

c) Calculate $\Delta\nu/\nu = \Delta\lambda/\lambda$ and the quality factor $\mathcal{Q}$ of the line (see Problem 1.2).

d) Calculate the energy E and the uncertainty ΔE for the emitted photons; express your results in eV.

Solutions

3.1

a) If λ is the wavelength in glass, owing to $n = \lambda_0/\lambda$, one has:

$$\varphi_L = \frac{2\pi}{\lambda} d - \frac{2\pi}{\lambda_0} d = \frac{2\pi(n-1)}{\lambda_0} d .$$

The requirement $\delta\varphi = 2\pi$ entails:

$$1 = (n-1)\frac{d}{\lambda_0}\left(\frac{1}{\cos\alpha} - 1\right) \;\Rightarrow\; 1 \simeq (n-1)\frac{d\alpha^2}{2\lambda_0} \;\Rightarrow\; \alpha \simeq 4.4°.$$

b) Let $E(x,t) = E_0 \cos(k\,x - \omega t)$ be the electric field of the incident wave (the polarization is not relevant). The wave arriving at the counter C_1 is

$$\frac{E_0}{\sqrt{2}}\big(a\cos(\omega t + \varphi_1) + b\cos(\omega t + \varphi_2)\big), \qquad \varphi_1 - \varphi_2 = \varphi$$

and, as a consequence, the intensity at C_1 is

$$I_1(\varphi) = \frac{1}{2}I_0\big(a^2 + b^2 + 2ab\cos\varphi\big) = \frac{1}{2}I_0\big(1 + 2ab\cos\varphi\big)$$

$$\frac{I_1^{max} - I_1^{min}}{I_1^{max} + I_1^{min}} = 2ab .$$

c) The intensity at C_2 can be simply calculated as:

$$I_2(\varphi) = I_0 - I_1(\varphi) = \frac{1}{2}I_0\big(1 - 2ab\cos\varphi\big) .$$

Therefore the wave arriving at C_2 is

$$\frac{E_0}{\sqrt{2}}\big(a\cos(\omega t + \psi_1) + b\cos(\omega t + \psi_2)\big), \qquad \psi_1 - \psi_2 = \varphi \pm \pi$$

i.e. there is an extra phase difference of $180°$ with respect to that between the two components arriving at C_1: this is due to the fact that along the path $s_1 \to s_2 \to s_4 \to C_2$ there occur three reflections, whereas along the path $s_1 \to s_3 \to s_4 \to C_2$ there occur one reflection and two transmissions; instead, in the case of the light arriving at C_1, there occur two reflections and one transmission along both paths, therefore in a reflection there is an additional phase change of $\pm 90°$ with respect to a transmission.

d) If at s_4 only the light from the path $s_1 \to s_3 \to s_4$ arrives, the same intensity $\frac{1}{2}a^2 I_0$ is measured by both counters.

3.2

a) The energy of a photon is $E_\gamma \simeq 12000/6000 = 2\,\text{eV} = 3.2 \times 10^{-12}\,\text{erg}$, so the number of photons per second is $N = 5 \times 10^{-3} \times 0.25/(3.2 \times 10^{-12}) = 4 \times 10^8$. One photon every 2.5×10^{-9} seconds enters the interferometer and each photon spends 2×10^{-9} seconds in it, so the average number of photons inside the interferometer is 0.8.

b) $\overline{N}_1 = \frac{1}{2}N(1 + 2ab) = 2 \times 10^8 \cdot 1.92 = 3.83 \times 10^8$
$\overline{N}_2 = N - \overline{N}_1 = 0.17 \times 10^8$.

The probability distribution relative to the counting of the counters is the binomial distribution with probability $p = \overline{N}_1/N = \frac{1}{2}(1 + 2ab)$ that each single photon be revealed by C_1 and probability $q = 1 - p = \frac{1}{2}(1 - 2ab)$ by C_2. Therefore, for each counter $\Delta N_i = \sqrt{Npq} = 4000$ and, as a consequence, $\Delta N_1/\overline{N}_1 = 10^{-5}$, $\Delta N_2/\overline{N}_2 = 2.35 \times 10^{-4}$.

3.3

a) The average length is $c \times \tau = 4.8\,\text{m}$.

b) Let l_1 be the length of the path $s_1 \to s_2 \to s_4 \to C_{1,2}$ (the one with the optical fiber) and l_2 that of the path $s_1 \to s_3 \to s_4 \to C_{1,2}$, $t_1 = l_1/c$ and $t_2 = l_2/c \ll t_1$. Let $t = 0$ be the instant when a photon enters the interferometer: one of the counters will click either in the interval of time $t_1, t_1 + \tau$ or in the interval $t_2, t_2 + \tau$, then the recording of the arrival is a measurement of the followed path: in this case there can be no interference, it is as if the other path would be inaccessible. As a consequence (see Problem 3.1d) each counter counts $\frac{1}{2}N(a^2 + b^2) = \frac{1}{2}N$ photons per second.

Experiments confirm this conclusion: it is known since the times of the first experiments on interference – then with no reference to the fact that light consists of photons – that, if the difference between the two paths of an interferometer exceeds the spatial coherence of the incident radiation (4.8 m in the present case), there is no interference.

From the point of view of the single photon, we can imagine the photon as a wave packet (of length 4.8 m) that is split into two packets (each of length 4.8 m) by the mirror s_1: they arrive at different times on the mirror s_4 and, as a consequence, cannot interfere. In addition – and this is the typically quantum feature of the problem – each photon is an indivisible entity: the fact that, for example, no counter clicks in the interval

$t_2, t_2 + \tau$ ($t_2 \ll t_1$) is, per se, a measurement that perturbs the state of the photon that, after this interval of time, can no longer be described as two packets along the two paths, but must be described as a single packet which is still going the long path. If instead in the interval $t_2, t_2 + \tau$ a counter clicks, the 'half packet' which was going the long path *is no longer there.* After we have recorded when a counter has clicked, the statement "the photon was, prior to the measurement, in one well determined path" is – in the framework of the "orthodox" Copenhagen interpretation of quantum mechanics – 'misleading and nonscientific' since it has no possibility of being verified.

3.4

a) If p is the momentum of the incident neutrons and p' is the momentum at height d, one has:

$$\frac{p^2}{2m_n} = \frac{p'^2}{2m_n} + m_n g d \quad \Rightarrow \quad p - p' \simeq \frac{m_n^2 g d}{p}$$

and, using the Broglie relation $p = h/\lambda = \hbar k$, one obtains $\Delta k \simeq \dfrac{m_n^2 g d}{\hbar^2 k}$.

b) $\varphi = kL - k'L = \Delta k \times L = \dfrac{m_n^2 g d L}{\hbar^2 k} = \dfrac{2\pi \lambda m_n^2 g A}{h^2} = 120 \, \text{radians}$

$A = d \times L$ is the area enclosed by $s_1 \to s_2 \to s_4 \to s_3 \to s_1$.

Indeed, to the first order in g, the result does not depend on the shape of the circuit $s_1 \to s_2 \to s_4 \to s_3 \to s_1$, but only on the enclosed area, as can be shown by observing that $\varphi = \hbar^{-1} \oint \vec{p} \cdot d\vec{q}$ and by using Stokes theorem: since only the horizontal parts of the circuit are relevant, we can define the vector field $\vec{p}(x, z)$ as $p_x(x, z) = (p^2 - 2m_n^2 g z)^{1/2}$, $p_z(x, z) = 0$. Then, paying attention to the sign of the circulation,

$$-\left(\text{curl}\, \vec{p} \right)_y = -\frac{\partial p_x}{\partial z} = \frac{m_n^2 g}{\sqrt{p^2 - 2m_n^2 g z}} = \frac{m_n^2 g}{p} + O(g^2) .$$

c) Let θ denote the angle by which the interferometer is rotated with respect to the vertical plane: one has $A \to A \cos\theta$, then $\varphi \to \varphi \cos\theta$. In the range $-30° \le \theta \le +30°$ the phase φ varies from $120 \times \cos 30° = 104$ to 120 and then again to 104: so there is an excursion of 32 radians and one observes $32/2\pi \simeq 5$ maxima ('fringes'). The result has been confirmed by several experiments performed between 1975 and 1987.

3.5

a) The Hamilton equations for the variables $q_1, q_2; p_1, p_2$ are:

$$\begin{cases} \dot{q}_1 = \dfrac{1}{m} p_1 + \omega q_2 \\ \dot{q}_2 = \dfrac{1}{m} p_2 - \omega q_1 \end{cases} \Rightarrow \begin{array}{l} p_1 = m\dot{q}_1 - m\omega q_2 \\ p_2 = m\dot{q}_2 + m\omega q_1 \end{array} \quad \begin{cases} \dot{p}_1 = \omega p_2 \\ \dot{p}_2 = -\omega p_1 \end{cases} \Rightarrow$$

$$\begin{cases} m\,\ddot{q}_1 = \dot{p}_1 + m\omega\,\dot{q}_2 = m\omega^2 q_1 + 2m\omega\,\dot{q}_2 \\ m\,\ddot{q}_2 = \dot{p}_2 - m\omega\,\dot{q}_1 = m\omega^2 q_2 - 2m\omega\,\dot{q}_1 \end{cases} \Rightarrow \ m\vec{a} = m\omega^2 \vec{r} - 2m\vec{\omega} \wedge \vec{v} \,.$$

b) Let us take the origin of the Cartesian axes in the point where the neutrons encounter the mirror s_1. To the first order in ω one has, in the several paths between the mirrors:

$s_1 \to s_2$: $\ x = \dot{x} = 0$, $k_x = -m_n \omega\, y/\hbar$; $\ k_y = \sqrt{2m_n E}/\hbar$

$s_2 \to s_4$: $\ y = l_y$, $\dot{y} = 0$, $k_y = +m_n \omega\, x/\hbar$; $\quad E = \dfrac{p_x^2}{2m_n} + \omega\, l_y p_x \quad \Rightarrow$

$$k_x = \frac{1}{\hbar}\left(\sqrt{m_n^2 \omega^2 l_y^2 + 2m_n E} - m_n \omega\, l_y \right) \simeq \frac{1}{\hbar}\left(\sqrt{2m_n E} - m_n \omega\, l_y \right)$$

$s_1 \to s_3$: $\ y = \dot{y} = 0$, $k_y = +m_n \omega\, x/\hbar$; $\ k_x = \sqrt{2m_n E}/\hbar$

$s_3 \to s_4$: $\ x = l_x$, $\dot{x} = 0$, $k_x = -m_n \omega\, y/\hbar$; $\ k_y \simeq \dfrac{1}{\hbar}\left(\sqrt{2m_n E} + m_n \omega\, l_x \right)$

c) $\varphi = \displaystyle\int_{\gamma_1} \vec{k}\cdot d\vec{l}_1 - \int_{\gamma_2} \vec{k}\cdot d\vec{l}_2 = \int_{s_1 \to s_2} k_y\, dy + \int_{s_2 \to s_4} k_x\, dx - \int_{s_1 \to s_3} k_x\, dx - \int_{s_3 \to s_4} k_y\, dy$

$$= \frac{1}{\hbar}\left[\left((l_x + l_y)\sqrt{2m_n E} - m_n \omega\, l_x l_y\right) - \left((l_x + l_y)\sqrt{2m_n E} + m_n \omega\, l_x l_y\right)\right]$$

$$= -\frac{2m_n \omega\, l_x l_y}{\hbar} = -\frac{2\cdot 1.7\times 10^{-24}\cdot 7.3\times 10^{-5}\cdot 9}{1.05\times 10^{-27}} = -2.1\,\text{rad} = -122°.$$

Alternatively: from $2E = m_n v^2 - m_n \omega^2 (x^2 + y^2)$, to first order in ω one has $v \simeq \sqrt{2E/m_n}$ therefore $p_1 = p_1^0 - m_n \omega\, y$, $p_2 = p_2^0 + m_n \omega\, x$ where p_1^0, $p_2^0 = m_n v$ or 0, depending on the orientation (x or y) of the path. In any case these terms can be omitted since they would give a contribution to φ of order 0 in ω. Then we can define the vector field $p_1(x, y) = -m_n \omega\, y$, $p_2(x, y) = +m_n \omega\, x$ and make use of Stokes' theorem, as in Problem 3.4.

3.6

a) It is known in diffraction theory that light is diffracted by the hole in the screen in a cone whose semi-aperture is $\theta \simeq \lambda/a$ (see the figure) that gives rise to a spot on the plate whose radius is $r \simeq \theta\, L = \lambda\, L/a$.

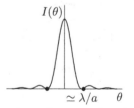

b) The single photon is absorbed by a single molecule and starts the photoreaction that involves a few molecules close to the point where it has been absorbed. So a practically pointlike stain can be seen on the plate. Only after many photons have reached the plate one can note that they are distributed with a density proportional to the intensity $I(\theta)$ of the figure above and give rise to the diffraction pattern of a).

c) We only know that the point where the photon arrived belongs to the diffraction pattern whose diameter is $2r$, so we can trace back the position of the source within the interval of amplitude $2\Delta = 2r\,D/L = 2\lambda D/a$, i.e. up to an uncertainty of the order of Δ.

If N photons arrive, the average of the positions they arrive at determines the position of the source up to the uncertainty $\Delta/\sqrt{N}$ (the root mean square of the distributions of the averages is $\Delta/\sqrt{N}$).

3.7

a) Assume the two slits are in the plane $x = 0$, parallel to the y axis and with $z_{1,2} = \pm d/2$; the screen is in the plane $x = D$, with $D \gg d$. The width of the central fringe is determined by the condition $|r_2 - r_1| = \lambda/2$ (see figure), where λ is the wavelength of the photons and

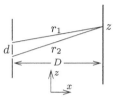

$$r_{1,2} = \sqrt{D^2 + (z \pm d/2)^2} \quad \Rightarrow \quad |r_2 - r_1| \simeq \frac{|z|\,d}{D}\,.$$

So the width of the central fringe is $a = \lambda D/d$.

Note that, when the dimension of the slits is small with respect to d, the width of the fringes is of the same order of the amplitude of the diffraction pattern relative to a slit of width d.

b) If the photons traveled along a straight line from the slits to the counters, the distance between the counters should be equal to $L\,d/D$, i.e. the two counters, seen from the hole, should make the angle d/D. However, due to diffraction, each photon that crosses the hole in the screen S propagates within a cone whose semi-angle is

$$\theta \simeq \frac{\lambda}{a} = \frac{d}{D}$$

so each photon can reach both the counter C_1 and C_2 independently of the slit it has crossed.

3.8

a) In a circularly polarized wave $\cos\vartheta = \sin\vartheta = 1/\sqrt{2}$ and $\varphi = \pm\pi/2$, so:

$$\hat{e}_{\sigma+} = \frac{1}{\sqrt{2}}(\hat{e}_1 + i\,\hat{e}_2)\,, \qquad \hat{e}_{\sigma-} = \frac{1}{\sqrt{2}}(\hat{e}_1 - i\,\hat{e}_2)$$

($\hat{e}_{\sigma\pm}$ respectively correspond to the right and left circular polarizations). Let $\hat{e}'_1$, $\hat{e}'_2$ two mutually orthogonal unit vectors in the plane orthogonal to the direction of propagation:

$$\begin{cases} \hat{e}'_1 = \cos\alpha\,\hat{e}_1 + \sin\alpha\,\hat{e}_2 \\ \hat{e}'_2 = -\sin\alpha\,\hat{e}_1 + \cos\alpha\,\hat{e}_2 \end{cases} \Rightarrow \begin{cases} \hat{e}_1 = \cos\alpha\,\hat{e}'_1 - \sin\alpha\,\hat{e}'_2 \\ \hat{e}_2 = \sin\alpha\,\hat{e}'_1 + \cos\alpha\,\hat{e}'_2 \end{cases} \Rightarrow$$

$$\frac{1}{\sqrt{2}}(\hat{e}_1 \pm i\,\hat{e}_2) = e^{i\alpha}\frac{1}{\sqrt{2}}(\hat{e}'_1 \pm i\,\hat{e}'_2)\;.$$

b) Since only the component of the electric field parallel to the transmission axis is transmitted, the fraction of the intensity that crosses the polaroid is $\cos^2\vartheta$ (Malus' law), therefore the probability that a single photon crosses the polaroid sheet is $\cos^2\vartheta$. If the photon is in a state of circular polarization it has probability $\frac{1}{2}$ to cross the sheet and emerges in the state of linear polarization parallel to the transmission axis.

c) The probability that a single linearly polarized photon crosses a polaroid sheet whose transmission axis makes the angle α with the direction of polarization of the photon is $\cos^2\alpha$, therefore:

$$N = 2:\;\; P = \left(\frac{1}{2}\right)^2 = \frac{1}{4}\;;$$

$$N = 90:\;\; P = \left(\cos^2(\pi/180)\right)^{90} \simeq \left(1 - \frac{1}{2}(\pi/180)^2\right)^{180}$$

$$\simeq 1 - 90\,(\pi/180)^2 = 0.97\;;$$

$$N \to \infty:\;\; P \to 1\;.$$

d) The probability that a circularly polarized photon crosses a polaroid sheet is $\frac{1}{2}$, in addition the emerging photon is linearly polarized in the direction of the transmission axis. As a consequence:

$$N = 2:\;\; P = \frac{1}{2}\times\frac{1}{2} = \frac{1}{4}\;;$$

$$N = 90:\;\; P = \frac{1}{2}\times\left(\cos(\pi/180)\right)^{89} \simeq 0.49\;;$$

$$N \to \infty:\;\; P \to \frac{1}{2}\;.$$

3.9

a) The light is (either right or left) circularly polarized: only in these two cases the amplitude of the wave transmitted by P_S is independent of the orientation of the transmission axis.

b) Since the polaroid sheet transmits a half of the intensity of the incident light, removing it causes only an increase by a factor 2 of the intensity of the interference pattern.

c) Putting $I_0 = cE_0^2/4\pi$ and $\varphi(x) = \varphi_A(x) - \varphi_B(x)$, the intensity on the plate is obtained by taking the time average:

$$I(x) \equiv \frac{c}{4\pi}\,\overline{E^2(x,t)} = I_0\overline{\left|\hat{e}_\alpha\cos\left(\omega t + \varphi_A(x)\right) + \hat{e}_\beta\cos\left(\omega t + \varphi_B(x)\right)\right|^2}$$

$$= I_0\left(1 + (\hat{e}_\alpha\cdot\hat{e}_\beta)\cos\varphi(x)\right)\quad\Rightarrow$$

$$I^{\max} = I_0\left(1 + \hat{e}_\alpha\cdot\hat{e}_\beta\right),\quad I^{\min} = I_0\left(1 - \hat{e}_\alpha\cdot\hat{e}_\beta\right)\;;$$

$$V \equiv \frac{I^{\max} - I^{\min}}{I^{\max} + I^{\min}} = \hat{e}_\alpha \cdot \hat{e}_\beta = \cos(\alpha - \beta) \geq 0 \,.$$

The visibility V is a maximum when the transmission axes of the polaroid sheets are parallel to each other ($V = 1$), is a minimum when they are orthogonal ($V = 0$): in the latter case there is no interference and one observes only the sum of intensities of the two diffraction patterns relative to the two slits.

d) The electric field on the plate L is polarized parallel to the direction of the transmission axis of P_L and the amplitude transmitted by P_L is

$$E_0 \Big(\cos(\alpha - \gamma) \cos\big(\omega t + \varphi_A(x)\big) + \cos(\beta - \gamma) \cos\big(\omega t + \varphi_B(x)\big) \Big) \quad \Rightarrow$$

$$I(x) = \frac{1}{2} I_0 \Big(\cos^2(\alpha - \gamma) + \cos^2(\beta - \gamma) + 2\cos(\alpha - \gamma) \cos(\beta - \gamma) \cos\varphi(x) \Big)$$

$$\Rightarrow \quad V = \frac{2\cos(\alpha - \gamma) \cos(\beta - \gamma)}{\cos^2(\alpha - \gamma) + \cos^2(\beta - \gamma)} \,.$$

V is a maximum and equals 1 when the two components coming from the two slits have the same amplitude: $\cos(\alpha - \gamma) \doteq \cos(\beta - \gamma)$, i.e. when the transmission axis of P_L is parallel to the bisector between $\hat{e}_\alpha$ and $\hat{e}_\beta$. If $\hat{e}_\alpha = \hat{e}_1$, $\hat{e}_\beta = \hat{e}_2$ and P_L is absent, then $V = 0$, as seen in c); whereas, if P_L is inserted with its transmission axis at $45°$ with respect to $\hat{e}_1$ and $\hat{e}_2$, the energy arriving at L is a half, but $V = 1$ and as a consequence an interference pattern with maximal visibility is observed.

3.10

a) No: for both right and left polarized photons the probability of being transmitted is $\frac{1}{2}$, regardless of the orientation of the transmission axis.

b) The quarter-wave plate transforms circularly polarized photons into linearly polarized photons:

$$\hat{e}_{\sigma+} \to \frac{1}{\sqrt{2}}(\hat{e}_1 - \hat{e}_2)\,, \qquad \hat{e}_{\sigma-} \to \frac{1}{\sqrt{2}}(\hat{e}_1 + \hat{e}_2)\,.$$

So, after one single photon has crossed the plate, it is sufficient to verify whether it crosses a polaroid sheet with the transmission axis parallel to $\hat{e}_1 + \hat{e}_2$ (or $\hat{e}_1 - \hat{e}_2$): if the photon crosses it, it was left polarized, if not it was right polarized.

c) Using part of the available photons and making measurements with a polaroid sheet, whose transmission axis makes the (arbitrary) angle α with respect to $\hat{e}_1$, it is possible to determine the transition probability $P(\hat{e}_{\vartheta\varphi} \to \hat{e}_\alpha)$, where $\hat{e}_\alpha = \cos\alpha\,\hat{e}_1 + \sin\alpha\,\hat{e}_2$ is the linear polarization state of the photons that cross the polaroid sheet. One has:

$$P(\alpha) \equiv P(\hat{e}_{\vartheta\varphi} \to \hat{e}_\alpha) = |\cos\alpha \cos\vartheta + \sin\alpha \sin\vartheta\, e^{i\varphi}|^2$$

$$= \cos^2\alpha \cos^2\vartheta + \sin^2\alpha \sin^2\vartheta + \frac{1}{2}\sin 2\alpha \sin 2\vartheta \cos\varphi\,.$$

$P(0)$ and $P(\pi/2)$ allow one to find $\cos^2\vartheta$ and $\sin^2\vartheta$ respectively. In this way four possible values for ϑ are left: as $\hat{e}_{\vartheta\varphi}$ may be determined up to a phase factor, one is allowed to choose $\cos\vartheta \geq 0$. The choice of $\sin\vartheta \geq 0$ is possible at the cost of redefining φ, yet unknown after these measurements. $P(\pi/4)$ provides the value of $\cos\varphi$. Only the sign of φ remains unknown: if another part of the available photons crosses a quarter-wave plate before impinging the polaroid sheet, then $\varphi \to \varphi + \pi/2 \Rightarrow \cos\varphi \to -\sin\varphi$ and, for example, $P(\pi/4)$ determines also the sign of φ. In conclusion: the state of the photons that have not been used is now completely determined.

3.11

a) In the basis $\hat{e}_1$, $\hat{e}_2$ the polarization vector $a\,\hat{e}_1 + b\,\hat{e}_2$ (a, $b \in \mathbb{C}$) is represented by the pair (a, b) and since

$$\begin{cases} \hat{e}_1 \to \cos\alpha\,\hat{e}_1 + \sin\alpha\,\hat{e}_2 \\ \hat{e}_2 \to \cos\alpha\,\hat{e}_2 - \sin\alpha\,\hat{e}_1 \end{cases} \Rightarrow U\begin{pmatrix}1\\0\end{pmatrix} = \begin{pmatrix}\cos\alpha\\\sin\alpha\end{pmatrix}; \quad U\begin{pmatrix}0\\1\end{pmatrix} = \begin{pmatrix}-\sin\alpha\\\cos\alpha\end{pmatrix}$$

$$\Rightarrow \quad U = \begin{pmatrix}\cos\alpha & -\sin\alpha\\\sin\alpha & \cos\alpha\end{pmatrix}.$$

b) The polarization states that remain unaltered are all and only the eigenvectors of the matrix U:

$$\begin{pmatrix}\cos\alpha & -\sin\alpha\\\sin\alpha & \cos\alpha\end{pmatrix}\begin{pmatrix}a\\b\end{pmatrix} = \lambda\begin{pmatrix}a\\b\end{pmatrix} \quad \Rightarrow \quad a = 1,\ b = \pm i\,.$$

They correspond to the polarization states $\hat{e}_1 \pm i\,\hat{e}_2$, namely to either right or left circularly polarized photons.

In analogy to a birefringent crystal, that exhibits two different refraction indices for lights linearly polarized in two mutually orthogonal directions, an optically active substance exhibits two different refraction indices for light that is either right or left circularly polarized.

c) The states of linear polarization $\hat{e}_1$ and $\hat{e}_2$: $\hat{e}_1 \to \hat{e}_1$, $\hat{e}_2 \to i\,\hat{e}_2$.

d) No, since in both cases the final state is uniquely determined by the initial state, contrary to what happens when a measurement is made: in the case of the plates the photons undergo a *causal* evolution to the final state, and such evolution is represented by a unitary transformation.

3.12

a) When the particles cross the screen, their y coordinate has an uncertainty $\Delta y = a$, so $\Delta p_y \simeq h/a$ and the uncertainty in the angle with respect to the direction of incidence (diffraction angle) is given by:

$$\theta \simeq \frac{\Delta p_y}{p} \simeq \frac{h}{a} \times \frac{\lambda}{h} = \frac{\lambda}{a}\,.$$

b) Although the y coordinate of the photons that have crossed the screen with the slits has an uncertainty of the order of d, Δp_y is of the order of h/a. In fact, as seen above, the photons come out from the slits within a cone of aperture $\theta \simeq \lambda/a$ then $\Delta p_y = p_x\, \theta \simeq h/a$ and $\Delta y\, \Delta p_y \simeq h\, d/a \gg h$. We shall meet again a similar situation ($\Delta p \gg h/\Delta x$) in the Problem 5.19.

3.13

a) From the geometry of the second figure in the text $\Delta L = \sqrt{2}\, \Delta q^{(\mathrm{m})}$.

b) If $\lambda = h/p$ is the wavelength of the photons, ΔL should not exceed λ, therefore $\Delta q^{(\mathrm{m})} < \lambda/\sqrt{2}$.

c) The recoil of the mirror can be detected if the momentum transferred by the photon to the mirror is greater then $\Delta p^{(\mathrm{m})}$: $\Delta p^{(\mathrm{m})} < 2p\, \sqrt{2}/2 = p\, \sqrt{2}$.

d) $\Delta q^{(\mathrm{m})}\, \Delta p^{(\mathrm{m})} < h$. Therefore, since the uncertainty relations hold also for macroscopic objects (as the mirror), the observation of the interference is incompatible with the knowledge of the way followed by the photon, as already seen in Problems 3.7 and 3.9.

3.14

a) $\Delta x \simeq c \times \tau$.

b) $\Delta p \simeq \dfrac{h}{\Delta x} = \dfrac{h}{c\,\tau}$; $\quad p = \dfrac{h}{\lambda} = \dfrac{h\,\nu}{c} \;\Rightarrow\; \Delta\nu = \dfrac{c\,\Delta p}{h} \simeq \dfrac{1}{\tau} = 6.2 \times 10^7\,\mathrm{Hz}$.

$\nu = \dfrac{c}{\lambda} \simeq 5 \times 10^{14}\,\mathrm{Hz}$.

c) $\dfrac{\Delta\lambda}{\lambda} = \dfrac{\Delta\nu}{\nu} = \dfrac{\lambda}{c}\, \Delta\nu \simeq \dfrac{\lambda}{c\,\tau} = 1.2 \times 10^{-7}$;

$Q \equiv 2\pi\,\nu \times \tau = 2\pi\,\dfrac{\nu}{\Delta\nu} \simeq 5 \times 10^7$.

d) $E = \dfrac{12400}{5890} = 2.11\,\mathrm{eV}$; $\quad \dfrac{\Delta E}{E} = \dfrac{\Delta\nu}{\nu} \quad\Rightarrow\quad \Delta E \simeq 2.5 \times 10^{-7}\,\mathrm{eV}$,

$\Rightarrow \quad \tau \times \Delta E \simeq h$.

4

States, Measurements and Probabilities

Superposition principle; observables; statistical mixtures; commutation relations.

Note. The states of photon rectilinear polarization denoted by $e_{1,2}$ in the previous chapter will from now on be denoted by $|\,e_{1,2}\,\rangle$; in the sequel we shall use the notation $|\,e_{1,2}\,\rangle$ for vectors in the complex Hilbert space, whereas $\hat{e}_{1,2}$ will stand for vectors in the 'physical' real configuration space.

4.1 Photons are sent on a screen in which two slits A and B have been made. Let $|\,A\,\rangle$ be (a vector that represents) the state of each photon crossing slit A (B is closed) and, vice versa, $|\,B\,\rangle$ (a vector ...) the state of each photon crossing slit B (A is closed).

a) How many states $|\,C\,\rangle = \alpha\,|\,A\,\rangle + \beta\,|\,B\,\rangle$ can be obtained as superpositions of $|\,A\,\rangle$ and $|\,B\,\rangle$, with α and β arbitrary complex numbers?

It is possible to modify the relative intensity of the light going through the two slits by putting a slab of material whose transparency is not 100% (an attenuator) in front of one of them.

b) An attenuator is placed in front of slit A and B is closed. What is the state of the photons crossing A?

It is possible to modify the relative phase of the light crossing the slits by putting a plate of transparent glass in front of one of them (a phase shifter: see Problem 3.1).

c) A phase shifter is placed in front of slit A and B is closed. What is the state of the photons going through A?

d) The vectors $|\,A\,\rangle$ and $\alpha\,|\,A\,\rangle$ represent the same state; the same is true for the vectors $|\,B\,\rangle$ and $\beta\,|\,B\,\rangle$. Do $\alpha\,|\,A\,\rangle + \beta\,|\,B\,\rangle$ and $|\,A\,\rangle + |\,B\,\rangle$ represent the same state?

e) How is it possible to realize (we mean experimentally, i.e. in the laboratory) the state $\alpha\,|\,A\,\rangle + \beta\,|\,B\,\rangle$ with α and β arbitrary complex numbers? What

© Springer International Publishing AG 2017
E. d'Emilio and L.E. Picasso, *Problems in Quantum Mechanics*,
UNITEXT for Physics, DOI 10.1007/978-3-319-53267-7_4

differences appear in the interference patterns produced on a screen by such photons as α and β vary?

4.2   A beam of photons all prepared in the same polarization state is available, but it is not known whether the state is $\cos\vartheta \,|\,e_1\,\rangle + \sin\vartheta\,e^{i\varphi}\,|\,e_2\,\rangle$ or $\cos\vartheta \,|\,e_1\,\rangle + \sin\vartheta\,|\,e_2\,\rangle$. (In other words ϑ is known, but φ is not known.)

a) Say whether it is possible to determine the polarization state of the photons by means of (possibly many) measurements performed with only a polaroid sheet.

b) Is it possible to determine the state of the photons by means of *only one measurement* of a suitable observable?

4.3   Birefringent crystals, optically active substances (see Problem 3.11) and – of course – detectors are available. Say how the following nondegenerate observables (i.e. instruments) relative to photon polarization states can be constructed.

a) The observable that has the states of rectilinear polarization $|\,e_1\,\rangle$ and $|\,e_2\,\rangle$ as eigenstates. Show that the orthogonality of the states $|\,e_1\,\rangle$ and $|\,e_2\,\rangle$ (as vectors of the Hilbert space: $\langle\,e_1 \mid e_2\,\rangle = 0$) follows from the possibility of constructing such an observable.

b) The observable that has the states $|\,e_{\sigma_\pm}\,\rangle = \frac{1}{\sqrt2}\big(|\,e_1\,\rangle \pm i\,|\,e_2\,\rangle\big)$ of circular polarization as eigenstates.

c) Determine the state orthogonal to $|\,e_{\vartheta\varphi}\,\rangle \equiv \cos\vartheta \,|\,e_1\,\rangle + \sin\vartheta\,e^{i\varphi}\,|\,e_2\,\rangle$ and construct the observable that has these two states as eigenstates.

4.4   Let $\xi^{\rm op}$ be the operator associated to the observable ξ.

a) Is it true that, if the observable ξ is measured on the system in the state $|\,A\,\rangle$, the state after the measurement is $|\,B\,\rangle = \xi^{\rm op}\,|\,A\,\rangle$?

b) Is the statement made above true (for any ξ) at least in the case when $|\,A\,\rangle$ is an eigenstate of ξ?

Now let ξ and η be two compatible observables and $|\,A\,\rangle$ the state of the system. In the first case ξ and then η are measured. In the second case, the system always being in the state $|\,A\,\rangle$, η and then ξ are measured.

c) Is it true, in general, that the same results are obtained in the two cases?

4.5   Consider, in a three-dimensional Hilbert space in which $|\,1\,\rangle$, $|\,2\,\rangle$, $|\,3\,\rangle$ form an orthonormal basis, i) the observable ξ that has $|\,1\,\rangle$ and $|\,2\,\rangle$ as eigenvectors both belonging to the eigenvalue ξ_1, and $|\,3\,\rangle$ belonging to the eigenvalue ξ_3; ii) the observable η whose eigenvectors are $|\,1\,\rangle$ corresponding to the eigenvalue η_1, $|\,2\,\rangle$ and $|\,3\,\rangle$ both belonging to the eigenvalue η_2.

a) Do bases exist consisting of simultaneous eigenvectors of the two observables? In the affirmative case, just one basis (up to multiples) or more than one? Are the observables compatible with each other?

Assume that the system is in the state $a\,|\,1\,\rangle + b\,|\,2\,\rangle + c\,|\,3\,\rangle$ (a, b, c being real numbers).

b) Consider all the possible pairs of results of measurements of ξ and η in the given order and calculate the probability of occurrence for each pair.

4.6   Consider the following four statistical mixtures consisting of $N \gg 1$ photons:

a) $N/2$ photons in the polarization state $|\,e_1\,\rangle$ and $N/2$ photons in the state $|\,e_2\,\rangle$; calculate the probabilities of finding the photons $i)$ in the state $\frac{1}{\sqrt{2}}(\,|\,e_1\,\rangle + |\,e_2\,\rangle)$; $ii)$ in the state $|\,e_{\sigma_+}\,\rangle = \frac{1}{\sqrt{2}}(\,|\,e_1\,\rangle + \mathrm{i}\,|\,e_2\,\rangle)$.

b) $N/2$ photons in the state of circular polarization $|\,e_{\sigma_+}\,\rangle$ and $N/2$ photons in the state $|\,e_{\sigma_-}\,\rangle$; calculate the probabilities of finding the photons $i)$ in the state $|\,e_1\,\rangle$; $ii)$ in the state $\frac{1}{\sqrt{2}}(\,|\,e_1\,\rangle + |\,e_2\,\rangle)$.

c) $N/2$ photons in the state $\cos\vartheta\,|\,e_1\,\rangle + \sin\vartheta\,\mathrm{e}^{\mathrm{i}\varphi}\,|\,e_2\,\rangle$ and $N/2$ photons in the state $-\sin\vartheta\,|\,e_1\,\rangle + \cos\vartheta\,\mathrm{e}^{\mathrm{i}\varphi}\,|\,e_2\,\rangle$;   calculate the probabilities of finding the photons $i)$ in the state $\frac{1}{\sqrt{2}}(\,|\,e_1\,\rangle + |\,e_2\,\rangle)$; $ii)$ in the state $|\,e_{\sigma_+}\,\rangle = \frac{1}{\sqrt{2}}(\,|\,e_1\,\rangle + \mathrm{i}\,|\,e_2\,\rangle)$.

d) $N/4$ photons in each of the following states: $|\,e_1\,\rangle,\ |\,e_2\,\rangle,\ \frac{1}{\sqrt{2}}(\,|\,e_1\,\rangle + |\,e_2\,\rangle),$ $\frac{1}{\sqrt{2}}(\,|\,e_1\,\rangle - |\,e_2\,\rangle)$;   calculate the probability of finding the photons in the state $|\,e_{\vartheta\varphi}\,\rangle = \cos\vartheta\,|\,e_1\,\rangle + \sin\vartheta\,\mathrm{e}^{\mathrm{i}\varphi}\,|\,e_2\,\rangle$.

e) Is it possible, by means of suitable measurements, to distinguish the four statistical mixtures a) to d)?

4.7   Let $|\,s_1\,\rangle$ and $|\,s_2\,\rangle$ be two orthogonal states of a system: $\langle\,s_1\,|\,s_1\,\rangle = \langle\,s_2\,|\,s_2\,\rangle = 1$, $\langle\,s_1\,|\,s_2\,\rangle = 0$. Consider the statistical mixture consisting of $N \gg 1$ systems in the states $|\,s_\varphi\,\rangle \equiv a\,|\,s_1\,\rangle + b\,\mathrm{e}^{\mathrm{i}\varphi}\,|\,s_2\,\rangle$, with a, b real and satisfying $a^2 + b^2 = 1$, and φ uniformly distributed on the interval $(0,\ 2\pi)$.

a) Show that the given statistical mixture is equivalent to the statistical mixture consisting of $N\,a^2$ systems in the state $|\,s_1\,\rangle$ and $N\,b^2$ systems in the state $|\,s_2\,\rangle$.

4.8   Consider the statistical mixture $\{\,|\,u_1\,\rangle, \nu_1;\ |\,u_2\,\rangle, \nu_2;\ \cdots\ |\,u_i\,\rangle, \nu_i;\ \cdots\}$ where ν_i is the fraction of systems 'prepared' in the state $|\,u_i\,\rangle$ $(\sum \nu_i = 1)$, and the vectors $|\,u_i\,\rangle$ are normalized but not necessarily orthogonal to one another.

a) Calculate the probability P_i of finding the system in the state $|u_i\rangle$. Does the equality $\sum_i P_i = 1$ hold?

Consider the Hermitian operator (*statistical operator* or *density matrix*):

$$\varrho \equiv \sum_i \nu_i E_{u_i}, \qquad E_{u_i} = |u_i\rangle\langle u_i|; \qquad \varrho = \varrho^\dagger.$$

b) Show that, for any observable ξ and for any orthonormal basis $|n\rangle$, $n = 1, 2, \cdots$, the mean value $\langle\!\langle \xi \rangle\!\rangle$ of ξ over the statistical mixture is given by:

$$\langle\!\langle \xi \rangle\!\rangle = \mathrm{Tr}\,(\varrho\,\xi) \equiv \sum_n \langle n \mid \varrho\,\xi \mid n \rangle.$$

c) Show that *i)* $\mathrm{Tr}\,\varrho = 1$; *ii)* the eigenvalues of ϱ are nonnegative; *iii)* $\mathrm{Tr}\,\varrho^2 \le 1$. Show that, if $\mathrm{Tr}\,\varrho^2 = 1$, then $\varrho^2 = \varrho$ and that, in the latter case, the statistical mixture is a pure state.

4.9   Consider the four statistical mixtures described in Problem 4.6:

$$\left\{ |e_1\rangle, \tfrac{1}{2};\ |e_2\rangle, \tfrac{1}{2} \right\}; \qquad \left\{ |e_{\sigma_+}\rangle, \tfrac{1}{2};\ |e_{\sigma_-}\rangle, \tfrac{1}{2} \right\};$$

$$\left\{ \cos\vartheta\,|e_1\rangle + \sin\vartheta\,e^{i\varphi}\,|e_2\rangle, \tfrac{1}{2};\ -\sin\vartheta\,|e_1\rangle + \cos\vartheta\,e^{i\varphi}\,|e_2\rangle, \tfrac{1}{2} \right\};$$

$$\left\{ |e_1\rangle, \tfrac{1}{4};\ |e_2\rangle, \tfrac{1}{4};\ \tfrac{1}{\sqrt{2}}(|e_1\rangle + |e_2\rangle), \tfrac{1}{4};\ \tfrac{1}{\sqrt{2}}(|e_1\rangle - |e_2\rangle), \tfrac{1}{4} \right\}.$$

a) Determine, for each of them, the corresponding statistical operator ϱ in the space $\mathcal{H}_2$ of the polarization states of photons. Calculate ϱ^2 and verify the inequality $\mathrm{Tr}\,\varrho^2 \le 1$.

b) Write the statistical operator corresponding to the pure state $|e_{\sigma_+}\rangle$ and verify the equality $\varrho^2 = \varrho$ is satisfied.

4.10   A photon crosses a semi-transparent mirror, whose reflection and transmission coefficients are equal. Let $|A\rangle$ represent the transmitted state, $|B\rangle$ the reflected state, $\langle A \mid A \rangle = 1$, $\langle B \mid B \rangle = 1$, $\langle A \mid B \rangle = 0$ (see figure).

a) Say whether the state of the emerging photon is described either by the pure state $\tfrac{1}{\sqrt{2}}(|A\rangle + |B\rangle)$ or by the statistical mixture $\left\{ |A\rangle, \tfrac{1}{2};\ |B\rangle, \tfrac{1}{2} \right\}$.

b) If a counter C, whose efficiency is 100%, is placed in the path of the reflected state, what is the state of the emerging photon in those cases when the counter does not reveal the photon?

Let us now consider the device consisting of two semi-transparent mirrors and the counter C as in the figure to the right.

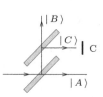

c) Write the state of a photon that emerges from the device when the counter does not click.

4.11 Consider Young's double-slit experiment with slits A and B and carried out with completely unpolarized monochromatic light. A polaroid sheet P_S, whose transmission axis is parallel to the plane containing the slits, is interposed between the source and the screen containing the slits.

S

L

a) How does the interference pattern change as the polaroid sheet is rotated around an axis orthogonal to the plane of the slits?

The polaroid sheet P_S is removed.

b) How does the interference pattern now change?

It is known (see Problems 3.7, 3.9, 3.13) that the occurrence of interference fringes is not compatible with establishing which of the two slits each photon comes from.
Consider now the following experiment: two shutters S_A and S_B are placed in front of the slits, with the feature that they are driven by a completely random process in such a way that when S_A is open S_B is closed and vice versa: any photon is allowed to cross just one of the two slits, but, due to randomness and the high speed of the commutation process, we do not know which one.

c) Say whether, under these conditions, interference fringes are observed.

4.12 The operators associated with the components of the angular momentum of a particle are:

$$L_x = y\,p_z - z\,p_y, \quad L_y = z\,p_x - x\,p_z, \quad L_z = x\,p_y - y\,p_x .$$

a) Show that L_x, L_y, L_z are Hermitian operators and, exploiting the formal properties of the commutators, calculate $[L_i,\, L_j]$ $(i, j = x, y, z)$.

b) Show that if a state $|\,s\,\rangle$ is an eigenstate of the operator $\vec{L}^{\,2} \equiv L_x^2 + L_y^2 + L_z^2$ belonging to the eigenvalue 0, then it is also a simultaneous eigenvector of L_x, L_y, L_z; determine, in the latter case, the corresponding eigenvalues. Vice versa: use the commutation relations among the components of angular momentum to show that a simultaneous eigenvector of L_x, L_y, L_z is also an eigenvector of $\vec{L}^{\,2}$ belonging to the eigenvalue 0.

c) By using the uncertainty relations $\Delta\xi\,\Delta\eta \ge \frac{1}{2}\overline{[\xi,\,\eta]}$ (the *Heisenberg–Robertson relations*), show that, given a component of the angular momentum and one of its eigenvectors, the mean values of the other two components (relative to orthogonal axes) in this state are vanishing.

4.13   Let A and B be two operators.

a) Explain why, in general, $e^{A+B} \neq e^A \, e^B$. In which cases does the equality sign hold?

Assume that the commutator between A and B is a c–number (more generally: $[A, B]$ commutes with both A and B); then the Baker–Campbell–Hausdorff identity:
$$e^A \, e^B = e^{A+B+\frac{1}{2}[A, B]}$$
holds.

b) Make the substitutions $A \to \lambda A$, $\; B \to \lambda B$ and verify that the Baker–Campbell–Hausdorff identity holds up to the second order in λ.

4.14   Consider a one-dimensional harmonic oscillator of mass m and angular frequency w. Let $|0\rangle$ be the ground state and $|1\rangle = \eta^\dagger |0\rangle$ the first excited state, with $\eta \equiv (2\pi w \hbar)^{-1/2}(p - i \, m w \, q)$.

a) Among the states that can be obtained as a superposition of $|0\rangle$ and $|1\rangle$, determine the one for which the mean value of q is a maximum and the one for which the mean value of p is a maximum.

b) Calculate the mean value of the energy H in the states found in a).

c) Calculate the mean values of q, of p and of the energy H in the statistical mixture $\{|0\rangle, \nu_0; \; |1\rangle, \nu_1\}$.

4.15   Consider a one-dimensional harmonic oscillator and let H be the Hamiltonian.

a) Calculate the commutator $[H, pq]$ and use the result to show that the mean values of kinetic energy and potential energy in the eigenstates of H are equal (virial theorem).

b) Calculate the mean values of kinetic energy and potential energy in the state obtained as a generic superposition of $|0\rangle$ and $|1\rangle = \eta^\dagger |0\rangle$.

c) Calculate the mean values of kinetic energy and potential energy in the state obtained as a generic superposition of $|0\rangle$ and $|2\rangle = \frac{1}{\sqrt{2}}(\eta^\dagger)^2 |0\rangle$.

Solutions

4.1

a) The vectors that can be obtained as linear combinations of $|A\rangle$ and $|B\rangle$, the coefficients being complex, are ∞^4, but, since vectors proportional to each other by a complex number represent the same state, the states are ∞^2, therefore the following writings are equivalent:

$$\alpha\,|A\rangle + \beta\,|B\rangle; \qquad |A\rangle + \gamma\,|B\rangle, \quad \gamma = \beta/\alpha;$$

$$\frac{1}{\sqrt{|\alpha|^2 + |\beta|^2}}\Big(|\alpha|\,|A\rangle + |\beta|\,e^{i\varphi}\,|B\rangle\Big), \qquad \varphi = \arg\beta - \arg\alpha$$

where, also in the last line, due to

$$\frac{|\alpha|^2}{|\alpha|^2 + |\beta|^2} + \frac{|\beta|^2}{|\alpha|^2 + |\beta|^2} = 1\,,$$

the independent real parameters are two.

b) The state of the photons is unchanged, therefore it is still represented by $|A\rangle$. Indeed, the diffraction pattern that can be observed – after the arrival of many photons – on a photographic plate posed after A is independent of the transparency of the attenuator: the photons arrive more or less rarely, but their state is the same in any event.

c) As above: the diffraction pattern does not depend on the phase.

d) The vectors $\alpha\,|A\rangle + \beta\,|B\rangle$ and $|A\rangle + |B\rangle$ are not proportional to each other (if $\alpha \neq \beta$), so they represent different states: think, for example, of the states of polarization $|e_1\rangle + |e_2\rangle$ and $|e_1\rangle + i\,|e_2\rangle$.

e) If the state is written as:

$$|\alpha|\,|A\rangle + |\beta|\,e^{i\varphi}\,|B\rangle, \qquad |\alpha|^2 + |\beta|^2 = 1$$

then $|\alpha|^2$ is the probability of finding the photon in the state $|A\rangle$, whereas $|\beta|^2$ is the probability to find it in the state $|B\rangle$. As a consequence, since the probability that a photon crosses the slit is proportional to the transparency of the slit, it is sufficient to put an attenuator in front of one slit. The phase difference φ may be realized by putting a phase shifter in front of one of the slits. As the transparency of the attenuator is varied, the visibility of the fringes varies accordingly (see Problem 3.1):

$$\frac{I^{\max} - I^{\min}}{I^{\max} + I^{\min}} = 2|\alpha\beta|$$

while varying φ gives rise to a translation of the whole interference pattern: for example, if $\varphi \to \varphi + \pi$, the figure is translated by half a fringe.

4.2

a) If the state is $\cos\vartheta \,|\,e_1\rangle + \sin\vartheta \,|\,e_2\rangle$, then the photons do not cross a polaroid sheet with its transmission axis orthogonal to the direction $\cos\vartheta\,\hat{e}_1 + \sin\vartheta\,\hat{e}_2$ (i.e parallel to the direction $-\sin\vartheta\,\hat{e}_1 + \cos\vartheta\,\hat{e}_2$), whereas if the state is $\cos\vartheta \,|\,e_1\rangle + \sin\vartheta\,e^{i\varphi}\,|\,e_2\rangle$, according to Malus' law, they do cross it with probability:

$$P = \left|\left(-\sin\vartheta\,\langle e_1| + \cos\vartheta\,\langle e_2|\right)\left(\cos\vartheta\,|\,e_1\rangle + \sin\vartheta\,e^{i\varphi}\,|\,e_2\rangle\right)\right|^2$$

$$= \left|(1 - e^{i\varphi})\sin\vartheta\,\cos\vartheta\right|^2 = \sin^2\frac{\varphi}{2}\,\sin^2 2\vartheta \ .$$

So, by making many measurements, it is possible to determine the state.

b) No: since the two states are not orthogonal to each other, no observable can give a result able to exclude with certainty one of the two states.

4.3

a) A birefringent crystal pinpoints two directions: $\hat{e}_1$ (optical axis) and $\hat{e}_2$, orthogonal to the former. Photons in the polarization state $|\,e_1\rangle$ that impinge the crystal emerge in the extraordinary ray in the same polarization state $|\,e_1\rangle$; photons that impinge, being in the polarization state $|\,e_2\rangle$, emerge in the ordinary ray, they too with unaltered polarization state. It

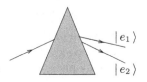

is convenient to separate the two rays: this can be achieved by means of a crystal with nonparallel faces (see the figure). Two detectors are then arranged in such a way as to distinguish the photons that emerge in the extraordinary ray from those that emerge in the ordinary ray. The eigenvalues of this observable are arbitrary, it is sufficient that the two detectors are identified on the display by two different numbers.

Since a measurement of such an observable on photons in the state $|\,e_1\rangle$ never gives the state $|\,e_2\rangle$ as a result, and vice versa, it follows that $P(|\,e_1\rangle \to |\,e_2\rangle) = 0$ and, as a consequence, $\langle e_1 \,|\, e_2\rangle = 0$, a result that was already implicit in the Malus' law.

b) One possible method consists in using an optically active medium (also in this case with nonparallel faces), that treats the circular polarization states in the same way as the birefringent crystal treats the mutually orthogonal linear polarizations (see Problem 3.11). Of course two detectors will be posed in the final part of the apparatus.

   Another method that, instead of the optically active medium, takes advantage of quarter-wave plates is obtained as a particular case of what is described in the next point c).

c) The state orthogonal to $\cos\vartheta\,|e_1\rangle + \sin\vartheta\,e^{i\varphi}\,|e_2\rangle$ is represented by $-\sin\vartheta\,|e_1\rangle + \cos\vartheta\,e^{i\varphi}\,|e_2\rangle$. By means of a birefringent crystal of suitable thickness ('$-\varphi/2\pi$ wavelength plate': see Problem 3.10) the two states are transformed into the states of linear polarization $\cos\vartheta\,|e_1\rangle + \sin\vartheta\,|e_2\rangle$ and $-\sin\vartheta\,|e_1\rangle + \cos\vartheta\,|e_2\rangle$; then a birefringent crystal with optical axis at an angle ϑ with respect to the direction $\hat{e}_1$ is placed (the states $\cos\vartheta\,|e_1\rangle + \sin\vartheta\,|e_2\rangle$ and $-\sin\vartheta\,|e_1\rangle + \cos\vartheta\,|e_2\rangle$ – and only they – are transmitted unchanged); finally a $+\varphi/2\pi$ wavelength plate, placed in order to restore the two initial states, completes the apparatus.

4.4

a) No! After a measurement the state is only statistically determined, instead $\xi^{\mathrm{op}}\,|A\rangle$ is a well defined vector.

b) What if $|A\rangle$ were an eigenstate of the observable ξ corresponding to the eigenvalue 0? In the latter case, if the statement were true, the state after the measurement should be represented by the null vector: no state corresponds to such a vector. The application of the operator ξ^{op} to a vector $|A\rangle$ only is a mathematical operation that is effected in the Hilbert space of states: no physical action in the laboratory corresponds to this.

c) If one is very lucky, yes ..., but in general this is not true. Indeed suppose, for example, that ξ is the (trivial) observable that only possesses the eigenvalue 1 (so that the operator associated to ξ is the identity): a measurement of ξ, made either before or after the measurement of η, does not change the state, therefore this is the same as not making the measurement at all; the question then is if two measurements of η always on the state $|A\rangle$ should necessarily give the same result: this in general is not true, as the result of the measurement of an observable is only statistically determined. Only if $|A\rangle$ is a simultaneous eigenstate of both ξ and η, the results of the measurements the two observables are (a priori determined and) independent of the order.

4.5

a) The vectors $|1\rangle$, $|2\rangle$, $|3\rangle$ are eigenvectors of both the observables that are, as a consequence, compatible with each other. There exist no other common bases: $|3\rangle$ must belong to the basis because it corresponds to a nondegenerate eigenvalue of ξ; the same is true for $|1\rangle$ that belongs to a nondegenerate eigenvalue of η; the basis is uniquely completed by the vector $|2\rangle$.

b) Normalizing the given vector to 1 yields:

$$|A\rangle = \frac{a|1\rangle + b|2\rangle + c|3\rangle}{\sqrt{a^2 + b^2 + c^2}}.$$

A measurement of ξ can give as a result either the degenerate eigenvalue ξ_1 or the nondegenerate eigenvalue ξ_3: in the former case, owing to von Neumann postulate (the "projection postulate"), the system is in the state obtained by projecting $|A\rangle$ onto the manifold generated by $|1\rangle$ and $|2\rangle$, i.e. in the state represented by the (normalized) vector

$$\frac{a|1\rangle + b|2\rangle}{\sqrt{a^2 + b^2}}$$

and the corresponding probability is

$$P = \left|\left(\frac{a\langle 1| + b\langle 2|}{\sqrt{a^2 + b^2}}\right)|A\rangle\right|^2 = \frac{a^2 + b^2}{a^2 + b^2 + c^2}$$

(in order to compute the probability, the vectors representing both the initial and final state must be normalized); in the case ξ_3 is the result of the measurement, the final state is $|3\rangle$ with probability $c^2/(a^2 + b^2 + c^2)$. Then η is measured and one must proceed in an analogous way. In the sequel the possible results of the various measurements, the corresponding state and the probability of the outcome are reported:

$$|A\rangle \xrightarrow{\xi} \begin{cases} \xi_1: \dfrac{a|1\rangle + b|2\rangle}{\sqrt{a^2 + b^2}}; \ P = \dfrac{a^2 + b^2}{a^2 + b^2 + c^2} \xrightarrow{\eta} \begin{cases} \eta_1: |1\rangle; \ P = \dfrac{a^2}{a^2 + b^2} \\[2mm] \eta_2: |2\rangle; \ P = \dfrac{b^2}{a^2 + b^2} \end{cases} \\[6mm] \xi_3: |3\rangle; \qquad\qquad P = \dfrac{c^2}{a^2 + b^2 + c^2} \xrightarrow{\eta} \eta_2: |3\rangle; \ P = 1. \end{cases}$$

In conclusion, normalizing $a^2 + b^2 + c^2 = 1$, the probabilities corresponding to the possible pairs of results are:

$$P(\xi_1, \eta_1) = a^2; \quad P(\xi_1, \eta_2) = b^2; \quad P(\xi_3, \eta_2) = c^2.$$

4.6

a) i):

$$P = \frac{1}{N}\left[\frac{N}{2}\left|\frac{1}{\sqrt{2}}\big(\langle e_1| + \langle e_2|\big)|e_1\rangle\right|^2 + \frac{N}{2}\left|\frac{1}{\sqrt{2}}\big(\langle e_1| + \langle e_2|\big)|e_2\rangle\right|^2\right] = \frac{1}{2}.$$

ii):

$$P = \frac{1}{N}\left[\frac{N}{2}\left|\frac{1}{\sqrt{2}}\big(\langle e_1| - i\langle e_2|\big)|e_1\rangle\right|^2 + \frac{N}{2}\left|\frac{1}{\sqrt{2}}\big(\langle e_1| - i\langle e_2|\big)|e_2\rangle\right|^2\right] = \frac{1}{2}.$$

b) i): $P = \dfrac{1}{2}$; ii): $P = \dfrac{1}{2}$.

c) i): $P = \dfrac{1}{2}\left[\left|\dfrac{1}{\sqrt{2}}\big(\langle e_1| + \langle e_2|\big)\Big(\cos\vartheta\,|e_1\rangle + \sin\vartheta\,\mathrm{e}^{\mathrm{i}\varphi}\,|e_2\rangle\Big)\right|^2\right.$

$$\left. + \left|\dfrac{1}{\sqrt{2}}\big(\langle e_1| + \langle e_2|\big)\Big(-\sin\vartheta\,|e_1\rangle + \cos\vartheta\,\mathrm{e}^{\mathrm{i}\varphi}\,|e_2\rangle\Big)\right|^2\right] = \dfrac{1}{2}.$$

ii): $P = \dfrac{1}{2}$.

d) $P = \dfrac{1}{N}\left[\dfrac{N}{4}\big|\langle e_{\vartheta\varphi}\,|\,e_1\rangle\big|^2 + \dfrac{N}{4}\big|\langle e_{\vartheta\varphi}\,|\,e_2\rangle\big|^2 + \dfrac{N}{4}\left|\langle e_{\vartheta\varphi}\,|\dfrac{1}{\sqrt{2}}\big(|e_1\rangle + |e_2\rangle\big)\right|^2\right.$

$$\left. + \dfrac{N}{4}\left|\langle e_{\vartheta\varphi}\,|\dfrac{1}{\sqrt{2}}\big(|e_1\rangle - |e_2\rangle\big)\right|^2\right] = \dfrac{1}{2}.$$

e) The four statistical mixtures are indistinguishable since for all of them the probability of finding photons in whatever state always is $1/2$.

4.7

a) The equivalence of the two statistical mixtures can be proven either show-ing that the transition probability to the generic state $|s\rangle$ is the same in both cases, or – equivalently – that the mean value of any observable is the same in the two mixtures. The transition probability to $|s\rangle$ is

$$P = \dfrac{1}{2\pi}\int |\langle s\,|\,s_\varphi\rangle|^2\mathrm{d}\varphi = \dfrac{1}{2\pi}\int \big|a\langle s\,|\,s_1\rangle + b\langle s\,|\,s_2\rangle\,\mathrm{e}^{\mathrm{i}\varphi}\big|^2\mathrm{d}\varphi$$

$$= a^2|\langle s\,|\,s_1\rangle|^2 + b^2|\langle s\,|\,s_2\rangle|^2$$

(the terms containing φ have vanishing integral) and equals the transi-tion probability to the state $|s\rangle$ in the case of the statistical mixture $\{\,|s_1\rangle,\ \nu_1 = a^2;\ \ |s_2\rangle,\ \nu_2 = b^2\,\}$.
As for the mean values, one has:

$$\langle\!\langle\xi\rangle\!\rangle = \dfrac{1}{2\pi}\int\Big(a^2\langle s_1|\,\xi\,|s_1\rangle + b^2\langle s_2|\,\xi\,|s_2\rangle + 2\Re e\,(ab\,\mathrm{e}^{\mathrm{i}\varphi}\langle s_1\,|\,\xi\,|\,s_2\rangle)\Big)\,\mathrm{d}\varphi$$

$$= a^2\langle s_1\,|\,\xi\,|\,s_1\rangle + b^2\langle s_2\,|\,\xi\,|\,s_2\rangle$$

equal to the mean value in the case of the mixture
$\{\,|s_1\rangle,\ \nu_1 = a^2;\ \ |s_2\rangle,\ \nu_2 = b^2\,\}$.

4.8

a) $P_i = \sum_j \nu_j |\langle u_i\,|\,u_j\rangle|^2$;

$$\sum_i P_i = \sum_{ij}\nu_j|\langle u_i\,|\,u_j\rangle|^2 \geq \sum_j \nu_j|\langle u_j\,|\,u_j\rangle|^2 = 1.$$

The equality $\sum_i P_i = 1$ holds only if $\langle u_i\,|\,u_j\rangle = \delta_{ij}$.

b) $\langle\langle\xi\rangle\rangle = \sum_i \nu_i \langle u_i \mid \xi \mid u_i \rangle = \sum_n \sum_i \nu_i \langle u_i \mid n \rangle\langle n \mid \xi \mid u_i \rangle$

$= \sum_{ni} \nu_i \langle n \mid \xi \mid u_i \rangle\langle u_i \mid n \rangle = \sum_{ni} \nu_i \langle n \mid \xi E_{u_i} \mid n \rangle = \mathrm{Tr}\,(\varrho\,\xi)\,.$

c) i) The first equality follows from $\sum_n \langle n \mid E_{u_i} \mid n \rangle = 1$ and $\sum_i \nu_i = 1$, or from $\langle\langle \mathbb{1} \rangle\rangle = 1$. ii) Note that $\varrho = \varrho^\dagger$. The eigenvalues of ϱ are nonnegative because for any state, and in particular if $\mid s \rangle$ is an eigenvector of ϱ, $\langle s \mid \varrho \mid s \rangle \geq 0$. iii) Finally, since the trace is invariant (i.e. it does not depend on the basis $\mid n \rangle$), it coincides with the sum of the eigenvalues, each of them being, as a consequence, ≤ 1; therefore the sum of their squares is ≤ 1 and equals 1 if and only if ϱ has only one nondegenerate eigenvalue equal to 1 and all the others are vanishing: in the latter case ϱ is a one-dimensional projector and, in conclusion, the statistical mixture is the pure state onto which ϱ projects. Alternatively:

$\mathrm{Tr}\,\varrho^2 = \sum_{ij}\sum_n \nu_i\nu_j\,\langle n \mid u_i \rangle\langle u_i \mid u_j \rangle\langle u_j \mid n \rangle = \sum_{ij}\nu_i\nu_j\,|\langle u_i \mid u_j \rangle|^2$

$\leq \sum_{ij}\nu_i\nu_j = \left(\sum_i \nu_i\right)^2 = 1$

where the equality sign holds if and only if $\langle u_i \mid u_j \rangle = 1$ for any i, j, i.e. if $\mid u_i \rangle = \mid u_j \rangle \equiv \mid u \rangle$ for any i, j, in which case $\varrho = \mid u \rangle\langle u \mid$ is a projector.

4.9

a) In the first three cases ϱ is a half of the sum of two projectors onto orthonormal vectors, i.e. ϱ is $\frac{1}{2}$ times the identity operator on $\mathcal{H}_2$: $\varrho = \frac{1}{2}\mathbb{1}_2$. In the fourth case $\varrho = \frac{1}{4}\mathbb{1}_2 + \frac{1}{4}\mathbb{1}_2 = \frac{1}{2}\mathbb{1}_2$.

As expected, the statistical operator is the same in all the four cases because, as seen in Problem 4.6, the four statistical mixtures are not operatively distinguishable. When, in a space of finite dimension, the statistical operator is a multiple of the identity, one says that the mixture is completely incoherent; in the case of the polarization states of photons one says that the light is unpolarized. One has:

$\varrho^2 = \frac{1}{4}\mathbb{1}_2 \quad \Rightarrow \quad \mathrm{Tr}\,\varrho^2 = \frac{1}{2}\,.$

b) $\varrho = \mid e_{\sigma_+} \rangle\langle e_{\sigma_+} \mid$. In terms of the vectors $\mid e_1 \rangle$, $\mid e_2 \rangle$:

$\varrho = \frac{1}{2}(\mid e_1 \rangle + \mathrm{i}\mid e_2 \rangle)(\langle e_1 \mid - \mathrm{i}\langle e_2 \mid) = \frac{1}{2}\mathbb{1}_2 + \frac{1}{2}\mathrm{i}(\mid e_2 \rangle\langle e_1 \mid - \mid e_1 \rangle\langle e_2 \mid)\,.$

$\varrho^2 = \mid e_{\sigma_+} \rangle\langle e_{\sigma_+} \mid e_{\sigma_+} \rangle\langle e_{\sigma_+} \mid = \mid e_{\sigma_+} \rangle\langle e_{\sigma_+} \mid = \varrho\,.$

4.10

a) The state of the emerging photon is a pure state. Indeed, it is possible to recombine the reflected component with the transmitted one in such a way as they can interfere (this is not possible when the state is a statistical mixture): it is sufficient to add two reflecting mirrors and a semi-transparent mirror to build up the Mach–Zehnder interferometer (see Problem 3.1).

b) The fact that the counter C does not record the arrival of the photon is, in any event, the result of a measurement: the measurement of the observable that gives "yes or no" as answer (dichotomic variable). Owing to the measurement postulate, the state is $|A\rangle$.

c) The state of the photon just before reaching the counter is

$$|X\rangle = \frac{1}{\sqrt{2}}|A\rangle + \frac{1}{2}|B\rangle + \frac{1}{2}|C\rangle .$$

Then, if the counter C does not record the arrival of the photon (answer 'no'), the state after the measurement is the projection of $|X\rangle$ onto the space orthogonal to $|C\rangle$, that is (N is the normalization factor)

$$N\left(\frac{1}{\sqrt{2}}|A\rangle + \frac{1}{2}|B\rangle\right) = \sqrt{\frac{2}{3}}|A\rangle + \sqrt{\frac{1}{3}}|B\rangle .$$

Applying the measurement postulate only to the part of the state reflected by the first mirror is wrong: in this case one would be led to the conclusion that $\frac{1}{2}(|B\rangle + |C\rangle) \to \frac{1}{\sqrt{2}}|B\rangle$ and that the state after the measurement should be $\frac{1}{\sqrt{2}}(|A\rangle + |B\rangle)$. If the latter conclusion were correct, one should observe, after letting many photons in the device, that equal numbers of photons emerge in the states $|A\rangle$ and $|B\rangle$. It is instead evident that 50% emerge in the state $|A\rangle$, 25% in the state $|B\rangle$ and 25% in the state $|C\rangle$.

4.11

a) Since the light is totally unpolarized, the photons are (as long as polarization is concerned) in a statistical mixture described, in the space $\mathcal{H}_2$ of the polarization states, by the statistical operator (see Problems 4.8 and 4.9) $\varrho = \frac{1}{2}\mathbb{1}_2$. As ϱ is a multiple of the identity, the mixture is equivalent to the mixture consisting of 50% of photons in whatever state of polarization and 50% in the orthogonal state: as a consequence, no matter how the system of photons has been prepared, it is in all respects equivalent to a statistical mixture in which 50% of the photons are linearly polarized parallel to the transmission axis $\hat{n}$ of P$_S$ and 50% in the direction orthogonal to $\hat{n}$. For such a collection of photons only 50% cross P$_S$, whatever the direction $\hat{n}$ may be. So the interference figure, independent of the polarization of the photons (see Problem 3.9), does not change when the direction $\hat{n}$ is varied.

b) Compared with the preceding case, the intensity of the interference pattern is doubled, all the other features remain unchanged. The fact the light is not polarized is not a problem, since "each photon interferes with itself".

c) The interference fringes are not observed, only the sum of the intensities of the diffraction patterns produced by the two slits is visible, since any photon that has crossed the screen with the slits is either in the state $|A\rangle$ or in the state $|B\rangle$ (see Problem 4.1), i.e. in the statistical mixture $\{|A\rangle, \nu_1 = \frac{1}{2}; |B\rangle, \nu_2 = \frac{1}{2}\}$, not in a coherent superposition. This means that the implication "ignorance of the crossed slit $\Rightarrow$ observability of the fringes" is false. In addition, the fact that the sequence of openings and closings of the shutters is 'practically' not knowable does not mean it is 'in principle' impossible to know.

4.12

a) Since $p_i = p_i^\dagger$, $q_i = q_i^\dagger$ and $[q_i , p_j] = 0$ for $i \neq j$, one has:

$$L_x^\dagger = (y\,p_z - z\,p_y)^\dagger = p_z y - p_y z = y\,p_z - z\,p_y = L_x$$

and analogously for the other components.

$$[L_x , L_y] = [(y\,p_z - z\,p_y) , (z\,p_x - x\,p_z)]$$
$$= [y\,p_z , z\,p_x] - [y\,p_z , x\,p_z] - [z\,p_y , z\,p_x] + [z\,p_y , x\,p_z] .$$

$$[y\,p_z , z\,p_x] = y\,[p_z , z\,p_x] + [y , z\,p_x]\,p_z$$
$$= y\,z\,[p_z , p_x] + y\,[p_z , z]\,p_x + z\,[y , p_x]\,p_z + [y , z]\,p_x p_z$$
$$= 0 - i\,\hbar\,y\,p_x + 0 + 0\,;$$

$$[y\,p_z , x\,p_z] = 0\,; \quad [z\,p_y , z\,p_x] = 0\,; \quad [z\,p_y , x\,p_z] = x\,[z , p_z]\,p_y = i\,\hbar\,x\,p_y$$

therefore:

$$[L_x , L_y] = i\,\hbar\,(x\,p_y - y\,p_x) = i\,\hbar\,L_z .$$

The others are obtained by cyclically permuting: $x \to y \to z \to x$:

$$[L_y , L_z] = i\,\hbar\,L_x\,; \qquad [L_z , L_x] = i\,\hbar\,L_y .$$

b) Since L_x, L_y, L_z are Hermitian operators, for any state $|\,A\,\rangle$ one has $\langle\,A\,|\,L_x^2\,|\,A\,\rangle = \langle\,A\,|\,L_x^\dagger L_x\,|\,A\,\rangle \geq 0$ because the last term is the square of the norm of the vector $L_x\,|\,A\,\rangle$; in addition, $\langle\,A\,|\,L_x^\dagger L_x\,|\,A\,\rangle = 0$ if and only if $L_x\,|\,A\,\rangle = 0$, i.e. $|\,A\,\rangle$ is an eigenstate of L_x belonging to the eigenvalue 0. As a consequence:

$$\langle\,s\,|\,L_x^2\,|\,s\,\rangle + \langle\,s\,|\,L_y^2\,|\,s\,\rangle + \langle\,s\,|\,L_z^2\,|\,s\,\rangle \equiv \langle\,s\,|\,\vec{L}^{\,2}\,|\,s\,\rangle = 0 \quad \Rightarrow$$

$$\langle\,s\,|\,L_x^2\,|\,s\,\rangle = \langle\,s\,|\,L_y^2\,|\,s\,\rangle = \langle\,s\,|\,L_z^2\,|\,s\,\rangle = 0 \quad \Rightarrow$$

$$L_x\,|\,s\,\rangle = L_y\,|\,s\,\rangle = L_z\,|\,s\,\rangle = 0 .$$

Vice versa, let

$$L_x\,|\,s\,\rangle = m_1\,|\,s\,\rangle\,; \qquad L_y\,|\,s\,\rangle = m_2\,|\,s\,\rangle\,; \qquad L_z\,|\,s\,\rangle = m_3\,|\,s\,\rangle$$

(m_1, m_2, m_3 are the eigenvalues), then from the commutation rules one has:

$$i\,\hbar\,L_z\,|\,s\,\rangle \equiv (L_x L_y - L_y L_x)\,|\,s\,\rangle = (m_1 m_2 - m_2 m_1)\,|\,s\,\rangle = 0 \quad \Rightarrow$$

$$L_z\,|\,s\,\rangle = 0 \Rightarrow m_3 = 0$$

and analogously $m_1 = m_2 = 0$, so $\vec{L}^{\,2}\,|\,s\,\rangle = 0$.

c) Let $L_z | m \rangle = m | m \rangle$; then in the state $| m \rangle$ $\Delta L_z = 0$, and

$$0 = \Delta L_z\, \Delta L_x \geq \frac{1}{2} |\langle m | [L_z, L_x] | m \rangle| = \frac{\hbar}{2} |\langle m | L_y | m \rangle|$$

$$\Rightarrow \langle m | L_y | m \rangle = 0$$

and likewise $\langle m | L_x | m \rangle = 0$.

4.13

a) The equality holds only if A and B commute. Indeed e^{A+B} is symmetric in A and B, whereas if A and B do not commute, $e^A e^B$ is not such: $e^A e^B \neq e^B e^A$.

b) Expanding and neglecting the terms of order λ^3 and higher:

$$e^{\lambda(A+B+\frac{\lambda}{2}[A,B])} \simeq 1 + \lambda(A+B) + \frac{\lambda^2}{2}\left(A^2 + AB + BA + B^2 + [A, B]\right)$$

$$= 1 + \lambda(A+B) + \frac{\lambda^2}{2}\left(A^2 + 2AB + B^2\right);$$

$$e^{\lambda A} e^{\lambda B} \simeq \left(1 + \lambda A + \frac{\lambda^2}{2}A^2\right)\left(1 + \lambda B + \frac{\lambda^2}{2}B^2\right)$$

$$\simeq 1 + \lambda(A+B) + \frac{\lambda^2}{2}\left(A^2 + 2AB + B^2\right).$$

Note that it has not been necessary to take advantage of the assumption that the commutator of A and B is a c–number: the assumption becomes necessary starting with the terms of order λ^3.

4.14

a) It is convenient (see Problem 4.1) to represent the generic state, obtained as superposition of $|0\rangle$ and $|1\rangle$, in the form $a|0\rangle + b e^{i\varphi}|1\rangle$, with a, b real numbers and $a^2 + b^2 = 1$. As the mean values of q and p in the eigenstates of the Hamiltonian are vanishing, one has:

$$\bar{q} = ab\left(e^{i\varphi}\langle 0 | q | 1 \rangle + e^{-i\varphi}\langle 1 | q | 0 \rangle\right)$$

$$\bar{p} = ab\left(e^{i\varphi}\langle 0 | p | 1 \rangle + e^{-i\varphi}\langle 1 | p | 0 \rangle\right)$$

$$q = -i\sqrt{\frac{\hbar}{2m\omega}}\,(\eta^\dagger - \eta), \qquad p = \sqrt{\frac{m\omega\hbar}{2}}\,(\eta^\dagger + \eta).$$

Therefore, since $\eta|0\rangle = 0$, $\eta^\dagger|0\rangle = |1\rangle$, $\eta|1\rangle = |0\rangle$, and $\langle 0 | \eta^\dagger = 0$, one has:

$$\langle 0 | q | 1 \rangle = i\sqrt{\frac{\hbar}{2m\omega}} = -\langle 1 | q | 0 \rangle; \quad \langle 0 | p | 1 \rangle = \sqrt{\frac{m\omega\hbar}{2}} = \langle 1 | p | 0 \rangle$$

$$\Rightarrow \quad \bar{q} = -2\sqrt{\frac{\hbar}{2m\omega}}\, ab\, \sin\varphi; \qquad \bar{p} = 2\sqrt{\frac{m\omega\hbar}{2}}\, ab\, \cos\varphi.$$

So $\bar{q}$ is a maximum for $a = b = \frac{1}{\sqrt{2}}$, $\varphi = -\pi/2$: $\frac{1}{\sqrt{2}}(|0\rangle - i\,|1\rangle)$, and $\bar{p}$ is a maximum for $a = b = \frac{1}{\sqrt{2}}$, $\varphi = 0$: $\frac{1}{\sqrt{2}}(|0\rangle + |1\rangle)$.

b) As $\langle 0 \mid H \mid 1 \rangle = \langle 1 \mid H \mid 0 \rangle = 0$, $\quad \overline{H} = \hbar\omega\left(\frac{1}{2}a^2 + \frac{3}{2}b^2\right)$. So, in both the states found in a), one has $\overline{H} = \hbar\omega$.

c) The mean values of q and p in the mixture are vanishing, as the "interference terms" between the states $|0\rangle$ and $|1\rangle$ now do not contribute to their calculation. The mean value of H is $\overline{H} = \hbar\omega\left(\frac{1}{2}\nu_0 + \frac{3}{2}\nu_1\right)$.

4.15

a) $\left[H, pq\right] = p\left[\dfrac{p^2}{2m}, q\right] + \left[\dfrac{m\omega^2}{2}\,q^2, p\right]q = -i\,\hbar\left(\dfrac{p^2}{m} - m\omega^2 q^2\right) = -2i\,\hbar(T - V)$

where T is the kinetic energy and V the potential energy. Take the mean value of both sides in an eigenstate of the energy:

$$-2i\,\hbar\Big(\langle E \mid T \mid E\rangle - \langle E \mid V \mid E\rangle\Big) = E\langle E \mid pq \mid E\rangle - \langle E \mid pq \mid E\rangle E = 0$$

$$\Rightarrow \langle E \mid T \mid E\rangle = \langle E \mid V \mid E\rangle = \tfrac{1}{2}E, \qquad \langle E \mid E\rangle = 1 .$$

b) As in Problem 4.14, let us represent the generic state, obtained as superposition of $|0\rangle$ and $|1\rangle$, in the form $a\,|0\rangle + b\,e^{i\varphi}\,|1\rangle$, with a, b real numbers and $a^2 + b^2 = 1$. As

$$\langle 0 \mid p^2 \mid 1\rangle \propto \langle 0 \mid (\eta + \eta^\dagger)^2 \mid 1\rangle = \langle 0 \mid \eta^2 + \eta\,\eta^\dagger + \eta^\dagger\eta + (\eta^\dagger)^2 \mid 1\rangle = 0$$

and $\langle 0 \mid q^2 \mid 1\rangle \propto \langle 0 \mid (\eta - \eta^\dagger)^2 \mid 1\rangle = 0$ too, one has:

$$\overline{T} = \Big(a\langle 0| + b\,e^{-i\varphi}\,\langle 1|\Big)\,T\,\Big(a|0\rangle + b\,e^{i\varphi}\,|1\rangle\Big)$$

$$= a^2\langle 0 \mid T \mid 0\rangle + b^2\langle 1 \mid T \mid 1\rangle = \hbar\omega\left(\frac{1}{4}a^2 + \frac{3}{4}b^2\right);$$

$$\overline{V} = a^2\langle 0 \mid V \mid 0\rangle + b^2\langle 1 \mid V \mid 1\rangle = \hbar\omega\left(\frac{1}{4}a^2 + \frac{3}{4}b^2\right).$$

c) $\langle 0 \mid p^2 \mid 2\rangle = \dfrac{m\omega\,\hbar}{2}\langle 0 \mid \eta^2 \mid 2\rangle = \dfrac{m\omega\,\hbar}{\sqrt{2}}\langle 2 \mid 2\rangle = \dfrac{m\omega\,\hbar}{\sqrt{2}} = \langle 2 \mid p^2 \mid 0\rangle$.

In the state $a\,|0\rangle + b\,e^{i\varphi}\,|2\rangle$, with a, b real numbers and $a^2 + b^2 = 1$, one has:

$$\overline{T} = \Big(a\langle 0| + b\,e^{-i\varphi}\,\langle 2|\Big)\,T\,\Big(a|0\rangle + b\,e^{i\varphi}\,|2\rangle\Big)$$

$$= \hbar\omega\left(\frac{1}{4}a^2 + \frac{5}{4}b^2 + \frac{\sqrt{2}}{2}\,ab\cos\varphi\right);$$

$$\overline{V} = \overline{H} - \overline{T} = \hbar\omega\left(\frac{1}{2}a^2 + \frac{5}{2}b^2\right) - \overline{T} = \hbar\omega\left(\frac{1}{4}a^2 + \frac{5}{4}b^2 - \frac{\sqrt{2}}{2}ab\cos\varphi\right).$$

5

Representations

Representations; unitary transformations; von Neumann theorem; coherent states; Schrödinger and momentum representations; degeneracy theorem.

5.1 In the space $\mathcal{H}_2$ of the polarization states of photons, consider the basis $|e_1\rangle$, $|e_2\rangle$ consisting of the vectors that represent two rectilinear states of polarization along two orthogonal directions. Determine the matrix that, in the above basis, represents the following observables (see Problem 4.3).

a) The observable Π that has $|e_1\rangle$, $|e_2\rangle$ as eigenstates corresponding to the eigenvalues $+1$ and -1.

b) The observable Π_ϑ whose eigenstates corresponding to the eigenvalues $+1$ and -1 are the *rectilinear* polarization state $\cos\vartheta\,|e_1\rangle + \sin\vartheta\,|e_2\rangle$ and the state orthogonal to it.

c) Is the relative phase between the basis vectors $|e_1\rangle$, $|e_2\rangle$ completely determined by the information contained in the previous question? Write the vectors relative to the states of rectilinear polarization in the basis $|\tilde{e}_1\rangle = e^{i\varphi_1}\,|e_1\rangle$, $|\tilde{e}_2\rangle = e^{i\varphi_2}\,|e_2\rangle$.

d) The observable Π_σ whose eigenstates corresponding to the eigenvalues $+1$ and -1 are the circular polarization states $|e_{\sigma\pm}\rangle = \frac{1}{\sqrt{2}}(|e_1\rangle \pm i\,|e_2\rangle)$.

e) The observable $\Pi_{\vartheta\varphi}$ whose eigenstates corresponding to the eigenvalues $+1$ and -1 are the state $\cos\vartheta\,|e_1\rangle + \sin\vartheta\,e^{i\varphi}\,|e_2\rangle$ and the state orthogonal to it.

5.2 Consider a statistical mixture of photons in which 70% are rectilinearly polarized along direction $\hat{e}_1$ and 30% are rectilinearly polarized along the orthogonal direction $\hat{e}_2$.

a) Write the statistical matrix, associated with the mixture, with respect to the basis $|e_1\rangle$, $|e_2\rangle$. Give a 'good' definition of the *degree of polarization* $\mathcal{P}$ of a mixture ($0 \le \mathcal{P} \le 1$), such that, in the case considered, its value is 0.4 (40%).

b) Always with respect to the basis $|e_1\rangle$, $|e_2\rangle$, write the statistical matrices that correspond to the pure states $|e_{\sigma\pm}\rangle$.

© Springer International Publishing AG 2017
E. d'Emilio and L.E. Picasso, *Problems in Quantum Mechanics*,
UNITEXT for Physics, DOI 10.1007/978-3-319-53267-7_5

c) Give an example of a statistical mixture prepared in a way different from the one given in the beginning: $\{|e_1\rangle, \nu_1 = 0.7 ; \ |e_2\rangle, \nu_2 = 0.3\}$, but equivalent to it.

Consider now the statistical mixture described, in the basis $|e_1\rangle$, $|e_2\rangle$, by the matrix:

$$\varrho = \begin{pmatrix} 0.5 & -0.2\,\mathrm{i} \\ +0.2\,\mathrm{i} & 0.5 \end{pmatrix} .$$

d) Calculate the degree of polarization of the mixture and determine the type of polarization (either rectilinear in some direction, or left/right circular, etc.).

5.3 Given a three-dimensional Hilbert space, consider the two observables ξ and η that, with respect to the basis $|1\rangle$, $|2\rangle$, $|3\rangle$, are represented by the matrices:

$$\xi \rightarrow \begin{pmatrix} \xi_1 & 0 & 0 \\ 0 & \xi_1 & 0 \\ 0 & 0 & \xi_3 \end{pmatrix}, \quad \xi_1 \neq \xi_3 ; \qquad \eta \rightarrow \begin{pmatrix} 0 & \gamma & 0 \\ \gamma^* & 0 & 0 \\ 0 & 0 & \eta_3 \end{pmatrix} .$$

a) Verify that the two observables are compatible with each other and find the basis with respect to which they are simultaneously diagonal.

b) Are the two observables a complete set of compatible observables?

5.4 Given the basis $|e_1\rangle$, $|e_2\rangle$ write the matrices that represent the unitary transformations corresponding to the following changes of basis:

a) from $|e_1\rangle$, $|e_2\rangle$ to $\cos\vartheta\,|e_1\rangle + \sin\vartheta\,|e_2\rangle$, $-\sin\vartheta\,|e_1\rangle + \cos\vartheta\,|e_2\rangle$;

b) from $|e_1\rangle$, $|e_2\rangle$ to $\frac{1}{\sqrt{2}}(|e_1\rangle \pm \mathrm{i}\,|e_2\rangle)$.

5.5 A diatomic molecule can capture an electron and we assume the state space of the electron is two-dimensional and generated by the orthogonal vectors $|1\rangle$, $|2\rangle$ that respectively represent the state of the electron captured by either the first or the second atom.

a) Write the matrix representing, with respect to the basis $|1\rangle$, $|2\rangle$, the most general electron Hamiltonian H and determine its eigenvalues.

b) Show that it is always possible to choose the phases of the basis vectors $|\tilde{1}\rangle = \mathrm{e}^{\mathrm{i}\varphi_1}\,|1\rangle$, $|\tilde{2}\rangle = \mathrm{e}^{\mathrm{i}\varphi_2}\,|2\rangle$ so that the matrix representing H in the new basis $|\tilde{1}\rangle$, $|\tilde{2}\rangle$ is real.

Assume that the molecule consists of two identical atoms. The equality of the atoms entails the invariance of H under the unitary transformation that exchanges the *states* (not necessarily the vectors) represented by $|\tilde{1}\rangle$ and $|\tilde{2}\rangle$:

$$\Pi\,|\tilde{1}\rangle = \alpha\,|\tilde{2}\rangle, \quad \Pi\,|\tilde{2}\rangle = \beta\,|\tilde{1}\rangle, \qquad |\alpha| = |\beta| = 1 .$$

c) The requirement that $\Pi^\dagger H \Pi = H$ (or equivalently, since $\Pi^\dagger = \Pi^{-1}$, that $\Pi H = H \Pi$) implies some restrictions both on the matrix that represents H in the basis $|\tilde{1}\rangle$, $|\tilde{2}\rangle$ and on α, β. Find these restrictions. Show that, provided Π is redefined by a multiplicative factor ($\Pi \to e^{i\chi}\Pi$), it is possible to have $\alpha = \beta = 1$.

Assume now that the molecule is triatomic (the three atoms not necessarily being equal) and that it can capture the electron in the three orthogonal states $|1\rangle$, $|2\rangle$, $|3\rangle$.

d) By suitably choosing the phases of the basis vectors, how many of the elements of the matrix representing H can be made real?

5.6 Consider the scale transformation $q \to \tilde{q} = \lambda q$, $p \to \tilde{p} = \lambda^{-1}p$ where λ is an arbitrary real parameter.

a) Verify that the transformation is a canonical transformation.

The Hamiltonian of a harmonic oscillator of mass m and angular frequency ω is

$$H = \frac{p^2}{2m} + \frac{1}{2}m\omega^2 q^2 .$$

b) Exploit the *von Neumann theorem* (there exists a unitary operator U implementing the transformation $q \to \tilde{q} \equiv \lambda q = U q U^{-1}$, $p \to \tilde{p} \equiv \lambda^{-1}p = U p U^{-1}$) to show that the two Hamiltonians H_1 and H_2:

$$H_1 = \frac{p^2}{2m_1} + \frac{1}{2}m_1\omega^2 q^2 , \qquad H_2 = \frac{p^2}{2m_2} + \frac{1}{2}m_2\omega^2 q^2$$

have the same eigenvalues. (As a consequence the eigenvalues of the Hamiltonian of a harmonic oscillator of given ω are independent of its mass.) Are the eigenvectors of H_1 and H_2 the same?

The Hamiltonian of the hydrogen atom is

$$H = \frac{\vec{p}^2}{2m} - \frac{e^2}{r} .$$

c) Show that the discrete eigenvalues of H depend on the charge e and the mass m only through the product me^4. Verify this is true for the energy levels given by the Bohr theory.

5.7 Consider a harmonic oscillator of mass m and angular frequency ω and the canonical transformation $q \to \tilde{q} = \Lambda p$, $p \to \tilde{p} = -\Lambda^{-1}q$, where Λ is an arbitrary real parameter (with dimensions $[\Lambda] = \mathrm{T}\,\mathrm{M}^{-1}$).

a) Show that there exist values of Λ such that the transformation $q \to \tilde{q}$, $p \to \tilde{p}$ is an invariance transformation for the Hamiltonian H, and find them.

b) Exploit the above invariance transformation to show that, for any eigenstate of H, the mean value of the kinetic energy equals the mean value of the potential energy.

c) Calculate the product $\Delta q \, \Delta p$ of the uncertainties of q and p in the eigenstates of H.

d) Show that the mean value of $qp + pq$ in any eigenstate of H is vanishing.

5.8 Let $U(a) = e^{-ipa/\hbar}$ be the operator that translates coordinates: $U(a) q U^{-1}(a) = q - a$, and $V(b) = e^{iqb/\hbar}$ be the operator that translates momenta: $V(b) p V^{-1}(b) = p - b$ (a and b being real numbers).

a) Show that $\tilde{p} \equiv a\,p - b\,q$ and $\tilde{q} \equiv \frac{1}{2}(a^{-1}q + b^{-1}p)$ are canonically conjugate variables.

b) Exploit the preceding result to show that the operators $U(a)V(b) = e^{-ipa/\hbar}e^{iqb/\hbar}$ and $W(a,b) \equiv e^{-i(pa-qb)/\hbar}$ induce the same canonical transformation and that, as a consequence, $U(a)V(b) = e^{i\varphi} W(a,b)$ (von Neumann). Use the Baker–Campbell–Hausdorff identity (see Problem 4.13) to calculate the phase factor $e^{i\varphi}$.

c) Find the unitary operator $G(v)$ that implements the Galilei transformation for a particle of mass m:
$$q \to \tilde{q} \equiv G(v)\, q\, G^{-1}(v) = q - vt\,, \quad p \to \tilde{p} \equiv G(v)\, p\, G^{-1}(v) = p - mv\,.$$

5.9 Given any one-dimensional system, consider the operator:
$$\eta_\lambda = \frac{1}{\sqrt{2\lambda\hbar}}(p - i\lambda q)\,, \qquad \lambda > 0\,.$$
From the theory of the harmonic oscillator we know that there exists a unique state $|0_\lambda\rangle$ such that $\eta_\lambda|0_\lambda\rangle = 0$. From now on we shall simply write $|0\rangle$, η instead of $|0_\lambda\rangle$, η_λ. Let
$$|\alpha\rangle \equiv V(b)\,U(a)\,|0\rangle\,, \qquad \alpha = \frac{b}{\sqrt{2\lambda\hbar}} - i\sqrt{\frac{\lambda}{2\hbar}}\,a \equiv \alpha_1 + i\alpha_2$$
where $U(a) = e^{-ipa/\hbar}$ and $V(b) = e^{iqb/\hbar}$ (see Problem 5.8) are the translation operators for coordinates and momenta respectively. (The states $|\alpha\rangle$ are named *coherent states*.)

a) Show that for any $\alpha \in \mathbb{C}$, $\eta|\alpha\rangle = \alpha|\alpha\rangle$.

b) Show that $\bar{q} = \langle\alpha\,|\,q\,|\,\alpha\rangle = a$, $\bar{p} = \langle\alpha\,|\,p\,|\,\alpha\rangle = b$ and that the states $|\alpha\rangle$ are minimum uncertainty states. (In the solution we will show that the converse also holds, namely that all minimum uncertainty states are coherent states.)

c) Determine the representatives $\langle n\,|\,\alpha\rangle$ of the vector $|\alpha\rangle$ in the basis $|n\rangle = \frac{1}{\sqrt{n!}}(\eta^\dagger)^n\,|0\rangle$ and the scalar product $\langle\alpha\,|\,\beta\rangle$ of two coherent states.

d) Show that there exists no vector orthogonal to all the $|\alpha\rangle$ vectors.

5.10 Consider a one-dimensional harmonic oscillator of mass m and angular frequency ω in the coherent state relative to the oscillator (i.e., see Problem 5.9, with $\lambda = m\omega$):

$$|a\rangle = V(b)\,U(a)\,|0\rangle, \qquad \alpha = \frac{b}{\sqrt{2m\omega\,\hbar}} - i\sqrt{\frac{m\omega}{2\hbar}}\,a$$

where $|0\rangle$ is the ground state of the oscillator.

a) Calculate the mean value $\overline{H}$ of the energy in the state $|a\rangle$, the degree of excitation of the oscillator, defined as $\overline{n} \equiv \overline{H}/\hbar\omega - \frac{1}{2}$, and the dispersion $\Delta n = \Delta H/\hbar\omega$.

b) Calculate the mean values of kinetic and potential energy in the state $|a\rangle$.

c) Calculate the uncertainties Δq and Δp in the state $|a\rangle$ and verify that $\alpha = \overline{p}/(2\Delta p) - i\,\overline{q}/(2\Delta q)$.

5.11 Consider a one-dimensional harmonic oscillator endowed with a charge e and subject to a uniform and constant electric field $\mathcal{E}$ oriented in the direction of the motion of the oscillator.

a) Write the Hamiltonian and find the energy levels of the system.

Assume the system is in the ground state. At a given instant the electric field is turned off and afterwards the energy of the oscillator is measured.

b) Calculate the probability of finding the eigenvalue $E_n = (n + \frac{1}{2})\hbar\omega$ of the Hamiltonian that is now the Hamiltonian of the free oscillator, i.e. of the oscillator in the absence of the electric field.

5.12 Consider a harmonic oscillator of mass m and angular frequency ω.

a) Calculate the mean value of q^6 in the ground state. (It may help to consider it as the squared norm of the vector $q^3\,|0\rangle$.) Show that the result implies that the probability of finding the oscillator out of the region accessible to a classical oscillator with the same energy is nonvanishing.

As for momentum, a similar result holds.

b) Find the interval of the allowed values of the momentum for a classical oscillator with energy $E_0 = \frac{1}{2}\hbar\omega$. By a wise use of the result of the previous question, show that for a quantum oscillator in the ground state the probability of finding the momentum out of that interval is nonvanishing.

5.13 Consider a particle in n dimensions.

a) Say what the dimensions (in terms of length L, mass M, time T) of normalized wavefunctions are in the Schrödinger representation.

Consider a particle in one dimension in the state $|A\rangle$ whose wavefunction is

$$\psi_A(x) = N\,e^{-a\,x^4}, \qquad a > 0$$

where N is the normalization coefficient.

b) Show that $N = C\,a^\gamma$, where C is a constant. Find the value of γ.

c) Show that, even not knowing the value of C, the mean value of q^4 in the state $|A\rangle$ can be explicitly calculated.

d) For what other values of n can the mean value of q^n in the state $|A\rangle$ be explicitly calculated?

5.14   Consider a harmonic oscillator of mass m and angular frequency ω.

a) Start from the equation $\eta\,|0\rangle = 0$ and find the normalized wavefunction $\varphi_0(k)$ for the ground state $|0\rangle$ in the momentum representation.

Given the normalized eigenfunctions of the Hamiltonian H in the Schrödinger representation:

$$\psi_n(x) = \frac{1}{\sqrt{2^n\,n!}}\left(\frac{m\omega}{\pi\hbar}\right)^{1/4} H_n(\sqrt{m\omega/\hbar}\,x)\,e^{-(m\omega/2\hbar)\,x^2}$$

it is possible to find the eigenfunctions of H in the momentum representation without resorting to the Fourier transform. To this end it is convenient to exploit the invariance of H under the transformation:

$$U\,q\,U^{-1} = -\frac{p}{m\omega}, \qquad U\,p\,U^{-1} = m\omega\,q\,.$$

b) Let $|x\rangle$ and $|k\rangle$ be the improper eigenvectors respectively of q and p normalized according to $\langle x' \mid x''\rangle = \delta(x' - x'')$, $\langle k' \mid k''\rangle = \delta(k' - k'')$. Show that:

$$U\,|k\rangle = \frac{1}{\sqrt{m\omega}}\,|x = k/m\omega\rangle\,.$$

In calculating the normalization factor $1/\sqrt{m\omega}$ it may help to recall the property $\delta(x/a) = |a|\,\delta(x)$ of the Dirac delta function.

c) Find the normalized eigenfunctions $\varphi_n(k) \equiv \langle k \mid n\rangle$ of the Hamiltonian in the momentum representation.

5.15

a) Calculate the mean value of $qp + pq$ in the coherent states $|\alpha\rangle$ defined in Problem 5.9.

b) Find the wavefunctions of coherent states in both the momentum and Schrödinger representations.

5.16

a) Show that the mean value of p in any state with real wavefunction $\psi(x)$ vanishes.

b) Calculate the mean value of p in the state described by the wavefunction $\psi(x) = e^{i\varphi(x)}\chi(x)$ with $\varphi(x)$ and $\chi(x)$ real functions.

c) Calculate the mean value of p in the state whose wavefunction is $\psi(x) = \chi(x)\,e^{ikx}$.

5.17 A free particle in one dimension $(H = p^2/2m)$ is in the state $|1\rangle$ whose normalized wavefunction is $\psi_1(x) = (\alpha/\pi)^{1/4}\,e^{-\alpha x^2/2}$.

a) Calculate the mean values of q^2, p^2 and p^4 and show that the odd powers of p have vanishing mean values.

b) Exploit the preceding results and calculate the mean values of p^2 and p^4 in the state $|2\rangle$ whose wavefunction is $\psi_2(x) = (\alpha/\pi)^{1/4}\,e^{-\alpha x^2/2}\,e^{ikx}$.

c) Calculate the energy uncertainty ΔE in the state $|2\rangle$ and show that when $\alpha \ll k^2$, $\Delta E/\overline{E} \simeq 2\,\Delta p/\overline{p}$ (as in classical physics, due to $E \propto p^2$).

5.18 The Schrödinger representation of an operator ξ is given by the function of two variables (actually it is a distribution) $\xi(x,\,y) \equiv \langle x\,|\,\xi\,|\,y\rangle$, where $|x\rangle$, $|y\rangle$ are the improper eigenvectors of position.

a) Given $|A\rangle \overset{\text{SR}}{\longrightarrow} \psi_A(x)$, find the wavefunction $\psi_B(x)$ of the state $|B\rangle = \xi\,|A\rangle$. Given $\xi \overset{\text{SR}}{\longrightarrow} \xi(x,\,y)$, what is the Schrödinger representation of the operator $\xi^\dagger$?

b) Find the Schrödinger representation of the projection operator $E_A = |A\rangle\langle A|$. Show that if $\chi(x)$ is an arbitrary normalized function, the operator $E_\chi \overset{\text{SR}}{\longrightarrow} E_\chi(x,\,y) = \chi(x)\,\chi^*(y)$ is a projection operator that projects onto a one-dimensional manifold.

c) If $|n\rangle \overset{\text{SR}}{\longrightarrow} \psi_n(x)$ is an orthonormal basis, write the completeness relation (or decomposition of the identity) $\sum_n |n\rangle\langle n| = \mathbb{1}$ in the Schrödinger representation.

The trace of an operator ξ (when it exists) is defined as (see Problem 4.8):
$$\text{Tr}\,\xi \equiv \sum_n \langle n\,|\,\xi\,|\,n\rangle$$
where $|n\rangle$ is an arbitrary orthonormal basis.

d) Show that $\text{Tr}\,\xi = \displaystyle\int_{-\infty}^{+\infty} \xi(x,\,x)\,dx$.

e) Given the projection operator $E_{\mathcal{V}}$ onto the manifold $\mathcal{V}$, show that $\text{Tr}\,E_{\mathcal{V}}$ equals the dimension of $\mathcal{V}$.

Consider the operator E whose Schrödinger representation is given by:

$$E(x, y) = \sqrt{\frac{\lambda}{\pi}}\, e^{-(\lambda/2)(x^2+y^2)}\left(1 + 2\lambda\, xy\right).$$

f) Show that E is a projection operator: $E^\dagger = E$, $E^2 = E$, calculate the dimension of the manifold onto which it projects and characterize it.

5.19 A particle in one dimension is in the state:

$$|A\rangle = |A_0\rangle + e^{i\varphi}\, U(a)\, |A_0\rangle$$

where $U(a) = e^{-ipa/\hbar}$ is the translation operator and $|A_0\rangle$ is the state with wavefunction $\psi_0(x) = (2\pi\Delta^2)^{-1/4}\, e^{-x^2/4\Delta^2}$, $\langle A_0 \mid A_0\rangle = 1$.

a) What condition must a and Δ satisfy in order that $\langle A_0 \mid U(a) \mid A_0\rangle$ be negligible? Calculate $\langle A_0 \mid U(a) \mid A_0\rangle$ for $a = 10\Delta$.

From now on we shall assume that $\langle A_0 \mid U(a) \mid A_0\rangle$ is negligible.

b) Determine the probability density $\rho(x)$ for the position of the particle. Within the approximation $\langle A_0 \mid U(a) \mid A_0\rangle \simeq 0$, is it possible to determine the phase φ by means of position measurements?

c) Determine the probability density $\tilde\rho(k)$ for the momentum of the particle.

d) Say what is the required precision for momentum measurements in order to distinguish the state $|A\rangle$ from the statistical mixture

$$\{\,|A_0\rangle,\ \nu_1 = \tfrac{1}{2};\ U(a)\,|A_0\rangle,\ \nu_2 = \tfrac{1}{2}\,\}.$$

5.20 Let $|A\rangle$ and $|B\rangle$ be two states whose wavefunctions in the Schrödinger representation are $\psi_A(x, y, z)$ and $\psi_B(x, y, z) = \psi_A^*(x, y, z)$. (Assume that $\psi_A(x, y, z)$ and $\psi_A^*(x, y, z)$ are not proportional to each other.)

a) Which, among the following observables, may have different mean value in the two states $|A\rangle$ and $|B\rangle$: $f(\vec{q})$; p_i $(i = 1, 2, 3)$; $\vec{p}^{\,2}$; $L_i \equiv (\vec{q} \wedge \vec{p})_i$?

b) Given the wavefunction $\varphi_A(\vec{k})$ of $|A\rangle$ in the momentum representation, find the wavefunction $\varphi_B(\vec{k})$ of $|B\rangle$.

5.21 A particle is in a state whose probability density for the position is

$$\rho(x) = |\psi(x)|^2 = \frac{N^2}{(x^2 + a^2)^2}.$$

a) Say whether the state of the particle is uniquely determined.

b) Is it possible to calculate the mean values of p and q in such a state?

5.22 It is known (see Problem 5.21) that the knowledge of either the probability density $\rho(x)$ for the position or, analogously, $\tilde\rho(k)$ for momentum is not sufficient to determine the state $|A\rangle$ of the particle, i.e. its wavefunction $\psi_A(x)$ and/or $\varphi_A(k)$. Establishing whether the knowledge of both probability densities is sufficient to uniquely identify the state of the particle is the purpose of this problem.

For a particle in one dimension, consider the states $|A\rangle$ and $|B\rangle$ of definite parity described by the wavefunctions (not proportional to each other):

$$|A\rangle \to \psi_A(x) = \pm\psi_A(-x), \qquad |B\rangle \to \psi_B(x) = \psi_A^*(x) = \pm\psi_B(-x).$$

a) Show that both $\rho_A(x) = \rho_B(x)$ and $\tilde\rho_A(k) = \tilde\rho_B(k)$.

Since $|A\rangle$ and $|B\rangle$ are different (by assumption $\psi_A(x)$ and $\psi_B(x)$ are linearly independent) there must exist observables whose mean values in the states $|A\rangle$ and $|B\rangle$ are different.

b) Say which, among the following observables, may have different mean values in the two states: $f(q)$, $g(p)$, $H = p^2/2m + V(q)$, $pq + qp$.

Let $\psi_A(x) = (a/\pi)^{1/4}\,e^{-(a+i\,b)\,x^2/2}$, $a > 0$, $b \in \mathbb{R}$ be the normalized wavefunction of $|A\rangle$.

c) Calculate the mean value of the observable $pq + qp$ in the states $|A\rangle$ and $|B\rangle$.

5.23 Consider a particle in one dimension and the canonical transformations generated by the family of unitary operators ($\alpha \in \mathbb{R}$):

$$U(\alpha) = e^{i\,\alpha(q\,p + p\,q)/2\hbar} \;:\quad \tilde q(\alpha) \equiv U(\alpha)\,q\,U^{-1}(\alpha), \; \tilde p(\alpha) \equiv U(\alpha)\,p\,U^{-1}(\alpha).$$

a) Show that:

$$\frac{d\,\tilde q(\alpha)}{d\alpha} = \tilde q(\alpha), \qquad \frac{d\,\tilde p(\alpha)}{d\alpha} = -\,\tilde p(\alpha)$$

and, by taking into account that $\tilde q(0) = q$, $\tilde p(0) = p$, explicitly determine $\tilde q(\alpha)$ and $\tilde p(\alpha)$.

b) Denoting by $|x\rangle$ and $|k\rangle$ the improper eigenvectors of q and p normalized according to $\langle x' \mid x''\rangle = \delta(x' - x'')$, $\langle k' \mid k''\rangle = \delta(k' - k'')$, show that:

$$U^\dagger(\alpha)\,|x\rangle = e^{\alpha/2}\,|e^\alpha x\rangle, \qquad U^\dagger(\alpha)\,|k\rangle = e^{-\alpha/2}\,|e^{-\alpha}k\rangle.$$

c) If $\psi_A(x) \equiv \langle x \mid A\rangle$ and $\varphi_A(k) \equiv \langle k \mid A\rangle$ stand for the normalized wavefunctions of the state $|A\rangle$ respectively in the Schrödinger and momentum representations, determine the wavefunctions of $|\widetilde A\rangle \equiv U(\alpha)|A\rangle$ in the two representations.

5.24   Consider a triatomic molecule consisting of three identical atoms placed at the vertices of an equilateral triangle. The molecule is twice ionized and the Hamiltonian of the 'third' electron in the field of the three ions is

$$H = \frac{\vec{p}^{\,2}}{2m} + V(|\vec{q} - \vec{a}\,|) + V(|\vec{q} - \vec{b}\,|) + V(|\vec{q} - \vec{c}\,|)$$

where the vectors $\vec{a}$, $\vec{b}$, $\vec{c}$ stand for the position of the three atoms with respect to the center of the molecule (see figure).

a) Taking the origin of the coordinates in the center of the molecule, write the canonical transformation of the variables $\vec{q}$, $\vec{p}$ corresponding to a rotation of $120°$ around the axis orthogonal to the plane containing the atoms and show that this transformation leaves H invariant.

b) Show that H is invariant also under the reflection with respect to the plane orthogonal to the molecule and containing $\vec{c}$ $(x \to -x,\ y \to y,\ z \to z)$. Do the $120°$ rotation and the reflection commute with each other?

c) May H have only nondegenerate eigenvalues?

Let $|1\rangle$, $|2\rangle$, $|3\rangle$, with wavefunctions respectively $\psi_1(\vec{r}) = \psi_0(|\vec{r} - \vec{a}\,|)$, $\psi_2(\vec{r}) = \psi_0(|\vec{r} - \vec{b}\,|)$, $\psi_3(\vec{r}) = \psi_0(|\vec{r} - \vec{c}\,|)$, the three particular states in which the third electron is bound to each of the three atoms. We assume (it is an approximation) that $|1\rangle$, $|2\rangle$, $|3\rangle$ are orthogonal to one another.

The operator that implements the $120°$ rotation, induces the following transformation:

$$U\,|1\rangle = |2\rangle, \quad U\,|2\rangle = |3\rangle, \quad U\,|3\rangle = |1\rangle$$

and the operator that implements the reflection $x \to -x,\ y \to y,\ z \to z$ induces:

$$I_x\,|1\rangle = |2\rangle, \quad I_x\,|2\rangle = |1\rangle, \quad I_x\,|3\rangle = |3\rangle .$$

d) Restrict to the subspace generated by the orthogonal vectors $|1\rangle, |2\rangle, |3\rangle$. Write, with respect to this basis, the matrix representing the most general electron Hamiltonian invariant under the transformations induced by U and I_x.

e) Find the eigenvalues of such a Hamiltonian.

f) Still restricting to the subspace generated by $|1\rangle$, $|2\rangle$, $|3\rangle$, find the simultaneous eigenvectors of H and I_x and those of H and U.

5.25   Consider an interferometer as that described in Problem 3.3 with the path $s_1 \to s_2 \to s_4$ much longer than the path $s_1 \to s_3 \to s_4$. The transmission and reflection coefficients of the semi-transparent mirror s_1 are equal, and so are the coefficients of s_4. Particles (e.g. neutrons) enter the interferometer one at a time. In the two figures the support of the particle wavefunction (i.e. the

regions where the wavefunction is nonvanishing) are drawn, respectively just before
and just after the crossing of s_1 .

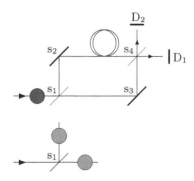

a) Consider the particle after the exit from
   the interferometer, but before reaching
   the detectors (i.e. in the two regions between s_4 and the detectors D_1 and D_2).
   How many regions R_i $(i = 1 \cdots ?)$ are
   there where the wavefunction is nonvanishing? Say where they are located.

Let $(t_1, t_1 + \tau)$ and $(t_2, t_2 + \tau)$ $(t_1 < t_2)$ be the (disjoint) time intervals in
which the detectors D_1 and D_2 (whose efficiency is assumed to be 100%) may
detect, *nondestructively*, the arrival of the particle.

b) Assume that the detector D_1 clicks (i.e. detects the particle) during the
   interval $(t_1, t_1 + \tau)$. How many distinct regions are there where the wavefunction, for $t > t_1 + \tau$, is nonvanishing?

c) Assume instead that neither detector clicks during the interval $(t_1, t_1 + \tau)$.
   Now how many distinct regions are there where the wavefunction, for
   $t_1 + \tau < t < t_2$, is nonvanishing?

d) If detector D_2 has been removed and detector D_1 does not click during
   the interval $(t_1, t_1 + \tau)$, how many are the distinct regions where the
   wavefunction, for $t_1 + \tau < t < t_2$, is nonvanishing? And if D_1 does not click
   even during the interval $t_2, t_2 + \tau$, in how many regions is the wavefunction
   nonvanishing for $t > t_2 + \tau$?

e) If, instead, counter D_2 is in place, but the observer does not read it, and
   counter D_1 does not click either before or after, what pieces of information
   are available about the state of the particle for $t_1 + \tau < t < t_2$ and for
   $t > t_2 + \tau$?

Solutions

5.1

a) $\Pi \rightarrow \begin{pmatrix} 1 & 0 \\ 0 & -1 \end{pmatrix}$.

b) The required matrix can be found either determining a, b, c such that:

$$\begin{pmatrix} a & b \\ b^* & c \end{pmatrix} \begin{pmatrix} \cos\vartheta \\ \sin\vartheta \end{pmatrix} = \begin{pmatrix} \cos\vartheta \\ \sin\vartheta \end{pmatrix} ; \quad \begin{pmatrix} a & b \\ b^* & c \end{pmatrix} \begin{pmatrix} -\sin\vartheta \\ \cos\vartheta \end{pmatrix} = - \begin{pmatrix} -\sin\vartheta \\ \cos\vartheta \end{pmatrix}$$

or writing the operator associated with the observable as the sum of the projectors onto the eigenvectors, times the corresponding eigenvalues ($+1$ and -1):

$$\Pi_\vartheta = (\cos\vartheta \, | e_1 \rangle + \sin\vartheta \, | e_2 \rangle) \times (\cos\vartheta \, \langle e_1 | + \sin\vartheta \, \langle e_2 |)$$
$$-(- \sin\vartheta \, | e_1 \rangle + \cos\vartheta \, | e_2 \rangle) \times (- \sin\vartheta \, \langle e_1 | + \cos\vartheta \, \langle e_2 |)$$

and taking its matrix elements ($\langle e_1 | \Pi_\vartheta | e_1 \rangle$, $\langle e_1 | \Pi_\vartheta | e_2 \rangle$, etc.):

$$\Pi_\vartheta \rightarrow \begin{pmatrix} \cos 2\vartheta & \sin 2\vartheta \\ \sin 2\vartheta & -\cos 2\vartheta \end{pmatrix} .$$

c) Yes, because we have stated that the vectors $\cos\vartheta \, | e_1 \rangle + \sin\vartheta \, | e_2 \rangle$ represent states of linear polarization: if $| e_1 \rangle \rightarrow e^{i\varphi_1} | e_1 \rangle$ and $| e_2 \rangle \rightarrow e^{i\varphi_2} | e_1 \rangle$, the representation of the states of linear polarization changes:

$$\cos\vartheta \, | e_1 \rangle + \sin\vartheta \, | e_2 \rangle = \cos\vartheta \, e^{-i\varphi_1} | \tilde{e}_1 \rangle + \sin\vartheta \, e^{-i\varphi_2} | \tilde{e}_2 \rangle .$$

d) $\Pi_\sigma \rightarrow \begin{pmatrix} 0 & -i \\ i & 0 \end{pmatrix}$.

e) $\Pi_{\vartheta\varphi} \rightarrow \begin{pmatrix} \cos 2\vartheta & \sin 2\vartheta \, e^{-i\varphi} \\ \sin 2\vartheta \, e^{i\varphi} & -\cos 2\vartheta \end{pmatrix}$.

5.2

a) $\varrho = \begin{pmatrix} 0.7 & 0 \\ 0 & 0.3 \end{pmatrix}$.

A 'good definition' of the degree of polarization must give the value 1 when all the photons are in the same polarization state, 0 when the mixture is totally unpolarized (see Problem 4.9). The given mixture consists of 30% of photons polarized in the direction $\hat{e}_1$, 30% in the direction $\hat{e}_2$ and 40% of photons still polarized in the direction $\hat{e}_1$: so 60% of the mixture is completely incoherent, 40% completely polarized: we can define $\mathcal{P}$ as the completely polarized fraction. $\mathcal{P}$ so defined equals the difference $p_1 - p_2$ between the eigenvalues of ϱ, where $p_1 \geq p_2$: indeed, since ϱ can always be diagonalized and $p_1 + p_2 = 1$, one has:

$$\varrho = \begin{pmatrix} p_1 & 0 \\ 0 & p_2 \end{pmatrix} = p_2 \mathbb{1} + (p_1 - p_2) \begin{pmatrix} 1 & 0 \\ 0 & 0 \end{pmatrix}, \qquad 0 \leq p_1 - p_2 \leq 1$$

that corresponds to decomposing the mixture in an incoherent part $(100 \times p_2\%)$ and a completely polarized part $(100 \times (p_1 - p_2)\%)$: the mixture is partially polarized in the state of polarization represented by the eigenstate of ϱ corresponding to p_1, i.e. the larger of the eigenvalues.

b) $\varrho^{(+)} = |e_{\sigma_+}\rangle\langle e_{\sigma_+}| \Rightarrow \varrho_{ij}^{(+)} = \langle e_i | e_{\sigma_+}\rangle\langle e_{\sigma_+} | e_j \rangle; \quad i, j = 1, 2 \quad \Rightarrow$

$$\varrho^{(+)} \to \frac{1}{2}\begin{pmatrix} 1 & -i \\ i & 1 \end{pmatrix}; \qquad \varrho^{(-)} \to \frac{1}{2}\begin{pmatrix} 1 & i \\ -i & 1 \end{pmatrix}.$$

c) A statistical mixture equivalent to that given in the text is, for example, the one consisting of 30% of photons in the state $|e_{\sigma_+}\rangle$, 30% in the state $|e_{\sigma_-}\rangle$ and 40% of photons in the state $|e_1\rangle$. In the basis $|e_1\rangle$, $|e_2\rangle$:

$$\varrho = 0.3 \times \frac{1}{2}\begin{pmatrix} 1 & -i \\ i & 1 \end{pmatrix} + 0.3 \times \frac{1}{2}\begin{pmatrix} 1 & i \\ -i & 1 \end{pmatrix} + 0.4 \times \begin{pmatrix} 1 & 0 \\ 0 & 0 \end{pmatrix} = \begin{pmatrix} 0.7 & 0 \\ 0 & 0.3 \end{pmatrix}.$$

More generally, it suffices to take a mixture – no matter how prepared – whose 60% is totally incoherent (this can be achieved in many ways) and for the remaining 40% consists of photons all in the state $|e_1\rangle$.

d) The eigenvalues of the matrix ϱ given in the text are 0.7 and 0.3, so the degree of polarization is 40%. The eigenvectors of the matrix are $(1, \pm i)$, i.e. the state $|e_{\sigma_+}\rangle$, belonging to the eigenvalue 0.7, and $|e_{\sigma_-}\rangle$, belonging to the eigenvalue 0.3, so the mixture is partially right circularly polarized.

5.3

a) The two matrices commute: indeed, the first is diagonal and, where the second consists of a 2×2 nondiagonal block, the first is a multiple of the identity. In order to find the basis in which they are simultaneously diagonal, it is sufficient to diagonalize the 2×2 block: since its trace is 0 and the determinant is $-|\gamma|^2$, the eigenvalues are $\pm|\gamma|$ and, putting $\gamma = |\gamma|\, e^{i\varphi}$, the eigenvectors are:

$$|\tilde{1}\rangle \equiv \frac{1}{\sqrt{2}}\Big(|1\rangle + e^{-i\varphi}\,|2\rangle\Big), \qquad |\tilde{2}\rangle \equiv \frac{1}{\sqrt{2}}\Big(|1\rangle - e^{-i\varphi}\,|2\rangle\Big).$$

In the basis $|\tilde{1}\rangle$, $|\tilde{2}\rangle$, $|\tilde{3}\rangle \equiv |3\rangle$ the representation of the two observables is

$$\xi \to \begin{pmatrix} \xi_1 & 0 & 0 \\ 0 & \xi_1 & 0 \\ 0 & 0 & \xi_3 \end{pmatrix} ; \qquad \eta \to \begin{pmatrix} |\gamma| & 0 & 0 \\ 0 & -|\gamma| & 0 \\ 0 & 0 & \eta_3 \end{pmatrix} .$$

b) The two observables form a complete set of compatible observables if there are no equal pairs of eigenvalues, i.e. if the pairs $(\xi_1, |\gamma|)$, $(\xi_1, -|\gamma|)$, (ξ_3, η_3) are all different, i.e. if $\gamma \neq 0$.

5.4

a) The columns of the transformation matrix are the vectors of the arrival basis expressed with respect to the initial basis:

$$U_1 = \begin{pmatrix} \cos \vartheta & -\sin \vartheta \\ \sin \vartheta & \cos \vartheta \end{pmatrix} .$$

b) $U_2 = \dfrac{1}{\sqrt{2}} \begin{pmatrix} 1 & 1 \\ i & -i \end{pmatrix} .$

5.5

a) The most general Hamiltonian is represented by the most general 2×2 Hermitian matrix:

$$H \to \begin{pmatrix} E_1 & a\,e^{i\varphi} \\ a\,e^{-i\varphi} & E_2 \end{pmatrix} , \qquad a \geq 0 .$$

The eigenvalues are:

$$E_\pm = \frac{1}{2}\left(E_1 + E_2 \pm \sqrt{(E_1 - E_2)^2 + 4a^2} \right) .$$

b) Since $\langle 1 | H | 2 \rangle = a\,e^{i\varphi}$, letting $|\tilde{1}\rangle = |1\rangle$, $|\tilde{2}\rangle = e^{-i\varphi}|2\rangle$, in the new basis $\langle \tilde{1} | H | \tilde{2} \rangle = a$ (real number):

$$H \to \begin{pmatrix} E_1 & a \\ a & E_2 \end{pmatrix} .$$

c) The unitary matrix that represents Π is $\begin{pmatrix} 0 & \beta \\ \alpha & 0 \end{pmatrix}$.

The condition $\Pi^\dagger\, H\, \Pi = H$ reads

$$\begin{pmatrix} 0 & \alpha^* \\ \beta^* & 0 \end{pmatrix}\begin{pmatrix} E_1 & a \\ a & E_2 \end{pmatrix}\begin{pmatrix} 0 & \beta \\ \alpha & 0 \end{pmatrix} = \begin{pmatrix} E_1 & a \\ a & E_2 \end{pmatrix}$$

$\Rightarrow E_1 = E_2$ and if $a \neq 0$: $\alpha = \beta \equiv e^{-i\chi}$.

Letting $\Pi \to e^{i\chi}\Pi$, the representation of the so redefined Π becomes

$$\Pi \to \begin{pmatrix} 0 & 1 \\ 1 & 0 \end{pmatrix} .$$

d) The diagonal elements of the matrix that represents H automatically are real; let $\langle 1 \mid H \mid 2 \rangle = a\,e^{i\varphi_2}$, $\langle 2 \mid H \mid 3 \rangle = b\,e^{i\varphi_3}$, $\langle 3 \mid H \mid 1 \rangle = c\,e^{i\varphi_1}$; since what really matters are the relative phases $\varphi_{12} \equiv \varphi_1 - \varphi_2$, φ_{23}, φ_{31}, and $\varphi_{12} + \varphi_{23} + \varphi_{31} = 0$, we can redefine the phases of only two vectors and then make only two matrix elements real; for example:

$$|\tilde{2}\rangle = e^{-i\varphi_2}\,|2\rangle; \quad |\tilde{3}\rangle = e^{-i(\varphi_2+\varphi_3)}\,|3\rangle \quad \Rightarrow$$

$$\langle \tilde{1} \mid H \mid \tilde{2} \rangle = a; \quad \langle \tilde{2} \mid H \mid \tilde{3} \rangle = b; \quad \langle \tilde{3} \mid H \mid \tilde{1} \rangle = c\,e^{i(\varphi_1+\varphi_2+\varphi_3)} \equiv c\,e^{i\varphi}$$

whence:

$$H \rightarrow \begin{pmatrix} E_1 & a & c\,e^{-i\varphi} \\ a & E_2 & b \\ c\,e^{+i\varphi} & b & E_3 \end{pmatrix}.$$

5.6

a) $[\tilde{p}, \tilde{q}] = [p, q]; \qquad \tilde{q}^{\dagger} = \tilde{q}, \quad \tilde{p}^{\dagger} = \tilde{p}$.

b) Let U be the operator that implements the transformation: $\tilde{q} = U q\,U^{-1}$, $\tilde{p} = U p\,U^{-1}$. One has:

$$U\left(\frac{p^2}{2m_1} + \frac{1}{2}m_1\omega^2 q^2\right)U^{-1} = \frac{p^2}{2m_1\lambda^2} + \frac{1}{2}m_1\lambda^2\,\omega^2 q^2 \,.$$

Putting $\lambda^2 = m_2/m_1$ one has $U H_1 U^{-1} = H_2$. As a consequence, H_1 and H_2, being unitarily equivalent, have the same eigenvalues. The eigenvectors are different: if $|E^{(1)}\rangle$ is an eigenvector of H_1, $|E^{(2)}\rangle \equiv U|E^{(1)}\rangle$ is an eigenvector of H_2, belonging to the same eigenvalue:

$$H_1|E^{(1)}\rangle = E|E^{(1)}\rangle \quad \Rightarrow \quad H_2 U\,|E^{(1)}\rangle = U H_1|E^{(1)}\rangle = E U\,|E^{(1)}\rangle \,.$$

c) In the case of the hydrogen atom one may reason in an analogous way: $H(m_1, e_1)$ and $H(m_2, e_2)$ are equivalent if it is possible to choose λ such that $m_1\lambda^2 = m_2$, $e_1^2/\lambda = e_2^2$, namely if $m_2/m_1 = e_1^4/e_2^4 \Rightarrow m_1 e_1^4 = m_2 e_2^4$. Alternatively: $E(m, e^2, \cdots) = E(m\lambda^2, e^2/\lambda, \cdots)$; since we can think of E as of a function of me^4 and of either e or of m, it follows that E must depend on the combination me^4, but not also on either e or m separately. According to the Bohr theory, the energy levels are $E_n = me^4/2\hbar^2 n^2$.

5.7

a) $H = \dfrac{p^2}{2m} + \dfrac{1}{2}m\omega^2 q^2 \rightarrow \widetilde{H} = \dfrac{1}{2m\Lambda^2}q^2 + \dfrac{m\omega^2\Lambda^2}{2}p^2$.

The requirement $\widetilde{H} = H$ enforces two conditions on Λ that can be simultaneously satisfied inasmuch as equivalent:

$$\frac{1}{m} = m\omega^2\Lambda^2, \qquad m\omega^2 = \frac{1}{m\Lambda^2} \quad \Rightarrow \quad \Lambda = \pm(m\omega)^{-1} \,.$$

b) Let U be the operator that implements the transformation:

$$q \to \tilde{q} \equiv U q U^{-1} = \frac{p}{m\omega}, \qquad p \to \tilde{p} \equiv U p U^{-1} = -m\omega q; \qquad U^{-1} = U^{\dagger}.$$

Since U commutes with H and H is nondegenerate, the eigenvectors of H are also eigenvectors of U:

$$U \mid E \rangle = e^{i\vartheta} \mid E \rangle; \qquad \langle E \mid U^{\dagger} \cdots U \mid E \rangle = \langle E \mid \cdots \mid E \rangle$$

therefore:

$$\langle E \mid f(q, p) \mid E \rangle = \langle E \mid U^{\dagger} U f(q, p) U^{\dagger} U \mid E \rangle = \langle E \mid U f(q, p) U^{\dagger} \mid E \rangle$$
$$= \langle E \mid f(\tilde{q}, \tilde{p}) \mid E \rangle$$

and in particular:

$$\langle E \mid \frac{p^2}{2m} \mid E \rangle = \langle E \mid \frac{\tilde{p}^2}{2m} \mid E \rangle = \langle E \mid \frac{1}{2} m\omega^2 q^2 \mid E \rangle .$$

c) According to the above result, the mean values of both the kinetic and the potential energy are a half of the total energy $(n + \frac{1}{2})\hbar\omega$. In addition, the mean values of q and p in the eigenstates of the Hamiltonian are vanishing, therefore:

$$(\Delta p)^2 = \overline{p^2} = \left(n + \frac{1}{2}\right) m\hbar\omega; \quad (\Delta q)^2 = \overline{q^2} = \left(n + \frac{1}{2}\right)\frac{\hbar}{m\omega} \qquad \Rightarrow$$

$$\Delta q \, \Delta p = \left(n + \frac{1}{2}\right)\hbar .$$

d) $\langle E_n \mid qp + pq \mid E_n \rangle = \langle E_n \mid \tilde{q}\tilde{p} + \tilde{p}\tilde{q} \mid E_n \rangle = -\langle E_n \mid pq + qp \mid E_n \rangle = 0 .$

5.8

a) $[\tilde{p}, \tilde{q}] = \frac{1}{2}[(ap - bq), (a^{-1}q + b^{-1}p)] = [p, q] .$

b) $U(a) V(b) q V^{-1}(b) U^{-1}(a) = U(a) q U^{-1}(a) = q - a ,$

$U(a) V(b) p V^{-1}(b) U^{-1}(a) = p - b .$

One has:

$W(a, b) = e^{-i\tilde{p}/\hbar} \quad \Rightarrow$

$W(a, b) \tilde{q} W^{-1}(a, b) = \tilde{q} - 1, \quad W(a, b) \tilde{p} W^{-1}(a, b) = \tilde{p}$

and, in addition,

$$q = a\tilde{q} - \frac{1}{2} b^{-1}\tilde{p}, \qquad p = \frac{1}{2} a^{-1}\tilde{p} + b\tilde{q} \quad \Rightarrow$$

$$W(a, b) q W^{-1}(a, b) = a(\tilde{q} - 1) - \frac{1}{2} b^{-1}\tilde{p} = q - a$$

and likewise:

$W(a, b) p W^{-1}(a, b) = p - b .$

From the identity $e^A e^B = e^{A+B+\frac{1}{2}[A, B]}$, and putting $A = -ipa/\hbar$, $B = iqb/\hbar$, one obtains:

$U(a) V(b) = W(a, b) e^{-iab/2\hbar} .$

c) $G(v) = e^{-iv(tp-mq)/\hbar} = e^{-imv^2t/2\hbar} e^{imvq/\hbar} e^{-ivtp/\hbar} .$

5.9

a) One has:

$$U^{-1}(a)\, \eta\, U(a) = \eta + i\alpha_2\,, \quad V^{-1}(b)\, \eta V(b) = \eta + \alpha_1 \quad \Rightarrow$$

$$\eta\,|\,\alpha\,\rangle = \eta\, V(b)\, U(a)\,|\,0\,\rangle = V(b)\, U(a)\, U^{-1}(a) V^{-1}(b)\, \eta\, V(b)\, U(a)\,|\,0\,\rangle$$
$$= V(b)\, U(a)\, (\eta + \alpha_1 + i\alpha_2)\,|\,0\,\rangle = \alpha\,|\,\alpha\,\rangle\,.$$

b) $\langle\,\alpha\,|\,q\,|\,\alpha\,\rangle = \langle\,0\,|\,U^{\dagger}(a) V^{\dagger}(b)\, q\, V(b)\, U(a)\,|\,0\,\rangle$

$$= \langle\,0\,|\,U^{-1}(a) V^{-1}(b)\, q\, V(b)\, U(a)\,|\,0\,\rangle = \langle\,0\,|\,U^{-1}(a)\, q\, U(a)\,|\,0\,\rangle$$
$$= \langle\,0\,|\,q + a\,|\,0\,\rangle = a$$

and in an analogous way $\langle\,\alpha\mid p\mid\alpha\,\rangle = b$. The uncertainties Δq and Δp in the state $|\,\alpha\,\rangle$ equal the uncertainties in the state $|\,0\,\rangle$, that is a minimum uncertainty state inasmuch as coinciding with the ground state of a harmonic oscillator, indeed:

$$\langle\,\alpha\,|\,(q - \bar{q})^2\,|\,\alpha\,\rangle = \langle\,\alpha\,|\,(q - a)^2\,|\,\alpha\,\rangle$$

$$= \langle\,0\,|\,U(-a) V(-b)\,(q - a)^2\, V(b)\, U(a)\,|\,0\,\rangle$$

$$= \langle\,0\,|\,\big((q + a) - a\big)^2\,|\,0\,\rangle = \langle\,0\,|\,q^2\,|\,0\,\rangle$$

and likewise $\langle\,\alpha\,|\,(p - b)^2\,|\,\alpha\,\rangle = \langle\,0\,|\,p^2\,|\,0\,\rangle$.

In order to demonstrate the converse, we may take advantage of the operators $U(a)$ and $V(b)$ and take ourselves back to the case of states $|\,s\,\rangle$ in which $\bar{p} = \bar{q} = 0$. Let λ be a real number; due to the commutation relations $[q,\,p] = i\hbar$ and $p + i\lambda q = (p - i\lambda q)^{\dagger}$, one has:

$$\langle\,s\,|\,(p + i\lambda q)(p - i\lambda q)\,|\,s\,\rangle = (\Delta p)^2 + \lambda^2 (\Delta q)^2 - \lambda\hbar \geq 0 \qquad \forall\,\lambda \in \mathbb{R}\,.$$

If (and only if) $\Delta p\,\Delta q = \tfrac{1}{2}\hbar$, then a value $\bar{\lambda}$ of λ exists such that the equality sign holds: indeed, the discriminant of the quadratic form in the variable λ is $\hbar^2 - 4(\Delta p)^2 (\Delta q)^2$: the latter is nonnegative if and only if it is vanishing, i.e. if $\Delta p\,\Delta q = \tfrac{1}{2}\hbar$; in this case:

$$\langle\,s\,|\,(p + i\bar{\lambda} q)(p - i\bar{\lambda} q)\,|\,s\,\rangle = 0 \quad \Leftrightarrow \quad (p - i\bar{\lambda} q)\,|\,s\,\rangle = 0\,.$$

c) $\langle\,n\,|\,\alpha\,\rangle = \dfrac{1}{\sqrt{n!}}\langle\,0\,|\,\eta^n\,|\,\alpha\,\rangle = \dfrac{\alpha^n}{\sqrt{n!}}\,\langle\,0\,|\,\alpha\,\rangle$

(thanks to the fact that $\eta^n\,|\,\alpha\,\rangle = \alpha^n\,|\,\alpha\,\rangle$); $\langle\,0\,|\,\alpha\,\rangle$ is determined – up to an irrelevant phase – by the normalization condition:

$$1 = \sum_n |\langle\,n\,|\,\alpha\,\rangle|^2 = |\langle\,0\,|\,\alpha\,\rangle|^2 \sum_n \frac{|\alpha|^{2n}}{n!} = |\langle\,0\,|\,\alpha\,\rangle|^2\, e^{|\alpha|^2} \quad \Rightarrow$$

$$\langle\,0\,|\,\alpha\,\rangle = e^{-\frac{1}{2}|\alpha|^2} \quad \Rightarrow \quad \langle\,n\,|\,\alpha\,\rangle = \frac{\alpha^n}{\sqrt{n!}}\, e^{-\frac{1}{2}|\alpha|^2}\,.$$

$$\langle\,\alpha\,|\,\beta\,\rangle = e^{-\frac{1}{2}(|\alpha|^2 + |\beta|^2)} \sum_n \frac{(\alpha^*\beta)^n}{n!} = e^{-\frac{1}{2}(|\alpha|^2 + |\beta|^2)}\, e^{\alpha^*\beta}\,.$$

d) Let $|A\rangle = \sum a_n |n\rangle$:

$$\langle A \mid \alpha \rangle = e^{-\frac{1}{2}|\alpha|^2} \sum \frac{a_n^* \alpha^n}{\sqrt{n!}} \, .$$

The series in the right hand side defines a holomorphic function of the complex variable α (the series has an infinite radius of convergence); so, if it is vanishing for any α (the vanishing on an infinite set of values of α with at least one finite accumulation point is sufficient), then $a_n = 0$ for any n, therefore $|A\rangle = 0$.

5.10

a) $\overline{H} = \langle \alpha \mid H \mid \alpha \rangle = \langle 0 \mid U^\dagger(a) V^\dagger(b) H \, V(b) \, U(a) \mid 0 \rangle$

$$= \langle 0 \mid \frac{1}{2m}(p+b)^2 + \frac{1}{2}m\omega^2(q+a)^2 \mid 0 \rangle = \frac{1}{2}\hbar\omega + \frac{b^2}{2m} + \frac{1}{2}m\omega^2 a^2$$

$$= \frac{1}{2}\hbar\omega + \hbar\omega|\alpha|^2 = \frac{1}{2}\hbar\omega + \frac{\bar{p}^2}{2m} + \frac{1}{2}m\omega^2\bar{q}^2, \quad \bar{p} = b, \, \bar{q} = a; \quad \bar{n} = |\alpha|^2.$$

Since $P_n \equiv P(|\alpha\rangle \rightarrow |n\rangle) = |\langle \alpha \mid n \rangle|^2 = e^{-|\alpha|^2} |\alpha|^{2n}/n!$ is the Poisson distribution, one has $\Delta n = \sqrt{\bar{n}} = |\alpha|$. Also the direct calculation is possible.

b) $\dfrac{\overline{p^2}}{2m} = \langle \alpha \mid \dfrac{p^2}{2m} \mid \alpha \rangle = \dfrac{1}{2m}\langle 0 \mid (p+b)^2 \mid 0 \rangle = \dfrac{1}{4}\hbar\omega + \dfrac{\bar{p}^2}{2m}$

$$\frac{1}{2}m\omega^2\overline{q^2} = \frac{1}{4}\hbar\omega + \frac{1}{2}m\omega^2 a^2 = \frac{1}{4}\hbar\omega + \frac{1}{2}m\omega^2\bar{q}^2 \, .$$

c) Due to the above result:

$$\frac{\overline{p^2}}{2m} - \frac{\bar{p}^2}{2m} = \frac{1}{4}\hbar\omega \quad \Rightarrow \quad \Delta p = \sqrt{\frac{m\omega\hbar}{2}}, \qquad \Delta q = \sqrt{\frac{\hbar}{2m\omega}}$$

so Δq and Δp are independent of α. As $\bar{p} = b$ and $\bar{q} = a$, one immediately verifies that $\alpha = \bar{p}/(2\Delta p) - \mathrm{i}\,\bar{q}/(2\Delta q)$.

5.11

a) The Hamiltonian of the oscillator in presence of the electric field is

$$H = \frac{p^2}{2m} + \frac{1}{2}m\omega^2 q^2 - e\,\mathcal{E}q = \frac{p^2}{2m} + \frac{1}{2}m\omega^2\left(q^2 - 2\frac{e\mathcal{E}}{m\omega^2}q\right)$$

$$= \frac{p^2}{2m} + \frac{1}{2}m\omega^2\left(q - \frac{e\mathcal{E}}{m\omega^2}\right)^2 - \frac{e^2\mathcal{E}^2}{2m\omega^2}$$

that, up to the last term which is a c–number, is unitarily equivalent to the Hamiltonian of the oscillator without field: putting $a = e\mathcal{E}/(m\omega^2)$ and $U(a)$ the translation operator, one has:

$$H = U(a)\left(\frac{p^2}{2m} + \frac{1}{2}m\omega^2 q^2 - \frac{e^2\mathcal{E}^2}{2m\omega^2}\right)U^{-1}(a) \quad \Rightarrow$$

$$E_n = \left(n + \tfrac{1}{2}\right)\hbar\omega - \frac{e^2\mathcal{E}^2}{2m\omega^2} \, .$$

b) If $|n\rangle$ stand for the eigenvectors of the Hamiltonian without field, the ground state of the oscillator with field is $|\tilde{0}\rangle = U(a)|0\rangle$, which is a coherent state with $\alpha = -i\sqrt{m\omega/2\hbar}\,a$; so, thanks to the result of Problem 5.9,

$$P(|\tilde{0}\rangle \to |n\rangle) = |\langle n\,|\,\alpha\rangle|^2 = e^{-(m\omega/2\hbar)\,a^2}\,\frac{1}{n!}\left(\frac{m\omega\,a^2}{2\hbar}\right)^n.$$

5.12

a) $q = -i\sqrt{\dfrac{\hbar}{2m\omega}}\,(\eta^\dagger - \eta)$; $\quad (\eta^\dagger - \eta)^3\,|0\rangle = \left((\eta^\dagger)^3 - \eta^\dagger\eta\,\eta^\dagger - \eta\,(\eta^\dagger)^2\right)|0\rangle$

(all the omitted terms, as e.g. $(\eta^\dagger)^2\eta$, give 0 on $|0\rangle$)

$$\cdots = \sqrt{3!}\,|3\rangle - 3\,|1\rangle \quad \Rightarrow \quad \langle 0\,|\,q^6\,|\,0\rangle = 15\left(\frac{\hbar}{2m\omega}\right)^3 = \frac{15}{8}\left(\frac{\hbar}{m\omega}\right)^3.$$

The classically allowed region is where $\frac{1}{2}m\omega^2 x^2 \le E_0 = \frac{1}{2}\hbar\omega$, namely $|x| \le x_0 \equiv \sqrt{\hbar/m\omega}$. If the probability of finding the particle out of the classically allowed region were zero, any power n of q should have a mean value smaller than x_0^n, in particular $\overline{q^6}$ should be smaller than $(\hbar/m\omega)^3$. The least value of n for which $\overline{q^n} > x_0^n$ is $n = 6$.

b) The interval classically allowed for the momentum is $|p| \le p_0 \equiv \sqrt{m\omega\,\hbar}$. In order to calculate $\overline{p^6}$ one may either proceed as above, or one may consider the unitary transformation:

$$q \to \tilde{q} \equiv U q\,U^{-1} = \frac{p}{m\omega}, \qquad p \to \tilde{p} \equiv U p\,U^{-1} = -m\omega\,q$$

which is an invariance transformation for the Hamiltonian (see Problem 5.7). As H is nondegenerate, $U\,|0\rangle = e^{i\vartheta}\,|0\rangle$, therefore:

$$\langle 0\,|\,p^6\,|\,0\rangle = \langle 0\,|\,U^\dagger\tilde{p}^6 U\,|\,0\rangle = \langle 0\,|\,\tilde{p}^6\,|\,0\rangle$$

$$= (m\omega)^6\langle 0\,|\,q^6\,|\,0\rangle = \frac{15}{8}(m\omega\,\hbar)^3.$$

5.13

a) $\displaystyle\int |\psi(x_1,\,\cdots\,x_n)|^2\,dx_1 \cdots dx_n = 1 \quad \Rightarrow \quad [\psi(x_1,\,\cdots\,x_n)] = L^{-n/2}.$

b) As a is the only dimensionful constant appearing in the wavefunction and $[a] = L^{-4}$, it follows that N is proportional to $a^{1/8}$. Alternatively, by means of the change of variable $y = a^{1/4}x$ in the normalization integral, one has:

$$\frac{1}{N^2} = \int_{-\infty}^{+\infty} e^{-2a\,x^4}\,dx = a^{-1/4}\int_{-\infty}^{+\infty} e^{-2y^4}\,dy \quad \Rightarrow \quad N \propto a^{1/8}.$$

c) $\displaystyle\overline{q^4} = N^2\int_{-\infty}^{+\infty} x^4 e^{-2a\,x^4}\,dx = -\frac{1}{2}N^2\,\frac{d}{da}\int_{-\infty}^{+\infty} e^{-2a\,x^4}\,dx = -\frac{1}{2}N^2\,\frac{d}{da}\frac{1}{N^2}$

$$= \frac{d\log N}{da} = \frac{1}{8\,a}.$$

d) $n = 4k$ with k positive integer. Indeed $d^k e^{-a x^4}/da^k = (-1)^k x^{4k} e^{-a x^4}$.

5.14

a) In the momentum representation $\left(p \to k, \quad q \to i\hbar \dfrac{d}{dk}\right)$:

$$\eta \to \frac{1}{\sqrt{2m\omega\hbar}}\left(k + m\omega\hbar\frac{d}{dk}\right)$$

therefore:

$$\left(\frac{d}{dk} + \frac{1}{m\omega\hbar}k\right)\varphi_0(k) = 0 \quad \Rightarrow \quad \varphi_0(k) = A e^{-k^2/(2m\omega\hbar)}$$

$$1 = |A|^2 \int_{-\infty}^{+\infty} e^{-k^2/(m\omega\hbar)}\,dk \quad \Rightarrow \quad \varphi_0(k) = \left(\pi m\omega\hbar\right)^{-1/4} e^{-k^2/(2m\omega\hbar)} .$$

b) $qU\,|\,k\,\rangle = U U^{-1} q U\,|\,k\,\rangle = U \dfrac{p}{m\omega}\,|\,k\,\rangle = \dfrac{k}{m\omega} U\,|\,k\,\rangle$

so $U\,|\,k\,\rangle$ is an improper eigenvector of q belonging to the (improper) eigenvalue $k/m\omega$: $U\,|\,k\,\rangle = B\,|\,x = k/m\omega\,\rangle$. The proportionality constant B may be determined from the normalization condition:

$$\langle\,k'\,|\,k''\,\rangle = \delta(k' - k'') = \langle\,k'\,|\,U^\dagger U\,|\,k''\,\rangle$$

$$= |B|^2\langle\,x' = k'/m\omega\,|\,x'' = k''/m\omega\,\rangle$$

$$= |B|^2\,\delta\!\left((k'-k'')/m\omega\right) = m\omega\,|B^2|\,\delta(k'-k'') \;\Rightarrow\; |B| = \frac{1}{\sqrt{m\omega}}\,.$$

c) Since the eigenvalues of the Hamiltonian are nondegenerate, $U\,|\,n\,\rangle = e^{i\vartheta}\,|\,n\,\rangle$, then:

$$\varphi_n(k) = \langle\,k\,|\,n\,\rangle = \langle\,k\,|\,U^\dagger U\,|\,n\,\rangle = \frac{e^{i\vartheta}}{\sqrt{m\omega}}\,\langle\,x = k/m\omega\,|\,n\,\rangle$$

$$= \frac{e^{i\vartheta}}{\sqrt{m\omega}}\,\psi_n(k/m\omega)$$

so, up to the phase factor $e^{i\vartheta}$,

$$\varphi_n(k) = \frac{1}{\sqrt{2^n\,n!}}\,(\pi m\omega\hbar)^{-1/4}\,H_n(k/\sqrt{m\omega\hbar})\,e^{-k^2/(2m\omega\hbar)}\,.$$

5.15

a) $\langle\,\alpha\,|\,qp+pq\,|\,\alpha\,\rangle = \langle\,0_\lambda\,|\,U^\dagger(a)V^\dagger(b)(qp+pq)\,V(b)\,U(a)\,|\,0_\lambda\,\rangle$

$$= \langle\,0_\lambda\,|\,(q+a)(p+b) + (p+b)(q+a)\,|\,0_\lambda\,\rangle = 2ab\,.$$

b) The wavefunction of the state $|\,0_\lambda\,\rangle$ in the Schrödinger representation is

$$|\,0\,\rangle \equiv |\,0_\lambda\,\rangle \;\xrightarrow{\text{SR}}\; \psi_0(x) = \left(\frac{\lambda}{\pi\hbar}\right)^{1/4} e^{-(\lambda/2\hbar)\,x^2}$$

so from the definition of $|\,\alpha\,\rangle$:

$$|\alpha\rangle = V(b)\,U(a)\,|0\rangle \qquad \alpha \equiv \alpha_1 + i\,\alpha_2\,; \qquad a = -\sqrt{\frac{2\hbar}{\lambda}}\,\alpha_2\,, \; b = \sqrt{2\lambda\,\hbar}\,\alpha_1$$

one has:

$$U(a)\,|0\rangle \xrightarrow{\text{SR}} \psi_0(x-a) \quad \Rightarrow$$

$$|\alpha\rangle \xrightarrow{\text{SR}} \psi_\alpha(x) = e^{i\,b\,x/\hbar}\,\psi_0(x-a) = \Big(\frac{\lambda}{\pi\,\hbar}\Big)^{1/4} e^{i\,b\,x/\hbar}\, e^{-(\lambda/2\hbar)\,(x-a)^2}\,.$$

Likewise, in the momentum representation:

$$|\alpha\rangle \xrightarrow{\text{MR}} \varphi_\alpha(k) = \big(\pi\lambda\,\hbar\big)^{-1/4} e^{i\,ab/\hbar}\, e^{-i\,k\,a/\hbar}\, e^{-(k-b)^2/(2\lambda\,\hbar)}\,.$$

5.16

a) If $\psi(x) = \psi^*(x)$, one has:

$$\overline{p} = -i\,\hbar\int_{-\infty}^{+\infty}\psi(x)\psi'(x)\,\mathrm{d}x = -\frac{1}{2}i\,\hbar\,\psi^2(x)\Big|_{-\infty}^{+\infty} = 0\,.$$

Alternatively: $\overline{p}$ must be a real number, whereas (if the integral exists) $-i\,\hbar\dots$ is imaginary, therefore $\overline{p}$ must vanish.

b) Putting

$$N^2 = \Big(\int_{-\infty}^{+\infty}|\psi(x)|^2\,\mathrm{d}x\Big)^{-1} = \Big(\int_{-\infty}^{+\infty}\chi^2(x)\,\mathrm{d}x\Big)^{-1}$$

one has:

$$\overline{p} = -i\,N^2\hbar\int_{-\infty}^{+\infty}\big(\chi(x)\chi'(x) + i\,\varphi'(x)\,\chi^2(x)\big)\,\mathrm{d}x = N^2\hbar\int_{-\infty}^{+\infty}\varphi'(x)\,\chi^2(x)\,\mathrm{d}x\,.$$

c) $\overline{p} = \hbar\,k\,.$

5.17

a) $\overline{q^2} = \sqrt{\dfrac{\alpha}{\pi}}\displaystyle\int_{-\infty}^{+\infty}x^2 e^{-\alpha\,x^2}\,\mathrm{d}x = -\sqrt{\dfrac{\alpha}{\pi}}\dfrac{\mathrm{d}}{\mathrm{d}\alpha}\int_{-\infty}^{+\infty}e^{-\alpha\,x^2}\,\mathrm{d}x = -\sqrt{\dfrac{\alpha}{\pi}}\dfrac{\mathrm{d}}{\mathrm{d}\alpha}\sqrt{\dfrac{\pi}{\alpha}}$

$$= \frac{1}{2\alpha}\,.$$

As $|1\rangle$ is a coherent state, it is of minimum uncertainty and, in addition, $\overline{p^2} = (\Delta p)^2$, $(\Delta q)^2 = \overline{q^2}$, therefore:

$$\overline{p^2} = \frac{\hbar^2}{4(\Delta q)^2} = \frac{1}{2}\hbar^2\alpha\,.$$

$$\overline{p^4} = \hbar^4\sqrt{\frac{\alpha}{\pi}}\int_{-\infty}^{+\infty}e^{-\alpha\,x^2/2}\frac{\mathrm{d}^4}{\mathrm{d}x^4}e^{-\alpha\,x^2/2}\,\mathrm{d}x = \hbar^4\sqrt{\frac{\alpha}{\pi}}\int_{-\infty}^{+\infty}\Big(\frac{\mathrm{d}^2}{\mathrm{d}x^2}e^{-\alpha\,x^2/2}\Big)^2\mathrm{d}x$$

$$= \hbar^4\sqrt{\frac{\alpha}{\pi}}\int_{-\infty}^{+\infty}(\alpha^4 x^4 - 2\alpha^3 x^2 + \alpha^2)e^{-\alpha\,x^2}\,\mathrm{d}x = \hbar^4(\alpha^4\,\overline{q^4} - 2\alpha^3\,\overline{q^2} + \alpha^2)$$

$$\overline{q^4} = \sqrt{\frac{\alpha}{\pi}} \int_{-\infty}^{+\infty} x^4 e^{-\alpha x^2} \, dx = \sqrt{\frac{\alpha}{\pi}} \frac{d^2}{d\alpha^2} \int_{-\infty}^{+\infty} e^{-\alpha x^2} \, dx = \sqrt{\frac{\alpha}{\pi}} \frac{d^2}{d\alpha^2} \sqrt{\frac{\pi}{\alpha}}$$

$$= \frac{3}{4\alpha^2} \quad \Rightarrow \quad \overline{p^4} = \frac{3}{4} \hbar^4 \alpha^2 \, .$$

If n is odd, $e^{-\alpha x^2/2} (d^n/dx^n) e^{-\alpha x^2/2}$ is an odd function and its integral vanishes.

b) The state $|2\rangle$ is obtained by applying the operator e^{iqk} (that translates the momentum operator) to the state $|1\rangle$. Therefore the mean value of p in $|2\rangle$ is $\bar{p} = \hbar k$ and, in general, the mean value of p^n equals the mean value of $(p + \bar{p})^n$ in the state $|1\rangle$; in conclusion, having in mind that the odd powers of p have vanishing mean value in the state $|1\rangle$, one has:

$$\overline{p^2} \equiv \langle 2 \,|\, p^2 \,|\, 2\rangle = \langle 1 \,|\, (p + \bar{p})^2 \,|\, 1\rangle = \langle 1 \,|\, p^2 + \bar{p}^2 \,|\, 1\rangle = \hbar^2 \Big(\frac{1}{2}\alpha + k^2\Big) \, .$$

Note that Δp is the same in the two states $|1\rangle$ and $|2\rangle$.

$$\overline{p^4} = \langle 1 \,|\, (p + \bar{p})^4 \,|\, 1\rangle = \langle 1 \,|\, p^4 + 6p^2\,\bar{p}^2 + \bar{p}^4 \,|\, 1\rangle$$

$$= \hbar^4 \Big(\frac{3}{4}\alpha^2 + 3\,\alpha\,k^2 + k^4\Big) \, .$$

c) In the sequel all the mean values are in the state $|2\rangle$.

$$(\Delta E)^2 \equiv \frac{1}{4m^2} \Big(\overline{p^4} - (\overline{p^2})^2\Big) = \frac{\hbar^4}{4m^2} \Big(\frac{1}{2}\alpha^2 + 2\,\alpha\,k^2\Big)$$

$$= \frac{\hbar^4\alpha^2}{8m^2} + \Big(\frac{1}{2m}\,2\bar{p}\,\Delta p\Big)^2 \, .$$

If $\alpha \ll k^2$, namely $\Delta p \ll \bar{p}$, then:

$$\Delta E \simeq \frac{1}{m}\bar{p}\,\Delta p, \quad \overline{E} \equiv \overline{H} = \frac{\overline{p^2}}{2m} = \frac{\bar{p}^2 + (\Delta p)^2}{2m} \simeq \frac{\bar{p}^2}{2m} \quad \Rightarrow \quad \frac{\Delta E}{E} \simeq 2\frac{\Delta p}{\bar{p}} \, .$$

5.18

a) From the completeness relations for the improper eigenvectors of the position operator:

$$\int_{-\infty}^{+\infty} |x\rangle \, dx \, \langle x| = \mathbb{1} \, ,$$

$$\psi_B(x) \equiv \langle x \,|\, \xi \,|\, A\rangle = \int_{-\infty}^{+\infty} \langle x \,|\, \xi \,|\, y\rangle \langle y \,|\, A\rangle \, dy = \int_{-\infty}^{+\infty} \xi(x, y)\,\psi_A(y)\, dy \, .$$

$$\xi^\dagger(x, y) \equiv \langle x \,|\, \xi^\dagger \,|\, y\rangle = \langle y \,|\, \xi \,|\, x\rangle^* = \xi^*(y, x) \, .$$

b) $E_A(x, y) = \langle x \,|\, A\rangle \langle A \,|\, y\rangle = \psi_A(x)\,\psi_A^*(y) \, .$

Vice versa, for any $|s\rangle \xrightarrow{\text{SR}} \psi_s(x)$:

$$E_\chi \,|\, s\rangle \xrightarrow{\text{SR}} \int_{-\infty}^{+\infty} E_\chi(x, y)\,\psi_s(y)\, dy = \chi(x) \int_{-\infty}^{+\infty} \chi^*(y)\,\psi_s(y)\, dy = c\,\chi(x)$$

where c is the scalar product between the state represented by the wave-function $\psi_s(x)$ and that represented by the wavefunction $\chi(x)$. Therefore $E_\chi = E_\chi^\dagger$ projects onto the state represented by $\chi(x)$.

c) $\mathbb{1} \xrightarrow{\text{SR}} \langle\, x \mid y\,\rangle = \delta(x-y) = \sum_n \langle\, x \mid n\,\rangle\langle\, n \mid y\,\rangle = \sum_n \psi_n(x)\,\psi_n^*(y)$.

d) $\mathrm{Tr}\,\xi = \sum_n \iint \langle\, n \mid x\,\rangle\langle\, x \mid \xi \mid y\,\rangle\langle\, y \mid n\,\rangle \, dx\, dy$

$= \iint \langle\, x \mid \xi \mid y\,\rangle \sum_n \langle\, y \mid n\,\rangle\langle\, n \mid x\,\rangle \, dx\, dy = \iint \langle\, x \mid \xi \mid y\,\rangle\langle\, y \mid x\,\rangle \, dx\, dy$

$= \iint \xi(x,\, y)\, \delta(x-y)\, dx\, dy = \int_{-\infty}^{+\infty} \xi(x,\, x)\, dx$.

e) E_V has only 1 and 0 as eigenvalues ($E_V^\dagger = E_V,\ E_V^2 = E_V$), the multiplic-ity of the eigenvalue 1 equals the dimension of V and the trace coincides with the sum of the eigenvalues.

f) $\xi = \xi^\dagger \iff \xi(x,\, y) = \xi^*(y,\, x)$ which holds for $E(x,\, y)$.

$$E^2(x,\, y) = \langle\, x \mid E^2 \mid y\,\rangle = \int_{-\infty}^{+\infty} \langle\, x \mid E \mid z\,\rangle\langle\, z \mid E \mid y\,\rangle \, dz$$

$$= \int_{-\infty}^{+\infty} E(x,\, z)\, E(z,\, y)\, dz = \frac{\lambda}{\pi}\, e^{-(\lambda/2)(x^2+y^2)} \int_{-\infty}^{+\infty} e^{-\lambda z^2}\left(1 + 4\lambda^2 x\, y\, z^2\right) dz$$

(the odd integrands have vanishing integral). As

$$\int_{-\infty}^{+\infty} e^{-\lambda z^2}\, dz = \sqrt{\frac{\pi}{\lambda}}\,, \qquad \int_{-\infty}^{+\infty} z^2\, e^{-\lambda z^2}\, dz = -\frac{d}{d\lambda}\sqrt{\frac{\pi}{\lambda}} = \frac{1}{2\lambda}\sqrt{\frac{\pi}{\lambda}}$$

$E^2(x,\, y) = E(x,\, y)$ follows.

$$\mathrm{Tr}\,E = \sqrt{\frac{\lambda}{\pi}} \int_{-\infty}^{+\infty} e^{-\lambda x^2}\left(1 + 2\lambda x^2\right) dx = \sqrt{\frac{\lambda}{\pi}}\left(\sqrt{\frac{\pi}{\lambda}} + 2\lambda\, \frac{1}{2\lambda}\sqrt{\frac{\pi}{\lambda}}\right) = 2\,.$$

$$E \mid s\,\rangle \xrightarrow{\text{SR}} \int_{-\infty}^{+\infty} E(x,\, y)\, \psi_s(y)\, dy = \left(\frac{\lambda}{\pi}\right)^{1/4} e^{-\lambda x^2/2}\left(c_1 + c_2\sqrt{2\lambda}\, x\right)$$

where c_1, c_2 are the scalar products of $\psi_s(x)$ with the normalized vectors

$$\left(\frac{\lambda}{\pi}\right)^{1/4} e^{-\lambda x^2/2}, \qquad \left(\frac{\lambda}{\pi}\right)^{1/4} \sqrt{2\lambda}\, x\, e^{-\lambda x^2/2}$$

that generate the two-dimensional manifold that is the image of E.

5.19

a) As $\psi_0(x)$ is a Gaussian function appreciably different from zero in a region whose amplitude is 4Δ and the wavefunction of $U(a) \mid A_0\,\rangle$ is $\psi_0(x-a)$, $\mid A_0\,\rangle$ and $U(a) \mid A_0\,\rangle$ are practically orthogonal if $a \gg \Delta$. If $a = 10\Delta$ one has:

$$\langle A_0 | U(a) | A_0 \rangle = \frac{1}{\sqrt{2\pi}\,\Delta} \int_{-\infty}^{+\infty} e^{-x^2/4\Delta^2}\, e^{-(x-a)^2/4\Delta^2}\, \mathrm{d}x$$

$$= \frac{1}{\sqrt{2\pi}\,\Delta} \int_{-\infty}^{+\infty} e^{-(x+a/2)^2/4\Delta^2}\, e^{-(x-a/2)^2/4\Delta^2}\, \mathrm{d}x = e^{-a^2/8\Delta^2} \simeq 4 \times 10^{-6}.$$

b) The normalized state $|A\rangle$, in the approximation $\langle A_0 | U(a) | A_0 \rangle \simeq 0$, is

$$\frac{1}{\sqrt{2}}\left(|A_0\rangle + e^{i\varphi} U(a) | A_0 \rangle \right)$$

with wavefunction:

$$\psi_A(x) = \frac{1}{(8\pi\Delta^2)^{1/4}}\left(e^{-x^2/4\Delta^2} + e^{i\varphi} e^{-(x-a)^2/4\Delta^2} \right)$$

(the approximation concerns the normalization coefficient). Taking into account that the product of $\psi_0(x)$ and $\psi_0(x-a)$ is negligible,

$$\rho(x) = \frac{1}{\sqrt{8\pi\Delta^2}}\left(e^{-x^2/2\Delta^2} + e^{-(x-a)^2/2\Delta^2} \right).$$

Since φ does not appear in the expression of $\rho(x)$, it is not possible to determine it by means of position measurements.

c) In the momentum representation (see Problem 5.14):

$$|A_0\rangle \to \varphi_0(k) = \left(\frac{2\Delta^2}{\pi\hbar^2}\right)^{1/4} e^{-k^2\Delta^2/\hbar^2}; \quad U(a)|A_0\rangle = e^{-ipa/\hbar}|A_0\rangle \;\Rightarrow$$

$$\varphi_A(k) = \frac{1}{\sqrt{2}}\left(\frac{2\Delta^2}{\pi\hbar^2}\right)^{1/4} e^{-k^2\Delta^2/\hbar^2}\left(1 + e^{i\varphi} e^{-ika/\hbar}\right) \;\Rightarrow$$

$$\tilde\rho(k) = \sqrt{\frac{2\Delta^2}{\pi\hbar^2}}\, e^{-2k^2\Delta^2/\hbar^2}\left(1 + \cos(ka/\hbar - \varphi)\right).$$

d) In order to distinguish the state $|A\rangle$ from the statistical mixture:
$$\{\, |A_0\rangle,\ \nu_1 = \tfrac{1}{2};\ U(a)|A_0\rangle,\ \nu_2 = \tfrac{1}{2}\,\}$$
it is necessary that the momentum measurements one performs be able to reveal the interference term $\cos(ka/\hbar - \varphi)$ that, on a period $2\pi\hbar/a$, has a vanishing average. Therefore the precision must be $\Delta k < 2\pi\hbar/a = h/a$ (see the figure). Note that $\tilde\rho(k)$ (not by any chance!) coincides with the interference pattern generated by two

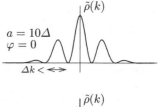

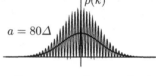

'Gaussian' slits of width $\simeq \Delta$ separated by a distance a: if $a \gg \Delta$, the fringes are too close to each other and if the resolving power of the device that measures the momentum is not sufficiently high, only the average intensity is observed, i.e. the thick curve in the second figure.

5.20

a) $\langle A \mid f(\vec{q}) \mid A \rangle = \displaystyle\int f(x,y,z) \, |\psi_A(x,y,z)|^2 \, dV = \langle B \mid f(\vec{q}) \mid B \rangle .$

$\langle A \mid p_i \mid A \rangle = -i\hbar \displaystyle\int \psi_A^*(x,y,z) \frac{\partial}{\partial x_i} \psi_A(x,y,z) \, dV .$

$\langle B \mid p_i \mid B \rangle = -i\hbar \displaystyle\int \psi_A(x,y,z) \frac{\partial}{\partial x_i} \psi_A^*(x,y,z) \, dV = -\langle A \mid p_i \mid A \rangle^*$

$\qquad\qquad = -\langle A \mid p_i \mid A \rangle .$

Likewise one may show that $\langle A \mid \vec{p}^{\,2} \mid A \rangle = \langle B \mid \vec{p}^{\,2} \mid B \rangle$ and that $\langle A \mid L_i \mid A \rangle = -\langle B \mid L_i \mid B \rangle$: in general, the mean value of all those observables that are not invariant under time reversal, such as $\vec{p}$ and $\vec{L}$, may be different in the two states.

b) $\varphi_B(\vec{k}) = \dfrac{1}{(2\pi\hbar)^{3/2}} \displaystyle\int e^{-i\,\vec{k}\cdot\vec{r}/\hbar} \, \psi_B(\vec{r}) \, dV = \dfrac{1}{(2\pi\hbar)^{3/2}} \displaystyle\int e^{-i\,\vec{k}\cdot\vec{r}/\hbar} \, \psi_A^*(\vec{r}) \, dV$

$\qquad = \dfrac{1}{(2\pi\hbar)^{3/2}} \left(\displaystyle\int e^{i\,\vec{k}\cdot\vec{r}/\hbar} \, \psi_A(\vec{r}) \, dV \right)^* = \varphi_A^*(-\vec{k}) .$

5.21

a) No: $\psi(x) = e^{i\varphi(x)} \sqrt{\rho(x)}$ with $\varphi(x)$ an arbitrary real function of x.

b) It is possible to calculate the mean value of q and, more generally, of the observables that are functions of q only. In the latter case the knowledge of $\rho(x)$ is sufficient, whereas it is not possible to calculate the mean value of p without explicitly knowing the function $\varphi(x)$.

5.22

a) $\rho_B(x) = |\psi_B(x)|^2 = |\psi_A(x)|^2 = \rho_A(x) .$

From Problem 5.20 one has, for the wavefunctions in the momentum representation,

$\varphi_B(k) = \varphi_A^*(-k) = \pm\varphi_A^*(k) \quad\Rightarrow\quad |\varphi_A(k)|^2 = |\varphi_B(k)|^2 .$

So not even the knowledge of both the probability densities for the position and the momentum determines the state in a unique way.

b) $\langle A \mid f(q) \mid A \rangle = \displaystyle\int_{-\infty}^{+\infty} \rho_A(x) f(x) \, dx = \int_{-\infty}^{+\infty} \rho_B(x) f(x) \, dx = \langle B \mid f(q) \mid B \rangle .$

Likewise $\langle A \mid g(p) \mid A \rangle = \langle B \mid g(p) \mid B \rangle$ and the same result holds for H too, since the mean value of the sum is the sum of the mean values. The mean value of the observable $pq + qp$ may instead be different in the two states: indeed, it cannot be expressed in terms of the position and momentum probability densities – see the next point.

c) One has:

$$\langle A \mid qp \mid A \rangle = -i\hbar \left(\frac{a}{\pi}\right)^{1/2} \int_{-\infty}^{+\infty} e^{-(a-ib)\,x^2/2}\, x \,\frac{d}{dx}\, e^{-(a+ib)\,x^2/2} dx$$

$$= \hbar\,(-b+i\,a)\left(\frac{a}{\pi}\right)^{1/2} \int_{-\infty}^{+\infty} x^2 e^{-a\,x^2} dx = \hbar\,(b-i\,a)\left(\frac{a}{\pi}\right)^{1/2} \frac{d}{da} \int_{-\infty}^{+\infty} e^{-a\,x^2} dx$$

$$= \hbar\,(b-i\,a)\left(\frac{a}{\pi}\right)^{1/2} \frac{d}{da}\left(\frac{\pi}{a}\right)^{1/2} = -\hbar\,\frac{b-i\,a}{2a}$$

therefore, as $\langle A \mid pq \mid A \rangle = \langle A \mid qp \mid A \rangle^*$ and $\psi_B(x)$ obtains from $\psi_A(x)$ by changing the sign of b:

$$\langle A \mid qp+pq \mid A \rangle = -\hbar\,\frac{b}{a}\,;\qquad \langle B \mid qp+pq \mid B \rangle = \hbar\,\frac{b}{a}\,.$$

5.23

a) Putting $G \equiv (qp+pq)/2\hbar$, as G commutes with $U(\alpha)$, one has $dU(\alpha)/d\alpha = i\,G\,U(\alpha) = i\,U(\alpha)\,G$ and, as a consequence,

$$\frac{d\tilde{q}(\alpha)}{d\alpha} = \frac{d}{d\alpha}\left(U(\alpha)\,q\,U^{-1}(\alpha)\right) = i\,U(\alpha)\,G\,q\,U^{-1}(\alpha) - i\,U(\alpha)\,q\,G\,U^{-1}(\alpha)$$

$$= i\,U(\alpha)\,[G,\,q]\,U^{-1}(\alpha) = U(\alpha)\,q\,U^{-1}(\alpha) = \tilde{q}(\alpha)$$

and analogously for $\tilde{p}(\alpha)$. The equations for $\tilde{q}(\alpha)$ and $\tilde{p}(\alpha)$ may be integrated as the differential equations for c–numeric functions (it is sufficient to take the matrix elements of both sides of the equations) and one obtains:

$$\tilde{q}(\alpha) = e^{\alpha}q\,,\qquad \tilde{p}(\alpha) = e^{-\alpha}p$$

i.e. the canonical transformation considered in Problem 5.6, in the case $\lambda > 0$.

b) If $q\,|\,x\rangle = x\,|\,x\rangle$, one has:

$$q\,U^{\dagger}(\alpha)\,|\,x\rangle = U^{\dagger}(\alpha)\,\tilde{q}(\alpha)\,|\,x\rangle = e^{\alpha}x\,U^{\dagger}(\alpha)\,|\,x\rangle \;\Rightarrow\; U^{\dagger}(\alpha)\,|\,x\rangle = c_x\,|\,e^{\alpha}x\rangle$$

where c_x is a proportionality factor that can be determined (up to a phase) from the normalization condition (see Problem 5.14):

$$\langle x' \mid x'' \rangle = \delta(x'-x'') = \langle x' \mid U(\alpha)\,U^{\dagger}(\alpha)\mid x'' \rangle = |c_x|^2\,\langle e^{\alpha}x' \mid e^{\alpha}x'' \rangle$$

$$= |c_x|^2\,\delta\big(e^{\alpha}(x'-x'')\big) = |c_x|^2\,e^{-\alpha}\delta(x'-x'') \;\Rightarrow\; |c_x| = e^{\alpha/2}\,.$$

The phase of c_x could depend on x. Analogously $U^{\dagger}(\alpha)\,|\,k\rangle = c_k\,|\,e^{-\alpha}k\rangle$ with $|c_k| = e^{-\alpha/2}$ and the phase of c_k could depend on k. A way to show that c_x and c_k are independent respectively of x and k is the following:

$$\langle k \mid U^{\dagger}(\alpha)\mid x\rangle = c_x\,\exp\left(-i\,k\,e^{\alpha}x/\hbar\right) = c_k\,\exp\left(-i\,e^{\alpha}k\,x/\hbar\right) \;\Rightarrow\; c_x = c_k\,.$$

c) $\psi_{\tilde{A}}(x) \equiv \langle x \mid U(\alpha)\mid A\rangle = e^{\alpha/2}\langle e^{\alpha}x \mid A\rangle = e^{\alpha/2}\,\psi_A(e^{\alpha}x)$.

$$\varphi_{\tilde{A}}(k) \equiv \langle k \mid U(\alpha)\mid A\rangle = e^{-\alpha/2}\langle e^{-\alpha}k \mid A\rangle = e^{-\alpha/2}\,\varphi_A(e^{-\alpha}k)\,.$$

5.24

a) One has
$$q_1' = q_1 \cos(2\pi/3) + q_2 \sin(2\pi/3), \quad q_2' = -q_1 \sin(2\pi/3) + q_2 \cos(2\pi/3),$$
$$q_3' = q_3, \quad \text{namely:}$$

$$q_i' = \sum_{j=1}^{3} \mathcal{R}_{ij} q_j, \quad \mathcal{R} = \begin{pmatrix} -\dfrac{1}{2} & \dfrac{\sqrt{3}}{2} & 0 \\ -\dfrac{\sqrt{3}}{2} & -\dfrac{1}{2} & 0 \\ 0 & 0 & 1 \end{pmatrix}; \quad \mathcal{R}^{-1} = \begin{pmatrix} -\dfrac{1}{2} & -\dfrac{\sqrt{3}}{2} & 0 \\ \dfrac{\sqrt{3}}{2} & -\dfrac{1}{2} & 0 \\ 0 & 0 & 1 \end{pmatrix}$$

and analogous formulae for the p_i. The kinetic energy is invariant under any rotation, and so it is in particular for the 120° rotation. Furthermore:

$$V(|\vec{q}\,' - \vec{a}\,|) = V(|\mathcal{R}\vec{q} - \mathcal{R}\mathcal{R}^{-1}\vec{a}\,|) = V(|\mathcal{R}(\vec{q} - \mathcal{R}^{-1}\vec{a}\,)|) = V(|\vec{q} - \mathcal{R}^{-1}\vec{a}\,|)$$
$$= V(|\vec{q} - \vec{b}\,|)$$

and likewise $V(|\vec{q}\,' - \vec{b}\,|) = V(|\vec{q} - \vec{c}\,|), \quad V(|\vec{q}\,' - \vec{c}\,|) = V(|\vec{q} - \vec{a}\,|).$

b) The kinetic energy is invariant under the reflection $p_x \to -p_x$; in addition:

$$V(|\vec{q} - \vec{a}\,|) \leftrightarrow V(|\vec{q} - \vec{b}\,|), \quad V(|\vec{q} - \vec{c}\,|) \to V(|\vec{q} - \vec{c}\,|).$$

Let U be the operator that implements the 120° rotation and I_x the one that implements the reflection. One has (sum over repeated indices understood):

$$U\,q_i\,U^{-1} = \mathcal{R}_{ij}\,q_j; \qquad I_x\,q_i\,I_x^{-1} = \mathcal{I}_{ij}\,q_j, \qquad \mathcal{I} = \begin{pmatrix} -1 & 0 & 0 \\ 0 & 1 & 0 \\ 0 & 0 & 1 \end{pmatrix}$$

$$I_x \left(U\,q_i\,U^{-1} \right) I_x^{-1} = \mathcal{R}_{ij}\,I_x\,q_j\,I_x^{-1} = \mathcal{R}_{ij}\,\mathcal{I}_{jk}\,q_k$$

$$U \left(I_x\,q_i\,I_x^{-1} \right) U^{-1} = \mathcal{I}_{ij}\,U\,q_j\,U^{-1} = \mathcal{I}_{ij}\,\mathcal{R}_{jk}\,q_k$$

and, since $\mathcal{R}\mathcal{I} \neq \mathcal{I}\mathcal{R}$ ($\mathcal{I}\mathcal{R}\mathcal{I}^{-1} = \mathcal{R}^{-1} \neq \mathcal{R}$), the two transformations do not commute with each other.

c) Since U and I_x commute with H, if H were nondegenerate its eigenvectors should be eigenvectors of both U and I_x as well; in this case U and I_x would have a complete set of simultaneous eigenvectors and should therefore commute. As a consequence H has degenerate eigenvalues (degeneracy theorem).

d) As $U\,H\,U^\dagger = H$, one has:

$$H_{11} \equiv \langle 1 \mid H \mid 1 \rangle = \langle 1 \mid U^\dagger U\,H\,U^\dagger U \mid 1 \rangle = \langle 2 \mid H \mid 2 \rangle \equiv H_{22} = H_{33}$$

$$H_{12} = H_{23} = H_{31}$$

furthermore $I_x\,H\,I_x^\dagger = H \Rightarrow H_{12} = H_{21}$, therefore all the nondiagonal matrix elements are equal to one another and real:

$$H \rightarrow \begin{pmatrix} E_0 & a & a \\ a & E_0 & a \\ a & a & E_0 \end{pmatrix}.$$

e) In order to determine the eigenvalues of H, we rewrite the matrix in the following way:

$$H \rightarrow \begin{pmatrix} E_0 & a & a \\ a & E_0 & a \\ a & a & E_0 \end{pmatrix} = (E_0 - a) \begin{pmatrix} 1 & 0 & 0 \\ 0 & 1 & 0 \\ 0 & 0 & 1 \end{pmatrix} + \begin{pmatrix} a & a & a \\ a & a & a \\ a & a & a \end{pmatrix}$$

whence, due to the fact that the characteristic of the last matrix is 1, two of its eigenvalues are vanishing and the third coincides with its trace, i.e. $3a$; in conclusion the eigenvalues of H are:

$$E_1 = E_2 = E_0 - a; \quad E_3 = E_0 + 2a .$$

f) The eigenvector of H corresponding to the eigenvalue E_3 is

$$| E_3 \rangle = \frac{1}{\sqrt{3}} (|1\rangle + |2\rangle + |3\rangle)$$

which is (better: must be) an eigenvector of U and I_x as well. All the vectors orthogonal to $|E_3\rangle$ are eigenvectors of H; the eigenvectors of I_x are $|1\rangle + |2\rangle + \alpha |3\rangle$, $|1\rangle - |2\rangle$ and, among them, those orthogonal to $|E_3\rangle$ are:

$$\frac{1}{\sqrt{6}} (|1\rangle + |2\rangle - 2|3\rangle); \quad \frac{1}{\sqrt{2}} (|1\rangle - |2\rangle).$$

The matrix that represents the operator U in the basis $|1\rangle$, $|2\rangle$, $|3\rangle$ is

$$U \rightarrow \begin{pmatrix} 0 & 0 & 1 \\ 1 & 0 & 0 \\ 0 & 1 & 0 \end{pmatrix}$$

and, since $U^3 = \mathbb{1}$, its eigenvalues are the cubic roots of 1 and the eigenvectors, besides the one already mentioned, are:

$$\frac{1}{\sqrt{3}} (|1\rangle + e^{2\pi i/3} |2\rangle + e^{4\pi i/3} |3\rangle); \quad \frac{1}{\sqrt{3}} (|1\rangle + e^{4\pi i/3} |2\rangle + e^{2\pi i/3} |3\rangle)$$

that are also eigenvectors of H.

5.25

a) The wavefunction is nonvanishing in four different regions (see the figure): $\psi = \psi_1' + \psi_2' + \psi_1'' + \psi_2''$, where the components ψ_1' and ψ_2' are generated by the component of the wavefunction that has gone the 'short' path, and ψ_1'' and ψ_2'' are generated by the component that has gone the 'long' path.

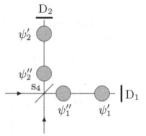

b) There survives only ψ_1' (downstream D_1): the other components "collapse".

c) There remain only the components ψ_1'' and ψ_2'' that have not yet reached the detectors.

d) In the interval $t_1 + \tau < t < t_2$ there remain the three components ψ_2', ψ_1'' and ψ_2''. Then, for $t > t_2 + \tau$, only ψ_2' and ψ_2'' survive.

e) In the interval $t_1 + \tau < t < t_2$ the state of the particle is ψ_2' if D_2 has clicked, otherwise it is $\psi_1'' + \psi_2''$. Since each detector clicks in one of the two time intervals with probability 25%, the state is ψ_2' in $\frac{1}{3}$ of the cases and $\psi_1'' + \psi_2''$ in $\frac{2}{3}$ of the cases (conditional probability), so the state of the system is the statistical mixture $\{\psi_2', \frac{1}{3}; \frac{1}{\sqrt{2}}(\psi_1'' + \psi_2''), \frac{2}{3}\}$. For $t > t_2 + \tau$, after having realized that D_1 has not clicked, the state of the system is the statistical mixture $\{\psi_2', \frac{1}{2}; \psi_2'', \frac{1}{2}\}$.

6

One-Dimensional Systems

Nondegeneracy theorem; variational method; rectangular potentials; transfer matrix and S-matrix; delta potentials; superpotential; completeness.

6.1 The Hamiltonian of a particle in one dimension is $H = p^2/2m + V(q)$. Assume that $V(q) = V(-q)$ and that H only has discrete eigenvalues. Let I be the operator that represents the spatial inversion: $I\,q\,I^{-1} = -q$, $I\,p\,I^{-1} = -p$.

a) May observables ξ exist that commute with H, but not with I?

b) What if H has continuous eigenvalues?

c) Let $|E_n\rangle$, $n = 0 \cdots$, be the eigenvectors of H corresponding to discrete eigenvalues. Say which of the matrix elements, relative to the operators q, p, q^2, p^2 between two such states, certainly vanish ("selection rules for parity").

6.2 The eigenvalue Schrödinger equation for a particle in one dimension is

$$-\frac{\hbar^2}{2m}\,\psi_E''(x) + V(x)\,\psi_E(x) = E\,\psi_E(x)\ .$$

One may have to solve the equation either analytically or numerically. In any event it is convenient to cast the equation in dimensionless form: to this end, if a is a length intrinsic to the problem – i.e. constructed from the constants that appear in the equation $(\hbar,\ m$ and those appearing in the expression for $V(x))$ – one sets $\xi = x/a$, $f_\epsilon(\xi) \equiv \psi_E(a\,\xi)$.

a) Write the equation for $f_\epsilon(\xi)$ in dimensionless form; show that the eigenvalues ϵ are determined in units of $\hbar^2/2ma^2$: $\epsilon \equiv E/(\hbar^2/2ma^2)$. Given that $f_\epsilon(\xi)$ is normalized $\left(\int |f_\epsilon(\xi)|^2\,d\xi = 1\right)$, say what the value of the constant C $(C > 0)$ must be such that $\psi_E(x) = C\,f_\epsilon(x/a)$ is normalized.

Consider, for the sake of concreteness, a particle of mass m subject to the potential:

$$V(x;\,\omega,\,g) = \frac{1}{2}m\omega^2 x^2 + g\,x^4,\qquad g > 0\ .$$

© Springer International Publishing AG 2017
E. d'Emilio and L.E. Picasso, *Problems in Quantum Mechanics*,
UNITEXT for Physics, DOI 10.1007/978-3-319-53267-7_6

b) Find the length a that can be made with $\hbar$, m, ω and the length b that can be made with $\hbar$, m, g. Write the dimensionless form of the Schrödinger equation in terms of either x/a and x/b. Say which of the two forms is more convenient in the case $m = 12\,\text{GeV}/\text{c}^2$, $\omega = 2\pi \times 10^{12}\,\text{s}^{-1}$, $g = 6.8 \times 10^{-2}\,\text{eV}/\text{Å}^4$.

c) Show that the Schrödinger equation for $f_\epsilon(\xi)$ is the eigenvalue equation $H \,|\, E \rangle = E \,|\, E \rangle$ in the Schrödinger representation for the canonical variables $\tilde{q} = q/a$, $\tilde{p} = a\,p$.

6.3   As in Problem 6.2, consider a particle of mass m subject to the potential:

$$V(x) = \tfrac{1}{2}m\omega^2 x^2 + g\,x^4$$

with $m = 12\,\text{GeV}/\text{c}^2$, $\omega = 2\pi \times 10^{12}\,\text{s}^{-1}$, $g = 6.8 \times 10^{-2}\,\text{eV}/\text{Å}^4$. Suppose one has to determine the eigenvalues E_n and the eigenfunctions $\psi_n(x)$ (or at least some of them) of the Hamiltonian H with the help of a computer.

a) Which initial conditions must be imposed at $x = 0$ if one is interested in the ground state? Which ones if one is interested in the first excited state?

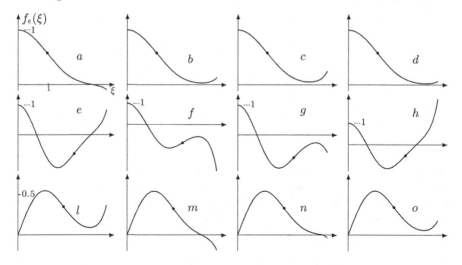

Unfortunately, in order to determine the eigenfunctions $\psi_n(x)$, it is not sufficient to assign the correct initial conditions, because we do not know the corresponding eigenvalues E_n. It is therefore necessary to proceed by trial and error, i.e. to solve the equation for several values of E until a solution, exhibiting as much as possible the properties of the eigenfunction one is after, is found. The graphs reproduced above correspond, not necessarily in the same order, to the following values of $\epsilon \equiv E/\hbar\omega$: 0.557, 0.558, 0.559, 0.560, 1.75, 1.76, 1.77,   1.78, 3.0, 3.1, 3.2, 3.3; the abscissa axis refers to the dimensionless variable (see Problem 6.2)

$\xi \equiv x/a = \sqrt{m\omega/\hbar}\,x$ and the unit on this axis is the same for all graphs, whereas the unit on the ordinate axis is given explicitly. The curves are drawn only for $0 \le \xi \le 3$: for $\xi \le 0$ they may be obtained from the fact that all the eigenfunctions have definite parity.

b) For each graph say to which eigenfunction it represents an approximation. To each graph associate the corresponding value of ϵ: to this end relevant pieces of information may be obtained from the curvature at the origin, from the position of the inflection points where $f_\epsilon(\xi) \ne 0$ (dotted on the graphs), from the number and position of the zeroes and, finally, from the behaviour of the curves in the region to the right of each graph. What is the best estimate one may obtain for the eigenvalues one is after?

6.4   It is known that for a particle in one dimension subject to the rectangular "potential well":

$$V_0(x) = \begin{cases} 0 & |x| > a \\ -V_0 & |x| < a \end{cases} \qquad V_0 > 0$$

there always exists at least one bound state with energy $E_0 < 0$.

Consider a particle subject to a negative potential: $V(x) < 0$ such that
$\lim_{x\to\pm\infty} V(x) = 0$ and for which there
exists a rectangular potential well $V_0(x)$
($V_0(x) \le 0$) of suitable depth and width
with the property $V_0(x) \ge V(x)$. Let
$\psi_0(x)$ be the wavefunction of the ground state of the particle subject to
$V_0(x)$ and let

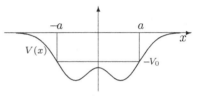

$$H = \frac{p^2}{2m} + V(x), \qquad H_0 = \frac{p^2}{2m} + V_0(x).$$

a) Show that the mean value of H in the state of wavefunction $\psi_0(x)$ is negative.

b) Show that H has at least one bound state whose energy is less or equal to the energy E_0^0 of the ground state of H_0.

6.5   Let $H_1 = p^2/2m + V_1(x)$ and $H_2 = p^2/2m + V_2(x)$ be two Hamiltonians with $V_{1,2}(x) \xrightarrow[|x|\to\infty]{} 0$ and $V_2(x) \ge V_1(x)$.

a) Show that if H_2 has a bound state, then also H_1 has at least one bound state.

Assume the potentials are even functions: $V_1(x) = V_1(-x)$, $V_2(x) = V_2(-x)$.

b) Exploit the above assumption to show that, if H_2 has two bound states, then also H_1 has at least two bound states.

Let us drop the assumption that V_1 and V_2 are even functions, but maintain the hypothesis that H_2 has two bound states $|E_0^{(2)}\rangle$, $|E_1^{(2)}\rangle$ whose energies are $E_0^{(2)} < E_1^{(2)} < 0$, and let $|E_0^{(1)}\rangle$ be the (existing!) ground state of H_1.

c) Show that the mean value of H_1 is negative in all the states of the vector space $\mathcal{V}$ generated by $|E_0^{(2)}\rangle$, $|E_1^{(2)}\rangle$, and – owing to the existence of a vector in $\mathcal{V}$ orthogonal to $|E_0^{(1)}\rangle$ – conclude that also H_1 has at least two bound states.

d) Generalize the result to the case of n bound states.

6.6   It is known that a particle in one dimension subject to a rectangular potential well of depth V_0 and width $2a$ admits as many bound states as the least integer greater or equal to $\sqrt{2mV_0 a^2/\hbar^2}\,/(\pi/2) = \sqrt{8mV_0 a^2/\pi^2 \hbar^2}$. By exploiting the results of Problem 6.5:

a) Find the minimum number n_b of bound states that the following potentials admit:

$$V(x) = -3\,\frac{\hbar^2}{mb^2}\,\mathrm{e}^{-x^2/b^2}; \qquad V(x) = -4\,\frac{\hbar^2}{mb^2}\,\mathrm{e}^{-x^2/b^2}.$$

b) Find the value of λ such that the potential:

$$V(x) = -\frac{\lambda}{x^2 + b^2}, \qquad \lambda > 0$$

admits at least N bound states.

c) Find the number of bound states admitted by the potential:

$$V(x) = -\frac{\lambda}{|x| + b}, \qquad \lambda > 0, \quad b > 0.$$

6.7   Assume the energy of the ground state of a harmonic oscillator of mass m and angular frequency ω is not known. Consider the states described by the following normalized wavefunctions ("trial functions"):

$$\psi(x;\,a) = \begin{cases} \sqrt{\dfrac{315}{256\,a^9}}\,(x^2 - a^2)^2 & |x| \le a \\ 0 & |x| \ge a. \end{cases} \qquad a > 0$$

a) Calculate the mean value $\overline{H}(a)$ of the Hamiltonian H in the states described by $\psi(x;\,a)$. To this end the following integrals are useful:

$$\int_{-1}^{+1} x^2(x^2 - 1)^2\,\mathrm{d}x = \frac{16}{105}; \qquad \int_{-1}^{+1} x^2(x^2 - 1)^4\,\mathrm{d}x = \frac{256}{3465}.$$

b) Find the value $\bar{a}$ of a for which the mean value of the Hamiltonian is a minimum. Say whether the mean value $\overline{H}(\bar{a})$ of H in the state described by the wavefunction $\psi(x;\,\bar{a})$ underestimates or, rather, overestimates of the (unknown) energy E_0 of the ground state of the oscillator. Calculate $\overline{H}(\bar{a})$.

c) Would one like to take advantage of the same technique (known as *variational method*) to estimate the energy of the first excited level of the oscillator, which properties should the trial functions possess in this case?

We now want to compare the wavefunction $\psi(x; \bar{a})$ found in b) with the wavefunction $\psi_0(x)$ of the ground state of the oscillator.

d) Knowing that $E_n = \left(n + \frac{1}{2}\right)\hbar\omega$ and setting $\psi(x; \bar{a}) = \alpha\,\psi_0(x) + \beta\,\widetilde{\psi}(x)$ (with $\widetilde{\psi}(x)$ and $\psi_0(x)$ orthogonal to each other and $|\alpha|^2 + |\beta|^2 = 1$), find an upper bound for $|\beta|$.

6.8 A particle of mass m is subject to the potential $V(x) = g\,x^4$, $g > 0$.

a) Determine α and n in such a way that $\exp(-\alpha|x|^n)$ represents the asymptotic behaviour of the eigenfunctions of the Hamiltonian H for $|x| \to \infty$.

b) How do the eigenvalues of H depend on g and m (see Problem 5.6)?

c) Exploit the variational method with the trial functions of Problem 6.7:

$$\psi(x; a) = \begin{cases} \sqrt{\dfrac{315}{256\,a^9}}\,(x^2 - a^2)^2 & |x| \le a \\ 0 & |x| \ge a \end{cases} \qquad a > 0$$

to find an upper bound for the energy of the ground state. To this end, in addition to those given in the text of Problem 6.7, the following integral is useful:

$$\int_{-1}^{+1} x^4(x^2 - 1)^4\, dx = \frac{256}{15015}\,.$$

6.9 A particle of mass m is subject to a potential that, in the region $x < 0$ is positive and large to the extent that it is legitimate to consider the limit $V(x) \to +\infty$ for $x < 0$: this amounts to constrain the particle to move only in the region $x \ge 0$. In the latter region the potential is

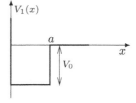

$$V_1(x) = \begin{cases} -V_0 & 0 < x < a \\ 0 & x \ge a\,. \end{cases} \qquad V_0 > 0$$

It is known that, in the above mentioned limit, the eigenfunctions of the Hamiltonian must vanish for $x = 0$: $\psi_E(0) = 0$ (see Problem 6.10).

a) Is it true that not only the eigenfunctions of the Hamiltonian, but *all* the wavefunctions $\psi(x)$ must vanish at $x = 0$?

b) We want now to establish whether the system always admits a bound state. To this end, assuming it exists, draw the corresponding wavefunction $\psi_{E_0}(x)$ in both regions $0 \le x \le a$ and $x \ge a$ and conclude that if $V_0 \le \pi^2\hbar^2/(8ma^2)$ there exist no bound states.

c) Show that the condition $V_0 > \pi^2\hbar^2/(8ma^2)$ is not only necessary but also sufficient for the existence of bound states.

d) Write the eigenfunctions of the Hamiltonian corresponding to the continuous spectrum.

e) Let us assume that in the case of a particle subject to the rectangular potential well of width $2a$: $V_2(x) = 0$ for $|x| > a$, $V_2(x) = -V_0$ for $|x| < a$ $(-\infty < x < +\infty)$ the system admits n bound states. How many are the bound states for the particle subject to the potential $V_1(x)$?

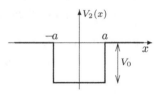

6.10   A particle of mass m is subject to the potential:

$$V(x) = \begin{cases} V_1 & x < 0 \\ -V_0 & 0 \le x \le a \\ 0 & x > a. \end{cases}$$

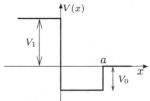

a) Show that in the limit $V_1 \to +\infty$ the eigenfunctions of the Hamiltonian vanish for $x \le 0$.

b) Is it true that not only the eigenfunctions of the Hamiltonian, but *all* the wavefunctions $\psi(x)$ must vanish for $x < 0$?

Assume that in the limit $V_1 = +\infty$ the system does not admit bound states: $\sqrt{2mV_0a^2/\hbar^2} < \pi/2$ (see Problem 6.9), whereas it is known that for $V_1 = 0$ the system always admits at least one bound state.

c) Determine $\overline{V}_1$ such that for $V_1 < \overline{V}_1$ the system admits at least one bound state.

6.11   Consider a particle subject to the rectangular potential well of width $2a$:

$$V(x) = \begin{cases} 0 & |x| \ge a \\ -V_0 & |x| < a. \end{cases}$$

Setting $\xi = ka$, $\eta = \kappa a$ where $\kappa = \sqrt{2m|E|/\hbar^2}$, $k = \sqrt{2m(V_0 - |E|)/\hbar^2}$ (note that, with respect to the case $V(x) = V_0$ for $|x| > a$, $V(x) = 0$ for $|x| < a$, ξ and η obtain through the substitution $E \to E + V_0$), the energies E_n of the bound states are determined by the systems of equations (the first one refers to the even states, the second one to the odd states):

$$\begin{cases} \eta = \xi \tan \xi \\ \xi^2 + \eta^2 = \dfrac{2mV_0a^2}{\hbar^2} \end{cases} \qquad \begin{cases} \eta = -\xi/\tan \xi \\ \xi^2 + \eta^2 = \dfrac{2mV_0a^2}{\hbar^2} \end{cases} \qquad \xi > 0, \quad \eta > 0.$$

a) Show that for every discrete eigenvalue E_n one has:

$$E_n(V_0, a) = -\frac{\hbar^2}{2ma^2} F_n(2mV_0 a^2/\hbar^2)$$

with (see Problem 6.5) $F_n'(x) > 0$ ($F_n(x)$ is not explicitly calculable).

b) Show that, both for $V_0 \to 0$ with fixed a, and for $a \to 0$ with fixed V_0,
$E_0(V_0, a) \approx -2mV_0^2 a^2/\hbar^2$.

c) Determine the asymptotic behaviour of $E_0(V_0, a)$ both for $V_0 \to \infty$ with fixed a, and for $a \to \infty$ with fixed V_0 .

d) Determine the limit of $E_0(V_0, a)$ for $V_0 \to \infty$ and $a \to 0$ with fixed $V_0 a = \lambda$.

e) Exploiting the behaviour of $F_0(x)$ for $x \to 0$ and for $x \to \infty$ found in b) and c), draw a qualitative graph for $|E_0|$ as a function of V_0, with a fixed.

f) Estimate, also with the aid of the graph, the energy of the ground state for
    i) an electron in a (one-dimensional) potential well of depth $V_0 = 1\,\mathrm{eV}$ and width $2a = 2\,\text{Å}$,   ii) a proton (mass $m_\mathrm{p} \simeq 1836\,m_\mathrm{e}$) in a well of depth $V_0 = 1\,\mathrm{MeV}$ and width $2a = 2 \times 10^{-12}\,\mathrm{cm}$.

6.12   Consider a particle subject to the potential:

$$V(x) = \begin{cases} 0 & x \le 0 \\ V_0 & x > 0 \end{cases}$$

with $V_0 > 0$ ("potential step").

a) Write the eigenfunctions of the Hamiltonian belonging to the eigenvalues $E < V_0$.

The transmission coefficient T and the reflection coefficient R are defined as the ratio of the transmitted and, respectively, the reflected flux (or density of probability current) to the incident flux.

b) Calculate the transmission and reflection coefficients for particles incoming from the region $x < 0$ with $E > V_0$ and verify they satisfy the condition demanded by flux conservation (continuity equation for the current).

c) Given the wavefunction for particles incoming from the region $x < 0$ with $E > V_0$, due to the reality of the Schrödinger equation it is possible to write another independent solution: find, among the linear combinations of these two solutions, the eigenfunction belonging to energy E relative to particles incoming from the region $x > 0$. Calculate the transmission and reflection coefficients and verify they are equal to those for particles incoming from the left ($x < 0$).

6.13 Consider the potential barrier:

$$V(x) = \begin{cases} 0 & x \le 0 \\ V_0 & 0 < x < a \\ 0 & x \ge a . \end{cases}$$

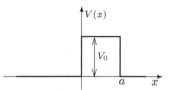

a) Determine the values of the energy E, for a particle of mass m, such that the reflected wave is absent (total transmission). Distinguish between the two cases $E \le V_0$ and $E > V_0$.

b) Do any values of the energy E exist for which the transmitted wave is absent (total reflection)?

6.14 Let $V(x)$ be a potential such that $V(x) = 0$ for both $x < x_1$ and $x > x_2$, being $x_1 < x_2$. In the external regions $x < x_1$ and $x > x_2$ the eigenfunctions $\psi_E(x)$ $(E > 0)$ of the Hamiltonian have the form:

$$\psi_E(x) = \begin{cases} \alpha e^{ikx} + \beta e^{-ikx} & x \le x_1 \\ \gamma e^{ikx} + \delta e^{-ikx} & x \ge x_2 . \end{cases}$$

Thanks to the linearity of the Schrödinger equation, the coefficients γ, δ depend linearly on α and β:

$$\begin{pmatrix} \gamma \\ \delta \end{pmatrix} = \begin{pmatrix} A & B \\ C & D \end{pmatrix} \begin{pmatrix} \alpha \\ \beta \end{pmatrix}, \qquad\qquad \begin{pmatrix} A & B \\ C & D \end{pmatrix} \equiv M$$

$\left(\text{the matrix } M \equiv M(k) \text{ is named } transfer \text{ } matrix \text{ for the potential } V(x)\right).$

a) Show that from the reality of the Schrödinger equation it follows that $D = A^*$, $C = B^*$.

b) Show that the continuity equation implies $\det M = 1$.

c) Determine the transmission and reflection coefficients for the potential $V(x)$ in terms of the matrix elements of M.

d) Given the transfer matrix $M(k)$ for the potential $V(x)$, find the transfer matrices $M_{\rm tr}(k)$ for the translated potential $V(x - a)$ and $M_{\rm rf}(k)$ for the reflected potential $V(-x)$.

6.15 Consider a potential $V(x)$ that is nonvanishing only for $x_1 \le x \le x_2$. In the one-dimensional case the *scattering matrix*, or *S-matrix*, allows one to determine the amplitudes of transmitted and reflected waves for particles incoming either from the left or from the right:

$$\psi_E^{(l)}(x) = \begin{cases} e^{ikx} + \beta e^{-ikx} & x \le x_1 \\ \gamma e^{ikx} & x \ge x_2 \end{cases} : \qquad S^\dagger \begin{pmatrix} 1 \\ 0 \end{pmatrix} = \begin{pmatrix} \gamma \\ \beta \end{pmatrix}$$

$$\psi_E^{(r)}(x) = \begin{cases} e^{-ikx} + \tilde\gamma e^{ikx} & x \ge x_2 \\ \tilde\beta e^{-ikx} & x \le x_1 \end{cases} : \qquad S^\dagger \begin{pmatrix} 0 \\ 1 \end{pmatrix} = \begin{pmatrix} \tilde\gamma \\ \tilde\beta \end{pmatrix}$$

(we have written $S^\dagger$ instead of S to stick to the use current in the literature).

a) Given the transfer matrix M (see Problem 6.14), find the S-matrix.

b) Show that the S-matrix is unitary.

c) Express the transmission and reflection coefficients in terms of the matrix elements of S.

6.16   Let $H = p^2/2\,m + V(q)$ be the Hamiltonian for a particle of mass m in one dimension. The particle is in an eigenstate of H whose probability density for the position is

$$\rho(x) = |\psi(x)|^2 = \frac{N^2}{(x^2 + a^2)^2}\ .$$

a) Say whether the state is uniquely determined and, in the affirmative case, write its wavefunction $\psi(x)$.

b) Calculate the mean values of p and q in this state.

c) Say whether the state is a minimum uncertainty state: $\Delta p\,\Delta q = \hbar/2$.

d) Make use of the Schrödinger equation to determine the potential $V(x)$.

e) Say whether there exists an odd eigenstate of H belonging to a discrete eigenvalue.

6.17   The eigenfunctions of the Hamiltonian H of a particle of mass m in one dimension satisfy the Schrödinger equation:

$$\left(-\frac{\mathrm{d}^2}{\mathrm{d}x^2} + U(x)\right)\psi(x) = \frac{2mE}{\hbar^2}\,\psi(x)\,, \qquad U(x) = \frac{2m}{\hbar^2}\,V(x)$$

with $V(x)$ unknown. It is known, instead, that the functions:

$$\psi_k(x) = \frac{i\,ka - \tanh(x/a)}{i\,ka + 1}\,\frac{e^{i\,k\,x}}{\sqrt{2\pi}}\,, \quad k > 0; \quad \int_{-\infty}^{+\infty} \psi_k^*(x)\,\psi_{k'}(x)\,\mathrm{d}x = \delta(k - k')$$

are eigenfunctions of H.

a) Determine the asymptotic behaviours of $\psi_k(x)$ for $x \to \pm\infty$, calculate the coefficients of reflection R and transmission T and verify that they satisfy $R + T = 1$.

b) Determine the potential $V(x)$ and say whether it admits bound states.

c) Show that $\psi_{-k}(x)$ is an eigenfunction of H belonging to the same eigenvalue E to which $\psi_k(x)$ belongs.

The following relation will be derived in the solution:

$$\int_{-\infty}^{+\infty} \psi_k(x)\,\psi_k^*(y)\,\mathrm{d}k = \delta(x - y) - \frac{1}{2a\,\cosh(x/a)\,\cosh(y/a)}\ .$$

d) Show that H has only one bound state and determine its wavefunction.

6.18 Consider a particle of mass m in one dimension subject to the attractive "delta potential" $V(x) = -\lambda\,\delta(x)$, $\lambda > 0$, where $\delta(x)$ is the Dirac delta function. This may be considered as the limit, for

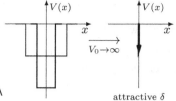

$$a \to 0, \qquad V_0 = \lambda/a \to +\infty, \qquad V_0 a = \lambda$$

of a rectangular potential well of depth V_0 and width a.

a) Taking the limit on the eigenfunctions $\phi_E(x)$ relative to the bound states of the rectangular potential well, show that the attractive delta potential has only one bound state. Explicitly calculate the corresponding eigenvalue E_0 of the energy and the eigenfunction $\psi_0(x)$; show in particular that $\psi_0(x)$ is continuous at $x = 0$, but its first derivative is discontinuous.

b) In the case the particle is an electron, what value must λ take so that $|E_0| = 1\,\mathrm{eV}$? Express your result in $\mathrm{eV}\times\text{Å}$.
   At what distance a from the origin $|\psi_0(a)|^2 \simeq 10^{-2}|\psi_0(0)|^2$?

It is possible to solve the Schrödinger equation with the attractive delta potential directly in the momentum representation.

c) Show that the Schrödinger equation in the momentum representation reads:
$$\left(\frac{p^2}{2m} - E\right)\varphi_E(p) = \frac{\lambda}{(2\pi\,\hbar)^{1/2}}\,\psi_E(0)\;.$$
   Write explicitly the functional dependence of φ_E on p and determine the energy E_0 of the bound state by exploiting the identity:
$$\psi_0(0) = (2\pi\,\hbar)^{-1/2}\int_{-\infty}^{+\infty}\varphi_0(p)\,\mathrm{d}p\;.$$
   Given $\varphi_0(p)$, use the Cauchy residue theorem to find again $\psi_0(x)$ in the coordinate representation.

It is also possible to solve the Schrödinger equation with the attractive delta potential directly in the coordinate representation.

d) Find the discontinuity of the logarithmic derivative of $\psi_E(x)$ at $x = 0$ by integrating once the Schrödinger equation in the interval $[-\epsilon, +\epsilon]$ and then taking the limit $\epsilon \to 0^+$. Use the result to find again the energy E_0 and the wavefunction $\psi_0(x)$ of the bound state.

6.19 Consider a particle of mass m in one dimension subject to the potential:

$$V(x) = -\lambda\,\delta(x) \qquad \begin{cases} \lambda < 0 & \text{repulsive potential} \\ \lambda > 0 & \text{attractive potential} \end{cases}$$

a) Determine the transfer matrix M for energies $E > 0$ (see Problem 6.14) and $S^\dagger$, where S is the scattering matrix (see Problem 6.15). Calculate the transmission coefficient $T(k)$ and the reflection coefficient $R(k)$.

b) Determine the eigenfunctions of the energy $\psi_k^{(l)}(x)$, $\psi_k^{(r)}(x)$ for $E > 0$ corresponding to the scattering states $|k^{sc}\rangle$ both for particles coming from the left ($x = -\infty$; $k > 0$) and from the right ($x = +\infty$; $k < 0$).

c) Determine the eigenfunctions of the energy with $E > 0$ relative to the states of definite parity.

d) Assume that the scattering states are normalized in such a way that:

$$\langle k'^{(r)} | k^{(r)} \rangle = \langle k'^{(l)} | k^{(l)} \rangle = \delta(k - k'), \qquad \langle k'^{(r)} | k^{(l)} \rangle = 0 .$$

Say what is the value of the following integral:

$$E(x, y) = \int_0^\infty \left(\psi_k^{(l)}(x)\, \psi_k^{(l)}(y)^* + \psi_k^{(r)}(x)\, \psi_k^{(r)}(y)^* \right) dk$$

both for $\lambda > 0$ and $\lambda < 0$, without calculating it.

6.20 Consider the Schrödinger equation for a particle of mass m subject to a potential $V(x) = 0$ for $x < x_1$ and for $x > x_2$, being $x_1 < x_2$. The transfer matrix M can be defined, thanks to the linearity of the Schrödinger equation, even for the solutions with $E < 0$:

$$f_E(x) = \begin{cases} \alpha e^{\kappa x} + \beta e^{-\kappa x} & x \le x_1 \\ \gamma e^{\kappa x} + \delta e^{-\kappa x} & x \ge x_2 \end{cases} \qquad \alpha, \beta, \gamma, \delta \in \mathbb{C}; \quad \kappa = \sqrt{2m|E|/\hbar^2}$$

independently of the fact that $f_E(x)$ is normalizable. It is useful to find the bound states.

a) Demonstrate that $M(\kappa)$ is a real matrix and that, thanks to the continuity equation, its determinant equals 1.

b) Find the condition on the elements of the matrix $M(\kappa)$ such that $f_E(x)$ is normalizable, namely that E is a discrete eigenvalue.

c) Assume that the transfer matrix $M(\kappa)$ for the potential $V(x)$ is known. Determine the transfer matrices $M_{tr}(\kappa)$ for the translated potential $V(x - a)$ and $M_{rf}(\kappa)$ for the reflected potential $V(-x)$.

d) Find the matrix M for the solutions with $E < 0$ in the case of the attractive potential $V(x) = -\lambda\, \delta(x)$ with $\lambda > 0$ and determine the energy of the bound state.

6.21 Let $V_1(x)$ be a potential vanishing for $x < x_1$ and $x > x_2$ ($x_1 < x_2$) and let M_1 be the relative transfer matrix (see Problems 6.14 and 6.20). Let $V_2(x) \equiv V_1(x - a)$, $a > x_2 - x_1$ be the potential obtained by translating $V_1(x)$ (V_1 and V_2 have disjoint supports).

a) Show that the transfer matrix M for a particle subject to the potential:

$$V(x) = V_1(x) + V_2(x)$$

is the product (in the suitable order) of the transfer matrices M_1 and M_2 relative to the single potentials $V_1(x)$ and $V_2(x)$.

Let us suppose that E_0 is the energy of a bound state of a particle subject to the only potential $V_1(x)$ (or $V_2(x)$).

b) Show that for large values of a, for any bound state relative to the single potential $V_1(x)$ (or $V_2(x)$) the particle subject to the potential $V(x) = V_1(x) + V_2(x)$ has two bound states with energies E_1 and E_2 whose distance decreases exponentially as a increases: exploit the condition that the existence of the bound state of energy E_0 enforces upon M_1 (see Problem 6.20) and the fact that $\det M_1 = 1$.

c) Show that, for large values of a, a similar result obtains for the potential $\widetilde{V}(x) = V_1(x) + V_1(a-x)$. What does the difference between the potentials $V_1(x) + V_1(a-x)$ and $V_1(x) + V_1(x-a)$ consist in?

6.22   Consider a particle of mass m in one dimension, subject to the potential:

$$V(x) = \frac{\hbar^2}{2m}\left(\phi'^2(x) - \phi''(x)\right)$$

($\phi(x)$ is called the 'superpotential').

a) Show that the Hamiltonian may be written in the form:

$$H = \frac{1}{2m}\left(p + i f(x)\right)\left(p - i f(x)\right)$$

with $f(x)$ a suitable real function, and that, as a consequence, H has nonnegative eigenvalues.

b) Show that if $E = 0$ is an eigenvalue of H corresponding to a bound state $|E=0\rangle$, then necessarily:

$$\left(p - i f(x)\right)|E=0\rangle = 0.$$

If the bound state $|E=0\rangle$ exists, what is its wavefunction $\psi_0(x)$? Can one claim the existence of the bound state with $E=0$ for whatever $\phi(x)$?

c) Find the superpotential $\phi(x)$ corresponding to the potential:

$$V(x) = \frac{\hbar^2}{2ma^2}\left(\tanh^2(x/a) - \frac{1}{\cosh^2(x/a)}\right)$$

and exploit the above result to find the energy and the wavefunction of the ground state for a particle subject to the potential:

$$\tilde{V}(x) = V(x) - \frac{\hbar^2}{2ma^2} = -\frac{\hbar^2}{ma^2\cosh^2(x/a)}.$$

d) If $\phi'(x) = (m\lambda/\hbar^2)\varepsilon(x)$ $(\varepsilon(x) = \pm 1$ for $x \gtrless 0)$, find the potential $V(x)$. Exploit the result to find the energy and the wavefunction of the bound state for a particle subject to the potential $\tilde{V}(x) = -\lambda\,\delta(x)$, $\lambda > 0$.

6.23   A particle of mass m in one dimension is subject to the attractive potential:

$$V(x) = -\lambda\left[\delta(x-a) + \delta(x+a)\right],\qquad \lambda > 0.$$

a) Show that there always exists at least one bound state and that – given the form the eigenfunctions of the energy may have for $E < 0$ – there may exist at most two of them.

As in the case of the rectangular potential well, the energies of the bound states are solutions of transcendental equations.

b) Find the equations that allow for the determination of the bound states and, possibly by means of a plot, show that the second bound state exists only if $\lambda > \hbar^2/2ma$.

c) Assuming the condition just written is fulfilled, show that the difference in energy between the two bound states tends exponentially to zero as $a \to \infty$; in particular specify when a may be considered large.

d) Draw an accurate plot of the wavefunctions relative to the two bound states.

6.24   The problem of finding the bound states of a particle subject to the potential of Problem 6.23:

$$V(x) = -\lambda\left[\delta(x+a) + \delta(x-a)\right],\qquad \lambda > 0$$

may be solved approximately by means of the variational method (see Problems 6.7 and 6.8), taking as trial functions the set of functions given by:

$$\psi(x;\,\theta) = N\left(\cos\theta\,\psi_1(x) + \sin\theta\,\psi_2(x)\right)$$

where $\psi_1(x) = \sqrt{\kappa_0}\,e^{-\kappa_0\,|x+a|}$ and $\psi_2(x) = \sqrt{\kappa_0}\,e^{-\kappa_0\,|x-a|}$ $(\kappa_0 = m\lambda/\hbar^2)$ respectively are the normalized eigenfunctions relative to the bound state (state $|1\rangle$) with $V_1(x) = -\lambda\,\delta(x+a)$ and to the bound state (state $|2\rangle$) with $V_2(x) = -\lambda\,\delta(x-a)$ (see Problems 6.18 and 6.19).

a) Calculate the scalar product $\langle 1\,|\,2\rangle$ and the normalization factor N.

b) Estimate, by means of the variational method, the energies of the ground and of the first excited state.

c) Say in which of the two cases – a either large or small – the result is more reliable. Compare the result with that of Problem 6.23.

d) Say whether by means of the wider choice of trial functions:

$$\psi(x;\,\theta,\varphi) = N\big(\cos\theta\,\psi_1(x) + \sin\theta\,e^{i\varphi}\,\psi_2(x)\big)$$

more accurate results for the energies of the ground and of the first excited states may be obtained.

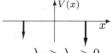

6.25   A particle of mass m in one dimension is subject to the asymmetric potential:

$$V(x) = V_1(x) + V_2(x) = -\lambda_1\,\delta(x+a) - \lambda_2\,\delta(x-a)\,, \qquad \lambda_2 > \lambda_1 > 0\,.$$

a) Making use of the transfer matrix M for the solutions with $E < 0$ (see Problems 6.20 and 6.21), find the equation that determines the bound states of the system.

b) Show that the ground state always exists, whereas the excited state exists only if $4a\,\kappa_2\,\kappa_1 > \kappa_2 + \kappa_1$, i.e. $\lambda_1\lambda_2/(\lambda_1 + \lambda_2) > \hbar^2/4ma$ ($\kappa_{1,2} = m\lambda_{1,2}/\hbar^2$).

c) Make a qualitative plot of the wavefunctions relative to the ground and to the (supposedly existing) excited state.

Consider now the case when $\lambda_2 = -\lambda_1 \equiv -\lambda$ (i.e. V_1 attractive, V_2 repulsive).

d) Show that there always exist only one bound state and make a qualitative plot of the corresponding wavefunction.

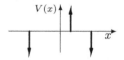

6.26   Consider a particle of mass m in one dimension subject to the potential:

$$V_3(x) = -\lambda\,\delta(x+a) + \lambda\,\delta(x-b) - \lambda\,\delta(x-a)$$

where $\lambda > 0$, $-a \le b \le a$.

a) Find the transfer matrix for the solutions with $E < 0$ and show that the energy of the ground state of the Hamiltonian H_3 is $-m\lambda^2/2\hbar^2$, i.e. the same as for the only bound state for a single attractive delta.

b) Show that the equation that determines the second bound state may be written in the form:

$$\cosh(2\kappa a) - \cosh(2\kappa b) = \frac{1}{2}\left(\frac{\kappa}{\kappa_0}\right)^2 e^{2\kappa a}\,, \qquad \kappa_0 = \frac{m\lambda}{\hbar^2}$$

and that it admits a solution if and only if $|b| < \bar{b} = a\,\sqrt{1 - \left(\dfrac{1}{2\kappa_0\,a}\right)^2}$.

It is known (see the solution of Problem 6.22) that the result of point a) extends to the potential ($a_1 < b_1 < a_2 < \cdots < b_n < a_{n+1}$):

$$V_{2n+1}(x) = -\lambda\sum_{i=1}^{n+1}\delta(x - a_i) + \lambda\sum_{i=1}^{n}\delta(x - b_i)\,, \qquad \lambda > 0\,, \qquad n \ge 0\,.$$

c) Show by induction the above result, making use of the transfer matrix.

Solutions

6.1

a) H commutes with I: $I H I^{-1} = H$. Indeed:

$$I V(q) I^{-1} = V(-q) = V(q); \quad I p^2 I^{-1} = I p I^{-1} I p I^{-1} = (-p)^2 = p^2 .$$

If an observable commuting with H but not with I existed, the Hamiltonian should have at least one degenerate eigenvalue (degeneracy theorem, see Problem 5.24), but in a one-dimensional problem the discrete eigenvalues of the Hamiltonian are all nondegenerate (nondegeneracy theorem).

b) Consider the example of the free particle: the Hamiltonian $H = p^2/2m$ commutes both with the momentum p and the space inversion I, but p and I do not commute with each other. Indeed in this case the eigenvalues are degenerate and the above argument fails.

c) The operators q and p are odd under space inversion ($I q I^{-1} = -q$, etc.) and this entails that q and p have vanishing matrix elements between states with the same parity $w = \pm 1$: $I \, | \, w \, \rangle = w \, | \, w \, \rangle$, independently of the fact that the latter may be eigenvectors of some Hamiltonian; indeed, due to $I^{-1} = I^\dagger$ and $\langle \, w \, | \, I^\dagger = w \, \langle \, w \, |$,

$$\langle \, w \, | \, q \, | \, w \, \rangle = \langle \, w \, | \, I^\dagger I \, q \, I^{-1} I \, | \, w \, \rangle = w^2 \langle \, w \, | \, I \, q \, I^{-1} \, | \, w \, \rangle$$

$$= -\langle \, w \, | \, q \, | \, w \, \rangle = 0 .$$

Therefore, since by the oscillation theorem (and the fact that H commutes with I) the eigenvectors of H are alternatively even, odd, even ..., the odd operators (such as q and p) may have nonvanishing matrix elements only between states $| \, E_n \, \rangle$ and $| \, E_m \, \rangle$, with $n + m$ odd.
Instead, q^2 and p^2 are even operators (i.e. they commute with I) and this entails that they have vanishing matrix elements between states with opposite parity.

6.2

a) As $\dfrac{d^2}{dx^2}\,\psi_E(x)=\dfrac{d^2}{dx^2}f_\epsilon(x/a)=\dfrac{1}{a^2}\dfrac{d^2}{d\xi^2}f_\epsilon(\xi)\,,$ one has:

$$-\frac{\hbar^2}{2ma^2}f_\epsilon''(\xi)+U(\xi)\,f_\epsilon(\xi)=E\,f_\epsilon(\xi)\,,\qquad U(\xi)\equiv V(a\,\xi)\,.$$

Dividing both sides by $\hbar^2/2ma^2$ (that has the dimensions of an energy) and putting $u(\xi)=(2ma^2/\hbar^2)\,U(\xi)$, $\epsilon=(2ma^2/\hbar^2)\,E$:

$$-f_\epsilon''(\xi)+u(\xi)\,f_\epsilon(\xi)=\epsilon\,f_\epsilon(\xi)$$

that is the Schrödinger equation in dimensionless form: it is particularly expressive inasmuch as the general features of the eigenfunctions corresponding to the lowest eigenvalues (maxima, zeroes, inflexion points, ...) show up in the region $|\xi|\lesssim1$.
If $f_\epsilon(\xi)$ is normalized (with respect to the measure $d\xi$), $\psi_E(x)=C\,f_\epsilon(x/a)$ is normalized (with respect to the measure dx) if:

$$1=C^2\int|f_\epsilon(x/a)|^2\,dx=C^2\int|f_\epsilon(\xi)|^2a\,d\xi=a\,C^2\quad\Rightarrow\quad C=\frac{1}{\sqrt a}\,.$$

Indeed, in whatever representation, the wavefunction has the dimensions of the reciprocal of the square root of the variable it depends on.

b) $a=\sqrt{\hbar/m\omega}$ and is the characteristic length intervening in the theory of the harmonic oscillator: note that the $\psi_n(x)$ given in the text of Problem 5.14 actually depend on the variable x/a.
Taking into account that g has the dimensions of (energy)/(length)^4, one finds that the only length one can construct out of $\hbar,\,m,\,g$ is $b=\left(\hbar^2/mg\right)^{1/6}$ whence $g=\hbar^2/mb^6$.
Putting $\xi=x/a$ one has:

$$u(\xi)=\frac{2ma^2}{\hbar^2}\left(\frac12 m\omega^2a^2\xi^2+g\,a^4\xi^4\right)=\xi^2+2\left(\frac{a}{b}\right)^6\xi^4\quad\Rightarrow$$

$$-f_\epsilon''(\xi)+\xi^2\,f_\epsilon(\xi)+2\left(\frac{a}{b}\right)^6\xi^4 f_\epsilon(\xi)=\epsilon\,f_\epsilon(\xi)\,.$$

Putting now $\xi=x/b$ (we use the same notation, but the meaning of ξ and ϵ are now different: $\epsilon=(2mb^2/\hbar^2)\,E$), one has:

$$u(\xi)=\frac{2mb^2}{\hbar^2}\left(\frac12 m\omega^2b^2\xi^2+g\,b^4\xi^4\right)=\left(\frac{b}{a}\right)^4\xi^2+2\,\xi^4\quad\Rightarrow$$

$$-f_\epsilon''(\xi)+\left(\frac{b}{a}\right)^4\xi^2\,f_\epsilon(\xi)+2\,\xi^4 f_\epsilon(\xi)=\epsilon\,f_\epsilon(\xi)\,.$$

In the proposed case, taking the eV and the Å as units,

$$\left(\frac{a}{b}\right)^6=\frac{\hbar\,g}{m^2\omega^3}=\frac{0.66\times10^{-15}\cdot6.8\times10^{-2}}{(144\times10^{18}/81\times10^{72})\times(2\pi)^3\times10^{36}}\simeq0.1$$

so the anharmonic term is of the same order of magnitude as the quadratic one $(g\,x^4 \simeq m\omega^2 x^2/2)$ for $x/a \simeq b^3/a^3 \simeq 3$: as the wavefunctions of the ground state and of the first excited states of the harmonic oscillator are appreciably different from zero for $x/a \lesssim 1$, where the anharmonic term still is negligible, these are – to a good approximation – the solutions of the Schrödinger equation and we must expect that the (positive: $g\,x^4 > 0$) corrections to the eigenvalues of the oscillator, due to the anharmonic term, are small.   Therefore it looks reasonable to assume the first form of the Schrödinger equation (the one in the variable x/a) as the more convenient when $a \ll b$, i.e. when the system basically is a harmonic oscillator with a small anharmonic correction $(\propto x^4)$; on the contrary the second form is convenient when $b \ll a$. This idea will be resumed and developed in the framework of the perturbation theory of the energy levels (Chapter 12).

c) $H = \dfrac{p^2}{2m} + V(q) = \dfrac{\tilde{p}^2}{2ma^2} + V(a\tilde{q})$

and, since in the Schrödinger representation for the variables $\tilde{q}\,,\ \tilde{p}\quad \tilde{q} \to \xi$,
$\tilde{p} \to -\mathrm{i}\hbar\dfrac{\mathrm{d}}{\mathrm{d}\xi}$, putting $\langle \xi \mid E \rangle = f_\epsilon(\xi)$ and $U(\xi) \equiv V(a\xi)$, one has:

$$H\,|\,E\,\rangle = E\,|\,E\,\rangle \xrightarrow{\ \mathrm{SR}\ } -\dfrac{\hbar^2}{2ma^2}\dfrac{\mathrm{d}^2}{\mathrm{d}\xi^2}f_\epsilon(\xi) + U(\xi)\,f_\epsilon(\xi) = E\,f_\epsilon(\xi)\,.$$

6.3

a) The ground state is even, so $\psi_E(0) \neq 0$, $\psi_E'(0) = 0$. As the Schrödinger equation is homogeneous, the value of $\psi_E(0)$, provided it is nonvanishing (as in the case of the excited even states), is arbitrary: since it is convenient to solve the Schrödinger equation in its dimensionless form:

$$-\frac{1}{2}f_\epsilon''(\xi) + \left(\frac{1}{2}\xi^2 + \left(\frac{a}{b}\right)^6\xi^4\right)f_\epsilon(\xi) = \epsilon\,f_\epsilon(\xi)\,, \qquad \left(\frac{a}{b}\right)^6 = 0.1$$

usually one puts $f_\epsilon(0) = 1$ (note that, comparing with the previous problem, we have introduced the factor $\frac{1}{2}$ so that the energies can be measured in units of $\hbar\omega$). These initial conditions are appropriate when one searches for eigenfunctions with an even n. Regarding the solutions with odd n, in particular for the first excited state: $\psi(0) = 0$, $\psi'(0) \neq 0$ and usually one puts $f_\epsilon'(0) = 1$.

b) Let us proceed by steps. The 12 values of ϵ form three well distinguished groups $(\,0.557,\ 0.558,\ 0.559,\ 0.560)$; $(\,1.75,\ 1.76,\ 1.77,\ 1.78\,)$; $(\,3.0,\ 3.1,\ 3.2,\ 3.3\,)$. Since the potential is that of a harmonic oscillator with a small anharmonic positive correction, we expect that the three groups approximate respectively the lowest energy level (with an energy $\epsilon_0 \gtrsim 0.5$), the first and the second excited level (with energies $\epsilon_1 \gtrsim 1.5$ and $\epsilon_2 \gtrsim 2.5$); clearly the numbers in the second group $(\epsilon = 1.75 \div 1.78)$ correspond (not in the same order) to the graphs l, m, n, o, because the latter are the only ones where $f_\epsilon(0) = 0$. In order to decide which group of graphs corresponds to the first and the third group of values of ϵ, note that, due to:

$$f_\epsilon''(\xi)/f_\epsilon(\xi) = 2\big(u(\xi) - \epsilon\big) \tag{1}$$

and $f_\epsilon(0) = 1$, the greater curvature in the origin (where $u(\xi) = 0$) corresponds to the greater value of ϵ: evidently, in spite of the different scale on the ordinate axis (indeed, a fortiori) the curves a, b, c, d have in the origin a smaller curvature than the figures e, f, g, h; so the former correspond to the set of values $0.557 \div 0.560$, the latter to the set $3.0 \div 3.3$. Two more elements that confirm this conclusion are: i) the position of the inflexion points where $f_\epsilon(\xi) \neq 0$: (1) entails that in such points $\epsilon = u(\xi)$ and, for the given $u(\xi)$, the more such inflexion points shift to the right, the higher the value of ϵ (the curves a, b, c, d have an inflexion point for $\xi \simeq 1$ whereas for the other curves it occurs for higher values of ξ); ii) number and position of the zeroes in the two groups of curves: all the curves of the second group have at least one 'stable' zero – indeed, at (practically) the same value of $\xi \simeq 1$ regardless of the four values of ϵ – as it must be for a wavefunction describing the second excited state; on the contrary, those of the first group have either no zeroes (curve a) or a 'fluctuating' zero, i.e. for $\xi \gg 1$ and whose position is strongly dependent on the value of ϵ (an intuition of this is suggested by looking at the behaviour of the graph a around its zero).

In order to complete the association graphs–values of ϵ, let us take advantage of the fact that the closer is ϵ to an eigenvalue, the 'later' $f_\epsilon(\xi)$ begins to diverge. In addition, the zeroes with $\xi > 0$ 'migrate' towards the left as ϵ grows: this is a consequence of the oscillation theorem and of the fact that, for given initial conditions, the solutions $f_\epsilon(\xi)$ depend continuously on ϵ: the eigenfunction f_n (of the Hamiltonian) has n zeroes and vanishes at $\xi = \pm\infty$ whereas f_{n+2} has two extra zeroes: as ϵ grows from ϵ_n to ϵ_{n+2}, the zeroes at infinity of f_n move with continuity towards the positions of the two extra zeroes of f_{n+2}. From this it follows that if ϵ has a value slightly higher than a given eigenvalue, $f_\epsilon(\xi)$ diverges with a given sign while, if ϵ has a value slightly lower than the same eigenvalue, $f_\epsilon(\xi)$ diverges with opposite sign. Using the above elements and denoting by ϵ_a, ϵ_b, $\cdots$ the value of ϵ corresponding to the graphs a, b, $\cdots$, one has that $(\epsilon_b, \epsilon_c, \epsilon_d) < \epsilon_0$ (absence of zeroes), $\epsilon_a > \epsilon_0$ (one zero); moreover c diverges earlier than b that diverges earlier than d, so $\epsilon_c < \epsilon_b < \epsilon_d$ and

$$\epsilon_c = 0.557\,, \ \epsilon_b = 0.558\,, \ \epsilon_d = 0.559\,, \ \epsilon_a = 0.560\,; \qquad 0.559 < \epsilon_0 < 0.560\,.$$

Likewise, the number of zeroes indicates that $(\epsilon_f, \epsilon_g) < \epsilon_2 < (\epsilon_e, \epsilon_h)$ and since f diverges earlier than g and h earlier than e,

$$\epsilon_f = 3.0\,, \ \epsilon_g = 3.1\,, \ \epsilon_e = 3.2\,, \ \epsilon_h = 3.3\,; \qquad 3.1 < \epsilon_2 < 3.2\,.$$

For the odd states:

$$\epsilon_l = 1.75\,, \ \epsilon_o = 1.76\,, \ \epsilon_n = 1.77\,, \ \epsilon_m = 1.78\,; \qquad 1.76 < \epsilon_1 < 1.77\,.$$

The values, accurately determined by means of a program of numeric integration, are:

$$\epsilon_0 = 0.559162; \qquad \epsilon_1 = 1.76950; \qquad \epsilon_2 = 3.13839.$$

6.4

a) Let $|E_0^0\rangle$ be the state represented by the wavefunction $\psi_0(x)$. One has:

$$\langle E_0^0 \mid H \mid E_0^0 \rangle = \langle E_0^0 \mid H_0 \mid E_0^0 \rangle + \langle E_0^0 \mid V(x) - V_0(x) \mid E_0^0 \rangle$$

$$= E_0^0 + \int |\psi_0(x)|^2 \left(V(x) - V_0(x) \right) \, dx$$

and since $V(x) - V_0(x) \le 0$ it follows that $\quad \langle E_0^0 \mid H \mid E_0^0 \rangle \le E_0^0 < 0$.

b) If H had no bound states, one should have only a continuous spectrum of energies $E \ge 0$ and the mean value of H in any state would be positive, in contrast with what has been found above. So there exists at least one bound state. If $|E_0\rangle$ stands for the ground state, one has:

$$E_0 \equiv \langle E_0 \mid H \mid E_0 \rangle \le \langle A \mid H \mid A \rangle \; \forall \, | A \rangle \; \Rightarrow \; E_0 \le \langle E_0^0 \mid H \mid E_0^0 \rangle \le E_0^0.$$

6.5

a) As $V_{1,2}(x) \xrightarrow[|x| \to \infty]{} 0$, the eigenvalues of H_1 and H_2 corresponding to the (possible) bound states are negative and the continuous spectra are $E \ge 0$. If H_1 had no bound states, the mean value of H_1 in any state should be positive, but the assumption $V_2(x) \ge V_1(x)$ entails that:

$$0 > E_0^{(2)} \equiv \langle E_0^{(2)} \mid H_2 \mid E_0^{(2)} \rangle \ge \langle E_0^{(2)} \mid H_1 \mid E_0^{(2)} \rangle$$

therefore H_1 must have at least one bound state with $E_0^{(1)} \le E_0^{(2)}$.

b) As both H_1 and H_2 commute with the space inversion operator, their eigenstates alternatively are even and odd; therefore the second bound state of H_2 is odd and – as a consequence – orthogonal to the ground state of H_1. So in the subspace of the states that are orthogonal to the ground state of H_1 there exists a state in which the mean value of H_1 is negative: it follows that in such a subspace H_1 must have at least one negative eigenvalue.

c) Let $|s\rangle = \alpha \,| E_0^{(2)} \rangle + \beta \,| E_1^{(2)} \rangle \quad (|\alpha|^2 + |\beta|^2 = 1)$. One has:

$$\langle s \mid H_1 \mid s \rangle \le \langle s \mid H_2 \mid s \rangle = |\alpha|^2 E_0^{(2)} + |\beta|^2 E_1^{(2)} \le E_1^{(2)} < 0.$$

If H_1 did not have the second bound state, it should have positive mean value in all the states orthogonal to $| E_0^{(1)} \rangle$, but in $\mathcal{V}$ there exists (at least) one vector $|\bar{s}\rangle$ orthogonal to $| E_0^{(1)} \rangle$ (the vector in $\mathcal{V}$ orthogonal to the projection of $| E_0^{(1)} \rangle$ on $\mathcal{V}$), so H_1 must have at least one more bound state $| E_1^{(1)} \rangle$. In addition:

$$E_1^{(1)} \le \langle \bar{s} \mid H_1 \mid \bar{s} \rangle \le E_1^{(2)} \quad \Rightarrow \quad E_1^{(1)} \le E_1^{(2)}.$$

d) One may proceed by induction: in the manifold generated by the n eigen-vectors of H_2 (on which H_1 is negative definite) there exists at least one vector orthogonal to the manifold generated by the $n-1$ eigenvectors of H_1: so, as before, one concludes that H_1 must have at least n bound states and that $E_n^{(1)} \le E_n^{(2)}$.

6.6

a) The potential well of width $2a$ inscribed in the Gaussian potential $V(x) = -\lambda e^{-x^2/b^2}$ has the depth $V_0 = \lambda e^{-a^2/b^2}$. In order to obtain the best estimate, we must determine a in such a way that $V_0 a^2$ be a maximum: this happens for $a = b$: so $V_0 a^2 = \lambda b^2/e$. The number of bound states of the potential well with $\lambda = 3\hbar^2/mb^2$ is the minimum integer greater or equal to:

$$\sqrt{(8m/\pi^2\hbar^2) \times (3\hbar^2/mb^2)\, b^2/e} = \sqrt{24/(\pi^2 \times e)} = 0.95 \quad \Rightarrow \quad n_b \ge 1.$$

Likewise, if $\lambda = 4\hbar^2/mb^2$,

$$\sqrt{(8m/\pi^2\hbar^2) \times (4\hbar^2/mb^2)\, b^2/e} = \sqrt{32/(\pi^2 \times e)} = 1.09 \quad \Rightarrow \quad n_b \ge 2.$$

b) As in point a) above, one must determine the maximum of the function $a^2|V(a)| = \lambda a^2/(a^2 + b^2)$: this function attains its maximum – equal to λ – for $a = \infty$, so it must happen that:

$$\frac{8m\lambda}{\pi^2\hbar^2} \ge (N-1)^2 \quad \Rightarrow \quad \lambda \ge (N-1)^2\, \frac{\pi^2\hbar^2}{8m}\,.$$

c) The function $\lambda a^2/(a+b)$ grows indefinitely as a grows, so the number of bound states is infinite for any $\lambda > 0$ and for any b.

6.7

a) It is convenient to calculate the mean value of p^2 as the squared norm of the wavefunction $-i\hbar\psi'(x; a)$. One has:

$$\overline{H}(a) = \frac{\hbar^2}{2m}\, \frac{315 \cdot 16}{256\, a^9} \int_{-a}^{+a} x^2(x^2 - a^2)^2\, \mathrm{d}x + \frac{1}{2}m\omega^2\, \frac{315}{256\, a^9} \int_{-a}^{+a} x^2(x^2 - a^2)^4\, \mathrm{d}x$$

and, by using the integrals provided in the text, one obtains:

$$\overline{H}(a) = \frac{3\hbar^2}{2ma^2} + \frac{m\omega^2 a^2}{22}\,.$$

b) The minimum of $\overline{H}(a)$ occurs for $\bar{a}^2 = \sqrt{33}\,\hbar/m\omega$ and $\overline{H}(\bar{a})$ is greater than E_0 because E_0 is the minimum eigenvalue. One has $\overline{H}(\bar{a}) \simeq 0.52\,\hbar\omega$: this value represents the best approximation by excess to $E_0 = 0.5\,\hbar\omega$ that can be obtained by means of the trial functions $\psi(x; a)$.

c) One must choose trial functions that are orthogonal to the (unknown) wavefunction of the ground state (but not to that – unknown – of the first excited state): since the Hamiltonian commutes with the space inversion $q \to -q, \ p \to -p$, owing to the oscillation theorem the wavefunction of the first excited state is odd and has only one zero: for instance, but not necessarily, one may choose as trial functions the normalized wavefunctions:

$$\psi_1(x; a) = \begin{cases} \sqrt{\dfrac{3465}{256 \, a^{11}}} \, x \, (x^2 - a^2)^2 & |x| \le a \\ 0 & |x| \ge a . \end{cases} \qquad a > 0$$

With the above choice the minimum mean value of H is $\overline{H} = \sqrt{33/13} \, \hbar \omega \simeq 1.59 \, \hbar \omega$, that represents the best approximation by excess of $E_1 = 1.5 \, \hbar \omega$ that can be obtained by means of the trial functions $\psi_1(x; a)$.

d) Since $\psi(x; \bar{a})$ is even, the expansion of $\widetilde{\psi}(x)$ in terms of the eigenfunctions of the Hamiltonian of the oscillator starts from $n = 2$, so:

$$\overline{H} = 0.52 \, \hbar \omega \ge |\alpha|^2 E_0 + |\beta|^2 E_2 = \left(1 - |\beta|^2\right) E_0 + |\beta|^2 E_2$$

$$= \frac{1}{2} \hbar \omega + 2 \hbar \omega \, |\beta|^2 \Rightarrow \quad |\beta|^2 \le 0.01 \quad \Rightarrow \quad |\beta| \le 0.1 .$$

6.8

a) The asymptotic expression of $\psi_E(x)$ solves the Schrödinger equation in which terms infinitesimal with respect to $x^4 \psi_E(x)$ are neglected:

$$\psi_E''(x) - \frac{2mg}{\hbar^2} x^4 \psi_E(x) \approx 0 .$$

Putting $\psi_E(x) \approx e^{-\alpha |x|^n}$, one has:

$$\left(\alpha^2 n^2 x^{2(n-1)} + O(x^{n-2}) - \frac{2mg}{\hbar^2} x^4\right) e^{-\alpha |x|^n} \approx 0 \ \Rightarrow \ n = 3 , \ \alpha = \sqrt{\frac{2mg}{9\hbar^2}}$$

(the solution with $\alpha = -\sqrt{2mg/9\hbar^2}$ diverges at infinity).

b) If $q \to \lambda q, \ p \to \lambda^{-1} p$, then $H \to p^2/2m\lambda^2 + g \lambda^4 q^4$; therefore, as the transformation is unitary, $E(m, g) = E(m\lambda^2, g \lambda^4) \Rightarrow E = E(g/m^2)$.

c) $\overline{H} = \dfrac{3}{2} \dfrac{\hbar^2}{ma^2} + \dfrac{3}{143} g \, a^4$

attains its minimum for $\bar{a}^6 = (143/2)\hbar^2/2mg$ and:

$$E_0 < \overline{H}(\bar{a}) = \frac{9}{2} \left(\frac{\hbar^4 g}{286 \, m^2}\right)^{1/3} \simeq 0.68 \left(\frac{\hbar^4 g}{m^2}\right)^{1/3}$$

(an accurate numerical calculation gives $E_0 = 0.67 \left(\hbar^4 g/m^2\right)^{1/3}$, so the variational calculation with the proposed trial functions gives the result with an error smaller than 1.5%).

6.9

a) False: every wavefunction $\psi(x)$ may be expressed as a superposition of eigenstates of the Hamiltonian, but the convergence of the superposition in the L^2 norm generally is not pointwise.

b) $\psi_0(x)$ is $\sin kx$ for $x < a$ and must connect with a decreasing exponential for $x > a$, in such a way as to keep the first derivative continuous: so one must have $ka > \pi/2$. Since $k = \sqrt{2m(V_0 - |E_0|)/\hbar^2}$, one has:

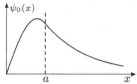

$$\frac{2mV_0a^2}{\hbar^2} \geq \frac{2m(V_0 - |E_0|)\,a^2}{\hbar^2} > \frac{\pi^2}{4} \quad \Rightarrow \quad V_0 > \frac{\pi^2\hbar^2}{8ma^2}\,.$$

c) Putting $\psi_{E_0}(x) = \sin kx$ for $0 \leq x \leq a$ and $\psi_{E_0}(x) = A\,e^{-\kappa(x-a)}$ for $x \geq a$, where $\kappa = \left(2m|E_0|/\hbar^2\right)^{1/2}$, the conditions of continuity of $\psi_{E_0}(x)$ and of its derivative (or, equivalently, of its logarithmic derivative) at $x = a$ give $\tan ka = -ka/\kappa a$; putting $\xi = ka$, $\eta = \kappa a$, the two equations:

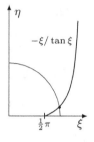

$$\eta = -\xi/\tan\xi\,, \quad \xi^2 + \eta^2 = \frac{2mV_0a^2}{\hbar^2}\,; \quad \xi \geq 0\,, \quad \eta \geq 0$$

follow. If $V_0 > \pi^2\hbar^2/(8ma^2)$, they always admit at least one solution with $\xi \equiv ka > \pi/2$.

d) The continuous spectrum is $E > 0$ and:

$$\psi_E(x) = \begin{cases} \sin k'x & 0 \leq x \leq a \\ \sin k'a\,\cos k(x-a) + \dfrac{k'}{k}\,\cos k'a\,\sin k(x-a) & x \geq a \end{cases}$$

where $k' = \sqrt{2m(V_0 + E)/\hbar^2}$, $k = \sqrt{2mE/\hbar^2}$.

e) The energy levels of the particle subject to the potential $V_1(x)$ ('half well') are all and only the energy levels corresponding to the odd states $\big(\psi_E(x) = -\psi_E(-x)\big)$ for the particle subject to the potential $V_2(x)$; therefore, if n is even, the number of energy levels is $n/2$, otherwise it is $(n-1)/2$.

6.10

a) As $V_1 \to +\infty$, we may assume $E < V_1$. If, for $x \leq 0$, we put $\psi_E(x) = e^{k_1 x}$, since $\psi_E(0) = 1$, instead of demonstrating that $\psi_E(x) \to 0$ for $x \leq 0$, we would demonstrate that $\psi_E(x) \to \infty$ for $x > 0$: this is the same thing, but it is unaesthetic. So we put:

$$\psi_E(x) = \begin{cases} A\,e^{k_1 x} & x \leq 0\,, \\ \sin(kx + \varphi) & 0 \leq x \leq a\,, \end{cases} \qquad \begin{aligned} k_1 &= \sqrt{2m(V_1 - E)/\hbar^2} \\ k &= \sqrt{2m(V_0 + E)/\hbar^2}\,. \end{aligned}$$

The conditions of continuity for $\psi_E(x)$ and for its derivative at $x = 0$ give:

$$A = \sin\varphi, \qquad \tan\varphi = \frac{k}{k_1} = \sqrt{\frac{V_0 + E}{V_1 - E}}$$

and, for $V_1 \to +\infty$, $\varphi \to 0$ and in conclusion $A \to 0$.

b) True: any superposition of functions that are identically vanishing for $x < 0$ converges to a function identically vanishing for $x < 0$ (but not necessarily so for $x = 0$: see Problem 6.9).

c) For the bound states one has (see Problem 6.9) $ka + \varphi > \pi/2$. If $\overline{V}_1$ is the maximum value of V_1 for which there occur bound states, when V_1 tends to $\overline{V}_1$ from below, the energy $E < 0$ of the bound state tends to 0 and the eigenstate becomes improper. For $E = 0$ the general solution of the Schrödinger equation in the region $x \geq a$ is $\alpha x + \beta$ and, since the wavefunction $\psi_{E=0}(x)$ must remain finite, it is constant and, as a consequence, $\psi'_{E=0}(a) = 0$, whence $ka + \varphi = \pi/2$. So, for $V_1 = \overline{V}_1$, one has $\sqrt{2mV_0a^2/\hbar^2} + \overline{\varphi} = \pi/2$, where $\tan\overline{\varphi} = \sqrt{V_0/\overline{V}_1}$, and in conclusion $\overline{V}_1 = V_0 \tan^2\sqrt{2mV_0a^2/\hbar^2}$.

6.11

a) The only parameter, that appears in the equations for the dimensionless variables ξ and η, is $2mV_0a^2/\hbar^2$: as a consequence, also the solutions of these equations may depend only on this parameter. Therefore:

$$E_n = -\frac{\hbar^2}{2m}\kappa_n^2 = -\frac{\hbar^2}{2m}\frac{\eta_n^2}{a^2} = -\frac{\hbar^2}{2ma^2}F_n(2mV_0a^2/\hbar^2).$$

The variational argument reported in Problem 6.5 shows that, as V_0 and/or a increase, E_n decreases (its absolute value increases): indeed, for example, if $-V_2 < -V_1$, the well $V_1(x)$ is contained in the well $V_2(x)$ and $E_n^{(2)} < E_n^{(1)}$, whence $F_n'(x) > 0$.

b) Both for $V_0 \to 0$ and $a \to 0$, $V_0a^2 \to 0$, therefore $\xi \to 0 \Rightarrow \tan\xi \approx \xi \Rightarrow \xi^2 \approx \eta$. As a consequence:

$$\eta^2 + \eta - \frac{2mV_0a^2}{\hbar^2} = 0 \Rightarrow \eta = \frac{1}{2}\left(-1 + \sqrt{1 + 8mV_0a^2/\hbar^2}\right) \approx \frac{2mV_0a^2}{\hbar^2}$$

$$\Rightarrow \quad E_0 \approx -\frac{2m}{\hbar^2}V_0^2a^2$$

(then $F_0(x) \approx x^2$ for $x \to 0$).

c) For the ground state one always has $\xi < \pi/2$. Both for $V_0 \to \infty$ and $a \to \infty$, $V_0a^2 \to \infty$, whence:

$$\xi \to \frac{\pi}{2} \quad \Rightarrow \quad E_0 \approx -V_0 + \frac{\hbar^2\pi^2}{8ma^2}$$

(so $F_0(x) \approx x - \pi^2/4$ for $x \to \infty$). In the case $a \to \infty$ one has $E_0 \to -V_0$, as it must be, but for $a = \infty$ the system has no longer bound states (free particle).

d) Putting $V_0 a = \lambda$, $V_0 a^2 \to 0$, then from b) one has $E_0 \to -2m\lambda^2/\hbar^2$.

e) The behaviour of E_0 as a function of V_0 is quadratic for $V_0 \approx 0$ and linear for $V_0 \approx \infty$: in the graph plotted in the figure the unit of energy, on both axes, is $\hbar^2/2ma^2$ and the graph itself has been obtained as the curve given in parametric form by the equations:

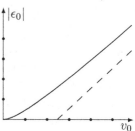

$$\begin{cases} v_0 = \xi^2 + (\xi \tan \xi)^2 & v_0 = V_0/(\hbar^2/2ma^2) \\ |\epsilon_0| = (\xi \tan \xi)^2 & \epsilon_0 = E_0/(\hbar^2/2ma^2) . \end{cases}$$

The dashed line is the asymptote $|E_0| = V_0 - (\hbar^2/2ma^2) \times \pi^2/4$ and the graph illustrates that it is approached rather slowly: the distance between the two curves decreases proportionally to $1/\sqrt{V_0}$.

f) In the case of the electron the unit of energy is $\hbar^2/2m_e\, a^2 = 3.8\,\text{eV}$ and, since $V_0 = 1\,\text{eV} < 3.8\,\text{eV}$,

$$E_0 \simeq -\frac{2m_e}{\hbar^2} V_0^2 a^2 = -\frac{1}{3.8} = -0.26\,\text{eV} \quad [-0.20\,\text{eV}]$$

(in parentheses we have reported the exact result up to 2 significant figures). In the case of the proton $\hbar^2/2m_p\, a^2 = 0.2\,\text{MeV} \ll 1\,\text{MeV}$, therefore:

$$E_0 \simeq -V_0 + \frac{\pi^2\hbar^2}{8m_p\, a^2} = -0.5\,\text{MeV} \quad [-0.77\,\text{MeV}] .$$

6.12

a) The eigenfunctions ψ_E of the Hamiltonian, relative to $0 < E < V_0$, have the form:

$$\psi_E(x) = \begin{cases} e^{ikx} + A\,e^{-ikx} & x \le 0 \\ B\,e^{-\kappa x} & x \ge 0 \end{cases} ; \quad k = \frac{1}{\hbar}\sqrt{2mE}, \quad \kappa = \frac{1}{\hbar}\sqrt{2m(V_0 - E)} .$$

The continuity conditions for the function and its derivative at the point $x = 0$ are:

$$\begin{cases} 1 + A = B \\ 1 - A = i\dfrac{\kappa}{k} B \end{cases} \Rightarrow A = \frac{1 - i\kappa/k}{1 + i\kappa/k}, \quad B = \frac{2}{1 + i\kappa/k} .$$

b) The eigenfunctions ψ_E of the Hamiltonian, relative to $E > V_0$, for particles that arrive from the region $x < 0$, have the form:

$$\psi_E(x) = \begin{cases} e^{ikx} + A\,e^{-ikx} & x \le 0 \\ B\,e^{ik'x} & x \ge 0 \end{cases} ; \quad k = \frac{1}{\hbar}\sqrt{2mE}, \quad k' = \frac{1}{\hbar}\sqrt{2m(E - V_0)} .$$

The continuity conditions for the function and its derivative at the point $x = 0$ are:

$$\begin{cases} 1 + A = B \\ 1 - A = \dfrac{k'}{k} B \end{cases} \Rightarrow \quad A = \dfrac{1 - k'/k}{1 + k'/k}, \quad B = \dfrac{2}{1 + k'/k}$$

and the reflection and transmission coefficients respectively are:

$$R = |A|^2 = \left(\dfrac{1 - k'/k}{1 + k'/k}\right)^2, \quad T = \dfrac{k'}{k}|B|^2 = \dfrac{4\,k'/k}{(1 + k'/k)^2} \ .$$

The conservation of the probability flux:

$$\Im m\left(\psi_E^*(x_1)\,\psi_E'(x_1)\right) = \Im m\left(\psi_E^*(x_2)\,\psi_E'(x_2)\right)$$

with $x_1 < 0$ and $x_2 > 0$ implies $\quad k\,|A|^2 + k'\,|B|^2 = k$, that is verified: $T + R = 1$. Note that, if $E \gg V_0$, the reflected wave is virtually absent: $T \simeq 1$, $R \simeq 0$.

c) Since the Schrödinger equation is real, if $\psi_E(x)$ is the solution determined above (particles coming from the region $x < 0$), also $\psi_E^*(x)$ is a solution. It follows that the solution $\widetilde{\psi}_E(x)$ one is after (particles coming from the region $x > 0$) is a linear combination of $\psi_E(x)$ and $\psi_E^*(x)$: $\widetilde{\psi}_E(x) = \alpha\,\psi_E(x) + \beta\,\psi_E^*(x)$. Having in mind that, for $E > V_0$, the coefficients A and B are real, one has:

$$\widetilde{\psi}_E(x) = \begin{cases} \widetilde{B}\,e^{-i k x} \\ e^{-i k' x} + \widetilde{A}\,e^{i k' x} \end{cases} = \begin{cases} (\alpha + \beta A)e^{i k x} + (\alpha A + \beta)e^{-i k x} & (x \le 0) \\ \alpha B\,e^{i k' x} + \beta B\,e^{-i k' x} & (x \ge 0) \end{cases}$$

$$\Rightarrow \quad \alpha + \beta A = 0, \quad \beta B = 1 \quad \Rightarrow \quad \alpha = -\dfrac{A}{B}, \quad \beta = \dfrac{1}{B} \quad \Rightarrow$$

$$\begin{cases} \widetilde{A} = \alpha B = -A \\ \widetilde{B} = \alpha A + \beta = \dfrac{1 - A^2}{B} = \dfrac{k'}{k} B \end{cases}$$

where the relationship $k\,|A|^2 + k'\,|B|^2 = k$ has been used. The conservation of the probability flux requires $k'\,|\widetilde{A}|^2 + k\,|\widetilde{B}|^2 = k'$, that is verified. Therefore:

$$R \equiv \left|\widetilde{A}\right|^2 = |A|^2, \quad T \equiv \dfrac{k}{k'}|\widetilde{B}|^2 = \dfrac{k'}{k}|B|^2 \ .$$

In conclusion: the reflection and transmission coefficients do not depend on the direction the particle comes from.

6.13

a) Let us first consider the case $0 < E < V_0$. If the reflected wave is absent, one must have:

$$\psi_E(x) = \begin{cases} e^{i k x} & x \le 0 \\ B\,e^{\kappa x} + B'\,e^{-\kappa x} & 0 \le x \le a \\ C\,e^{i k (x-a)} & x \ge a \end{cases}$$

$$k = (2mE/\hbar^2)^{1/2}, \quad \kappa = \left(2m(V_0 - E)/\hbar^2\right)^{1/2}.$$

The continuity conditions at $x = 0$ and $x = a$ are:

$$\begin{cases} 1 = B + B' \\ i\,(k/\kappa) = B - B' \end{cases} \qquad \begin{cases} C = B\,e^{\kappa a} + B'\,e^{-\kappa a} \\ C = -i\,(\kappa/k)\left(B\,e^{\kappa a} - B'\,e^{-\kappa a}\right) \end{cases} \Rightarrow$$

$$\begin{cases} C = (B + B')\cosh \kappa a + (B - B')\sinh \kappa a = \cosh \kappa a + i\,(k/\kappa)\sinh \kappa a \\ C = -i\,(\kappa/k)\left[(B - B')\cosh \kappa a + (B + B')\sinh \kappa a\right] \\ \quad = \cosh \kappa a - i\,(\kappa/k)\sinh \kappa a \end{cases}$$

since $\sinh \kappa a \neq 0$, for $E < V_0$ total transmission cannot occur. The same result obtains also for $E = V_0$: notice that in this case $\psi(x) = B + B'x/a$ for $0 \le x \le a$.

Let us now consider the case $E > V_0$. Also for $0 \le x \le a$ the wavefunction is a linear combination of exponentials of imaginary argument $e^{\pm i\,k'\,x}$. The continuity conditions now are $\left(k' = [2m(E - V_0)/\hbar^2]^{1/2}\right)$:

$$\begin{cases} 1 = B + B' \\ (k/k') = B - B' \end{cases} \qquad \begin{cases} C = B\,e^{i k'a} + B'\,e^{-i k'a} \\ C = (k'/k)\left(B\,e^{i k'a} - B'\,e^{-i k'a}\right) \end{cases}$$

$$\Rightarrow \quad \begin{cases} C = \cos k'a + i\,(k/k')\sin k'a \\ C = \cos k'a + i\,(k'/k)\sin k'a \end{cases} \quad \Rightarrow \quad \sin k'a = 0 \Rightarrow k'a = n\,\pi \quad \Rightarrow$$

$$E = V_0 + \frac{n^2 h^2}{8ma^2}, \quad n \ge 1.$$

In any event, much as in Problem 6.12, if $E \gg V_0$, $C \simeq e^{i k a}$ and the reflected wave is practically absent.

b) Total reflection is never possible: $\psi_E(x)$ would be vanishing for $x > a$ and, as a consequence, on the whole real line.

6.14

a) If $\psi_E(x)$ solves the Schrödinger equation, so does $\psi_E^*(x)$; then $\forall\, \alpha,\ \beta$ one must have:

$$\begin{pmatrix} \delta^* \\ \gamma^* \end{pmatrix} = \begin{pmatrix} A & B \\ C & D \end{pmatrix}\begin{pmatrix} \beta^* \\ \alpha^* \end{pmatrix} \quad \Rightarrow \quad \begin{pmatrix} \delta \\ \gamma \end{pmatrix} = \begin{pmatrix} A^* & B^* \\ C^* & D^* \end{pmatrix}\begin{pmatrix} \beta \\ \alpha \end{pmatrix} \Rightarrow$$

$$\begin{pmatrix} 0 & 1 \\ 1 & 0 \end{pmatrix}\begin{pmatrix} \gamma \\ \delta \end{pmatrix} = \begin{pmatrix} A^* & B^* \\ C^* & D^* \end{pmatrix}\begin{pmatrix} 0 & 1 \\ 1 & 0 \end{pmatrix}\begin{pmatrix} \alpha \\ \beta \end{pmatrix} \quad \Rightarrow$$

$$\begin{pmatrix} 0 & 1 \\ 1 & 0 \end{pmatrix}\begin{pmatrix} A & B \\ C & D \end{pmatrix} = \begin{pmatrix} A^* & B^* \\ C^* & D^* \end{pmatrix}\begin{pmatrix} 0 & 1 \\ 1 & 0 \end{pmatrix} \quad \Rightarrow \quad M = \begin{pmatrix} A & B \\ B^* & A^* \end{pmatrix}.$$

b) The continuity equation demands that $|\alpha|^2 - |\beta|^2 = |\gamma|^2 - |\delta|^2$. Putting:

$$u \equiv \begin{pmatrix} \alpha \\ \beta \end{pmatrix}; \quad v \equiv \begin{pmatrix} \gamma \\ \delta \end{pmatrix} = M\,u; \quad \Sigma \equiv \begin{pmatrix} 1 & 0 \\ 0 & -1 \end{pmatrix}$$

one has:

$$v^\dagger \Sigma v = u^\dagger \Sigma u \quad \Rightarrow \quad M^\dagger \Sigma M = \Sigma$$

and, by using the expression for M obtained above, one obtains $\det M = |A|^2 - |B|^2 = 1$.

c) For particles incoming from the region $x < x_1$ one must have $\delta = 0$. Putting $\alpha = 1$:

$$\begin{pmatrix} \gamma \\ 0 \end{pmatrix} = \begin{pmatrix} A & B \\ B^* & A^* \end{pmatrix} \begin{pmatrix} 1 \\ \beta \end{pmatrix} \Rightarrow \begin{cases} A + B\beta = \gamma \\ B^* + A^*\beta = 0 \end{cases} \Rightarrow \beta = -\frac{B^*}{A^*}, \quad \gamma = \frac{1}{A^*}$$

then:

$$T \equiv |\gamma|^2 = \frac{1}{|A|^2}, \quad R \equiv |\beta|^2 = \frac{|B|^2}{|A|^2}.$$

For particles incoming from the region $x > x_2$ one must have $\alpha = 0$. Putting $\delta = 1$:

$$\begin{pmatrix} A & B \\ B^* & A^* \end{pmatrix} \begin{pmatrix} 0 \\ \tilde{\beta} \end{pmatrix} = \begin{pmatrix} \tilde{\gamma} \\ 1 \end{pmatrix} \Rightarrow \tilde{\beta} = \frac{1}{A^*}, \quad \tilde{\gamma} = \frac{B}{A^*}$$

$$T \equiv |\tilde{\beta}|^2 = \frac{1}{|A|^2}, \quad R \equiv |\tilde{\gamma}|^2 = \frac{|B|^2}{|A|^2}.$$

As already remarked in Problem 6.12, the reflection and transmission coefficients do not depend on the direction the incoming particles come from.

d) If, with the translated potential $V(x-a)$ (transfer matrix M_{tr}):

$$\alpha e^{ikx} + \beta e^{-ikx} \to \gamma e^{ikx} + \delta e^{-ikx},$$

by the change of variable $x = y + a$ one has that, with the potential $V(y)$ (then transfer matrix M):

$$\alpha e^{ika} e^{iky} + \beta e^{-ika} e^{-iky} \to \gamma e^{ika} e^{iky} + \delta e^{-ika} e^{-iky} \quad \Rightarrow$$

$$\begin{pmatrix} \gamma e^{ika} \\ \delta e^{-ika} \end{pmatrix} = M \begin{pmatrix} \alpha e^{ika} \\ \beta e^{-ika} \end{pmatrix} \quad \Rightarrow$$

$$\begin{pmatrix} e^{ika} & 0 \\ 0 & e^{-ika} \end{pmatrix} \begin{pmatrix} \gamma \\ \delta \end{pmatrix} = M \begin{pmatrix} e^{ika} & 0 \\ 0 & e^{-ika} \end{pmatrix} \begin{pmatrix} \alpha \\ \beta \end{pmatrix} \quad \Rightarrow$$

$$M_{tr} = \begin{pmatrix} e^{-ika} & 0 \\ 0 & e^{ika} \end{pmatrix} M \begin{pmatrix} e^{ika} & 0 \\ 0 & e^{-ika} \end{pmatrix} = \begin{pmatrix} A & B e^{-2ika} \\ B^* e^{2ika} & A^* \end{pmatrix}.$$

If, with the potential $V(-x)$ (transfer matrix M_{rf}):

$$\alpha e^{ikx} + \beta e^{-ikx} \to \gamma e^{ikx} + \delta e^{-ikx}$$

by the change of variable $x = -y$ one has that, with the potential $V(y)$ (then transfer matrix M):

$$\delta e^{iky} + \gamma e^{-iky} \to \beta e^{iky} + \alpha e^{-iky} \quad \Rightarrow \quad \begin{pmatrix} \beta \\ \alpha \end{pmatrix} = M \begin{pmatrix} \delta \\ \gamma \end{pmatrix} \quad \Rightarrow$$

$$\begin{pmatrix} \delta \\ \gamma \end{pmatrix} = M^{-1} \begin{pmatrix} \beta \\ \alpha \end{pmatrix} \quad \Rightarrow \quad \begin{pmatrix} 0 & 1 \\ 1 & 0 \end{pmatrix} \begin{pmatrix} \gamma \\ \delta \end{pmatrix} = M^{-1} \begin{pmatrix} 0 & 1 \\ 1 & 0 \end{pmatrix} \begin{pmatrix} \alpha \\ \beta \end{pmatrix} \quad \Rightarrow$$

$$M_{rf} = \begin{pmatrix} 0 & 1 \\ 1 & 0 \end{pmatrix} M^{-1} \begin{pmatrix} 0 & 1 \\ 1 & 0 \end{pmatrix} = \begin{pmatrix} A & -B^* \\ -B & A^* \end{pmatrix}$$

and, as a consequence, $V(x) = V(-x)$ entails $B = -B^*$.

6.15

a) As in point c) of the preceding problem:

$$M \begin{pmatrix} 1 \\ \beta \end{pmatrix} \equiv \begin{pmatrix} A & B \\ B^* & A^* \end{pmatrix} \begin{pmatrix} 1 \\ \beta \end{pmatrix} = \begin{pmatrix} \gamma \\ 0 \end{pmatrix} \quad \Rightarrow \quad \beta = -\frac{B^*}{A^*}, \quad \gamma = \frac{1}{A^*}$$

$$\Rightarrow \quad S^\dagger \begin{pmatrix} 1 \\ 0 \end{pmatrix} = \frac{1}{A^*} \begin{pmatrix} 1 \\ -B^* \end{pmatrix};$$

$$M \begin{pmatrix} 0 \\ \tilde\beta \end{pmatrix} = \begin{pmatrix} \tilde\gamma \\ 1 \end{pmatrix} \quad \Rightarrow \quad \tilde\beta = \frac{1}{A^*}, \quad \tilde\gamma = \frac{B}{A^*} \quad \Rightarrow \quad S^\dagger \begin{pmatrix} 0 \\ 1 \end{pmatrix} = \frac{1}{A^*} \begin{pmatrix} B \\ 1 \end{pmatrix}$$

$$\Rightarrow \quad S^\dagger = \begin{pmatrix} \gamma & \tilde\gamma \\ \beta & \tilde\beta \end{pmatrix} = \frac{1}{A^*} \begin{pmatrix} 1 & B \\ -B^* & 1 \end{pmatrix}; \quad S = \frac{1}{A} \begin{pmatrix} 1 & -B \\ B^* & 1 \end{pmatrix}.$$

b) $$S S^\dagger = \frac{1}{|A|^2} \begin{pmatrix} 1 & -B \\ B^* & 1 \end{pmatrix} \begin{pmatrix} 1 & B \\ -B^* & 1 \end{pmatrix} = \frac{1}{|A|^2} \begin{pmatrix} 1+|B|^2 & 0 \\ 0 & 1+|B|^2 \end{pmatrix}$$

$$= \begin{pmatrix} 1 & 0 \\ 0 & 1 \end{pmatrix}$$

($|A|^2 - |B|^2 = 1$ has been used, see Problem 6.14).

c) $T = |S_{11}|^2 = |S_{22}|^2;$ $R = |S_{12}|^2 = |S_{21}|^2;$ $T + R = 1$ follows from the unitarity of the S-matrix.

6.16

a) The wavefunction is normalizable, so the eigenstate belongs to an eigenvalue of the discrete spectrum; since (in the one-dimensional case) the discrete eigenvalues of the Hamiltonian are nondegenerate and the Schrödinger operator is real, the corresponding eigenfunction must be real up to a complex factor of modulus 1 independent of x. In this case, contrary to what has been seen in Problem 5.21, the state is uniquely determined:

$$\psi(x) = \frac{N}{x^2 + a^2}.$$

As $\psi(x)$ has no zeroes, it corresponds to the ground state $|E_0\rangle$.

b) $\langle E_0 | p | E_0 \rangle = 0$ because $\psi(x)$ is real (see Problem 5.16), or else because $\psi(x)$ is even (see Problem 6.1); $\langle E_0 | q | E_0 \rangle = 0$ because $\psi(x)$ is even.

c) No: the minimum uncertainty states have $e^{ikx} e^{-\alpha x^2}$, with α and k real numbers (see Problem 5.15), as wavefunctions.

d) From the Schrödinger equation one obtains:

$$V(x) = E_0 + \frac{\hbar^2}{2m} \frac{\psi''(x)}{\psi(x)} = E_0 + \frac{\hbar^2}{m} \frac{3x^2 - a^2}{(x^2 + a^2)^2} = V(-x).$$

Note that the potential tends to E_0 for $x \to \pm\infty$. (One may take advantage of the fact that the potential – and the energy levels – are defined up to an additive constant to put $E_0 = 0$).

e) What has been found above shows that the energy E_0 of the ground state is adherent to the continuum: as a consequence, there is no room left for further bound states.

6.17

a) As $\tanh x \to \pm 1$ for $x \to \pm\infty$, one has:

$$\psi_k(x) \approx \frac{e^{ikx}}{\sqrt{2\pi}}, \quad x \to -\infty; \qquad \psi_k(x) \approx \frac{ika-1}{ika+1}\frac{e^{ikx}}{\sqrt{2\pi}}, \quad x \to +\infty.$$

The first equation implies that $R = 0$ (absence of reflection for any k); from the second one has:

$$T(k) = \left|\frac{ika-1}{ika+1}\right|^2 = 1.$$

b) The explicit calculation gives:

$$\psi_k''(x) = -k^2\,\psi_k(x) - \frac{2}{a^2\cosh^2(x/a)}\,\psi_k(x)$$

$$\Rightarrow \quad U(x)\,\psi_k(x) = \psi_k''(x) + \frac{2mE}{\hbar^2}\,\psi_k(x)$$

$$= \left(\frac{2mE}{\hbar^2} - k^2 - \frac{2}{a^2\cosh^2(x/a)}\right)\psi_k(x)$$

in which the coefficient of $\psi_k(x)$ is independent of both E and k only if $E - \hbar^2 k^2/2m$ is a constant; putting such a constant equal to zero, one has:

$$V(x) = -\frac{\hbar^2}{ma^2\cosh^2(x/a)}.$$

It is always possible to inscribe a rectangular well $V_0(x)$ in the potential in such a way that $V(x) \le V_0(x)$ for any x. As the potential well has at least one bound state, one is entitled to state (see Problem 6.4) that $V(x)$ possesses at least one bound state as well.

c) One has $\psi_{-k}(x) \propto \psi_k(-x)$ and, as the potential is even, also $\psi_k(-x)$ is an eigenfunction of H belonging to the same eigenvalue $E = \hbar^2 k^2/2m$ the eigenfunction $\psi_k(x)$ belongs to.

d) Let us calculate in detail the integral (put $p = ka$, $\xi = x/a$, $\eta = y/a$):

$$\int_{-\infty}^{+\infty}\psi_k(x)\,\psi_k^*(y)\,\mathrm{d}k = \int_{-\infty}^{+\infty}\frac{e^{ik(x-y)}}{2\pi}\left[\frac{(ka)^2 + \tanh(x/a)\tanh(y/a)}{(ka)^2+1}\right.$$

$$\left. +\,ika\,\frac{\big(\tanh(x/a)-\tanh(y/a)\big)}{(ka)^2+1}\right]\mathrm{d}k$$

$$= \delta(x-y) + \frac{1}{2\pi a}\int_{-\infty}^{+\infty}e^{ip(\xi-\eta)}\frac{(-1+\tanh\xi\,(\tanh\eta+ip)-ip\tanh\eta)}{p^2+1}\,\mathrm{d}p.$$

We now use the following relation (that can be derived by use of the residue theorem):

$$\int_{-\infty}^{+\infty} e^{i\alpha p}\, \frac{F(p)}{p^2+1}\, dp = \frac{2\pi}{2} \left[\theta(\alpha)\, e^{-\alpha}\, F(i) + \theta(-\alpha)\, e^{+\alpha}\, F(-i) \right]$$

then:

$$\frac{1}{2\pi a} \int_{-\infty}^{+\infty} \cdots = \frac{1}{2a} \Big[\theta(\xi-\eta) e^{-(\xi-\eta)} \left(-1 + \tanh\xi \tanh\eta - \tanh\xi + \tanh\eta \right)$$

$$+ \theta(\eta-\xi) e^{+(\xi-\eta)} \left(-1 + \tanh\xi \tanh\eta + \tanh\xi - \tanh\eta \right) \Big]$$

$$= \frac{1}{2a \cosh\xi \, \cosh\eta} \times \Big[\theta(\xi-\eta) e^{-(\xi-\eta)} \left(-\cosh\xi \cosh\eta + \sinh\xi \sinh\eta \right.$$

$$-\sinh\xi \cosh\eta + \cosh\xi \sinh\eta \right) + \theta(\eta-\xi) e^{+(\xi-\eta)} \left(-\cosh\xi \cosh\eta \right.$$

$$+ \sinh\xi \sinh\eta + \sinh\xi \cosh\eta - \cosh\xi \sinh\eta \right) \Big]$$

$$= -\frac{1}{2a \cosh\xi \, \cosh\eta} \left(\theta(\xi-\eta) + \theta(\eta-\xi) \right) = -\frac{1}{2a \cosh\xi \, \cosh\eta}$$

so that, in conclusion:

$$\int_{-\infty}^{+\infty} \langle x \mid k \rangle \langle k \mid y \rangle \, dk \equiv \int_{-\infty}^{+\infty} \psi_k(x)\, \psi_k^*(y)\, dk$$

$$= \delta(x-y) - \frac{1}{2a \cosh(x/a)\, \cosh(y/a)} \; .$$

The presence of the last term is a consequence of the fact that the set $\{\psi_k(x), k \in \mathbb{R}\}$ is not complete, namely of the existence of at least one bound state. Since the projector $\mathcal{E}(x,y) \equiv 1/\big(2a\cosh(x/a)\,\cosh(y/a)\big)$ is in the factorized form $\chi(x)\,\chi^*(y)$ (see Problem 5.18), there is only one bound state represented by the wavefunction:

$$\psi_B(x) = \frac{1}{\sqrt{2a}\cosh(x/a)} \; .$$

6.18

a) The even bound states of the rectangular potential well centered in the origin are represented by the wavefunctions:

$$\phi_E(x) = \begin{cases} \cos(ka/2)\, e^{\kappa(x+a/2)} & x \le -a/2 \\ \cos kx & |x| \le a/2 \\ \cos(ka/2)\, e^{-\kappa(x-a/2)} & x \ge +a/2 \end{cases}$$

(in which, thanks to the homogeneity of the Schrödinger equation, one has chosen to normalize $\phi_E(x)$ by the condition $\phi_E(0) = 1$ where – owing to the continuity of ϕ_E' at $x = a/2$ – the parameters:

$$\kappa = \sqrt{2m(-E)/\hbar^2}\,, \qquad\qquad k = \sqrt{2m(V_0+E)/\hbar^2}$$

fulfill the condition $\kappa = k \tan(ka/2)$.

Since $k^2 \propto (V_0 - |E|)$ and $|E| < V_0$, k^2a remains finite in the limit we are interested in, and $ka \to 0$, so $\tan(ka/2) \approx ka/2$ and $\kappa \approx k^2a/2$ remains finite. Since $\kappa \propto \sqrt{|E|} \Rightarrow Ea \to 0$ and

$$\kappa \approx \frac{k^2a}{2} \to \kappa_0 \equiv \frac{m\lambda}{\hbar^2} \quad \Rightarrow \quad E \to E_0 = -\frac{m\lambda^2}{2\hbar^2}.$$

If λ is replaced by 2λ, the above result reproduces what has been found in Problem 6.11d, where indeed the width of the well is $2a$.

One has, therefore, one even bound state represented by the wavefunction:

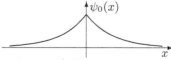

$$\psi_0(x) = e^{-\kappa_0|x|}$$

that is continuous at $x = 0$, but has a discontinuous derivative:

$$\psi_0'(0^+) - \psi_0'(0^-) = -2\kappa_0 = -\frac{2m\lambda}{\hbar^2}\psi_0(0).$$

There are no odd bound states: indeed, for a rectangular potential well of width a there are odd states only if $2mV_0a^2/\hbar^2 > \pi^2$, but $V_0a^2 \to 0$ (on the other hand, if $\psi(x)$ is odd and continuous, $\delta(x)\psi(x) = 0$).

b) $\lambda = \sqrt{2\hbar^2|E_0|/m} = 6.26 \times 10^{-20}$ erg cm $= 3.9$ eV Å .

$\kappa_0 = \sqrt{2m|E_0|/\hbar^2} \simeq 0.5 \times 10^8$ cm^{-1}; $\qquad \kappa_0 a = \ln 10 \Rightarrow a \simeq 4.6$ Å .

c) The eigenvalue equation:

$$\left(\frac{p^2}{2m} + V(q)\right)|E\rangle = E|E\rangle$$

in the momentum representation can be obtained by multiplying on the left by $\langle p|$ and by exploiting the completeness relation $\int_{-\infty}^{+\infty}|p'\rangle\,dp'\langle p'| = \mathbb{1}$:

$$\left(\frac{p^2}{2m} - E\right)\varphi_E(p) = -\int_{-\infty}^{+\infty}\langle p|V(q)|p'\rangle\varphi_E(p')\,dp'$$

$$\langle p|V(q)|p'\rangle = \int\langle p|V(q)|x\rangle\langle x|p'\rangle\,dx = \frac{1}{2\pi\hbar}\int e^{i(p'-p)x/\hbar}V(x)\,dx$$

(remember that $\langle x|p\rangle = e^{ipx/\hbar}/\sqrt{2\pi\hbar}$ and $\int e^{ipx/\hbar}\,dx = 2\pi\hbar\,\delta(p)$).

If $V(x) = -\lambda\,\delta(x)$ one has $\langle p|V(q)|p'\rangle = -\lambda/2\pi\hbar$ and:

$$\left(\frac{p^2}{2m} - E\right)\varphi_E(p) = \frac{\lambda}{\sqrt{2\pi\hbar}} \times \frac{1}{\sqrt{2\pi\hbar}}\int_{-\infty}^{+\infty}\varphi_E(p')\,dp' = \frac{\lambda}{\sqrt{2\pi\hbar}}\psi_E(0).$$

The above equation can be obtained in an alternative way, by multiplying the Schrödinger equation:

$$\left[-\frac{\hbar^2}{2m}\frac{d^2}{dx^2} - \lambda\,\delta(x)\right]\psi_E(x) = E\,\psi_E(x)$$

by $e^{-ipx/\hbar}/\sqrt{2\pi\hbar}$ and performing the integration in dp. Since we are after the bound state $-E = |E_0|$, one has:

$$\varphi_0(p) = \frac{\lambda}{\sqrt{2\pi\,\hbar}} \frac{1}{p^2/2m + |E_0|} \, \psi_0(0) \; .$$

The above equation provides the functional dependence of φ_0 on p: note that φ_0, as well as ψ_0, is real and even. By integrating both sides, using the identity in the text and $\int 1/(x^2 + 1)\,\mathrm{d}x = \arctan(x)$, one obtains:

$$1 = \frac{\lambda}{2\pi\,\hbar} \int_{-\infty}^{+\infty} \frac{2m}{p^2 + 2m|E_0|} \, \mathrm{d}p = \frac{\lambda}{\hbar}\sqrt{\frac{m}{2|E|}} \quad \Rightarrow \quad E_0 = -\frac{m\lambda^2}{2\hbar^2} \; .$$

Putting $\psi_0(0) = 1$ and $\kappa_0 = \sqrt{2m|E_0|/\hbar^2} = m\lambda/\hbar^2$, one has:

$$\psi_0(x) = \langle\, x \mid E_0 \,\rangle = \int_{-\infty}^{+\infty} \langle\, x \mid p \,\rangle \, \varphi_0(p)\, \mathrm{d}p = \frac{2m\lambda}{2\pi\,\hbar} \int_{-\infty}^{+\infty} \frac{1}{p^2 + \hbar^2\kappa_0^2} \, \mathrm{e}^{\,\mathrm{i}\,p\,x/\hbar}\, \mathrm{d}p \; .$$

The integration can be performed by aid of the residue theorem; considering p as a complex variable, note that the integrand has two poles on the imaginary axis: $p_\pm = \pm\,\mathrm{i}\,\hbar\,\kappa_0$.

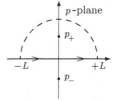

For $x > 0$ one may choose, as integration path in the p complex plane, the segment $[-L, +L]$ of the real axis joined with the counterclockwise oriented semicircle of radius L in the upper half plane: in the limit $L \to \infty$ the semicircle gives a vanishing contribution. Only the upper pole contributes to the sum of residues:

$$\psi_0(x) = \frac{\hbar\,\kappa_0}{\pi} \times 2\pi\,\mathrm{i}\, \frac{\mathrm{e}^{\,\mathrm{i}\,p_+ x/\hbar}}{p_+ - p_-} = \mathrm{e}^{-\kappa_0 x}, \qquad x > 0 \; .$$

In the case $x < 0$ the semicircle is taken in the lower half plane and the path is oriented clockwise: only p_- contributes and $\psi_0(x) = \mathrm{e}^{\kappa_0 x}$.

d) The Schrödinger equation is

$$\psi_E''(x) = -\frac{2m}{\hbar^2}\left(E + \lambda\,\delta(x)\right)\psi_E(x) \; .$$

As ψ_E, being continuous, is bounded in a neighborhood of $x = 0$, the term proportional to E gives a contribution $O(\epsilon)$, so in the limit $\epsilon \to 0^+$, one obtains:

$$\psi_E'(0^+) - \psi_E'(0^-) = -\frac{2m\lambda}{\hbar^2}\,\psi_E(0) \; .$$

The result is in agreement with what has been found in point a) and holds independently of the sign of λ.

In order to find the bound state(s), one puts (in the case of even states) $\psi_E(x) = \mathrm{e}^{-\kappa_0|x|}$ with $\kappa_0 = (2m|E_0|/\hbar^2)^{1/2}$ unknown; the discontinuity of the derivative requires $2\kappa_0 = 2m\lambda/\hbar^2$, therefore $\lambda > 0$ (attractive case) and $E_0 = -m\lambda^2/2\hbar^2$. For the odd states $\psi_E(0) = 0$: the $\psi_E(x)$ is therefore continuous with continuous derivative at $x = 0$ and, as a consequence, satisfies the Schrödinger equation for the free particle that has no bound states. On the other hand, the only odd and normalizable wavefunction compatible with the Schrödinger equation is $\psi_E(x) = \varepsilon(x)\,e^{-\kappa_0|x|}$ ($\varepsilon(x)$ is the sign function), that is not continuous at $x = 0$.

6.19

a) For any $E > 0$ put $k = \sqrt{2mE/\hbar^2}$ and:

$$\psi_k(x) = \begin{cases} \alpha\,e^{ikx} + \beta\,e^{-ikx} & x < 0 \\ \gamma\,e^{ikx} + \delta\,e^{-ikx} & x > 0 . \end{cases}$$

The conditions:

$$\psi_k(0^+) = \psi_k(0^-), \qquad \psi_k'(0^+) - \psi_k'(0^-) = -(2m\lambda/\hbar^2)\,\psi_k(0)$$

(see Problem 6.18) respectively take the form:

$$\begin{cases} \gamma + \delta = \alpha + \beta \\ ik(\gamma - \delta) - ik(\alpha - \beta) = -\dfrac{2m\lambda}{\hbar^2}(\alpha + \beta) . \end{cases}$$

Solving with respect to the unknown γ and δ, they give:

$$\begin{pmatrix} \gamma \\ \delta \end{pmatrix} = \begin{pmatrix} 1 + i\dfrac{\kappa_0}{k} & i\dfrac{\kappa_0}{k} \\ -i\dfrac{\kappa_0}{k} & 1 - i\dfrac{\kappa_0}{k} \end{pmatrix} \begin{pmatrix} \alpha \\ \beta \end{pmatrix}; \quad \begin{pmatrix} \alpha \\ \beta \end{pmatrix} = \begin{pmatrix} 1 - i\dfrac{\kappa_0}{k} & -i\dfrac{\kappa_0}{k} \\ i\dfrac{\kappa_0}{k} & 1 + i\dfrac{\kappa_0}{k} \end{pmatrix} \begin{pmatrix} \gamma \\ \delta \end{pmatrix}$$

where the position $\kappa_0 = m\lambda/\hbar^2$ has been made and the sign of κ_0 is the same as that of λ. In this way:

$$M = \begin{pmatrix} \dfrac{k + i\kappa_0}{k} & i\dfrac{\kappa_0}{k} \\ -i\dfrac{\kappa_0}{k} & \dfrac{k - i\kappa_0}{k} \end{pmatrix}; \qquad S^\dagger = \begin{pmatrix} \dfrac{k}{k - i\kappa_0} & \dfrac{i\kappa_0}{k - i\kappa_0} \\ \dfrac{i\kappa_0}{k - i\kappa_0} & \dfrac{k}{k - i\kappa_0} \end{pmatrix}.$$

The transmission and reflection coefficients:

$$T(k) = \frac{1}{|M_{11}|^2} \equiv |S_{11}|^2 = \frac{k^2}{k^2 + \kappa_0^2}, \quad R(k) = \left|\frac{M_{12}}{M_{11}}\right|^2 \equiv |S_{12}|^2 = \frac{\kappa_0^2}{k^2 + \kappa_0^2}$$

do not depend on the sign of λ.

b) The scattering state with source at left ($x = -\infty$; $k > 0$) is the state characterized by $\alpha = 1$, $\delta = 0$. The coefficients β and γ are obtained from:

$$S^\dagger \begin{pmatrix} 1 \\ 0 \end{pmatrix} = \begin{pmatrix} \gamma \\ \beta \end{pmatrix} \quad \Rightarrow \quad \beta = \frac{i\kappa_0}{k - i\kappa_0}, \qquad \gamma = \frac{k}{k - i\kappa_0}$$

therefore:

$$\psi_k^{(1)}(x) = N \begin{cases} e^{ikx} + \dfrac{i\kappa_0}{k - i\kappa_0}\,e^{-ikx} & x \le 0 \\ \dfrac{k}{k - i\kappa_0}\,e^{ikx} & x \ge 0 . \end{cases}$$

Likewise:

$$\psi_k^{(\mathrm{r})}(x) = N \begin{cases} \dfrac{k}{k - \mathrm{i}\,\kappa_0}\,\mathrm{e}^{-\mathrm{i}\,k\,x} & x \leq 0 \\[2mm] \dfrac{\mathrm{i}\,\kappa_0}{k - \mathrm{i}\,\kappa_0}\,\mathrm{e}^{\mathrm{i}\,k\,x} + \mathrm{e}^{-\mathrm{i}\,k\,x} & x \geq 0 \,. \end{cases}$$

Since $\psi_k^{(\mathrm{r})}(x) = \psi_k^{(\mathrm{l})}(-x)$, the normalization factor N is the same in the last two formulae.

c) Thanks to $\psi_k^{(\mathrm{r})}(x) = \psi_k^{(\mathrm{l})}(-x)$, the states with definite parity are given by:

$$\psi_k^+(x) = \frac{1}{\sqrt{2}}\left(\psi_k^{(\mathrm{l})}(x) + \psi_k^{(\mathrm{r})}(x)\right)$$

$$= \frac{N}{\sqrt{2}(k - \mathrm{i}\,\kappa_0)} \begin{cases} (k - \mathrm{i}\,\kappa_0)\,\mathrm{e}^{\mathrm{i}\,k\,x} + (k + \mathrm{i}\,\kappa_0)\,\mathrm{e}^{-\mathrm{i}\,k\,x} & x \leq 0 \\ (k + \mathrm{i}\,\kappa_0)\,\mathrm{e}^{\mathrm{i}\,k\,x} + (k - \mathrm{i}\,\kappa_0)\,\mathrm{e}^{-\mathrm{i}\,k\,x} & x \geq 0 \end{cases}$$

$$= \frac{N\sqrt{2}}{(k - \mathrm{i}\,\kappa_0)} \begin{cases} k \cos kx + \kappa_0 \sin kx & x \leq 0 \\ k \cos kx - \kappa_0 \sin kx & x \geq 0 \,. \end{cases}$$

The case of the odd states $\psi_k^-(x)$ is more interesting: indeed, since they vanish at the origin, they have continuous derivative and, as a consequence, they coincide with the analogous states of the free particle (they do not 'feel' the $\delta(x)$):

$$\psi_k^-(x) = N\sqrt{2}\,\sin kx \,.$$

d) If the scattering states are normalized as in the text of the problem, in the attractive case – where one bound state $|\,\kappa_0\,\rangle$ exists (see Problem 6.18) – the completeness relation reads:

$$\int_0^\infty \left(|\,k^{(\mathrm{l})}\,\rangle\langle\,k^{(\mathrm{l})}\,| + |\,k^{(\mathrm{r})}\,\rangle\langle\,k^{(\mathrm{r})}\,| \right) \mathrm{d}k + |\,\kappa_0\,\rangle\langle\,\kappa_0\,| = \mathbb{1}$$

that, by multiplying on the left by $\langle\,x\,|$ and on the right by $|\,y\,\rangle$, becomes:

$$E(x,\,y) + \psi_0(x)\,\psi_0^*(y) = \delta(x - y) \;\Rightarrow\; E(x,y) = \delta(x - y) - \kappa_0\,\mathrm{e}^{-\kappa_0(|x|+|y|)}.$$

In the repulsive case the scattering states are a complete set (the bound state is missing), then $E(x,y) = \delta(x - y)$.

For the sake of completeness, we calculate the normalization factor N in such a way that the orthogonality relations given in the text hold true. To this end we will take advantage of the fact that (in the sense of distributions) the following prescriptions are valid:

$$\int_0^{\pm\infty} \mathrm{e}^{\mathrm{i}\,\alpha\,x}\,\mathrm{d}x \equiv \lim_{\epsilon \to 0^+} \int_0^{\pm\infty} \mathrm{e}^{\mathrm{i}\,(\alpha \pm \mathrm{i}\,\epsilon)\,x}\,\mathrm{d}x = \mathrm{i}\left(\mathcal{P}\frac{1}{\alpha} \mp \mathrm{i}\,\pi\,\delta(\alpha)\right) \;\Rightarrow$$

$$\int_{-\infty}^{+\infty} \mathrm{e}^{\mathrm{i}\,\alpha\,x}\,\mathrm{d}x = 2\pi\,\delta(\alpha)$$

where $\mathcal{P}$ stands for Cauchy principal value.

We will limit ourselves to considering the states $| k^{(1)} \rangle$ (in the other cases the calculations go along the same lines):

$$\langle k'^{(1)} | k^{(1)} \rangle = N^2 \int_{-\infty}^{0} \left(e^{ik'x} + \frac{i\kappa_0}{k' - i\kappa_0} e^{-ik'x} \right)^* \left(e^{ikx} + \frac{i\kappa_0}{k - i\kappa_0} e^{-ikx} \right) dx$$

$$+ N^2 \int_{0}^{+\infty} \left(\frac{k'}{k' - i\kappa_0} e^{ik'x} \right)^* \left(\frac{k}{k - i\kappa_0} e^{ikx} \right) dx$$

$$= N^2 \int_{-\infty}^{0} \left(e^{i(k-k')x} + \frac{i\kappa_0 e^{-i(k+k')x}}{k - i\kappa_0} + \frac{-i\kappa_0 e^{i(k+k')x}}{k' + i\kappa_0} \right.$$

$$\left. + \frac{\kappa_0^2 e^{-i(k-k')x}}{(k' + i\kappa_0)(k - i\kappa_0)} \right) dx + N^2 \int_{0}^{+\infty} \frac{k k' e^{i(k-k')x}}{(k' + i\kappa_0)(k - i\kappa_0)} dx .$$

In the last but one integral let us perform the replacement $x \to -x$ and, as a consequence, the integration goes from 0 to $+\infty$:

$$\langle k'^{(1)} | k^{(1)} \rangle = N^2 \int_{-\infty}^{0} \left(e^{i(k-k')x} + \frac{i\kappa_0 e^{-i(k+k')x}}{k - i\kappa_0} + \frac{-i\kappa_0 e^{i(k+k')x}}{k' + i\kappa_0} \right) dx$$

$$+ N^2 \int_{0}^{+\infty} \frac{\kappa_0^2 + k k'}{(k' + i\kappa_0)(k - i\kappa_0)} e^{i(k-k')x} dx .$$

In the first term we extend the integration up to $+\infty$ and, accordingly, subtract $N^2 \int_{0}^{+\infty} e^{i(k-k')x} dx$: as $\int_{-\infty}^{+\infty} e^{i(k-k')x} dx = 2\pi \delta(k - k')$, there remains to show that:

$$\int_{-\infty}^{0} \left(\frac{i\kappa_0}{k - i\kappa_0} e^{-i(k+k')x} + \frac{-i\kappa_0}{k' + i\kappa_0} e^{i(k+k')x} \right) dx$$

$$+ \int_{0}^{+\infty} \left(\frac{\kappa_0^2 + k k'}{(k' + i\kappa_0)(k - i\kappa_0)} - 1 \right) e^{i(k-k')x} dx = 0 .$$

The first integral, calculated according to the prescription given above, equals:

$$\frac{i\kappa_0}{k - i\kappa_0} \times \frac{1}{-i(k + k')} + \frac{-i\kappa_0}{k' + i\kappa_0} \times \frac{1}{i(k + k')} = -\frac{\kappa_0}{(k' + i\kappa_0)(k - i\kappa_0)}$$

(the term with $\delta(k+k')$ vanishes since $k, k' > 0 \Rightarrow k+k' > 0$); the second can be calculated in the same way: also in this case the term containing $\delta(k - k')$ vanishes because the coefficient of $e^{i(k-k')x}$ vanishes for $k = k'$; the result is the opposite of the preceding one; so, in conclusion:

$$\langle k'^{(1)} | k^{(1)} \rangle = 2\pi N^2 \delta(k - k') \quad \Rightarrow \quad N = \frac{1}{\sqrt{2\pi}} .$$

6.20

a) Since the Schrödinger equation is real, if $\psi_E(x)$ is a solution, also:

$$\psi_E^*(x) = \begin{cases} \alpha^* e^{\kappa x} + \beta^* e^{-\kappa x} & x \le x_1 \\ \gamma^* e^{\kappa x} + \delta^* e^{-\kappa x} & x \ge x_2 \end{cases}$$

is a solution, therefore:

$$\begin{pmatrix} \gamma \\ \delta \end{pmatrix} = \begin{pmatrix} A & B \\ C & D \end{pmatrix} \begin{pmatrix} \alpha \\ \beta \end{pmatrix} \quad \Rightarrow \quad \begin{pmatrix} \gamma^* \\ \delta^* \end{pmatrix} = \begin{pmatrix} A & B \\ C & D \end{pmatrix} \begin{pmatrix} \alpha^* \\ \beta^* \end{pmatrix}$$

$$\Rightarrow \quad M(\kappa) = M^*(\kappa)$$

$(A, B, C, D$ are functions of κ). The continuity equation:

$$\frac{d}{dx} \Im m \left(\psi_E^*(x) \, \psi_E'(x) \right) = 0$$

$$\Rightarrow \quad \Im m \left(\psi_E^*(x_1) \, \psi_E'(x_1) \right) = \Im m \left(\psi_E^*(x_2) \, \psi_E'(x_2) \right)$$

holds for all the solutions of the Schrödinger equation regardless of the value of E, and entails that $\Im m(\alpha^* \beta) = \Im m(\gamma^* \delta)$; from this $A D - B C = 1$ follows.

b) E is an eigenvalue of the Hamiltonian if there exists the solution with $\beta = \gamma = 0$, so one must have:

$$\begin{pmatrix} 0 \\ \delta \end{pmatrix} = M(\kappa) \begin{pmatrix} \alpha \\ 0 \end{pmatrix} \quad \Rightarrow \quad \alpha \, A(\kappa) = 0 \; \Rightarrow \; A(\kappa) = 0 \,.$$

c) The transfer matrix relative to the translated potential $V(x - a)$ can be obtained as in Problem 6.14, provided $i k$ is replaced by κ:

$$M_{\mathrm{tr}}(\kappa) = \begin{pmatrix} A & B \, e^{-2\kappa a} \\ C \, e^{+2\kappa a} & D \end{pmatrix} .$$

As for the reflected potential $V(-x)$, in the same way as in Problem 6.14:

$$M_{\mathrm{rf}}(\kappa) = \begin{pmatrix} 0 & 1 \\ 1 & 0 \end{pmatrix} M^{-1} \begin{pmatrix} 0 & 1 \\ 1 & 0 \end{pmatrix} = \begin{pmatrix} A & -C \\ -B & D \end{pmatrix} .$$

d) $V(x) = -\lambda \, \delta(x)$: it is necessary, as in Problem 6.19, to enforce that $\psi_E(x)$ be continuous at $x = 0$ and that its derivative has the discontinuity $-(2m\lambda/\hbar^2) \, \psi_E(0)$ (see Problem 6.18). The solution can be obtained from the solution relative to the case $E > 0$ (see Problem 6.19) by means of the substitution $i k \to \kappa$ so that, putting $\kappa_0 = m\lambda/\hbar^2$, one has:

$$M(\kappa) = \begin{pmatrix} 1 - \dfrac{\kappa_0}{\kappa} & -\dfrac{\kappa_0}{\kappa} \\ \dfrac{\kappa_0}{\kappa} & 1 + \dfrac{\kappa_0}{\kappa} \end{pmatrix} .$$

The condition $A(\kappa) = 0$ for the existence of a bound state implies that:

$$\kappa = \kappa_0 \quad \Rightarrow \quad E_0 = -\frac{\hbar^2 \kappa_0^2}{2m} = -\frac{m\lambda^2}{2\hbar^2} \,.$$

Note that, in the case $\lambda < 0$ (repulsive potential), as by definition $\kappa > 0$, the condition $\kappa = \kappa_0 < 0$ cannot be fulfilled.

6.21

a) We will limit ourselves to consider the case of solutions of the Schrödinger equation with $E > 0$, the case $E < 0$ being analogous. Let $\psi_1(x)$ be a solution of the Schrödinger equation with only V_1 as potential. One has:

$$\psi_1(x) = \begin{cases} \alpha\,e^{ikx} + \beta\,e^{-ikx} & x < x_1 \\ \gamma\,e^{ikx} + \delta\,e^{-ikx} & x > x_2 \end{cases} \qquad \begin{pmatrix} \gamma \\ \delta \end{pmatrix} = M_1 \begin{pmatrix} \alpha \\ \beta \end{pmatrix}.$$

Let us now consider that solution $\psi_2(x)$ of the Schrödinger equation with only V_2 as potential and that for $x < x_1 + a$ is given by:

$$\psi_2(x) = \gamma\,e^{ikx} + \delta\,e^{-ikx} \qquad x < x_1 + a$$

(γ and δ, determined by $\psi_1(x)$, are the 'initial conditions' for $\psi_2(x)$, in the same way as α and β are the initial conditions for $\psi_1(x)$). One has:

$$\psi_2(x) = \lambda\,e^{ikx} + \mu\,e^{-ikx} \qquad x > x_2 + a; \qquad \begin{pmatrix} \lambda \\ \mu \end{pmatrix} = M_2 \begin{pmatrix} \gamma \\ \delta \end{pmatrix}.$$

The function $\psi(x)$ that coincides with $\psi_1(x)$ for $x < x_1 + a$ and with $\psi_2(x)$ for $x > x_2$ ($x_1 < x_2 < x_1 + a < x_2 + a$: between x_2 and $x_1 + a$ the functions $\psi_1(x)$ and $\psi_2(x)$ coincide by construction) satisfies the Schrödinger equation with potential $V(x) = V_1(x) + V_2(x)$ on the whole real axis, and is the unique solution relative to $V(x)$ that, for $x < x_1$, is given by $\alpha\,e^{ikx} + \beta\,e^{-ikx}$. Therefore:

$$\begin{pmatrix} \lambda \\ \mu \end{pmatrix} = M_2 \begin{pmatrix} \gamma \\ \delta \end{pmatrix}; \qquad \begin{pmatrix} \gamma \\ \delta \end{pmatrix} = M_1 \begin{pmatrix} \alpha \\ \beta \end{pmatrix} \quad \Rightarrow \quad \begin{pmatrix} \lambda \\ \mu \end{pmatrix} = M_2 M_1 \begin{pmatrix} \alpha \\ \beta \end{pmatrix}$$

and, as a consequence, $M = M_2 M_1$.

b) Let us consider the case of solutions with $E < 0$. Putting $\kappa = \sqrt{2m|E|/\hbar^2}$, one has (see Problem 6.20):

$$M_1 = \begin{pmatrix} A & B \\ C & D \end{pmatrix}, \quad M_2 = \begin{pmatrix} A & B\,e^{-2\kappa a} \\ C\,e^{2\kappa a} & D \end{pmatrix};$$

$$M = \begin{pmatrix} A^2 + BC\,e^{-2\kappa a} & * \\ * & * \end{pmatrix}.$$

We have emphasized only the element M_{11} because the bound states are determined (see Problem 6.20) by the equation $M_{11} = 0$:

$$A^2(\kappa) + B(\kappa)\,C(\kappa)\,e^{-2\kappa a} = 0 .$$

The existence of the bound state E_0 for the single potential guarantees that there exists κ_0 such that $A(\kappa_0) = 0$. For $a = \infty$ the equation takes the form $A^2(\kappa) = 0$ that has the twofold solution $\kappa = \kappa_0$: there are two states (for example, the particle bound to either the right well or the left well) that have the same energy. For a large, putting $\kappa = \kappa_0 + \delta\kappa$, one has:

$$\left[A'(\kappa_0)\right]^2 (\delta\kappa)^2 \simeq -B(\kappa_0)\,C(\kappa_0)\,e^{-2\kappa_0 a}$$

and, owing to $\det M_1 = 1$ and $A(\kappa_0) = 0$, one has $B(\kappa_0)\,C(\kappa_0) = -1$, therefore:

$$(\delta\kappa)^2 \simeq \frac{1}{\left[A'(\kappa_0)\right]^2}\,e^{-2\kappa_0 a} \quad \Rightarrow \quad \delta\kappa \simeq \pm \frac{1}{|A'(\kappa_0)|}\,e^{-\kappa_0 a} \quad \Rightarrow$$

$$E_2 - E_1 = -\frac{\hbar^2}{2m}\left((\kappa_0 - \delta\kappa)^2 - (\kappa_0 + \delta\kappa)^2\right) \simeq 4\,\frac{\hbar^2 \kappa_0\,\delta\kappa}{2m} = \frac{4|E_0|\,e^{-\kappa_0 a}}{\kappa_0|A'(\kappa_0)|} .$$

c) The potential $V_1(a - x)$ is obtained by $V_1(x)$ performing first a translation of length a $(V_1(x) \to V_1(x - a))$, then a reflection with respect to the point $x = a$: $x - a \to -(x - a)$. Then $V_1(a - x)$ has the same bound states as $V_1(x)$. Now, for large values of a $V_1(x)$ and $V_1(a - x)$ have disjoint supports, so:

$$M_1 = \begin{pmatrix} A & B \\ C & D \end{pmatrix}, \quad M_2 = \begin{pmatrix} A & -C\,e^{-2\kappa a} \\ -B\,e^{2\kappa a} & D \end{pmatrix}; \quad M_{11} = A^2 - C^2 e^{-2\kappa a}$$

$$M_{11} = 0 \quad \Rightarrow \quad (\delta\kappa)^2 \simeq \left[\frac{C(\kappa_0)}{A'(\kappa_0)}\right]^2 e^{-2\kappa_0 a} \quad \Rightarrow \quad \delta\kappa \simeq \pm \frac{C(\kappa_0)}{A'(\kappa_0)} e^{-\kappa_0 a}$$

with $C(\kappa_0) \neq 0$ thanks to $\det M = 1$ and $A(\kappa_0) = 0$.

6.22

a) Owing to $[p, f(x)] = -i\,\hbar\,f'(x)$,

$$(p + i f(x))(p - i f(x)) = p^2 + f^2(x) - \hbar f'(x) = p^2 + \hbar^2 (\phi'^2(x) - \phi''(x))$$

$$\Rightarrow \quad f(x) = \hbar\,\phi'(x), \quad H = \frac{1}{2m}(p + i\,\hbar\,\phi'(x))(p - i\,\hbar\,\phi'(x)) .$$

As $(p + i\,\hbar\,\phi'(x))(p - i\,\hbar\,\phi'(x)) = \eta^\dagger \eta$, where $\eta \equiv p - i\,\hbar\,\phi'(x)$, the mean value of H in any state is ≥ 0, then the eigenvalues of H are ≥ 0.

b) $H \mid E = 0 \rangle = 0 \; \Rightarrow \; \eta^\dagger \eta \mid E = 0 \rangle = 0 \; \Rightarrow \; \langle E = 0 \mid \eta^\dagger \eta \mid E = 0 \rangle = 0$

$\Rightarrow \quad \| \eta \mid E = 0 \rangle \| = 0 \quad \Rightarrow \quad \eta \mid E = 0 \rangle = 0 .$

So, if $\mid E = 0 \rangle$ exists, $\psi_0(x)$ is a solution of the equation:

$$\left(-i\hbar\frac{d}{dx} - i\hbar\,\phi'(x)\right) \psi_0(x) = 0 \quad \Rightarrow \quad \psi_0(x) = e^{-\phi(x)} .$$

In order that $e^{-\phi(x)}$ be in L^2, i.e. that $\mid E = 0 \rangle$ existed as a bound state, it is necessary (but not sufficient) that $\phi(x)$ tended to $+\infty$ for $x \to \pm\infty$ (for example, if $\phi(x) \approx \ln(\ln |x|)$, $e^{-\phi(x)} \approx 1/\ln|x|$ is not in L^2).
The above conclusion shows that, although the solutions of the Schrödinger equation:

$$\left(-i\hbar\frac{d}{dx} + i\hbar\,\phi'(x)\right)\left(-i\hbar\frac{d}{dx} - i\hbar\,\phi'(x)\right)u(x) = 0$$

are ∞^2, if $e^{-\phi(x)}$ is not in L^2, no other solution is in L^2: this may be explicitly verified, by determining – e.g. through the method of reduction of the degree (i.e. by means of the substitution $u(x) = v(x)\,e^{-\phi(x)}$) – all the solutions of the above equation.

c) Putting $g(x) = \phi'(x)$, in general in order to find $\phi(x)$ given $V(x)$, it is necessary to solve the Riccati equation:

$$g'(x) - g^2(x) = -\frac{2m}{\hbar^2}V(x) .$$

In our case it is evident that:

$$\phi'(x) = \frac{1}{a}\tanh(x/a) \quad \Rightarrow \quad \phi(x) = \frac{1}{a}\int_0^x \tanh(x'/a)\,\mathrm{d}x' = \ln\cosh(x/a)\;.$$

As $\mathrm{e}^{-\phi(x)} = 1/\cosh(x/a)$ is in L^2, $E = 0$ is an eigenvalue for the Hamiltonian with potential $V(x)$, so if the potential is $\widetilde{V}(x) = V(x) - \hbar^2/2ma^2$ the energy of the ground state $\big($wavefunction $1/\cosh(x/a)\big)$ is $E_0 = -\hbar^2/2ma^2$, in agreement with the results of Problem 6.14.

d) As $\varepsilon^2(x) = 1$ and $\varepsilon'(x) = 2\,\delta(x)$,

$$V(x) = \frac{\hbar^2}{2m}\left(\frac{m^2\lambda^2}{\hbar^4} - 2\frac{m\lambda}{\hbar^2}\,\delta(x)\right) = \frac{m\lambda^2}{2\hbar^2} - \lambda\,\delta(x)\;.$$

The Hamiltonian with potential $V(x)$ possesses the eigenvalue $E = 0$ (the corresponding wavefunction is $\mathrm{e}^{-m\lambda|x|/\hbar^2}$ which is L^2), so the Hamiltonian with potential $\widetilde{V}(x) = V(x) - m\lambda^2/2\hbar^2 = -\lambda\,\delta(x)$, possesses the eigenvalue $E_0 = -m\lambda^2/2\,\hbar^2$, in agreement with the results obtained in Problem 6.18.

By aid of the same technique it is straightforward to show that the potential:

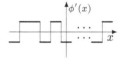

$$V(x) = -\lambda\sum_{i=0}^{2n}(-)^i\,\delta(x - a_i)\,, \qquad n \in \mathbb{N}, \quad a_i \in \mathbb{R}$$

no matter what the values of the a_i's are, has the ground state with energy $E_0 = -m\lambda^2/2\,\hbar^2$ equal to that of the single attractive δ: it suffices to take $\phi'(x) = (m\lambda/\hbar^2)\,g(x)$, with:

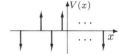

$$g(x) = \begin{cases} -1 & x < a_0 \\ +1 & a_{2i} < x < a_{2i+1} \quad i = 0\ldots n-1 \\ -1 & a_{2i-1} < x < a_{2i} \quad i = 1\ldots n \\ +1 & x > a_{2n}\;. \end{cases}$$

If instead the number of δ is even (always alternating attractive and repulsive), $g(x)$ has the same value ± 1 for both $x < a_{\min}$ and $x > a_{\max}$, then $\mathrm{e}^{-\phi(x)}$ diverges either for $x \to +\infty$ or for $x \to -\infty$: this means that the energy of the ground state (if it exists) is greater (i.e. less negative) than the energy of the ground state of the single attractive delta.

6.23

a) The existence of at least one bound state is guaranteed by the fact that the given Hamiltonian is smaller than the Hamiltonian with one single attractive δ (supported e.g. in $x = -a$), in the sense that the difference equals $-\lambda\,\delta(x - a)$, whose mean value is ≤ 0 in any state: from Problem 6.5 it follows that $E_0 \leq -m\lambda^2/2\hbar^2$, which is the energy of the bound state with one single delta.

In order to establish that there are no more than two bound states, let us note that the eigenfunctions of the Hamiltonian, with $E < 0$, are – in the three regions $x < -a$, $|x| < a$, $x > a$ – linear combinations of $e^{\kappa x}$ and $e^{-\kappa x}$, with $\kappa = \left(2m|E|/\hbar^2\right)^{1/2}$; therefore, as H commutes with the space inversion, all have the form:

$$\psi_E^+(x) = \begin{cases} B\,e^{\kappa x} \\ A\cosh\kappa x \,, \\ B\,e^{-\kappa x} \end{cases} \qquad \psi_E^-(x) = \begin{cases} C\,e^{\kappa x} \\ D\sinh\kappa x \\ -C\,e^{-\kappa x} \end{cases} \qquad \begin{array}{l} x \le -a \\ |x| \le a \\ x \ge a \end{array} \qquad (1)$$

respectively for the even and the odd states: the first has no zeroes, so the only even state is the ground state; the second (if it exists) has one zero: there (possibly) exists only one excited bound state.

b) In order to determine the bound states, it is necessary to enforce the continuity at $x = a$ for the $\psi_E(x)$ given by (1) (or equivalently at $x = -a$) and the discontinuity $\Delta\,\psi_E'(a)$ of the first derivative, given by $(-2m\lambda/\hbar^2)\,\psi_E(a)$ (see Problem 6.18):

$$\begin{cases} B\,e^{-\kappa a} = A\cosh\kappa a \\ B\,\kappa\,e^{-\kappa a} + A\,\kappa\,\sinh\kappa a = \dfrac{2m\lambda}{\hbar^2}\,A\cosh\kappa a \end{cases} \Rightarrow$$

$$\kappa\left(\cosh\kappa a + \sinh\kappa a\right) = 2\,\frac{m\lambda}{\hbar^2}\cosh\kappa a \qquad \text{(even state)};$$

$$\begin{cases} D\sinh\kappa a = -C\,e^{-\kappa a} \\ D\,\kappa\cosh\kappa a - C\,\kappa\,e^{-\kappa a} = \dfrac{2m\lambda}{\hbar^2}\,D\sinh\kappa a \end{cases} \Rightarrow$$

$$\kappa\left(\cosh\kappa a + \sinh\kappa a\right) = 2\,\frac{m\lambda}{\hbar^2}\sinh\kappa a \qquad \text{(odd state)}.$$

One of the several ways the equations relative to the bound states can be rewritten is the following:

$$e^{-2\kappa a} = \frac{\hbar^2}{m\lambda\,a}\,\kappa a - 1 \qquad \text{(even state)}$$

$$e^{-2\kappa a} = 1 - \frac{\hbar^2}{m\lambda\,a}\,\kappa a \qquad \text{(odd state)}\,.$$

$$(2)$$

The first of the above equations always possesses one solution $\kappa_+ > \kappa_0$, which in the figure is represented by the point of intersection between the exponential and the straight line with positive slope; the second equation admits one solution $\kappa_- < \kappa_0$ only in the case when the angular coefficient of the straight line in the right hand side is greater than the value of the derivative at the origin of the exponential: $-\hbar^2/m\lambda\,a > -2$, which is the condition given in the text. The intersection point between the two straight lines corresponds to $\kappa_0 = m\lambda/\hbar^2$, that is the value of κ giving the bound state for the single delta: it is evident that the (negative) energy corresponding to the odd solution (if it exists) is higher (i.e. less negative) than that of the single delta ($\kappa_- < \kappa_0$).

c) We find again in this particular case the general result we established in Problem 6.21. In the limit $a \to \infty$, i.e. widely separated wells, both the solutions $\kappa_\pm$ of the two equations tend to κ_0, that corresponds to the bound state of a single delta: there are two states (e.g. the particle is bound to either the right or the left well) that have the same energy. The finite value of a causes the two states to be not perfectly degenerate. When a is large – better: when $\kappa_0 a \equiv m\lambda a/\hbar^2 \gg 1$ – approximate solutions of the equations (2) may be found by expanding around κ_0: putting $\kappa_\pm = \kappa_0 + \epsilon_\pm$, one has:

$$\begin{cases} e^{-2\kappa_0 a} \approx \dfrac{\kappa_0 + \epsilon_+}{\kappa_0} - 1 = \dfrac{\epsilon_+}{\kappa_0} \\ e^{-2\kappa_0 a} \approx 1 - \dfrac{\kappa_0 + \epsilon_-}{\kappa_0} = -\dfrac{\epsilon_-}{\kappa_0} \end{cases} \qquad \Rightarrow \qquad \epsilon_\pm \approx \pm\kappa_0\, e^{-2\kappa_0 a}$$

and the energy difference is

$$E_- - E_+ = -\frac{\hbar^2}{2m}(\kappa_-^2 - \kappa_+^2) \simeq \frac{\hbar^2}{2m}2\kappa_0 \times 2\kappa_0\, e^{-2\kappa_0 a} = 4\,|E_0|e^{-2\kappa_0 a}\,.$$

The result is in agreement with what has been found in Problem 6.21, since $|A'(\kappa_0)| = 1/\kappa_0$; note that, in the present case, there appears the exponential $e^{-2\kappa_0 a}$ instead of $e^{-\kappa_0 a}$: indeed the two deltas separate from each other symmetrically and their distance is $2a$.

d)

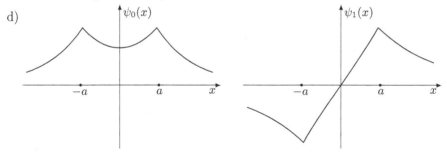

The two wavefunctions are equally normalized; $\kappa_0 a$ has been taken equal to 0.88 ($> \frac{1}{2}$), the even wavefunction corresponds to $\kappa_+ = 1.14\,\kappa_0$ and the odd one to $\kappa_- = 0.72\,\kappa_0$: by looking at both graphs, it appears that the exponential of the odd function decays more slowly than the one of the even function. It is also evident that if the particle is in an eigenstate of the energy, it is no longer bound to one of the two wells, but simultaneously to both.

6.24

a) $\langle 1\,|\,2 \rangle = \displaystyle\int_{-\infty}^{+\infty} \psi_1(x)\,\psi_2(x)\,dx$

$= \kappa_0 \left(\displaystyle\int_{-\infty}^{-a} e^{2\kappa_0 x}\,dx + 2a\,e^{-2\kappa_0 a} + \int_a^\infty e^{-2\kappa_0 x}\,dx \right) = (1 + 2\kappa_0 a)\,e^{-2\kappa_0 a}\,.$

$$1 = N^2 \left(\cos^2 \theta \langle 1 \mid 1 \rangle + \sin^2 \theta \langle 2 \mid 2 \rangle + 2 \sin \theta \cos \theta \langle 1 \mid 2 \rangle \right)$$

$$\Rightarrow \quad N^2 = \frac{1}{1 + \sin 2\theta \left(1 + 2\kappa_0 a \right) e^{-2\kappa_0 a}} \; .$$

b) $\overline{H}(\theta) = N^2 \left(\cos^2 \theta \langle 1 \mid H \mid 1 \rangle + \sin^2 \theta \langle 2 \mid H \mid 2 \rangle + 2 \sin \theta \cos \theta \langle 1 \mid H \mid 2 \rangle \right).$

Calling $E^{(0)} = -\hbar^2 \kappa_0^2 / 2m = -m\lambda^2 / 2\hbar^2$ the energy of the ground state with one single delta, and having in mind that:

$$H = \left(\frac{p^2}{2m} - \lambda \delta(x + a) \right) - \lambda \delta(x - a) = \left(\frac{p^2}{2m} - \lambda \delta(x - a) \right) - \lambda \delta(x + a)$$

one has:

$$\langle 1 \mid H \mid 1 \rangle = E^{(0)} - \lambda \langle 1 \mid \delta(x - a) \mid 1 \rangle = E^{(0)} - \lambda \kappa_0 \, e^{-4\kappa_0 a} = \langle 2 \mid H \mid 2 \rangle$$

$$\langle 1 \mid H \mid 2 \rangle = E^{(0)} \langle 1 \mid 2 \rangle - \lambda \langle 1 \mid \delta(x + a) \mid 2 \rangle = E^{(0)} \langle 1 \mid 2 \rangle - \lambda \kappa_0 \, e^{-2\kappa_0 a}$$

$$= E^{(0)} (1 + 2\kappa_0 a) \, e^{-2\kappa_0 a} - \lambda \kappa_0 \, e^{-2\kappa_0 a} = E^{(0)} (3 + 2\kappa_0 a) \, e^{-2\kappa_0 a}$$

so ($\lambda \kappa_0 = -2 E^{(0)}$)

$$\overline{H}(\theta) = E^{(0)} \, \frac{1 + 2 \, e^{-4\kappa_0 a} + \sin 2\theta \, (3 + 2\kappa_0 a) \, e^{-2\kappa_0 a}}{1 + \sin 2\theta \, (1 + 2\kappa_0 a) \, e^{-2\kappa_0 a}}$$

that is a minimum for $\sin 2\theta = 1$, therefore the (variational) energy of the ground state is

$$E_0^{(v)} = E^{(0)} \left(1 + 2 \, e^{-2\kappa_0 a} \, \frac{1 + e^{-2\kappa_0 a}}{1 + (1 + 2\kappa_0 a) \, e^{-2\kappa_0 a}} \right) .$$

The state that minimizes $\overline{H}(\theta)$ is

$$\mid E_0^{(v)} \rangle = \frac{\mid 1 \rangle + \mid 2 \rangle}{\sqrt{2} \left(1 + (1 + 2\kappa_0 a) \, e^{-2\kappa_0 a} \right)}$$

that is even; as a consequence, the variational state that corresponds to the first excited state (if it exists) is the odd one:

$$\mid E_1^{(v)} \rangle = \frac{\mid 1 \rangle - \mid 2 \rangle}{\sqrt{2} \left(1 - (1 + 2\kappa_0 a) \, e^{-2\kappa_0 a} \right)}$$

corresponding to $\sin 2\theta = -1$ (i.e. to the maximum of $\overline{H}(\theta)$) and, consequently:

$$E_1^{(v)} = E^{(0)} \left(1 - 2 \, e^{-2\kappa_0 a} \, \frac{1 - e^{-2\kappa_0 a}}{1 - (1 + 2\kappa_0 a) \, e^{-2\kappa_0 a}} \right) .$$

Notice that $E_1^{(v)} < 0$ for $\kappa_0 a \gtrsim 1.34 > 0.5$.

c) Certainly the result is more reliable for large values of a: indeed, in the limit $a = \infty$, $\psi_1(x)$ and $\psi_2(x)$ are exact eigenfunctions of H. For $\kappa_0 a \gg 1$, up to terms of the order $e^{-4\kappa_0 a}$:

$$E_0^{(v)} \approx E^{(0)} \left(1 + 2 \, e^{-2\kappa_0 a} \right), \qquad E_1^{(v)} \approx E^{(0)} \left(1 - 2 \, e^{-2\kappa_0 a} \right)$$

and, from (2) of Problem 6.23 (again up to terms of order $e^{-4\kappa_0 a}$), one has:

$$e^{-2\kappa_0 a} = \frac{\kappa_+}{\kappa_0} - 1 \;\Rightarrow\; \kappa_+ = \kappa_0\left(1 + e^{-2\kappa_0 a}\right) \;\Rightarrow\; E_+ \approx E^{(0)}\left(1 + 2\,e^{-2\kappa_0 a}\right)$$

(the index $+$ identifies the even state); likewise:

$$e^{-2\kappa_0 a} = 1 - \frac{\kappa_-}{\kappa_0} \;\Rightarrow\; \kappa_- = \kappa_0\left(1 - e^{-2\kappa_0 a}\right) \;\Rightarrow\; E_- \approx E^{(0)}\left(1 - 2\,e^{-2\kappa_0 a}\right).$$

Note instead that, for $a \to 0$ $E_0^{(v)} \to 3E^{(0)}$, whereas the exact result is $4E^{(0)}$, since $V(x) \to -2\lambda\,\delta(x)$.

d) No: the minimum and the maximum of $\overline{H}(\theta, \varphi)$ in the set of functions $\psi(x; \theta, \varphi)$ respectively are the minimum and the maximum eigenvalue of the restriction of the Hamiltonian H to the two-dimensional subspace $\mathcal{V}$ generated by the functions $\psi_1(x)$ and $\psi_2(x)$. H commutes with the space inversion I and $\mathcal{V}$ is invariant under I, so in $\mathcal{V}$ the eigenstates of H have definite parity: that is why they are exactly the states we have found with the trial functions $\psi(x; \theta)$.

6.25

a) The transfer matrix M is the product of the transfer matrices M_1 and M_2, relative to the potentials $V_1(x)$ and $V_2(x)$, given by (see Problem 6.20):

$$M_1(\kappa) = \begin{pmatrix} 1 - \dfrac{\kappa_1}{\kappa} & -\dfrac{\kappa_1}{\kappa}\,e^{2\kappa a} \\[2mm] \dfrac{\kappa_1}{\kappa}\,e^{-2\kappa a} & 1 + \dfrac{\kappa_1}{\kappa} \end{pmatrix}, \qquad M_2(\kappa) = \begin{pmatrix} 1 - \dfrac{\kappa_2}{\kappa} & -\dfrac{\kappa_2}{\kappa}\,e^{-2\kappa a} \\[2mm] \dfrac{\kappa_2}{\kappa}\,e^{2\kappa a} & 1 + \dfrac{\kappa_2}{\kappa} \end{pmatrix}$$

and (see Problem 6.21):

$$\left[M(\kappa)\right]_{11} = \left[M_2(\kappa) \times M_1(\kappa)\right]_{11} = \left(1 - \frac{\kappa_2}{\kappa}\right)\left(1 - \frac{\kappa_1}{\kappa}\right) - \frac{\kappa_2\,\kappa_1}{\kappa^2}\,e^{-4\kappa a}.$$

The bound states are determined by the equation (see Problem 6.20) $\left[M(\kappa)\right]_{11} = 0$:

$$\left(1 - \frac{\kappa_2}{\kappa}\right)\left(1 - \frac{\kappa_1}{\kappa}\right) - \frac{\kappa_2\,\kappa_1}{\kappa^2}\,e^{-4\kappa a} = 0. \tag{1}$$

b) Let us rewrite the equation for the bound states as:

$$(\kappa - \kappa_2)(\kappa - \kappa_1)\,e^{4\kappa a} = \kappa_2\,\kappa_1, \qquad \kappa > 0.$$

The function in the left hand side takes the value $\kappa_2\,\kappa_1$ for $\kappa = 0$, is negative for $\kappa_1 < \kappa < \kappa_2$ and grows indefinitely for $\kappa > \kappa_2$: one always has a solution for $\kappa > \kappa_2$, whereas the solution for $\kappa < \kappa_1$ exists only in the case the derivative at the origin is positive, as in the figure:

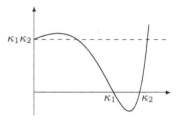

$$4a\,\kappa_2\,\kappa_1 > \kappa_2 + \kappa_1 \quad \Rightarrow \quad \frac{\lambda_1 \lambda_2}{\lambda_1 + \lambda_2} > \frac{\hbar^2}{4ma}.$$

Note that this condition takes, for $\lambda_2 = \lambda_1$, the form found in Problem 6.23 for the existence of the second bound state in the case of the symmetric potential.

c) The two curves reported in the figure below accurately reproduce the two wavefunctions in the case $\kappa_2 a = 1$, $\kappa_1 a = 0.75$: in this case there are two bound states, the first with $\kappa a = 1.04$, the second (the excited state) with $\kappa a = 0.57$.

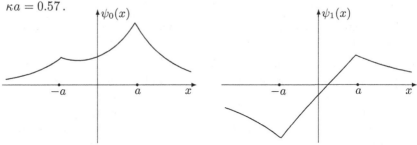

Obviously the two wavefunctions no longer have definite parity: in the ground state (that has a lower energy) the particle is – as expected – mainly concentrated around the more attractive well, while in the excited state it is more concentrated around the less attractive well: this fact can be understood both because the excited state has a greater energy and because its wavefunction must be orthogonal to that of the ground state.

d) The equation for the bound states is obtained from equation (1) by putting $\kappa_1 = \kappa_0$, $\kappa_2 = -\kappa_0$:

$$(\kappa + \kappa_0)(\kappa - \kappa_0) + \kappa_0^2\, e^{-4\kappa a} = 0 \quad \Rightarrow$$

$$\kappa^2 = \kappa_0^2\left(1 - e^{-4\kappa a}\right), \qquad \kappa > 0$$

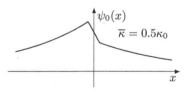

that always admits one and only one solution $\overline{\kappa} < \kappa_0$.

6.26

a) Putting $\varrho = \kappa_0/\kappa$, the transfer matrices $M_-(a)$ and $M_-(-a)$ relative to $-\lambda\,\delta(x - a)$ and $-\lambda\,\delta(x + a)$ are (see Problem 6.20):

$$M_-(a) = \begin{pmatrix} 1 - \varrho & -\varrho\, e^{-2\kappa a} \\ \varrho\, e^{+2\kappa a} & 1 + \varrho \end{pmatrix}; \qquad M_-(-a) = \begin{pmatrix} 1 - \varrho & -\varrho\, e^{+2\kappa a} \\ \varrho\, e^{-2\kappa a} & 1 + \varrho \end{pmatrix}$$

and that relative to the repulsive delta $+\lambda\,\delta(x - b)$ is obtained from $M_-(a)$ by substituting $\varrho \to -\varrho$, $a \to b$:

$$M_+(b) = \begin{pmatrix} 1 + \varrho & \varrho\, e^{-2\kappa b} \\ -\varrho\, e^{+2\kappa b} & 1 - \varrho \end{pmatrix}$$

so that the overall transfer matrix is $M_3 = M_-(a)\, M_+(b)\, M_-(-a)$. The existence of bound states requires (see Problem 6.20) the vanishing of:

$$[M_3]_{11} = [[M_-(a) M_+(b)] \times M_-(-a)]_{11}$$
$$= [M_-(a) M_+(b)]_{11} [M_-(-a)]_{11} + [M_-(a) M_+(b)]_{12} [M_-(-a)]_{21}.$$

Note that both $[M_-(-a)]_{11}$ and

$$[M_-(a) M_+(b)]_{12} = (1 - \varrho)(\varrho e^{-2\kappa b} - \varrho e^{-2\kappa a})$$

contain the factor $1 - \varrho$, so $\kappa = \kappa_0$ is a solution and $E = -\hbar^2 \kappa_0^2 / 2m = -m\lambda^2 / 2\hbar^2$ is one of the eigenvalues of H_3. In order to establish that it is the minimum eigenvalue, i.e. the one corresponding to the maximum κ, one may proceed in two ways:

1st way. Let us explicitly write the condition $[M_3]_{11} = 0$, omitting the factor $1 - \varrho$ relative to the solution $\kappa = \kappa_0$:

$$1 - \varrho^2 (1 - 2e^{-2\kappa a} \cosh(2\kappa b) + e^{-4\kappa a}) = 0 \quad \Rightarrow$$
$$\kappa^2 = \kappa_0^2 (1 - 2e^{-2\kappa a} \cosh(2\kappa b) + e^{-4\kappa a}) < \kappa_0^2 (1 - e^{-2\kappa a})^2 < \kappa_0^2.$$

2nd way. Let us show that the wavefunction corresponding to $\kappa = \kappa_0$ has no zeroes. Let us recall that the notation $\begin{pmatrix} \alpha \\ \beta \end{pmatrix}$ means $\alpha e^{\kappa x} + \beta e^{-\kappa x}$; by replacing $\varrho = 1$ in the transfer matrices one finds, proceeding from the left to the right and denoting by $*$ the irrelevant matrix elements:

$$\begin{cases} \begin{pmatrix} 1 \\ 0 \end{pmatrix} & x \le -a \\[2mm] \begin{pmatrix} 0 & * \\ e^{-2\kappa_0 a} & * \end{pmatrix} \begin{pmatrix} 1 \\ 0 \end{pmatrix} = \begin{pmatrix} 0 \\ e^{-2\kappa_0 a} \end{pmatrix} & -a \le x \le b \\[2mm] \begin{pmatrix} * & e^{-2\kappa_0 b} \\ * & 0 \end{pmatrix} \begin{pmatrix} 0 \\ e^{-2\kappa_0 a} \end{pmatrix} = \begin{pmatrix} e^{-2\kappa_0 (a+b)} \\ 0 \end{pmatrix} & b \le x \le a \\[2mm] \begin{pmatrix} 0 & * \\ e^{+2\kappa_0 a} & * \end{pmatrix} \begin{pmatrix} e^{-2\kappa_0 (a+b)} \\ 0 \end{pmatrix} = \begin{pmatrix} 0 \\ e^{-2\kappa_0 b} \end{pmatrix} & x \ge a \end{cases}$$

which consists of a sequence of alternatively growing and decreasing exponentials, all with positive coefficients: so the wavefunction characterized by $\kappa = \kappa_0$ never vanishes.

b) Multiplying both sides of the equation found above:

$$\kappa^2 = \kappa_0^2 (1 - 2e^{-2\kappa a} \cosh(2\kappa b) + e^{-4\kappa a})$$

by $e^{2\kappa a}$, one obtains:

$$\cosh(2\kappa a) - \cosh(2\kappa b) = \frac{1}{2} \left(\frac{\kappa}{\kappa_0} \right)^2 e^{2\kappa a}.$$

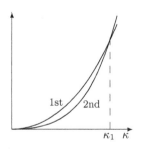

Both curves defined by the functions in the left and right hand sides 1st and 2nd curve respectively) of the above equation, start from the origin, are growing and concave upwards; the 2nd curve is higher than the 1st one for high κ.

The condition that guarantees the existence of one and only one intersection point is that, at $\kappa = 0$, the curvature of the left hand side be greater than that of the right hand side:

$$\frac{1}{2}\left[(2a)^2 - (2b)^2\right] > \frac{1}{2\,\kappa_0^2} \quad \Rightarrow \quad |b| < a\sqrt{1 - \left(\frac{1}{2\kappa_0 a}\right)^2}.$$

In the equation that defines the second eigenvalue and in the condition that guarantees its existence we observe the following features: i) both are even in b;   ii) for $b = 0$ both give results already known for the case of the first excited (odd) state in the case of two attractive symmetric deltas (see Problem 6.23): an odd – therefore vanishing in $x = 0$ – wavefunction is 'insensitive' to the delta in the origin (regardless of its being either attractive or repulsive);   iii) the more $|b|$ grows, the closer to the origin the abscissa κ_1 of the intersection point: correspondingly the higher the energy $E_1 = -\hbar^2 \kappa_1^2/2\,m$. When $|b|$ reaches the critical value $\bar{b}$, the second state is no longer bound; the energy of the ground state is independent of b (and of a too) and, consistently, for $b = \pm a$ the repulsive delta cancels one of the attractive ones and the remaining delta has only one bound state (see Problems 6.18 and 6.21).

c) We have already noted that both $\left[M_-\right]_{11}$ and $\left[M_3\right]_{11}$ contain the factor $1 - \varrho$. Let us assume that this be true for the transfer matrix relative to the first $2j - 1$ delta, with $j \geq 1$, i.e. that:

$$\left[M_{2j-1}\right]_{11} = \left[\left[M_-(a_1)M_+(b_1)\right]\left[M_-(a_2)M_+(b_2)\right]\cdots M_-(a_j)\right]_{11} \equiv (1 - \varrho)X.$$

Adding one more pair of deltas to the right, one has:

$$\left[M_{2j+1}\right]_{11} = \left[M_{2j-1} \times \left[M_+(b_j)M_-(a_{j+1})\right]\right]_{11}$$
$$= (1 - \varrho)\,X\left[M_+(b_j)M_-(a_{j+1})\right]_{11}$$
$$+ \left[M_{2j-1}\right]_{12}(1 - \varrho)\,\varrho\left(e^{2\kappa\,a_{j+1}} - e^{2\kappa\,b_j}\right)$$

that still contains the factor $1 - \varrho$. So $\varrho = 1$, namely $\kappa = \kappa_0$, certainly is a solution of the eigenvalue equation $\left[M_{2n+1}\right]_{11} = 0$. There remains to demonstrate that this solution represents the ground state. One proceeds as in point b), realizing that for $\varrho = 1$ one still has a sequence of alternatively growing and decreasing exponentials of real argument, all with coefficients of the same sign: so the wavefunction belonging to κ_0 has no zeroes.

7

Time Evolution

Time evolution in the Schrödinger and Heisenberg pictures; classical limit; time reversal; interaction picture; sudden and adiabatic approximations.

Note. *Notation:* $|A, t\rangle$ *represents at time t the state that at time $t = 0$ is represented either by the vector $|A, 0\rangle$ or simply $|A\rangle$.*

7.1 Consider a one-dimensional harmonic oscillator, whose angular frequency is ω, and, at time $t = 0$, the states $|A, 0\rangle = a|0\rangle + b\,e^{i\varphi}|1\rangle$ and $|B, 0\rangle = c|0\rangle + d\,e^{i\varphi}|2\rangle$ (a, b, c, d are real numbers $\neq 0$ such that $a^2 + b^2 = c^2 + d^2 = 1$, and $|n\rangle = e^{i\,\alpha_n}\frac{1}{\sqrt{n!}}\,(\eta^\dagger)^n\,|0\rangle$ represent the eigenstates of the Hamiltonian).

a) Show that the evolution of the states $|A, t\rangle$ and $|B, t\rangle$ is periodic. Find the corresponding periods.

b) Choose the phases α_n of the vectors $|n\rangle$ in such a way that the corresponding wavefunctions in the Schrödinger representation are real. Having made this choice, find the time dependence of the mean values of the observables: q, p, the total energy H, the kinetic energy and the potential energy in the states $|A, t\rangle$ and $|B, t\rangle$.

c) Set $a = b$, $\varphi = 0$. Make a qualitative plot of the wavefunctions corresponding to the states $|A, 0\rangle$ and $|A, \tau/2\rangle$, where τ is the period of the state $|A, t\rangle$. (The wavefunctions $\psi_n(x)$ are given in the text of the Problem 5.14, where $H_0(\xi) = 1$, $H_1(\xi) = 2\,\xi$.)

7.2 A particle of mass m is constrained in a segment of length a (infinite rectangular potential well).

a) Find the period of the time evolution of the state $|A, 0\rangle = \alpha|1\rangle + \beta|2\rangle$, where $|n\rangle$ are the eigenstates of the Hamiltonian and (as customary in this case) $|1\rangle$ represents the ground state.

b) Demonstrate that the time evolution of any nonstationary state is periodic.

c) Find the period of the state $|B, 0\rangle = \alpha|n\rangle + \beta|n+1\rangle$, when $n \gg 1$.

© Springer International Publishing AG 2017
E. d'Emilio and L.E. Picasso, *Problems in Quantum Mechanics*,
UNITEXT for Physics, DOI 10.1007/978-3-319-53267-7_7

d) Consider a state whose wavefunction $\psi(x, 0)$ at time $t = 0$ has its support in the left half of the segment of length a. Do instants exist such that the probability of finding the particle in the right half of the segment is equal to 1?

Consider a state superposition of the states $|n\rangle$, with $n \simeq \bar{n} \gg 1$.

e) Show that, under these conditions, the period of the state is given, to a good approximation, by the period of the classical motion of a particle of energy $E_{\bar{n}}$. Estimate $\bar{n}$ when m is the mass of the hydrogen atom, $a = 1\,\mathrm{cm}$, and the particle has an energy of the order of the thermal energy at the room temperature T_0.

7.3 In the case of a particle constrained in a segment (Problem 7.2) it has been shown that the time evolution period of states obtained as superposition of states with large quantum number n: $n \simeq \bar{n} \gg 1$ ("quasi-classical" states of energy $\simeq E_{\bar{n}}$) coincides (to a good approximation) with the period of the classical motion of a particle of energy $E_{\bar{n}}$.

a) Show that, in general, the period of the (quantum) evolution of quasi-classical states with energy $\simeq E_{\bar{n}}$ is given by $\tau = h/(\partial E_n/\partial n)|_{n=\bar{n}}$.

According to Bohr's theory, the energy levels of the hydrogen atom are $E_n = -e^2/(2n^2 a_{\mathrm{B}})$ where $n^2 a_{\mathrm{B}}$ is the radius of the orbit of the electron with energy E_n ($a_{\mathrm{B}} = \hbar^2/m_{\mathrm{e}} e^2$).

b) Verify that the period of the classical motion of the electron along a circular orbit of radius $n^2 a_{\mathrm{B}}$ coincides with the period of the quantum evolution of the quasi-classical states with energy $\simeq E_n$.

Let us consider a particle in one dimension, whose classical motion takes place in a finite region and is periodic. Let E be the energy of the particle and $A(E)$ the area of the surface in the phase space enclosed by the curve defined by the equation $H(q, p) = E$. The period of the classical motion of the particle is given by the relation (derived in the solution) $\tau_{\mathrm{cl}} = \mathrm{d}A(E)/\mathrm{d}E$.

c) Demonstrate the equality between the period of the classical motion and that of the quasi-classical states exploiting the expression for τ_{cl} and Bohr quantization rule (see Problem 2.6).

7.4 The Hamiltonian of a particle of mass m in one dimension is

$$H = \frac{p^2}{2m} + V(x) + V(a - x)$$

where $V(x)$ is an attractive potential with compact support and a is large enough so that $V(x)$ and $V(a - x)$ have disjoint supports (double well potential: see Problem 6.21). Let us assume, in addition, that $V(x)$ has only one bound state whose energy is $E_0 \equiv -\hbar^2 \kappa_0^2/2m$.

a) Show that H commutes with the space inversion operator $I_{\bar{x}}$, the inversion being performed with respect to a suitable point $\bar{x}$. If I_0 is the space inversion operator with respect to the origin ($I_0\, q\, I_0^{-1} = -q$, $I_0\, p\, I_0^{-1} = -p$), how is the operator $I_{\bar{x}}$ expressed in terms of I_0?

If $\kappa_0 a$ is large enough (distant wells) the Hamiltonian H has two bound states with energies E_1, E_2 (see Problem 6.21). Let $E_1 < E_2$.

b) Having suitably chosen the phases of the vectors $|E_1\rangle$ and $|E_2\rangle$ (see Problem 5.5), determine the states $|L\rangle$ and $|R\rangle$, among the superpositions of $|E_1\rangle$ and $|E_2\rangle$, in which the mean value of q is respectively a minimum and a maximum.

At time $t = 0$ the particle is 'localized in the left well', i.e. it is in the state $|L\rangle$ in which the mean value of q is a minimum.

c) Find the state $|L, t\rangle$ at time t, show that the state $|L, t\rangle$ evolves in a periodic way and find its period. Find the instant when, for the first time, the particle is localized in the right well. Does this result have a classical analogue?

d) Let us assume that we do not know $|E_1\rangle$ and $|E_2\rangle$, but that we know the bound state $|E_0, l\rangle$ of the single left well $V(x)$. How is the bound state $|E_0, r\rangle$ of the single right well $V(a - x)$ obtained? What is the best approximation for the eigenstates $|E_1\rangle$ and $|E_2\rangle$ of H in terms of $|E_0, l\rangle$ and $|E_0, r\rangle$?

7.5   Consider a one-dimensional harmonic oscillator of mass m and angular frequency ω that, at time $t = 0$, is in the coherent state (see Problem 5.10):

$$|\alpha, t = 0\rangle \equiv |\alpha\rangle = V(b)\, U(a)\, |0\rangle, \qquad \alpha = \frac{b}{\sqrt{2m\omega\,\hbar}} - i\sqrt{\frac{m\omega}{2\hbar}}\, a$$

($U(a)$ and $V(b)$ respectively are the translation operators for the coordinate and momentum and $|0\rangle$ is the ground state of the oscillator).

a) Find the state $|\alpha, t\rangle$ of the system at time t.

b) Find the mean values $\bar{q}(t)$ and $\bar{p}(t)$ in the state $|\alpha, t\rangle$ making use of both the Schrödinger and the Heisenberg pictures for the time evolution.

c) Show that the uncertainties Δq and Δp do not depend on time. Is this result true regardless of the state of the oscillator at time $t = 0$?

d) Calculate the mean values of kinetic energy and potential energy in the state $|\alpha, t\rangle$.

7.6   A free particle of mass m in one dimension is, at $t = 0$, in the state $|A, 0\rangle$ represented by the wavefunction $\psi_A(x, 0) = e^{-x^2/4a^2}\, e^{i k x}$.

a) Find the condition that a must fulfill so that $\bar{p} \gg \Delta p$. If the particle is a neutron (mass $m = 1.7 \times 10^{-24}$ g) with average kinetic energy $\bar{E} = 1\,\text{MeV}$, find the numerical value of a in the case $\bar{p} = 10^6 \Delta p$.

b) Write the Heisenberg equations for the operators $p(t)$ and $q(t)$ and integrate them with the initial conditions that derive from the choice $U(t = 0) = \mathbb{1}$, $U(t)$ being the time evolution operator.

c) Calculate the width $\Delta q(t)$ of the wave packet at time t.

d) Find the distance d that the neutron must go, starting at $t = 0$, in order that the width Δq be doubled. What is the width of the wave packet after the neutron has gone a distance $L = 1\,\text{m}$?

In order to find the wavefunction at time t when $\bar{p} \gg \Delta p$, one may take advantage of the identity (formally corresponding to the expansion of p^2 around $\bar{p} \times \mathbb{1}$):

$$H = \frac{p^2}{2m} = \frac{\bar{p}^2}{2m} + \frac{\bar{p}}{m}(p - \bar{p}) + \frac{1}{2m}(p - \bar{p})^2 = -\frac{\bar{p}^2}{2m} + \frac{\bar{p}}{m}p + \frac{1}{2m}(p - \bar{p})^2$$

and neglect the last term.

e) If $\psi(x, 0)$ is the wavefunction at $t = 0$, find, in the above mentioned approximation, the wavefunction at time t. What is the difference between the so obtained wavefunction and the exact solution $\psi(x, t)$ of the Schrödinger equation?

7.7   Consider a free particle of mass m in one dimension in the state:

$$|A, t\rangle \xrightarrow{\text{SR}} \psi_A(x, t) = \left(\frac{8a^2}{\pi}\right)^{1/4} \frac{e^{-x^2/(4a^2 + 2\mathrm{i}\hbar t/m)}}{\sqrt{4a^2 + 2\mathrm{i}\hbar t/m}}.$$

a) Verify that $\psi_A(x, t)$ is a solution of the Schrödinger equation:

$$-\frac{\hbar^2}{2m}\psi_A''(x, t) = \mathrm{i}\hbar\frac{\partial \psi_A(x, t)}{\partial t}.$$

b) Find the probability density $\rho_A(x, t)$ and show that $\forall x \; \rho_A(x, t) \to 0$ for $t \to \infty$.

c) Calculate the uncertainties Δq and Δp in the state $|A, t\rangle$.

d) Find the wavefunction $\varphi_A(p', t)$ of the state $|A, t\rangle$ in the momentum representation.

Consider the state $|B, t = 0\rangle \xrightarrow{\text{SR}} \psi_B(x, 0) = \left(2\pi a^2\right)^{-1/4} e^{-x^2/4a^2} e^{\mathrm{i}kx}$.

e) Find $\psi_B(x, t)$ and $\rho_B(x, t)$.

7.8   A particle of mass m in one dimension is subject to a constant external force (electric field, gravity ...), whose potential is $V(q) = -\gamma q$. At time $t = 0$ the particle is in the normalized state $|s\rangle$.

a) Write the Heisenberg equations for the operators $p(t)$ and $q(t)$ and integrate them.

b) Calculate the mean values:

$$\bar{p}(t) = \langle\, s \mid p(t) \mid s \,\rangle, \qquad \bar{q}(t) = \langle\, s \mid q(t) \mid s \,\rangle.$$

c) Calculate the uncertainties $\Delta q(t)$, $\Delta p(t)$ and compare the result with that for the free particle ($\gamma = 0$).

d) Let $\varphi_s(p', 0) = \langle\, p' \mid s, 0 \,\rangle$ be the wavefunction in the momentum representation relative to the state $\mid s \,\rangle$. Find explicitly $\varphi_s(p', t)$. For this purpose exploit the Baker–Campbell–Hausdorff formula:

$$e^A\, e^B = \exp\left(A + B + \tfrac{1}{2}[A,\, B] + \tfrac{1}{12}[A,[A,\, B]] + \tfrac{1}{12}[B,[B,\, A]] \right)$$

that extends the formula given in Problem 4.13 to the case when the commutator $[A,\, B]$ is not a c–number, but such are the double commutators $[[A,\, B],\, A]$, and $[[A,\, B],\, B]$: set $A = (a\,p^2 + b\,p)$, $B = c\,q$ and find a, b, c in such a way that, up to a phase factor, $e^A\, e^B$ coincides with the time evolution operator.

7.9 In the experiment of neutron interferometry described in Problem 3.4 neutrons of wavelength $\lambda = 1.4\,\text{Å}$ are used. They undergo Bragg reflections at an angle of $22°$ off the silicon crystals of the interferometer and the reflected neutrons have an angular dispersion ("Bragg window") $\Delta\theta \simeq 10^{-6}\,\text{rad}$.

a) Find $\Delta\lambda/\lambda$ (use Bragg relation).

Let us assume that the wavefunctions of the neutrons in the path between two consecutive reflections is

$$\psi(x, y, z) = e^{-\beta\,(x^2 + y^2)}\, e^{-\alpha\, z^2}\, e^{i\,k\,z} \qquad \text{with } \beta \ll \alpha.$$

b) Calculate the degree of monochromaticity $\Delta E/E$ of the reflected neutrons and the dimension of the wave packet in the direction of propagation (longitudinal dimension).

The neutrons that have entered the interferometer propagate, both horizontally and vertically, over distances d of the order of 5 cm between two adjacent crystals.

c) Find the longitudinal dimension of the packets after they have gone the distance d.

d) Calculate the amount by which the center of mass of the packet falls down in the horizontal paths ($m_n = 1.7 \times 10^{-24}\text{g}$).

Inside the cathode ray tube of a TV set the electrons are accelerated by means of potential differences of the order of $10^4\,\text{V}$ applied between a grid and the cathode. The electrons pass through the holes in the grid, whose dimensions are of the order of 10^{-2} cm, and go a distance L of the order of 10 cm, before reaching the screen.

e) Give an order of magnitude for the dimension orthogonal to the propagation direction (transverse dimension) of the wave packets arriving at the screen.

7.10 In classical physics the equations of motion for particles subject to forces that are conservative and independent of velocities are invariant under time reversal: this means that, if $x(t)$, $x \equiv (x_1, \cdots, x_n)$ is the solution of the equations of motion with the initial conditions $x(0) = x_0$, $\dot{x}(0) = v_0$, then $\tilde{x}(t) \equiv x(-t)$ is the solution of the equations of motion with initial conditions $\tilde{x}(0) = x_0$, $\dot{\tilde{x}}(0) = -v_0$.

Let $\psi(x, t)$ be a solution of the Schrödinger equation for the time evolution for a particle subject to the potential $V(x)$.

a) Say whether $\psi^*(x, t)$ and/or $\psi(x, -t)$ are solutions of the Schrödinger equation.

b) Verify that $\tilde{\psi}(x, t) \equiv \psi^*(x, -t)$ is a solution of the Schrödinger equation.

c) Let $\bar{q}(t)$, $\bar{p}(t)$ be the mean values of q and p in the state represented by the wavefunction $\psi(x, t)$. Calculate the mean values of q and p in the state represented by the wavefunction $\tilde{\psi}(x, t)$.

d) Is the *time reversal* transformation, that in the Schrödinger picture associates the state $\tilde{\psi}(x, t)$ to the state $\psi(x, t)$, a unitary transformation?

e) Express the time reversal transformation in the momentum representation.

7.11 Let $V(t)$ be a family of unitary operators that depend on time t and let H be the Hamiltonian of the system. To any vector $|A, t\rangle$ that solves the Schrödinger equation:

$$i\hbar \frac{d}{dt} |A, t\rangle = H |A, t\rangle$$

we associate the vector $|\tilde{A}, t\rangle = V^\dagger(t) |A, t\rangle$.

a) Find $\widetilde{H}$ such that the Schrödinger equation for $|\tilde{A}, t\rangle$ reads:

$$i\hbar \frac{d}{dt} |\tilde{A}, t\rangle = \widetilde{H} |\tilde{A}, t\rangle$$

and show that $\widetilde{H} = \widetilde{H}^\dagger$.

b) Write the unitary operator $\widetilde{U}(t)$ that evolves the vectors $|\tilde{A}, t\rangle$ in time:

$$|\tilde{A}, t\rangle = \widetilde{U}(t) |\tilde{A}, 0\rangle$$

and show that it satisfies the equation:

$$i\hbar\,\frac{d}{dt}\,\widetilde{U}(t) = \widetilde{H}\,\widetilde{U}(t)\,.$$

Let now $H(t) = H_0 + H'(t)$ with H_0 independent of time ('free Hamiltonian'). The picture for time evolution that is obtained by setting $V(t) = U_0(t) \equiv e^{-i\,H_0\,t/\hbar}$, is called *interaction picture* and $H'(t)$ the *interaction Hamiltonian*.

c) Find $\widetilde{H}$. If $H_0 = p^2/2m$ and $H' \equiv H'(q,\,p,\,t)$, explicitly write $\widetilde{H}$.

In the interaction picture the evolution of states is given by:

$$|\,\tilde{A},\,t\,\rangle = \widetilde{U}(t)\,|\,\tilde{A},\,0\,\rangle = \widetilde{U}(t)\,|\,A,\,0\,\rangle\,.$$

d) If the states evolve as in the interaction picture, how must the observables evolve in order to obtain the correct evolution of the mean values: $\overline{\xi}(t) = \langle\,A,\,t\,|\,\xi\,|\,A,\,t\,\rangle$?

7.12   The Hamiltonian of a particle in one dimension is $H = p^2/2m - \gamma\,q$.

a) Determine $\widetilde{H}$ in the interaction picture (see Problem 7.11).

b) Write $\widetilde{U}(t)$ in the form:

$$\widetilde{U}(t) = e^{i\gamma p t^2/2m\hbar}\,e^{i\gamma q t/\hbar}\,e^{-i\alpha(t)/\hbar}$$

and find $\alpha(t)$.

c) Write the time evolution operator $U(t)$ as $U_0(t)\,\widetilde{U}(t)$ and compare it with that found in Problem 7.8.

7.13   A system endowed with a magnetic moment, subject to both a static magnetic field $\vec{B}_0$ and to a magnetic field rotating in the plane orthogonal to $\vec{B}_0$, may, in some cases (spin $\frac{1}{2}$ system), be described as a system that has only two independent states (*two-level system*) whose Hamiltonian is

$$H(t) = H_0 + H_1(t) = \begin{pmatrix} \frac{1}{2}E_0 & 0 \\ 0 & -\frac{1}{2}E_0 \end{pmatrix} + \begin{pmatrix} 0 & a\,e^{-i\,\omega_0\,t} \\ a\,e^{i\,\omega_0\,t} & 0 \end{pmatrix}\,.$$

As the Hamiltonian depends on time and $[H(t_1),\,H(t_2)] \neq 0$, the time evolution operator $U(t) \neq \exp(-i\int_0^t H(t')\,dt')$. Therefore we take advantage of the method introduced in Problem 7.11: to any vector $|\,A,\,t\,\rangle$ that solves the Schrödinger equation:

$$i\hbar\,\frac{d}{dt}\,|\,A,\,t\,\rangle = H(t)\,|\,A,\,t\,\rangle$$

we associate the vector $|\,\tilde{A},\,t\,\rangle = V^\dagger(t)\,|\,A,\,t\,\rangle$, where:

$$V(t) = \begin{pmatrix} e^{-i\,\omega_0\,t/2} & 0 \\ 0 & e^{i\,\omega_0\,t/2} \end{pmatrix}\,.$$

a) Find $\widetilde{H}$ such that $i\hbar \dfrac{d}{dt}|\tilde{A}, t\rangle = \widetilde{H}|\tilde{A}, t\rangle$.

b) Show that if $M \equiv \alpha\, \Sigma$, M and Σ being 2×2 matrices with α real and $\Sigma^2 = \mathbb{1}$, then:

$$e^{iM} \equiv \sum_n \frac{(i\alpha\,\Sigma)^n}{n!} = \mathbb{1} \times \cos\alpha + i\,\Sigma \times \sin\alpha \equiv \cos\alpha + i\,\Sigma\,\sin\alpha \ .$$

Exploit this result to find the unitary operator $\widetilde{U}(t)$ that evolves in time the vectors $|\tilde{A}, t\rangle$: $|\tilde{A}, t\rangle = \widetilde{U}(t)|\tilde{A}, 0\rangle$.

Let $|-\rangle$ and $|+\rangle$ respectively be the eigenstates of H_0 corresponding to the eigenvalues $-\frac{1}{2}E_0$, $+\frac{1}{2}E_0$ and assume that at $t = 0$ the system is in the state $|-\rangle$.

c) Calculate the probability $P_{+-}(t)$ to find the system in the state $|+\rangle$ for $t > 0$. Find the maximum of $P_{+-}(t)$. Under which conditions does such maximum equal 1 ?

7.14   A particle of mass m in one dimension is subject to the impulsive force $f(t) = \gamma\,\delta(t)$, where $\delta(t)$ is the Dirac delta function.

a) Write the Heisenberg equations for the operators $p(t)$ and $q(t)$ and integrate them with the initial condition $U(t = 0^-) = \mathbb{1}$.

b) Find the time evolution operator from the instant $t = 0^-$ to the instant $t = 0^+$.

c) Assume that $\psi(x, 0^-)$ is the wavefunction of the particle at time $t = 0^-$. Find the wavefunction $\psi(x, 0^+)$.

7.15   This problem is preparatory to Problem 7.16.   Let us consider a *classical* one-dimensional harmonic oscillator of mass m and angular frequency ω, subject to the impulsive force $f(t) = \gamma\,\delta(t)$. For $t < 0$ the oscillator is at rest.

a) Write the equation of motion for the oscillator and integrate it.

b) Find the energy of the oscillator for $t > 0$.

Let us now assume that the oscillator, at rest for $t < 0$, is subject to an external (nonimpulsive) force $F(t)$ nonvanishing only for $0 < t < \tau$. Let us denote by $D_r(t)$ the solution found in a) in the case $\gamma = 1$:

$$m\left(\frac{d^2}{dt^2} + \omega^2\right)D_r(t) = \delta(t) \ .$$

c) Verify that:

$$x(t) = \int_0^\infty D_r(t - t')\, F(t')\, dt'$$

is the solution of the equation of motion in presence of the force $F(t)$ corresponding to the given initial conditions.

d) Write the explicit form of the *retarded Green function* $D_r(t - t')$ and of its derivative $\dot{D}_r(t - t')$.

e) Write the solution of the equation of motion with generic initial conditions $x(0) = x_0$, $\dot{x}(0) = v_0$ (namely the general solution).

7.16 A one-dimensional harmonic oscillator of mass m and angular frequency ω is in the ground state $|0\rangle$. At time $t = 0$ the oscillator is subject to the impulsive force $f(t) = \gamma \delta(t)$.

a) Find the state of the oscillator for $t > 0$.

b) Calculate, for $t > 0$, the mean value of the energy and the probability of finding the energy eigenvalue E_n.

Assume now that the oscillator is subject to an external nonimpulsive force $F(t)$ such that $F(t) = 0$ for $t < 0$ and $t > \tau$. Again, for $t < 0$ the oscillator is in the ground state.

c) Write the Heisenberg equations for $q(t)$ and $p(t)$ and integrate them: since they are nonhomogeneous linear equations, take as a particular solution the c–number solution given in the text of Problem 7.15.

d) Calculate $\Delta q(t)$, $\Delta p(t)$ and their product. Exploit the result to show that for every t the state of the oscillator is a coherent state $|\alpha(t)\rangle$ (see Problem 7.5) and find the expression for $\alpha(t)$ (see Problem 5.10).

e) Calculate, for $t > \tau$, the mean value of the energy in the case $F(t) = F_0 > 0$ for $0 < t < \tau$.

7.17 A one-dimensional harmonic oscillator of mass m and angular frequency ω is, for $t \leq 0$, in the ground state $|0\rangle$. Starting at $t = 0$ the center of oscillation is moved according to the time law $\xi(t)$, where $\xi(t) = 0$ for $t \leq 0$, $\xi(t) = \xi_0$ for $t \geq \tau$.

The purpose of the problem is to study the time evolution of the state of the oscillator in the two extremal cases when the center of oscillation is displaced in a time interval either very short or very long with respect to the time characteristic of the oscillator, namely its period $2\pi/\omega$.

a) Write the Hamiltonian $H(t)$ of the oscillator, say which are its eigenvalues, write the wavefunction $\psi_0^{(t)}(x, t)$ corresponding to the minimum eigenvalue of $H(t)$. Say whether $\psi_0^{(t)}(x, t)$ is a solution of the time dependent Schrödinger equation:

$$H\left(-i\hbar\frac{\partial}{\partial x}, x, t\right)\psi_0^{(t)}(x, t) = i\hbar\frac{\partial}{\partial t}\psi_0^{(t)}(x, t).$$

b) Integrate the Heisenberg equations for $q(t)$ and $p(t)$ and exploit the result of Problem 7.16 to find the state $|\alpha(t)\rangle$ of the oscillator for $t \geq 0$.

Let us assume that the center of oscillation is displaced from $x = 0$ to $x = \xi_0$ in a time interval $\tau \ll \omega^{-1}$, so that $\xi(t)$ may be approximated with the step function:

$$\xi(t) = \begin{cases} 0 & t \leq 0 \\ \xi_0 & t > 0 \end{cases}$$

(*sudden approximation*).

c) Find the state $|\alpha(t)\rangle$ of the oscillator for $t \geq 0$ and show that the time evolution of the state is continuous even for $t = 0$. Calculate $\bar{q}(t)$ for $t \geq 0$ and show that it oscillates harmonically around ξ_0.

We now want to study the time evolution of the system in the opposite case when the function $\xi(t)$ varies very slowly (*adiabatic approximation*): let us assume that $\xi(t) = f(t/\tau)$, f being a continuous and differentiable function such that $f(0) = 0$, $f(1) = \xi_0$.

d) Show that in the limit $\tau \to \infty$ ($\xi(t)$ switches from 0 to ξ_0 in an infinite time) the state $|\alpha(t)\rangle$ of the oscillator coincides at any time with the ground state of $H(t)$ (a suggestion useful to this purpose: before taking the limit $\tau \to \infty$, express $\alpha(t)$ in terms of the derivative f' of the function f by performing a suitable integration by parts).

e) Verify in addition that, in the above limit, the wavefunction $\psi_0^{(t)}(x, t)$ is a solution of the time dependent Schrödinger equation.

Solutions

7.1

a) $|A,t\rangle = a\,e^{-i\frac{1}{2}\omega t}|0\rangle + b\,e^{i\varphi}e^{-i\frac{3}{2}\omega t}|1\rangle$

$|B,t\rangle = c\,e^{-i\frac{1}{2}\omega t}|0\rangle + d\,e^{i\varphi}e^{-i\frac{5}{2}\omega t}|2\rangle$.

The condition of periodicity is that, for the first time after the time τ, the *state* (not necessarily the vector) becomes equal to the state at time $t = 0$: in the case of the state $|A,t\rangle$ this requires that:

$$e^{-i\frac{1}{2}\omega\tau} = e^{-i\frac{3}{2}\omega\tau} \quad\Rightarrow\quad e^{-i\omega\tau} = 1 \;\Rightarrow\; \tau = \frac{2\pi}{\omega}$$

whereas in the case of the state $|B,t\rangle$:

$$e^{-i\frac{1}{2}\omega\tau} = e^{-i\frac{5}{2}\omega\tau} \quad\Rightarrow\quad e^{-2i\omega\tau} = 1 \;\Rightarrow\; \tau = \frac{\pi}{\omega}$$

so the period of the state $|A,t\rangle$ is equal to that of the classical oscillator, while the period of the state $|B,t\rangle$ is half of the classical one.
Note that $|A,\tau\rangle = -|A,0\rangle$, $\quad|B,\tau\rangle = -i\,|B,0\rangle$ so, if mistakenly the periodicity of the vector were required, in the first case the result would be wrong by a factor 2: $|A,2\tau\rangle = |A,0\rangle$, in the second by a factor 4.

b) A choice of the phases, that makes the wavefunctions of the stationary states in the Schrödinger representation real, is

$$|n\rangle = \frac{(-i)^n}{\sqrt{n!}}(\eta^\dagger)^n|0\rangle$$

which is different by a factor $(-i)^n$ from that used in Problems 4.14 and 4.15; it gives rise to the wavefunctions reported in the text of Problem 5.14. With the above choice of the phases and modifying accordingly the results of Problems 4.14 and 4.15, one has:

$$\langle 0\,|\,q\,|\,1\rangle = \sqrt{\frac{\hbar}{2m\omega}} = \langle 1\,|\,q\,|\,0\rangle; \quad \langle 0\,|\,p\,|\,1\rangle = -i\sqrt{\frac{m\omega\,\hbar}{2}} = -\langle 1\,|\,p\,|\,0\rangle$$

$$\langle 0 \mid q^2 \mid 2 \rangle = \frac{\sqrt{2}}{2} \frac{\hbar}{m\omega} ; \qquad \langle 0 \mid p^2 \mid 2 \rangle = -\frac{\sqrt{2}}{2} m\omega \hbar$$

the last following from $\langle 0 \mid H \mid 2 \rangle = 0$. In addition, either by direct calculation or by exploiting the selection rule on the space inversion (see Problem 6.1):

$$\langle 0 \mid p \mid 2 \rangle = \langle 0 \mid q \mid 2 \rangle = \langle 0 \mid p^2 \mid 1 \rangle = \langle 0 \mid q^2 \mid 1 \rangle = 0 .$$

In the state $\mid A, t \rangle$:

$$\bar{q}(t) = 2ab \cos(\varphi - \omega t) \langle 0 \mid q \mid 1 \rangle = 2ab \sqrt{\frac{\hbar}{2m\omega}} \cos(\varphi - \omega t)$$

$$\bar{p}(t) = 2ab \sqrt{\frac{m\omega \hbar}{2}} \sin(\varphi - \omega t) .$$

The mean value of the total energy H does not depend on time (H is a constant of motion!); the mean values of the kinetic energy T and of the potential V do not depend on time because their matrix elements between $\mid 0 \rangle$ and $\mid 1 \rangle$ are vanishing.
In the state $\mid B, t \rangle$ the mean values of q and p and, of course, of the Hamiltonian H do not depend on time, whereas:

$$\bar{T}(t) = c^2 \langle 0 \mid T \mid 0 \rangle + d^2 \langle 2 \mid T \mid 2 \rangle + 2cd \cos(\varphi - 2\omega t) \langle 0 \mid T \mid 2 \rangle$$

$$= \hbar\omega \left(\frac{1}{4}c^2 + \frac{5}{4}d^2 - \frac{\sqrt{2}}{2} cd \cos(\varphi - 2\omega t) \right)$$

$$\bar{V}(t) = \bar{H} - \bar{T}(t) = \hbar\omega \left(\frac{1}{4}c^2 + \frac{5}{4}d^2 + \frac{\sqrt{2}}{2} cd \cos(\varphi - 2\omega t) \right) .$$

c) $\psi_A(x,0) = \frac{1}{\sqrt{2}} (\psi_0(x) + \psi_1(x)) \propto (1 + \sqrt{2}\,\xi) e^{-\xi^2/2} , \qquad \xi = \sqrt{m\omega/\hbar}\, x$

$$\psi_A(x,\tau/2) = \frac{1}{\sqrt{2}} \left(e^{-i\pi/2}\, \psi_0(x) + e^{-3i\pi/2}\psi_1(x) \right) \propto \frac{1}{\sqrt{2}} (\psi_0(x) - \psi_1(x))$$

$$= \frac{1}{\sqrt{2}} (\psi_0(x) + \psi_1(-x)) = \psi_A(-x,0) \propto (1 - \sqrt{2}\,\xi) e^{-\xi^2/2} .$$

The state with wavefunction $\psi_A(x,0)$ is, among the states that are superpositions of $\mid 0 \rangle$ and $\mid 1 \rangle$, the one in which the mean value of q is a maximum (see Problem 4.14, having in mind the different convention for the vectors $\mid n \rangle$); after half a period the state is obtained by space inversion from the state at time $t = 0$, therefore $\bar{q}(\tau/2) = -\bar{q}(0)$.

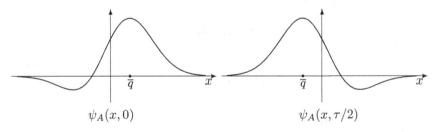

$$\psi_A(x,0) \qquad\qquad\qquad\qquad \psi_A(x,\tau/2)$$

7.2

a) The eigenvalues of the energy of a particle confined in a segment of length
a (particle in an infinite square well) are $E_n = n^2 E_1$, $E_1 = h^2/8ma^2$.
The period of the *state* (not of the vector!) is

$$\tau = \frac{h}{E_2 - E_1} = \frac{h}{3E_1} = \frac{8ma^2}{3h}.$$

b) Let

$$|A,t\rangle = \sum_n a_n\, e^{-i\, E_n\, t/\hbar}\, |n\rangle = e^{-i\, E_{n_0}\, t/\hbar} \sum_n a_n\, e^{-i\, (n^2 - n_0^2) E_1\, t/\hbar}\, |n\rangle$$

where n_0 is an arbitrary value of n such that $a_{n_0} \neq 0$. Since the factor
$e^{-i\, E_{n_0}\, t/\hbar}$ can be omitted and, as a consequence, the coefficient of $|n_0\rangle$
does no longer depend on t, the condition $e^{-i\, (n^2 - n_0^2) E_1 \tau/\hbar} = 1$ for any
n must hold and this, *in general*, is possible only if $\tau = h/E_1$: in some
particular cases (depending on which a_n are nonvanishing) the period may
be a submultiple of h/E_1.   For example, the state:

$$|A,t\rangle = \alpha\, e^{-i\, E_1\, t/\hbar}\, |1\rangle + \beta\, e^{-i\, E_2\, t/\hbar}\, |2\rangle + \gamma\, e^{-i\, E_3\, t/\hbar}\, |3\rangle$$

has a period h/E_1 while, as seen in a), the state $|A,0\rangle = \alpha\,|1\rangle + \beta\,|2\rangle$
evolves with a period three times smaller.

c) The period of the state $|B,0\rangle = \alpha\,|n\rangle + \beta\,|n+1\rangle$ is

$$\tau = \frac{h}{E_{n+1} - E_n} = \frac{h}{(2n+1)E_1} \simeq \frac{h}{2n\, E_1}.$$

d) Let us take the origin of the x axis in the center of the segment and let
$\psi_n(x)$ be the eigenfunctions of the Hamiltonian: $\psi_n(-x) = (-1)^n \psi_n(x)$.
One has:

$$\psi(x,\,0) = \sum_n c_n \psi_n(x) \quad \Rightarrow$$

$$\psi(x,\,t) = \sum_n c_n\, e^{-i\, E_n t/\hbar}\, \psi_n(x) = \sum_n c_n\, e^{-i\, E_1 n^2 t/\hbar}\, \psi_n(x)\,.$$

The state is periodic with period $\tau = h/E_1$; after half a period:

$$\psi(x,\,\tau/2) = \sum_n c_n\, e^{-i\pi n^2}\, \psi_n(x) = \sum_n c_n\, (-1)^n\, \psi_n(x)$$

$$= \sum_n c_n \psi_n(-x) = \psi(-x,\,0)$$

i.e. the particle is in the right half of the segment. So, for $t = (k + \frac{1}{2})\tau$
(k integer) the probability of finding the particle in the right half of the
segment is 1.

e) $\quad |A,t\rangle = e^{-i\, \bar{n}^2\, E_1\, t/\hbar} \sum_{n \simeq \bar{n}} a_n\, e^{-i\, (n^2 - \bar{n}^2) E_1\, t/\hbar}\, |n\rangle\,.$

If $n \simeq \bar{n}$ and $\bar{n} \gg 1$, all the differences $n^2 - \bar{n}^2$ are, to a good approximation, integer multiples of $2\bar{n}$: indeed $n^2 - \bar{n}^2 = (n+\bar{n})(n-\bar{n}) \simeq 2\bar{n}\,(n-\bar{n})$.
Therefore the period is $\tau \simeq h/2\bar{n}\, E_1$.   The classical period is

$$\tau_{\rm cl} = 2a/v = \sqrt{\frac{2ma^2}{E_{\bar n}}} = \sqrt{\frac{2ma^2}{\bar n^2 E_1}} = \sqrt{\frac{h^2}{4\,\bar n^2 E_1^2}} = \frac{h}{2\,\bar n\, E_1} \; .$$

$$E_{\bar n} \simeq k_{\rm B} T_0 \simeq 0.025\,{\rm eV}\,; \quad E_1 \simeq 2 \times 10^{-18}\,{\rm eV} \quad \Rightarrow \quad \bar n \simeq 1.1 \times 10^8 .$$

7.3

a) Let

$$|A,t\,\rangle = \sum_{n \simeq \bar n} a_n\, {\rm e}^{-{\rm i}\,E_n\, t/\hbar}\,|\,n\,\rangle = {\rm e}^{-{\rm i}\,E_{\bar n}\, t/\hbar} \sum_{n \simeq \bar n} a_n\, {\rm e}^{-{\rm i}\,(E_n - E_{\bar n})\, t/\hbar}\,|\,n\,\rangle$$

(the factor ${\rm e}^{-{\rm i}\,E_{\bar n}\, t/\hbar}$ is irrelevant). Since $n \simeq \bar n \gg 1$:

$$E_n - E_{\bar n} \simeq \left(\frac{\partial E_n}{\partial n}\right)_{n=\bar n} \times (n - \bar n) \quad \Rightarrow \quad \tau \simeq \frac{h}{(\partial E_n/\partial n)_{n=\bar n}} \; .$$

b) If v is the velocity of the electron in the circular orbit of radius $n^2 a_{\rm B}$,

$$\tau_{\rm cl} = \frac{2\pi\, n^2 a_{\rm B}}{v} = \frac{2\pi\, n^2 a_{\rm B}}{\sqrt{e^2/(m_{\rm e}\, n^2 a_{\rm B})}} = \frac{2\pi\, n^2 a_{\rm B}}{\sqrt{e^4/n^2 \hbar^2}} = \frac{2\pi\, \hbar\, n^3 a_{\rm B}}{e^2} = \frac{h}{\partial E_n/\partial n} \; .$$

c) We have seen above that the period of the (quantum) evolution of the quasi–classical states with energy $\simeq E_n$ is $\tau = h/(\partial E_n/\partial n)$. Thanks to Problem 2.6, we know that the curve $\gamma(E_n)$ in the phase space, on which $H(q,p) = E_n$, encloses the area:

$$A_n = \oint_{\gamma(E_n)} p\, {\rm d}q$$

then, due to the Bohr–Sommerfeld quantization rule $A_n = n\,h$, and if $\Delta E_n \equiv E_{n+1} - E_n \ll E_n$, one has:

$$\tau_{\rm cl} = \frac{{\rm d}A(E)}{{\rm d}E} \simeq \frac{A(E_{n+1}) - A(E_n)}{E_{n+1} - E_n} = \frac{h}{E_{n+1} - E_n} \simeq \frac{h}{\partial E_n/\partial n} \; .$$

Let us demonstrate the relation $\tau_{\rm cl} = {\rm d}A(E)/{\rm d}E$.

Let $\gamma(E)$ and $\gamma(E + \Delta E)$ be the two curves in the phase space on which respectively $H(q,p) = E$ and $H(q,p) = E + \Delta E$ (ΔE infinitesimal). The distance between the two curves is $\Delta E/|\nabla H|$, where ∇H is the gradient of $H(q,p)$, namely the vector whose components are $(\partial H/\partial q,\ \partial H/\partial p)$; then, if ${\rm d}\vec\gamma$ is an element of line on $\gamma(E)$:

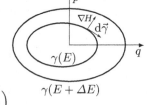

$${\rm d}\vec\gamma = (\dot q\, {\rm d}t,\ \dot p\, {\rm d}t) = \left(\frac{\partial H}{\partial p}\, {\rm d}t,\ -\frac{\partial H}{\partial q}\, {\rm d}t\right),$$

the oriented area enclosed by the two curves is given by:

$$A(E + \Delta E) - A(E) = \oint_{\gamma(E)} \frac{\Delta E}{|\nabla H|}\left({\rm d}\vec\gamma \wedge \frac{\nabla H}{|\nabla H|}\right)$$

(in two dimensions the external product between two vectors $\vec u \wedge \vec v \equiv u_1 v_2 - u_2 v_1$ is a scalar); in addition ${\rm d}\vec\gamma \wedge \nabla H = |\nabla H|^2\, {\rm d}t$, so:

$$A(E + \Delta E) - A(E) = \Delta E \times \int_{\gamma(E)} dt = \tau_{\text{cl}} \, \Delta E \quad \Rightarrow \quad \tau_{\text{cl}} = \frac{dA(E)}{dE} \; .$$

7.4

a) The Hamiltonian is invariant under the space inversion with respect to the point $\bar{x} = a/2$:

$$\begin{cases} (q - \tfrac{1}{2}a) \to -(q - \tfrac{1}{2}a) \\ p \to -p \end{cases} \quad \Leftrightarrow \quad \begin{cases} q \to -q + a \\ p \to -p \end{cases} \quad \Rightarrow$$

$$V(x) + V(a - x) \to V(a - x) + V(x) \; .$$

Let $U(a) = e^{-i p a/\hbar}$ denote the translation operator: $U(-a) \, q \, U^{-1}(-a) = q + a$; then one has $I_{\bar{x}} = I_0 \, U(-a)$ $\left(\text{or also } I_{\bar{x}} = U(a) \, I_0 \text{ as well as other equivalent expressions} \right)$, indeed:

$$I_0 \, U(-a) \, q \, U^{-1}(-a) \, I_0^{-1} = I_0 \, (q + a) \, I_0^{-1} = -q + a \, ,$$

$$I_0 \, U(-a) \, p \, U^{-1}(-a) \, I_0^{-1} = I_0 \, p \, I_0^{-1} = -p \; .$$

b) The states $| E_1 \rangle$ (ground) and $| E_2 \rangle$ (the first excited) respectively are even and odd under $I_{\bar{x}}$. Putting $\tilde{q} = q - \tfrac{1}{2}a$, one has $\langle E_1 | \tilde{q} | E_1 \rangle = \langle E_2 | \tilde{q} | E_2 \rangle = 0$ (selection rule on parity: see Problem 6.1), then the mean value of $\tilde{q}$ in the state $\alpha | E_1 \rangle + \beta | E_2 \rangle$ $\left(|\alpha|^2 + |\beta|^2 = 1 \right)$ has the value $2 \, \mathfrak{Re} \left(\alpha^* \beta \langle E_1 | \tilde{q} | E_2 \rangle \right)$. Since, in analogy with Problem 5.5, we are allowed to choose the phases of $| E_1 \rangle$ and $| E_2 \rangle$ in such a way that $\langle E_1 | \tilde{q} | E_2 \rangle$, if nonvanishing (see below), is real and positive, the mean value of $\tilde{q}$, as well as of q, is either a maximum or a minimum respectively for $\alpha = \pm \beta = 1/\sqrt{2}$. Then:

$$| L \rangle = \frac{1}{\sqrt{2}} \big(| E_1 \rangle - | E_2 \rangle \big), \qquad \langle L | q | L \rangle = \frac{a}{2} - \langle E_1 | \tilde{q} | E_2 \rangle$$

$$| R \rangle = \frac{1}{\sqrt{2}} \big(| E_1 \rangle + | E_2 \rangle \big), \qquad \langle R | q | R \rangle = \frac{a}{2} + \langle E_1 | \tilde{q} | E_2 \rangle \; .$$

The matrix element $\langle E_1 | \tilde{q} | E_2 \rangle$ is nonvanishing: indeed $\psi_{E_1}(x)$ is real and has no zeroes, i.e. it has a constant sign, $\psi_{E_2}(x)$ has its only zero in $x = \tfrac{1}{2}a$, so also the product $\psi_{E_1}(x) \, (x - \tfrac{1}{2}a) \, \psi_{E_2}(x)$ has constant sign and its integral is nonvanishing.

c) $| L, t \rangle = \dfrac{1}{\sqrt{2}} \, e^{-i \, E_1 t/\hbar} \big(| E_1 \rangle - e^{-i \, (E_2 - E_1) t/\hbar} \, | E_2 \rangle \big)$

so the period of the state is $\tau = h/(E_2 - E_1)$; after half a period the particle is in the state $| R \rangle$ and then it keeps on oscillating between the two wells. The result has no classical analogue: a particle localized in one of the two wells has a negative energy and, as a consequence, cannot cross the classically forbidden region that separates them. When $\kappa_0 \, a \to \infty$, $\tau \propto e^{\kappa_0 a} \to \infty$ (see Problem 6.21).

d) $| E_0, r \rangle = I_{\bar{x}} | E_0, l \rangle$.

The state $|E_1\rangle$ is even under $I_{\bar{x}}$, whereas $|E_2\rangle$ is odd; in the subspace generated by $|E_0,l\rangle$ and $|E_0,r\rangle$, the only states that are respectively even and odd are:

$$N_{\pm}\big(|E_0,l\rangle \pm |E_0,r\rangle\big);\qquad N_{\pm} = \big(2 \pm 2\langle E_0,l\,|\,E_0,r\rangle\big)^{-1/2}$$

with $N_{\pm}$ standing for normalization factors. In the limit $a \to \infty$, $E_1 = E_2 = E_0$ (see Problem 6.21), whence all the linear combinations of $|E_0,l\rangle$ and $|E_0,r\rangle$ are exact eigenstates of H, i.e. stationary states.

7.5

a) In Problem 5.9 we have established that $\langle n\,|\,\alpha\rangle = \dfrac{\alpha^n}{\sqrt{n!}}\,e^{-\frac{1}{2}|\alpha|^2}$ so, apart from the irrelevant phase factor $e^{-i\frac{1}{2}\omega t}$,

$$|\alpha,t\rangle = e^{-\frac{1}{2}|\alpha|^2}\sum_n \frac{\alpha^n}{\sqrt{n!}}\,e^{-in\omega t}\,|n\rangle$$

$$= e^{-\frac{1}{2}|\alpha|^2}\sum_n \frac{\big(\alpha\,e^{-i\omega t}\big)^n}{\sqrt{n!}}\,|n\rangle = |\alpha\,e^{-i\omega t}\rangle .$$

As a consequence, the coherent states remain such during the time evolution: the parameter $\alpha(t) \equiv \alpha\,e^{-i\omega t}$ uniformly moves clockwise on the circumference of radius $|\alpha|$ in the complex plane.

An equivalent way to express the above result is the following: for a classical oscillator, interpret the phase space of the rescaled variables $\alpha_1 \equiv p/\sqrt{2m\omega\hbar}$ and $\alpha_2 \equiv -\sqrt{m\omega/2\hbar}\,q$ as the Gauss plane of the complex variable $\alpha = \alpha_1 + i\alpha_2$ and to any classical state, associated to the (complex) coordinate α, associate the (quantum) coherent state $|\alpha\rangle$: then one has the commutative diagram represented in the figure, where in the horizontal lines one has the association classical state $\to$ coherent state and the vertical lines represent the time evolution.

b) Since (see Problem 5.9) for a coherent state $|\beta\rangle$:

$$\langle\beta\,|\,p\,|\,\beta\rangle = \sqrt{2m\omega\hbar}\;\Re e\,\beta,\qquad \langle\beta\,|\,q\,|\,\beta\rangle = -\sqrt{2\hbar/m\omega}\;\Im m\,\beta,$$

in the Schrödinger picture one has:

$$\bar{p}(t) = \langle\alpha,t\,|\,p\,|\,\alpha,t\rangle = \sqrt{2m\omega\hbar}\;\Re e\big(\alpha\,e^{-i\omega t}\big) = b\cos\omega t - m\omega\,a\sin\omega t$$

$$\bar{q}(t) = \langle\alpha,t\,|\,q\,|\,\alpha,t\rangle = -\sqrt{\frac{2\hbar}{m\omega}}\;\Im m\big(\alpha\,e^{-i\omega t}\big) = a\cos\omega t + \frac{b}{m\omega}\sin\omega t .$$

In the Heisenberg picture, where $\overline{\xi(t)} = \langle A,0\,|\,\xi(t)\,|\,A,0\rangle \equiv \bar{\xi}(t)$,

$$p(t) = p\cos\omega t - m\omega\,q\sin\omega t,\qquad q(t) = q\cos\omega t + \frac{p}{m\omega}\sin\omega t \quad \Rightarrow$$

$$\bar{p}(t) = \langle\alpha,0\,|\,p(t)\,|\,\alpha,0\rangle = \bar{p}\cos\omega t - m\omega\,\bar{q}\sin\omega t$$

$$\bar{q}(t) = \langle\alpha,0\,|\,q(t)\,|\,\alpha,0\rangle = \bar{q}\cos\omega t + \frac{\bar{p}}{m\omega}\sin\omega t$$

and, due to $\bar{p} \equiv \langle \alpha, 0 \mid p \mid \alpha, 0 \rangle = b$, $\bar{q} \equiv \langle \alpha, 0 \mid q \mid \alpha, 0 \rangle = a$, the previous result is recovered.

The coherent states are therefore the best candidates to represent the states of a classical oscillator through the identification $q_{cl} = \bar{q}$, $p_{cl} = \bar{p}$, $E_{cl} = \overline{E} \equiv \overline{H}$, and vice versa: indeed, in the first place, since they are states of minimum uncertainty, they satisfy the requisite that the fluctuations of the values that q and p may take around their mean values, are 'small' (and, as we shall see in the next point, this remains true even in the course of the time evolution); in the second place, since for the classical oscillator $E_{cl} \gg \hbar \omega$ and from Problem 5.10 we know that for the coherent states $\Delta E \propto \sqrt{\bar{n}}$, for the quasi-classical states one has $\Delta E / \overline{E} = 1/\sqrt{\bar{n}} \approx 0$; finally, both the classical and the quantum schemes provide the same results for the time evolution: $q_{cl}(t) = \bar{q}(t)$, $p_{cl}(t) = \bar{p}(t)$.

c) From Problem 5.10 we know that in any coherent state, and in particular in the state $\mid \alpha, t \rangle$, for any t:

$$\Delta p = \sqrt{\frac{m \omega \hbar}{2}}, \qquad \Delta q = \sqrt{\frac{\hbar}{2 m \omega}}.$$

The independence of Δq and Δp of time, obviously true also for the stationary states, is not true e.g. for the state $\mid 0 \rangle + \mid 1 \rangle$ (see Problem 7.1).

d) In a coherent state $\overline{p^2}/2m \equiv (\Delta p)^2/2m + \bar{p}^2/2m = \frac{1}{4} \hbar \omega + \bar{p}^2/2m$, whence:

$$\langle \alpha, t \mid \frac{p^2}{2m} \mid \alpha, t \rangle = \frac{1}{4} \hbar \omega + \frac{1}{2m}(b \cos \omega t - m \omega a \sin \omega t)^2$$

$$\langle \alpha, t \mid \frac{1}{2} m \omega^2 q^2 \mid \alpha, t \rangle = \frac{1}{4} \hbar \omega + \frac{1}{2} m \omega^2 (a \cos \omega t + \frac{b}{m \omega} \sin \omega t)^2.$$

7.6

a) The state $\mid A, 0 \rangle$ is a minimum uncertainty state (it is a coherent state, see Problem 5.9) and $\Delta q = a$, therefore $\Delta p = \hbar/2a$ and, in addition, $\bar{p} = \hbar k$ so that:

$$\bar{p} \gg \Delta p \quad \Rightarrow \quad k \gg \frac{1}{2a}. \quad \text{i.e.} \quad a \gg \frac{1}{2k}.$$

The mean energy of the neutron is

$$\overline{E} = \frac{\bar{p}^2}{2m} + \frac{(\Delta p)^2}{2m} \simeq \frac{\bar{p}^2}{2m} = \frac{\hbar^2 k^2}{2m} \quad \Rightarrow \quad k \simeq \sqrt{2m \overline{E}/\hbar^2} = 2.2 \times 10^{12} \, \text{cm}^{-1}$$

$$k = 10^6 \times \frac{1}{2a} \quad \Rightarrow \quad a = \Delta q \simeq 2.3 \times 10^{-7} \, \text{cm}.$$

b) The Hamiltonian is $H = p^2/2m$ and the Heisenberg equations are:

$$\dot{q}(t) = \frac{i}{\hbar} [H, q(t)] = \frac{p(t)}{m}, \qquad \dot{p}(t) = \frac{i}{\hbar} [H, p(t)] = 0.$$

Since $p(t) \equiv U^\dagger(t) p U(t)$, $q(t) \equiv U^\dagger(t) q U(t)$ and $U(t = 0) = \mathbb{1}$, one has $p(0) = p$, $q(0) = q$ where p and q are the usual momentum and position operators (indeed those in the Schrödinger picture). The second Heisenberg equation states that p is a constant of motion, i.e. $p(t) = p(0) \equiv p$. By substituting the latter into the equation for q and integrating – with the initial condition $q(0) = q$ – one has:

$$q(t) = q + \frac{p}{m} t$$

that is formally identical with the classical equation for the uniform motion, the only difference being in that it is an *operator* equation.

c) In order to obtain $\Delta q(t)$, take the square of $q(t)$ paying attention to the order of the factors:

$$q^2(t) = q^2 + \frac{1}{m}(pq + qp)t + \frac{p^2}{m^2}t^2$$

therefore:

$$\left(\Delta q(t)\right)^2 \equiv \overline{q(t)^2} - \overline{q(t)}^2 = (\Delta q)^2 + \frac{1}{m}\left(\overline{qp} + \overline{pq} - 2\bar{q}\bar{p}\right)t + \frac{(\Delta p)^2}{m^2}t^2$$

and, since in a coherent state $\overline{qp} + \overline{pq} = 2\bar{q}\bar{p}$ (see Problem 5.15),

$$\left(\Delta q(t)\right)^2 = (\Delta q)^2 + \frac{(\Delta p)^2}{m^2}t^2 = a^2 + \frac{\hbar^2}{4m^2a^2}t^2 .$$

d) $\Delta q(t) = 2\,\Delta q\,, \quad t = d\,\dfrac{m}{p} \quad \Rightarrow \quad d = \sqrt{3}\,\dfrac{\bar{p}}{\Delta p}\,\Delta q \simeq 0.4\,\text{cm}\,.$

$$t = L\,\frac{m}{\bar{p}} \quad \Rightarrow \quad \Delta q(t) \simeq \frac{\hbar L m}{2\bar{p}\,a} = L\,\frac{\Delta p}{\bar{p}} = 10^{-4}\,\text{cm}\,.$$

e) $|A, t\rangle = e^{-i\,H\,t/\hbar}\,|A, 0\rangle \simeq e^{i\,\bar{p}^2\,t/2m\hbar}\,e^{-i\,p\,\bar{p}\,t/m\hbar}\,|A, 0\rangle\,.$

Since the term $e^{-i\,\bar{p}^2\,t/2m\hbar}$ is an irrelevant phase factor and $e^{-i\,p\,\bar{p}\,t/m\hbar}$ is the operator that translates the coordinate q by $\bar{p}t/m$ then, up to a phase factor,

$$\psi(x, t) \simeq \psi(x - v\,t, 0)\,, \qquad v \equiv \frac{\bar{p}}{m}\,.$$

The approximation we have made allows one to correctly describe the motion of the center of mass of the wave packet, but it neglects the motion in the center-of-mass frame, i.e. the spreading, so it is tenable until $t \lesssim m\,\Delta q/\Delta p$, namely when the packet has gone a distance $d \lesssim \Delta q \times (\bar{p}/\Delta p)$.

7.7

a) Put $\gamma(t) \equiv 4a^2 + 2i\,\hbar t/m$. One has

$$i\hbar\,\frac{\partial}{\partial t}\left(\frac{e^{-x^2/\gamma(t)}}{\sqrt{\gamma(t)}}\right) = -\frac{2\hbar^2}{m}\left(\frac{x^2}{\gamma^{5/2}} - \frac{1}{2\gamma^{3/2}}\right)e^{-x^2/\gamma} - \frac{\hbar^2}{2m}\frac{\partial^2}{\partial x^2}\left(\frac{e^{-x^2/\gamma(t)}}{\sqrt{\gamma(t)}}\right)$$

$$= -\frac{\hbar^2}{2m}\left(-\frac{2}{\gamma^{3/2}} + \frac{4x^2}{\gamma^{5/2}}\right)e^{-x^2/\gamma}$$

so the Schrödinger equation is satisfied.

b) $|\psi_A(x, t)|^2 \equiv \psi_A^*(x, t)\, \psi_A(x, t) = \left(\dfrac{8\,a^2}{\pi}\right)^{1/2} \dfrac{e^{-2a^2 x^2/(4a^4 + \hbar^2 t^2/m^2)}}{2\sqrt{4a^4 + \hbar^2 t^2/m^2}}$

and putting $\alpha(t) \equiv 2a^2/(4a^4 + \hbar^2 t^2/m^2)$:

$$|\psi_A(x, t)|^2 = \left(\dfrac{\alpha(t)}{\pi}\right)^{1/2} e^{-\alpha(t)\, x^2}$$

one realizes that $\psi_A(x, t)$ is normalized (see Problem 5.17), whence:

$$\rho_A(x, t) = |\psi_A(x, t)|^2 = \left(\dfrac{2\,a^2}{\pi}\right)^{1/2} \dfrac{e^{-2a^2 x^2/(4a^4 + \hbar^2 t^2/m^2)}}{\sqrt{4a^4 + \hbar^2 t^2/m^2}} \;.$$

It is evident that $\rho_A(x, t) \to 0$ for $t \to \infty$, even if $\int \rho_A(x, t)\, \mathrm{d}x = 1 \; \forall\, t$.

c) By exploiting the expression of $|\psi_A(x, t)|^2$ in terms of $\alpha(t)$ and thanks to Problem 5.17, one sees that:

$$(\Delta q)^2 = \dfrac{1}{2\alpha(t)} = a^2 + \dfrac{\hbar^2 t^2}{4m^2 a^2} \;.$$

As the particle is free, Δp does not depend on t, so we calculate it at time $t = 0$:

$$\psi_A(x, 0) = \left(2\pi\, a^2\right)^{-1/4} e^{-x^2/4a^2} \quad \Rightarrow \quad (\Delta p)^2 = \dfrac{\hbar^2}{4a^2} \;.$$

So $\psi_A(x, t)$ is the free evolution of a state that at time $t = 0$ is represented by a Gaussian wave packet.

d) By suitably changing notation $\varphi_A(p', 0)$ may be obtained from Problem 5.15 (or also 5.14): $\varphi_A(p', 0) = (2a^2/\pi\, \hbar^2)^{1/4}\, e^{-a^2 p'^2/\hbar^2}$, therefore:

$$\varphi_A(p', t) = e^{-i\, p'^2 t/2m\hbar}\, \varphi_A(p', 0) = \left(\dfrac{2a^2}{\pi\, \hbar^2}\right)^{1/4} e^{-(4a^2 + 2i\, \hbar\, t/m)\, p'^2/4\hbar^2} \;.$$

e) The state $|\, B, 0\,\rangle$ is obtained by applying the operator $e^{i\, q\, k}$, that translates p, to the state $|\, A, 0\,\rangle$; so, if $U(t) \equiv e^{-i\, H\, t/\hbar}$ is the time evolution operator,

$$|\, B, t\,\rangle = U(t)\, |\, B, 0\,\rangle = U(t)\, e^{i\, q\, k}\, |\, A, 0\,\rangle = U(t)\, e^{i\, q\, k}\, U^\dagger(t)\, U(t)\, |\, A, 0\,\rangle$$

$$= e^{i\, k\, q(-t)}\, |\, A, t\,\rangle = e^{i\, k\, (q - p\, t/m)}\, |\, A, t\,\rangle$$

$$= e^{-i\, \hbar\, k^2 t/2m}\, e^{i\, k\, q}\, e^{-i\, k\, p\, t/m}\, |\, A, t\,\rangle$$

the Baker–Campbell–Hausdorff identity (see Problems 4.13 and 5.8) having been used in the last step. Therefore:

$$|\, B, t\,\rangle \xrightarrow{\text{SR}} e^{-i\, \hbar\, k^2 t/2m}\, e^{i\, k\, x}\, \psi_A(x - v\, t, t), \qquad v \equiv \hbar\, k/m = \overline{p}/m \;.$$

$$\rho_B(x, t) = \rho_A(x - v\, t, t) = \left(\dfrac{2\,a^2}{\pi}\right)^{1/2} \dfrac{e^{-2a^2 (x - v\, t)^2/(4a^4 + \hbar^2 t^2/m^2)}}{\sqrt{4a^4 + \hbar^2 t^2/m^2}}$$

i.e. a Gaussian wave packet that, as t grows, widens and lowers (but $\int \rho_B(x, t)\, \mathrm{d}x = 1$), while its center of mass moves with velocity v.

7.8

a) The Hamiltonian is $H = p^2/2m - \gamma q$ and the Heisenberg equations of motion are:

$$\dot{q}(t) = \frac{i}{\hbar}[H, q(t)] = \frac{p(t)}{m}, \qquad\qquad \dot{p}(t) = \frac{i}{\hbar}[H, p(t)] = \gamma.$$

The second equation, supplemented with the initial condition $p(0) = p$, gives:

$$p(t) = p + \gamma \mathbb{1} t.$$

By substituting the latter into the equation for q and integrating with the initial condition $q(0) = q$, one has:

$$q(t) = q + \frac{p}{m} t + \frac{1}{2}\frac{\gamma}{m} \mathbb{1} t^2.$$

The above operator equations are formally identical with the classical equations of the uniformly accelerated motion. The identity operator $\mathbb{1}$ is usually omitted.

b) $\bar{p}(t) = \bar{p} + \gamma t, \qquad \bar{q}(t) = \bar{q} + \frac{\bar{p}}{m} t + \frac{1}{2}\frac{\gamma}{m} t^2.$

c) In order to obtain $\Delta q(t)$, $\Delta p(t)$, take the squares of the operators $p(t)$ and $q(t)$:

$$p^2(t) = p^2 + 2\gamma p t + \gamma^2 t^2$$

$$q^2(t) = q^2 + \frac{1}{m}(pq + qp) t + \frac{\gamma}{m} q t^2 + \frac{p^2}{m^2} t^2 + \frac{\gamma}{m^2} p t^3 + \frac{1}{4}\left(\frac{\gamma}{m}\right)^2 t^4$$

whence:

$$\left(\Delta p(t)\right)^2 \equiv \overline{p(t)^2} - \overline{p(t)}^2 = (\Delta p)^2$$

$$\left(\Delta q(t)\right)^2 = (\Delta q)^2 + \frac{1}{m}(\overline{qp} + \overline{pq} - 2\bar{q}\bar{p}) t + \frac{(\Delta p)^2}{m^2} t^2.$$

In the calculation of $\Delta q(t)$ and $\Delta p(t)$, note the cancellation of all the terms containing γ: in this way the result is the same as for the free particle (see Problem 7.6); in particular, Δp stays constant and Δq, asymptotically for $t \to \pm\infty$, grows linearly with t. The difference between the evolution in presence of an external field and the free one shows up in the mean values (see point b) and in the moments (of q and p) of order higher than the second where the cancellations no longer take place.

d) With $A = ap^2 + bp$, $B = cq$ one has:

$$[A, B] = -i\hbar (2acp + bc), \quad [[A, B], A] = 0, \quad [[A, B], B] = (-i\hbar)^2 2ac^2$$

whence, collecting all the c–numbers in ϑ,

$$e^A e^B = e^{i\vartheta} \exp(ap^2 + bp + cq - i\hbar acp).$$

Putting $\quad a = -it/2m\hbar, \quad b = i\hbar ac, \quad c = i\gamma t/\hbar, \quad$ and therefore:

$$A = -\frac{it}{2m\hbar}(p^2 - \gamma t\, p), \qquad B = \frac{it\,\gamma}{\hbar}\, q$$

one has:

$$e^{-iHt/\hbar} = e^{-i\vartheta}\, \exp\left(-\frac{it}{2m\hbar}(p^2 - \gamma t\, p)\right)\, \exp\left(\frac{it\,\gamma}{\hbar}\, q\right).$$

In the momentum representation $q \to i\hbar\, d/dp'$ so that the operator $e^{i\gamma t q/\hbar} \to e^{-\gamma t\, d/dp'}$ applied to the wavefunction $\varphi_s(p', 0)$ provides the Taylor expansion of $\varphi_s(p' - \gamma t, 0)$. In conclusion:

$$\varphi_s(p', t) = e^{-i(p'^2 t - \gamma\, p'\, t^2)/2m\hbar}\, \varphi_s(p' - \gamma t, 0).$$

7.9

a) Thanks to the Bragg condition $2d\sin\theta = n\lambda$, one has:

$$2d\cos\theta\, \Delta\theta = n\, \Delta\lambda \quad \Rightarrow \quad \frac{\Delta\lambda}{\lambda} = \frac{\Delta\theta}{\tan\theta} \simeq 2.5 \times 10^{-6}.$$

b) As $\Delta p_z/\bar{p}_z = \Delta\lambda/\lambda \;\Rightarrow\; \Delta p_z \ll \bar{p}_z$ and in addition $\Delta p_x = \Delta p_y \ll \Delta p_z$ ($\beta \ll \alpha$), one has (see Problem 5.17):

$$E \simeq \frac{\bar{p}_z^2}{2m_n}, \qquad \Delta E \simeq \frac{\bar{p}_z\, \Delta p_z}{m_n} \quad \Rightarrow \quad \frac{\Delta E}{E} = 2\frac{\Delta p_z}{\bar{p}_z} = 2\frac{\Delta\lambda}{\lambda} \simeq 5 \times 10^{-6}.$$

The state of the neutrons is a state of minimum uncertainty, so:

$$\Delta z = \frac{\hbar}{2\Delta p_z} = \frac{\hbar}{2p_z}\frac{p_z}{\Delta p_z} = \frac{\lambda}{4\pi} \times \frac{\lambda}{\Delta\lambda} \simeq 4.5 \times 10^{-4}\, \text{cm} .$$

c) We have seen in Problem 7.8 that the dependence on time of the width of a wave packet in presence of gravity is identical with that of a free particle. As a consequence, for both the horizontal and the vertical propagation, one has (see Problem 7.6):

$$\left(\Delta z(t)\right)^2 = (\Delta z)^2 + \frac{(\Delta p)^2}{m_n^2}\, t^2 = (\Delta z)^2 + \frac{\hbar^2 t^2}{4m_n^2(\Delta z)^2}, \qquad t = \frac{d\, m_n}{\bar{p}_z} \quad \Rightarrow$$

$$\left(\Delta z(t)\right)^2 = (\Delta z)^2 + \left(\frac{\lambda\, d}{4\pi\, \Delta z}\right)^2 = (\Delta z)^2\left(1 + \frac{1}{16\pi^2}\frac{\lambda^2}{(\Delta z)^2} \times \frac{d^2}{(\Delta z)^2}\right) \quad \Rightarrow$$

$$\Delta z(t) \simeq \Delta z\, (1 + 3.8 \times 10^{-4}) \simeq \Delta z .$$

Even in the more realistic case the wavefunction is not Gaussian, but has the same characteristic dimensions, the conclusion that the spreading inside the interferometer is irrelevant is tenable.

d) $\delta h = \dfrac{1}{2}gt^2 = \dfrac{1}{2}g\left(\dfrac{d\, m_n\, \lambda}{h}\right)^2 = 1.6 \times 10^{-7}\, \text{cm} .$

e) The energy of the electrons is $E \simeq 10^4\, \text{eV}$, so (see Problem 2.14) their de Broglie wavelength is $\lambda \simeq 12.4 \times 10^{-2}\, \text{Å}$. Thanks to what has been seen above, the spreading does not depend on the mass but on the wavelength, the initial width of the packet and the distance it has gone: if $\Delta y \simeq 10^{-2}\, \text{cm}$ is the transverse width of the packet when it has crossed the grid, its width on reaching the screen is given by:

$$\Delta y(t) = \Delta y \left(1 + \frac{1}{16\pi^2} \frac{\lambda^2}{(\Delta y)^2} \times \frac{L^2}{(\Delta y)^2}\right)^{1/2} \simeq \Delta y \left(1 + 10^{-10}\right)^{1/2} \simeq \Delta y \, .$$

7.10

a) The Schrödinger equation (that, for the sake of simplicity, we write in only one dimension) is

$$-\frac{\hbar^2}{2m} \psi''(x, t) + V(x)\psi(x, t) = i\hbar \frac{\partial \psi(x, t)}{\partial t} \, .$$

Neither $\psi^*(x, t)$ nor $\psi(x, -t)$ solve it:

$$-\frac{\hbar^2}{2m} \psi^{*\prime\prime}(x, t) + V(x)\psi^*(x, t) = -i\hbar \frac{\partial \psi^*(x, t)}{\partial t} \, .$$

$$-\frac{\hbar^2}{2m} \psi''(x, -t) + V(x)\psi(x, -t) = i\hbar \left(\frac{\partial \psi(x, \tau)}{\partial \tau}\right)_{\tau=-t} = -i\hbar \frac{\partial \psi(x, -t)}{\partial t} \, .$$

b) It follows from the above equations: in the right hand side there is a double change of sign.

c) If $\psi(x, 0) \to \psi^*(x, 0)$, the mean value of q does not change, whereas the mean value of p changes its sign (see Problem 5.20), therefore:

$$\psi(x, t) \to \tilde\psi(x, t) \equiv \psi^*(x, -t) \quad \Rightarrow \quad \bar q(t) \to \bar q(-t), \quad \bar p(t) \to -\bar p(-t) \, .$$

d) The scalar product of two states is not invariant under time reversal:

$$\int \tilde\psi_1^*(x, t)\, \tilde\psi_2(x, t)\, dx = \left(\int \psi_1^*(x, t)\, \psi_2(x, t)\, dx\right)^*$$

so time reversal is not a unitary transformation and there exists no (linear) operator that implements it; indeed, if $|A\rangle \to |\tilde A\rangle$, $|B\rangle \to |\tilde B\rangle$, then:

$$|C\rangle = \alpha\,|A\rangle + \beta\,|B\rangle \quad \Rightarrow \quad |\tilde C\rangle = \alpha^*\,|\tilde A\rangle + \beta^*\,|\tilde B\rangle \, .$$

Time reversal is said to be an antiunitary transformation.

e) From Problem 5.20: $\varphi(k, t) \to \varphi^*(-k, -t)$.

7.11

a) Multiplying both sides of the Schrödinger equation by $V^\dagger(t)$ one has:

$$V^\dagger(t)\,H\,V(t)\,V^\dagger(t)\,|A, t\rangle = i\hbar V^\dagger(t)\frac{d}{dt}\Big(V(t)\,V^\dagger(t)\,|A, t\rangle\Big) \quad \Rightarrow$$

$$V^\dagger(t)\,H\,V(t)\,|\tilde A, t\rangle = i\hbar\Big(V^\dagger(t)\frac{d}{dt}V(t)\Big)\,|\tilde A, t\rangle + i\hbar\frac{d}{dt}\,|\tilde A, t\rangle$$

and putting:

$$\tilde H = V^\dagger(t)\,H\,V(t) - i\hbar V^\dagger(t)\frac{d}{dt}V(t)$$

one has:

$$i\hbar\frac{d}{dt}\,|\tilde A, t\rangle = \tilde H\,|\tilde A, t\rangle \, .$$

Taking the derivative of both sides of $V^\dagger(t)\,V(t)=1$:

$$-i\hbar V^\dagger(t)\frac{d}{dt}V(t)=i\hbar\left(\frac{d}{dt}V^\dagger(t)\right)V(t)=\left(-i\hbar V^\dagger(t)\frac{d}{dt}V(t)\right)^\dagger.$$

Therefore:

$$\tilde H^\dagger=V^\dagger(t)\,H\,V(t)+i\hbar\left(\frac{d}{dt}V^\dagger(t)\right)V(t)=\tilde H\;.$$

b) If $|A,\,t\rangle=U(t)\,|A,\,0\rangle$, one has:

$$|\tilde A,\,t\rangle=V^\dagger(t)\,U(t)\,V(0)\,V^\dagger(0)\,|A,\,0\rangle=V^\dagger(t)\,U(t)\,V(0)\,|\tilde A,\,0\rangle\quad\Rightarrow$$
$$\tilde U(t)=V^\dagger(t)\,U(t)\,V(0)\;.$$

Finally:

$$i\hbar\frac{d}{dt}\,|\tilde A,\,t\rangle=\tilde H\,|\tilde A,\,t\rangle\quad\Rightarrow\quad i\hbar\frac{d}{dt}\,\tilde U(t)\,|\tilde A,\,0\rangle=\tilde H\,\tilde U(t)\,|\tilde A,\,0\rangle$$

and, since the vector $|\tilde A,\,0\rangle$ is arbitrary, the equation given in the text follows.

c) $\tilde H=H_0+U_0^\dagger H'U_0-H_0=U_0^\dagger H'U_0\;.$

$U_0^\dagger H'(q,\,p,\,t)\,U_0=H'(U_0^\dagger q\,U_0,\,U_0^\dagger p\,U_0,\,t)\;.$

If $H_0=p^2/2m$ (free particle Hamiltonian):

$$U_0^\dagger(t)\,p\,U_0(t)=p,\quad U_0^\dagger(t)\,q\,U_0(t)=q+\frac{p}{m}t\;\Rightarrow\;\tilde H=H'(q+pt/m,\,p,\,t)\;.$$

d) $\bar\xi(t)\equiv\langle A,\,0\,|\,U^\dagger(t)\,\xi\,U(t)\,|\,A,\,0\rangle$

$$=\langle A,\,0\,|\,\tilde U^\dagger(t)\,U_0^\dagger(t)\,\xi\,U_0(t)\,\tilde U(t)\,|\,A,\,0\rangle$$
$$=\langle\tilde A,\,t\,|\,U_0^\dagger(t)\,\xi\,U_0(t)\,|\,\tilde A,\,t\rangle$$
$$\Rightarrow\quad|A,\,0\rangle\to|\tilde A,\,t\rangle,\quad\xi\to U_0^\dagger(t)\,\xi\,U_0(t)\;.$$

So, in the interaction picture, the states evolve with $\tilde U(t)$, the observables with $U_0(t)$.

7.12

a) $\tilde H=-\gamma\left(q+\frac{p}{m}t\right)\;.$

b) One must have $i\hbar\,d\tilde U/dt=\tilde H\,\tilde U$. Attention must be payed to the order of noncommuting factors:

$$i\hbar\frac{d}{dt}\,\tilde U(t)=-\frac{\gamma pt}{m}\tilde U(t)-e^{i\gamma pt^2/2m\hbar}\gamma q\,e^{i\gamma qt/\hbar}e^{-i\alpha(t)/\hbar}+\dot\alpha(t)\tilde U(t)$$

$$=\left(-\frac{\gamma pt}{m}+\dot\alpha\right)\tilde U(t)-e^{i\gamma pt^2/2m\hbar}\gamma q\,e^{-i\gamma pt^2/2m\hbar}\tilde U(t)$$

$$=\left(-\frac{\gamma pt}{m}-\gamma\left(q+\frac{\gamma t^2}{2m}\right)+\dot\alpha\right)\tilde U(t)\quad\Rightarrow\quad\alpha(t)=\frac{\gamma^2 t^3}{6m}\;.$$

c) $U(t) = U_0(t)\,\tilde U(t) = \mathrm{e}^{-\mathrm{i}\,\alpha(t)/\hbar}\,\mathrm{e}^{-\mathrm{i}\,p^2 t/2m\hbar}\,\mathrm{e}^{\mathrm{i}\,\gamma\,p\,t^2/2m\hbar}\,\mathrm{e}^{\mathrm{i}\,\gamma\,q\,t/\hbar}$

$\qquad = \mathrm{e}^{-\mathrm{i}\,\alpha(t)/\hbar}\,\mathrm{e}^{-\mathrm{i}\,t(p^2-\gamma\,t\,p)/2m\hbar}\,\mathrm{e}^{\mathrm{i}\,\gamma\,q\,t/\hbar}$.

7.13

a) From Problem 7.11 one has:

$$\tilde H = V^\dagger(t)\,H\,V(t) - \mathrm{i}\,\hbar\,V^\dagger(t)\,\frac{\mathrm{d}}{\mathrm{d}t}V(t) = \begin{pmatrix} \frac{1}{2}(E_0 - \hbar\omega_0) & a \\ a & -\frac{1}{2}(E_0 - \hbar\omega_0) \end{pmatrix}.$$

b) As $\Sigma^{2n} = \mathbb{1}$ and $\Sigma^{2n+1} = \Sigma$, the formula given in the text obtains by separately collecting the even and odd terms appearing in the series that defines $\mathrm{e}^{\mathrm{i}\,M}$.
   As $\tilde H$ does not depend on time, $\tilde U(t) = \mathrm{e}^{-\mathrm{i}\,\tilde H\,t/\hbar}$. One has:

$$\tilde H = \sqrt{\tfrac{1}{4}(E_0 - \hbar\omega_0)^2 + a^2}\,\Sigma\,, \qquad \Sigma = \begin{pmatrix} \sin\phi & \cos\phi \\ \cos\phi & -\sin\phi \end{pmatrix}$$

$$\sin\phi = \frac{\tfrac{1}{2}(E_0 - \hbar\omega_0)}{\sqrt{\tfrac{1}{4}(E_0 - \hbar\omega_0)^2 + a^2}}\,, \qquad \cos\phi = \frac{a}{\sqrt{\tfrac{1}{4}(E_0 - \hbar\omega_0)^2 + a^2}}$$

$$\tilde U(t) = \cos\omega t - \mathrm{i}\,\Sigma\,\sin\omega t\,, \qquad \omega \equiv \hbar^{-1}\sqrt{\tfrac{1}{4}(E_0 - \hbar\omega_0)^2 + a^2}\,.$$

c) Let $U(t)$ denote the time evolution operator. Then:

$$P_{+-}(t) = \big|\langle +\,|\,U(t)\,|-\rangle\big|^2.$$

From Problem 7.11 we know that $U(t) = V(t)\,\tilde U(t)\,V^\dagger(0) = V(t)\,\tilde U(t)$, and, due to $V^\dagger(t)\,|+\rangle = \mathrm{e}^{\mathrm{i}\,\omega_0\,t/2}\,|+\rangle$, one has:

$$P_{+-}(t) = \big|\langle +\,|\,\tilde U(t)\,|-\rangle\big|^2 = \cos^2\phi\,\sin^2\omega t\,; \qquad 0 \le P_{+-}(t) \le \cos^2\phi\,.$$

$P_{+-}(t)$ reaches the value 1 when the resonance condition $\omega_0 = E_0/\hbar$ is fulfilled.

7.14

a) Also in this case (as in Problems 7.6 and 7.8) the Heisenberg equations are formally identical with the classical equations:

$$\dot q(t) = \frac{p(t)}{m}\,, \qquad \dot p(t) = \gamma\,\delta(t) \quad \Rightarrow$$

$$p(t) = \begin{cases} p & \text{for } t < 0 \\ p + \gamma & \text{for } t > 0 \end{cases} \qquad q(t) = \begin{cases} q + \dfrac{p}{m}\,t & \text{for } t < 0 \\ q + \dfrac{p+\gamma}{m}\,t & \text{for } t > 0\,. \end{cases}$$

b) Let U stand for the (unitary) time evolution operator from time $t = 0^-$ to time $t = 0^+$, then:

$$U^\dagger q\,U = q\,, \quad U^\dagger p\,U = p + \gamma \quad \Rightarrow \quad U = \mathrm{e}^{\mathrm{i}\,\gamma\,q/\hbar}\,.$$

c) For the wavefunction one has:

$$\psi(x, 0^+) = U\,\psi(x, 0^-) = e^{i\gamma\,x/\hbar}\,\psi(x, 0^-)\,.$$

The result is the quantum version of the impulse–momentum theorem of classical mechanics: the effect of an impulsive force is that the particle does not change its position $\big($and also $\rho(x, 0^-) = \rho(x, 0^+)\big)$, but its momentum changes by γ.

7.15

a) The equation of motion is

$$\ddot{x} + \omega^2 x = \frac{\gamma}{m}\,\delta(t)\,.$$

The impulsive force $f(t)$ transfers the momentum $\int f(t)\,dt = \gamma$ to the oscillator, therefore the initial conditions at time $t = 0^+$ are:

$$x(0^+) = 0\,, \quad \dot{x}(0^+) = \frac{\gamma}{m} \quad \Rightarrow \quad x(t) = \begin{cases} 0 & t \le 0 \\ \dfrac{\gamma}{m\omega}\,\sin\omega t & t \ge 0\,. \end{cases}$$

b) As the oscillation amplitude is $A = \gamma/m\omega$, the energy is

$$E = \frac{1}{2}m\omega^2 A^2 = \frac{\gamma^2}{2m}\,.$$

c) The equation of motion and the initial conditions are:

$$\ddot{x} + \omega^2 x = \frac{F(t)}{m}\,; \qquad x(0) = 0\,, \quad \dot{x}(0) = 0\,.$$

As $D_r(t)$ satisfies the equation:

$$\ddot{D}_r(t) + \omega^2 D_r(t) = \frac{1}{m}\,\delta(t) \;\Rightarrow\; \ddot{D}_r(t - t') + \omega^2 D_r(t - t') = \frac{1}{m}\,\delta(t - t') \;\Rightarrow$$

$$\ddot{x}(t) + \omega^2 x(t) = \frac{1}{m}\int_0^\infty \delta(t - t')\,F(t')\,dt' = \frac{F(t)}{m}\,.$$

Since $D_r(t) = 0$ for $t \le 0$, i.e. $D_r(t - t') = 0$ for $t \le t'$, the initial conditions are satisfied, so that:

$$x(t) = \int_0^\infty D_r(t - t')\,F(t')\,dt' = \int_0^t D_r(t - t')\,F(t')\,dt' \quad \Rightarrow \quad x(0) = 0$$

$$\dot{x}(t) = \int_0^\infty \dot{D}_r(t - t')\,F(t')\,dt' = \int_0^t \dot{D}_r(t - t')\,F(t')\,dt' \quad \Rightarrow \quad \dot{x}(0) = 0\,.$$

d) From point a):

$$D_r(t - t') = \begin{cases} 0 & t \le t' \\ \dfrac{1}{m\omega}\,\sin\omega\,(t - t') & t \ge t' \end{cases}$$

$$\dot{D}_r(t - t') = \begin{cases} 0 & t \le t' \\ \dfrac{1}{m}\,\cos\omega\,(t - t') & t \ge t'\,. \end{cases}$$

e) It is sufficient to add, to the general solution of the homogenous equation, a particular solution, e.g. that given in the text:

$$x(t) = x_0 \cos \omega t + \frac{v_0}{\omega} \sin \omega t + \int_0^t D_r(t - t') F(t') dt'.$$

7.16

a) As in the case of the free particle (see Problem 7.14) $q(t = 0^+) = q$, $p(t = 0^+) = p + \gamma$, so the state at time $t = 0^+$ is $e^{i\gamma q/\hbar} |0\rangle$ that is the coherent state (see Problem 7.5) with $\alpha = \gamma/\sqrt{2m\omega\hbar}$. As a consequence, for $t > 0$ the state is $|\alpha e^{-i\omega t}\rangle$.

b) From Problem 5.10 and with $\alpha = \gamma/\sqrt{2m\omega\hbar}$, one has:

$$\overline{H} = \frac{1}{2}\hbar\omega + \hbar\omega |\alpha|^2 = \frac{1}{2}\hbar\omega + \frac{\gamma^2}{2m};$$

$$P(|\alpha e^{-i\omega t}\rangle \to |E_n\rangle) = P(|\alpha\rangle \to |E_n\rangle) = |\langle n | \alpha\rangle|^2 = \frac{|\alpha|^{2n}}{n!} e^{-|\alpha|^2}.$$

c) The Heisenberg equations are:

$$\dot{q} = \frac{p}{m}, \qquad \dot{p} = -m\omega^2 q + F(t), \quad \Rightarrow \quad \ddot{q} + \omega^2 q = \frac{1}{m} F(t).$$

The latter can be solved as in the classical case: to the general solutions of the homogeneous equation $A \cos \omega t + B \sin \omega t$ (where A and B are now operators), a particular solution of the complete equation must be added, for example the (c–numeric) one given in the text of Problem 7.15: then the initial conditions $q(0) = q$, $m\dot{q}(0) \equiv p(0) = p$ are imposed:

$$q(t) = q \cos \omega t + \frac{p}{m\omega} \sin \omega t + \frac{1}{m\omega} \int_0^t \sin \omega(t - t') F(t') dt'$$

$$p(t) = m\dot{q}(t) = p \cos \omega t - m\omega q \sin \omega t + \int_0^t \cos \omega(t - t') F(t') dt'.$$

d) The state at time $t = 0$ is the ground state $|0\rangle$ of the oscillator with no external force, so:

$$\left(\Delta q(t)\right)^2 \equiv \langle 0 | q(t)^2 | 0\rangle - \langle 0 | q(t) | 0\rangle^2 = (\Delta q)^2 \cos^2 \omega t + \frac{(\Delta p)^2}{m^2\omega^2} \sin^2 \omega t$$

$$\left(\Delta p(t)\right)^2 = (\Delta p)^2 \cos^2 \omega t + m^2\omega^2(\Delta q)^2 \sin^2 \omega t.$$

All the linear terms in $\overline{q}$ and $\overline{p}$ cancel (indeed they are vanishing since $\overline{q} = \overline{p} = 0$); the term containing $\overline{qp} + \overline{pq}$ is vanishing (see Problem 5.7) as $\langle 0 | qp + pq | 0\rangle = 0$; furthermore, in the state $|0\rangle$, $(\Delta q)^2 = \hbar/2m\omega$, $(\Delta p)^2 = m\omega\hbar/2$, then:

$$\Delta q(t) = \sqrt{\frac{\hbar}{2m\omega}}, \qquad \Delta p(t) = \sqrt{\frac{m\omega\hbar}{2}} \quad \Rightarrow \quad \Delta q(t)\,\Delta p(t) = \frac{1}{2}\hbar.$$

So at any time the state is a minimum uncertainty state, i.e. (see Problem 5.9) it is a coherent state $|\alpha(t)\rangle$. In addition (see Problem 5.10):

$$a(t) = \frac{\overline{p}(t)}{2\Delta p} - i\frac{\overline{q}(t)}{2\Delta q} \quad \Rightarrow \quad a(t) = \frac{1}{\sqrt{2m\omega\hbar}}\int_0^t e^{-i\omega(t-t')} F(t')\, dt'\,.$$

e) As $F(t) = 0$ for $t > \tau$, one has:

$$a(t \geq \tau) = \frac{e^{-i\omega t}}{\sqrt{2m\omega\hbar}}\int_0^\tau e^{i\omega t'} F(t')\,dt' = \frac{F_0}{\sqrt{2m\omega\hbar}}e^{-i\omega t}\frac{1}{i\omega}\left(e^{i\omega\tau}-1\right) \quad \Rightarrow$$

$$|a(t \geq \tau)|^2 = \frac{2F_0^2}{m\omega^3\hbar}\sin^2\frac{\omega\tau}{2} \quad \Rightarrow$$

$$\overline{H} = \frac{1}{2}\hbar\omega + |a|^2\,\hbar\omega = \frac{1}{2}\hbar\omega + \frac{2F_0^2}{m\omega^2}\sin^2\frac{\omega\tau}{2}\,.$$

7.17

a) The Hamiltonian is

$$H(t) = \frac{p^2}{2m} + \frac{1}{2}m\omega^2\big(q-\xi(t)\big)^2 \quad \Rightarrow \quad \begin{cases} H(t \leq 0) = \dfrac{p^2}{2m} + \dfrac{1}{2}m\omega^2 q^2 \\[2mm] H(t \geq \tau) = \dfrac{p^2}{2m} + \dfrac{1}{2}m\omega^2(q-\xi_0)^2. \end{cases}$$

$H(t)$ obtains from $H(0)$ by means of a translation, therefore the eigenvalues of $H(t)$ are $E_n = (n + \frac{1}{2})\hbar\omega$, independent of t and:

$$\psi_0^{(t)}(x,\,t) = \left(\frac{m\omega}{\pi\hbar}\right)^{1/4} e^{-(m\omega/2\hbar)\,(x-\xi(t))^2}\,e^{-i\frac{1}{2}\omega t}$$

$$H\!\left(-i\hbar\frac{\partial}{\partial x},\,x,\,t\right)\psi_0^{(t)}(x,\,t) = \frac{1}{2}\,\hbar\omega\,\psi_0^{(t)}(x,\,t)\,.$$

$$i\hbar\frac{\partial}{\partial t}\,\psi_0^{(t)}(x,\,t) = \left(\frac{1}{2}\,\hbar\omega + i\,m\omega\,\dot\xi(t)\,\big(x-\xi(t)\big)\right)\psi_0^{(t)}(x,\,t)$$

$$\neq \frac{1}{2}\,\hbar\omega\,\psi_0^{(t)}(x,\,t)\,.$$

whence $\psi_0^{(t)}(x,\,t)$ would not solve the time dependent Schrödinger equation even in the case the center of oscillation should move with uniform motion.

b) As the Hamiltonian differs by the c–number constant $\frac{1}{2}m\omega^2\xi(t)^2$ from the Hamiltonian relative to an oscillator subject to the external force $F(t) = m\omega^2\xi(t)$, the Heisenberg equation of motion for $q(t)$ and $p(t)$ are the same as in Problem 7.16 (provided the substitution $F(t) \to m\omega^2\xi(t)$ is made), so:

$$q(t) = q\cos\omega t + \frac{p}{m\omega}\sin\omega t + \omega\int_0^t \sin\omega(t-t')\,\xi(t')\,dt'$$

$$p(t) = m\,\dot q(t) = p\cos\omega t - m\omega q\sin\omega t + m\omega^2\int_0^t \cos\omega(t-t')\,\xi(t')\,dt'$$

and, thanks to the results of Problem 7.16, the state of the oscillator for $t \geq 0$ is the coherent state $|a(t)\rangle$ relative to the oscillator with the origin as center of oscillation, and:

$$\alpha(t) = \frac{m\omega^2}{\sqrt{2m\omega\,\hbar}}\, e^{-i\omega t} \int_0^t e^{i\omega t'}\, \xi(t')\, dt'\,.$$

c) The function $\xi(t)$ is discontinuous in $t = 0$ but bounded, therefore the integral $\int_0^t e^{i\omega t'}\, \xi(t')\, dt'$ is a continuous function. One has:

$$\alpha(t) = \frac{m\omega^2\,\xi_0}{\sqrt{2m\omega\,\hbar}}\, e^{-i\omega t} \int_0^t e^{i\omega t'}\, dt' = -i\sqrt{\frac{m\omega}{2\hbar}}\, \xi_0 \left(1 - e^{-i\omega t}\right).$$

From Problem 5.9 one has:

$$\alpha(t) = \frac{\overline{p}(t)}{2\Delta p} - i\,\frac{\overline{q}(t)}{2\Delta q} \quad\Rightarrow\quad \overline{q}(t) = -2\Delta q \times \Im m\,\alpha(t) = \xi_0(1 - \cos\omega t)$$

so, for $t \geq 0$, the oscillator oscillates around the new center of oscillation ξ_0 with amplitude ξ_0.

d) Let us perform the integration by parts:

$$\alpha(t) = \frac{m\omega^2}{\sqrt{2m\omega\,\hbar}}\, e^{-i\omega t} \left(\frac{e^{i\omega t}}{i\omega}\, \xi(t) - \int_0^t e^{i\omega t'}\, \dot{\xi}(t')\, dt'\right).$$

One has:

$$\int_0^t e^{i\omega t'}\, \dot{\xi}(t')\, dt' = \int_0^t e^{i\omega t'}\, \frac{1}{\tau}\, f'(t'/\tau)\, dt' = \int_0^{t/\tau} e^{i\omega \tau s}\, f'(s)\, ds$$

that, for any fixed t, tends to 0 for $\tau \to \infty$. Therefore:

$$\alpha(t) = -i\sqrt{\frac{m\omega}{2\hbar}}\, \xi(t)$$

that corresponds to the state described by the wavefunction $\psi_0^{(t)}(x,\,t)$ (see Problem 5.15).

e) Thanks to what we have seen in point a) and owing to $\dot{\xi}(t) = (1/\tau)f'(t/\tau)$, one has that, in the limit $\tau \to \infty$, $\dot{\xi}(t) \to 0$, and, as a consequence, $\psi_0^{(t)}(x,\,t)$ satisfies the time dependent Schrödinger equation.

8

Angular Momentum

Orbital angular momentum: states with $l = 1$ and representations; rotation operators; spherical harmonics; tensors and states with definite angular momentum ($l = 1$, $l = 2$).

Note. *It is known that the eigenvalues of a component of the angular momentum are $m\hbar$ and those of $\vec{L}^2$ are $l(l+1)\hbar^2$ with m, l integers: it is customary to refer to them simply as m and l: for example, if $l = 2$ we shall say that $\vec{L}^2$ has the eigenvalue 2, not $6\hbar^2 = 2(2+1)\hbar^2$. In addition, the eigenvalues of a component different from L_z, e.g. L_x, will be denoted by a subscript to the letter m, e.g. m_x.*

8.1 Consider a particle in three dimensions.

a) By exploiting the Schrödinger Representation for the angular momentum $\vec{L} \xrightarrow{\text{SR}} -i\,\hbar\,\vec{r} \wedge \nabla$, explicitly show that the states whose wavefunctions depend only on r, $|s\rangle \xrightarrow{\text{SR}} f(r)$, are eigenstates of all the components of $\vec{L}$. What is the value of $\vec{L}^2$ in such states?

b) Use the commutation rules $[L_i,\, q_j] = i\,\hbar\,\epsilon_{ijk}\,q_k$ to show that each state $|\mathcal{P}_i\rangle = q_i\,|s\rangle$ whose wavefunction is $x_i\,f(r)$, $i = 1, 2, 3$, is an eigenstate of the component of the angular momentum carrying the same index.

c) Calculate $\sum_i \big[L_i,\, [L_i,\, q_j]\big]$ and use the result to show that the states represented by the wavefunctions $x_j\,f(r)$, and their linear combinations, all are eigenstates of $\vec{L}^2$.

d) Write the commutation rules between L_z and $q_1 \pm i\,q_2$ and show that the states $|\mathcal{P}_\pm\rangle \xrightarrow{\text{SR}} (x \pm i\,y)\,f(r)$, $|\mathcal{P}_0\rangle \xrightarrow{\text{SR}} z\,f(r)$ are eigenstates of L_z. Which are the eigenvalues of L_x, L_y, L_z in the linear span of the vectors $|\mathcal{P}_i\rangle$?

e) Say whether the states with wavefunctions $\psi_1(x, y, z) = x\,f(r)$, $\psi_2(x, y, z) = y\,f(r)$, $\psi_3(x, y, z) = z\,f(r)$, are mutually orthogonal and equally normalized. Find the condition $f(r)$ must fulfill so that they are normalized to 1. Are the states represented by the wavefunctions $\psi_\pm(x, y, z) = (x \pm i\,y)\,f(r)$, $\psi_0(x, y, z) = z\,f(r)$, equally normalized?

© Springer International Publishing AG 2017
E. d'Emilio and L.E. Picasso, *Problems in Quantum Mechanics*,
UNITEXT for Physics, DOI 10.1007/978-3-319-53267-7_8

f) Write equally normalized wavefunctions for the eigenstates of L_x and of L_y belonging to the linear span of the vectors $|\mathcal{P}_i\rangle$.

8.2 Assume that the states $|\mathcal{P}_i\rangle \xrightarrow{\text{SR}} x_i\,f(r)$, $i = 1, 2, 3$ (see Problem 8.1) are normalized.

a) Find the matrices that represent L_x, L_y, L_z in states with $l = 1$ in the basis $|\mathcal{P}_i\rangle$ (*Cartesian basis*): to this purpose use the commutation rules among the components of $\vec{L}$ and those of $\vec{q}$ to calculate the matrix elements of L_x, L_y, L_z. Does the representation of $\vec{L}$ depend on the radial function $f(r)$?

b) Still restricting to states with $l = 1$, find the representation of L_x, L_y, L_z in the basis consisting of the eigenvectors of L_z (*spherical basis*):

$$|m = \pm 1\rangle \equiv e^{i\alpha_\pm}\,|\mathcal{P}_\pm\rangle \xrightarrow{\text{SR}} e^{i\alpha_\pm}\,\frac{x \pm i\,y}{\sqrt{2}}\,f(r),$$

$$|m = 0\rangle \equiv |\mathcal{P}_0\rangle \xrightarrow{\text{SR}} z\,f(r)$$

where $e^{i\alpha_\pm}$ are arbitrary phase factors.

Let $|l, m\rangle$ be a simultaneous eigenvector of $\vec{L}^2$ belonging to the eigenvalue $l(l+1)\,\hbar^2$ and of L_z belonging to the eigenvalue $m\hbar$ and let $L_\pm \equiv L_x \pm i\,L_y$.

c) It is known that $L_\pm\,|l, m\rangle \propto |l, m \pm 1\rangle$: calculate the absolute value of the proportionality coefficient in such a way that, if $|l, m\rangle$ is normalized, also the vectors $|l, m \pm 1\rangle$ are normalized. Choose the phases of the vectors $|l, m\rangle$ so that, in this basis, L_x is represented by a matrix whose elements are real and positive. Given this choice of the phases, find the representation of L_x, L_y, L_z in the states with $l = 1$.

d) Does a 3×3 matrix exist that both commutes with all the matrices found in either a) or in b), and that is not a multiple of the identity? Does a 3×3 matrix exist that commutes with only two of them?

8.3 Let $\vec{L}$ be the angular momentum of a particle. In the Schrödinger representation, with wavefunctions expressed in terms of polar coordinates r, θ, ϕ, one has $L_z \rightarrow -i\hbar\,\partial/\partial\phi$. Let $U(\alpha) \equiv e^{-i\alpha\,L_z/\hbar}$.

a) Calculate the action of $U(\alpha)$ on the wavefunctions in the Schrödinger representation, expressed both in polar and in Cartesian coordinates.

b) Find the representation of $U(\alpha)$ on the states with $l = 1$ in the Cartesian basis $|\mathcal{P}_i\rangle$ (see Problem 8.2): *i)* by exploiting the result of the previous question; *ii)* calculating L_z^2, L_z^3, $\cdots$ and resumming the series $U(\alpha) = \sum_n(-i\alpha/\hbar)^n\,L_z^n/n!$; *iii)* starting from the matrix that represents $U(\alpha)$ in the spherical basis $|\mathcal{P}_{\pm,0}\rangle$ where L_z is diagonal and making use of the matrix that allows one to transform the Cartesian basis into the spherical basis.

c) Take advantage of the result of question a) and calculate $U(\alpha)\, q_i\, U^{-1}(\alpha)$, $i = 1,\, 2,\, 3$ (it may be convenient to consider the operators $q_1 \pm i q_2$).

We want to generalize the result found in c) to the case of a (total, intrinsic, ...) angular momentum, for which not necessarily the Schrödinger representation exists, and to arbitrary vector operators. Let $\vec{J}$ be defined by the commutation rules $[J_i,\, J_j] = i\hbar\, \epsilon_{ijk}\, J_k$ and let $\vec{V}$ be a vector operator: $[J_i,\, V_j] = i\hbar\, \epsilon_{ijk}\, V_k$.

d) Let $|\, m,\, \cdots\rangle$ be an eigenvector of J_3 belonging to the eigenvalue $m\hbar$. Demonstrate that $(V_1 + i\, V_2)\, |\, m,\, \cdots\rangle$ (if nonvanishing) is an eigenvector of J_3 belonging to the eigenvalue $(m+1)\, \hbar$.

e) Let $U(\alpha) = e^{-i\alpha\, J_3/\hbar}$; use the result of the previous question and the completeness of the eigenvectors of J_3 to show that the transformation law for vector operators under rotations: $V_i' \equiv U(\alpha)\, V_i\, U^{-1}(\alpha)$, is the same as that found in c) for $\vec{q}$.

8.4 Let $L_n \equiv \sum L_i\, \hat{n}_i$ be whatever component of the angular momentum ($\hat{n}$ is a real unit vector).

a) Demonstrate, either by using the matrices found in Problem 8.2 or by a more general argument, that on the states with angular momentum $l = 1$ the identity:

$$L_n^3 - L_n = L_n\, (L_n + 1)\, (L_n - 1) = 0$$

holds.

b) Let $|\, A\rangle$ be an eigenstate of L_n . Show, making use of the commutation rules of the angular momentum, that the mean values in $|\, A\rangle$ of the components of $\vec{L}$ along the directions orthogonal to $\hat{n}$ are vanishing.

Let $|\, A\rangle \xrightarrow{\text{SR}} \psi_A(x, y, z) = (\alpha_1\, x + \alpha_2\, y + \alpha_3\, z)\, f(r)$, with $\langle\, A\, |\, A\rangle = 1$.

c) Calculate the mean values of L_x, L_y, L_z in the state $|\, A\rangle$ in the case $\alpha_1,\, \alpha_2,\, \alpha_3$ are real numbers.

d) Calculate the mean values of L_x, L_y, L_z in the state $|\, A\rangle$ in the case $\alpha_1,\, \alpha_2,\, \alpha_3$ are complex numbers.   Show that it is possible to write $\overline{L}_i = i\hbar\, \epsilon_{ijk}\, \alpha_j\, \alpha_k^* \equiv i\hbar\, [\vec{\alpha} \wedge \vec{\alpha}^{\,*}]_i$.

e) Show that, if all the mean values of L_x, L_y, L_z in the state $|\, A\rangle$ are vanishing, then $|\, A\rangle$ is an eigenvector of a suitable component L_n of the angular momentum.

8.5 Let $|\, A\rangle$ be a normalized eigenvector of a component $L_n = \sum L_i\, \hat{n}_i$ of the angular momentum, belonging to an eigenvalue $\lambda \neq 0$: $L_n\, |\, A\rangle = \lambda\, |\, A\rangle$.

a) Show that $\langle A \mid \vec{L} \mid A \rangle = \lambda \, \hat{n}$ (take advantage of the fact – see Problem 8.4 – that the mean values of the components of $\vec{L}$ orthogonal to $\hat{n}$ are vanishing).

Let now $\mid A \rangle$ be a normalized eigenvector of $\vec{L}^2$ belonging to the eigenvalue $l(l+1)\,\hbar^2$.

b) Show that, if $\langle A \mid \vec{L} \mid A \rangle = \pm \hbar l \, \hat{n}$ $(\hat{n}^2 = 1)$, then $\mid A \rangle$ is an eigenvector of L_n.

c) Calculate $\langle A \mid \vec{L} \mid A \rangle$ in the state of a particle whose normalized wavefunction is $\psi_A(x, y, z) = (x \, \cos\gamma + i y \, \sin\gamma)\, f(r)$ and say whether for any value of γ it is an eigenvector of a suitable component L_{n_γ} of $\vec{L}$.

d) Demonstrate that the states with angular momentum $l = 1$: $\mid A \rangle \overset{\text{SR}}{\longrightarrow}$ $\psi_A(x, y, z) = (\alpha_1\, x + \alpha_2\, y + \alpha_3\, z)\, f(r)$ that are eigenstates of some component L_n of $\vec{L}$ belonging to a nonvanishing eigenvalue $(\pm \hbar)$, are all and only those for which $\alpha_1^2 + \alpha_2^2 + \alpha_3^2 = 0$ $(\alpha_i \in \mathbb{C})$; show that $n_i = i\,\epsilon_{ijk}\,\alpha_j\,\alpha_k^*$.

8.6   The spherical harmonics $Y_{l,m}(\theta,\ \phi)$, if expressed in terms of the Cartesian coordinates, may be written as $r^{-l} \times$ (homogeneous polynomial of degree l in x, y, z). The orthonormality of spherical harmonics is expressed by:

$$\int Y_{l',m'}^*(\theta,\ \phi)\, Y_{l,m}(\theta,\ \phi)\, d\Omega = \delta_{l\,l'}\ \delta_{m\,m'}\ , \qquad d\Omega \equiv \sin\theta\, d\theta\, d\phi\ .$$

a) Is it true that all the states represented by the wavefunctions:

$$\psi(x, y, z) = \frac{1}{r^2} \times \text{(homogeneous polynomial of degree 2)} \times f(r)$$

are eigenstates of the angular momentum with $l = 2$?
Are all the states, represented by the wavefunctions:

$$\psi(x, y, z) = \frac{1}{r} \times \text{(homogeneous polynomial of degree 1)} \times f(r)\,,$$

eigenstates of $\vec{L}^2$ belonging to $l = 1$?

b) Only two among the following functions are (nonnormalized) spherical harmonics. Which ones?

$$\cos^2\theta\, e^{2i\,\phi}, \quad \sin^2\theta\, e^{2i\,\phi}, \quad \sin\theta \cos\theta\, e^{2i\,\phi}, \quad \sin\theta \cos\theta\, e^{i\,\phi}.$$

c) Write the most general homogeneous polynomial of degree 2 that, multiplied by a radial function, gives rise to states with $L_z = 0$. Exploit the orthogonality of spherical harmonics with different values of l to find the one belonging to $l = 2$.

d) Make use of the space inversion with respect to the plane $y = 0$ to show that, up to a phase factor, $Y_{l,-m}(\theta,\ \phi) = Y_{l,m}(\theta,\ -\phi)$ and write, both in polar and in Cartesian coordinates, all the normalized spherical harmonics $Y_{l=2,m}(\theta,\ \phi)$.

8.7 Consider the vector space $\mathcal{H}^{(n)}$ spanned by the wavefunctions:
$$\psi(x,y,z) = \frac{1}{r^n} \times \text{(homogeneous polynomial of degree } n \text{ in } x, y, z) \times f(r)$$
with assigned $f(r)$.

a) Show that $\mathcal{H}^{(n)}$ is invariant under the rotation operators $e^{-i\alpha \vec{L}\cdot\hat{n}}$ and, therefore, under $\vec{L}$. Show that $\mathcal{H}^{(n)} \supset \mathcal{H}^{(n-2)} \supset \mathcal{H}^{(n-4)} \cdots$.

b) Find the maximum eigenvalues $m_{\max}$ of L_z and $l_{\max}$ of $\vec{L}^2$ on the states belonging to $\mathcal{H}^{(n)}$. Find $l_{\min}$. Which are the eigenvalues of $\vec{L}^2$ restricted to the space $\mathcal{H}^{(n)}$?

c) Write, both in Cartesian and polar coordinates, the (nonnormalized) spherical harmonics $Y_{l_{\max},l_{\max}}$ and $Y_{l_{\max},l_{\max}-1}$.

d) How many linearly independent homogeneous polynomials of degree n give rise to states with $m = l_{\max} - 2$? How many 'have' $m = l_{\max} - 3$? How many have $m = l_{\max} - 4$?

8.8 Consider for a particle in three dimensions the operators in the Schrödinger representation:
$$\xi_i: \quad x, \ y, \ z; \quad x^2, \ y^2, \ z^2, \ xy, \ x^2 - y^2, \ x^2 + y^2$$
and let $|m\rangle$ be the eigenvectors of L_z.

a) For each of the above operators say which conditions must be fulfilled by $\Delta m \equiv m'' - m'$ so that the matrix elements $\langle m'' \mid \xi_i \mid m' \rangle$ may be nonvanishing (*selection rules* on m).

Let $|l\rangle$ be the eigenvectors of $\vec{L}^2$.

b) Use the result of Problem 8.7 and show that:
$$q_i \mid l \rangle = \alpha \mid l+1 \rangle + \beta \mid l-1 \rangle + \gamma \mid l-3 \rangle + \cdots$$
with α, β, γ, $\cdots$ not necessarily all different from 0.

c) Take advantage of the preceding result and of the identity:
$$\langle l'' \mid q_i \mid l' \rangle = \langle l' \mid q_i \mid l'' \rangle^*$$
and find the selection rule on l: $\Delta l \equiv l'' - l' = \pm 1$.

d) Determine, in a way analogous to the preceding one, the selection rules on l for all the operators ξ_i.

e) Show that the same selection rules on m and on l are obtained if x, y, z are replaced by p_x, p_y, p_z: to this end one may exploit the fact that the transformation $q_i \to \Lambda p_i$, $p_i \to -\Lambda^{-1} q_i$ is a canonical transformation (see Problem 5.7); if, instead, the direct calculation is carried out, it may be convenient to write $|m\rangle \xrightarrow{\text{SR}} f(\rho, z) e^{im\phi}$, where $\rho = \sqrt{x^2 + y^2}$.

8.9 Consider the states described by the following wavefunctions:

$$\psi_1(x,y,z) = (z+a)\, f(r), \quad \psi_2(x,y,z) = (x^2+y^2)\, f(r),$$
$$\psi_3(x,y,z) = (x^2-y^2)\, f(r), \quad \psi_4(x,y,z) = (x^2+y^2-\tfrac{2}{3}r^2)\, f(r)\,.$$

a) For each of the above wavefunctions, say which are the possible results of a measurement of L_z and, for each possible result, write the (not necessarily normalized) wavefunction after the measurement.

b) For each of the above wavefunctions, say which are the possible results of a measurement of $\vec{L}^{\,2}$ and, for each possible result, write the (not necessarily normalized) wavefunction after the measurement.

c) Calculate the probabilities of the possible results of a measurement of $\vec{L}^{\,2}$ in the case of the state represented by the wavefunction $\psi_2(x,y,z)$.

8.10 A particle is in a state that is a simultaneous eigenstate of $\vec{L}^{\,2}$ with eigenvalue l, and of L_z with eigenvalue m.

a) Calculate the mean value of L_x^2.

b) Verify the preceding result by calculating the probabilities for each eigenvalue of L_x, in the particular case $l=1$, $m=1$.

8.11 Consider the state $|\, \mathcal{D}_{1,2} \,\rangle \equiv |\, \mathcal{D}_{2,1} \,\rangle \xrightarrow{\text{SR}} x\, y\, f(r)$.

a) Say whether it is an eigenstate of L_x, L_y, and/or of L_z.

b) Are the states $|\, \mathcal{D}_{1,2} \,\rangle$, $|\, \mathcal{D}_{2,3} \,\rangle$, $|\, \mathcal{D}_{3,1} \,\rangle$ eigenstates of $\vec{L}^{\,2}$? Are they linearly independent? In the affirmative case, are they orthogonal?

Consider the three states:

$$|\, \mathcal{D}_{1,1} \,\rangle \xrightarrow{\text{SR}} (x^2 - \tfrac{1}{3}r^2)\, f(r),$$
$$|\, \mathcal{D}_{2,2} \,\rangle \xrightarrow{\text{SR}} (y^2 - \tfrac{1}{3}r^2)\, f(r),$$
$$|\, \mathcal{D}_{3,3} \,\rangle \xrightarrow{\text{SR}} (z^2 - \tfrac{1}{3}r^2)\, f(r)\,.$$

c) Say whether they are eigenstates of $\vec{L}^{\,2}$ and whether they are linearly independent.

d) Denote by $U(\alpha) = e^{-i\alpha\, L_z/\hbar}$ the operator implementing the rotation of the angle α around the z axis (see Problem 8.3). Find $U(\alpha)|\, \mathcal{D}_{i,j} \,\rangle$, $i,j = 1,\,2,\,3$ and show (or verify) that the six states $|\, \mathcal{D}_{i,j} \,\rangle = |\, \mathcal{D}_{j,i} \,\rangle$ transform as the components of a rank 2 tensor:

$$U(\alpha)|\, \mathcal{D}_{i,j} \,\rangle = \sum_{kl} R_{ik}(\alpha)\, R_{jl}(\alpha)\, |\, \mathcal{D}_{k,l} \,\rangle, \quad R(\alpha) = \begin{pmatrix} \cos\alpha & \sin\alpha & 0 \\ -\sin\alpha & \cos\alpha & 0 \\ 0 & 0 & 1 \end{pmatrix}.$$

Solutions

8.1

a) As $\nabla f(r) \propto \vec{r}$ (think e.g. of the central force fields), and $\vec{r} \wedge \vec{r} = 0$, one has that $L_i \,|\, s \rangle = 0 \;\; \forall\, i \quad \Rightarrow \quad L_i^2 \,|\, s \rangle = 0 \quad \Rightarrow \quad \vec{L}^2 \,|\, s \rangle = 0$.

b) As $L_i \,|\, s \rangle = 0$ and $[L_i \,,\, q_i] = 0$, one has:

$$L_i \,|\, \mathcal{P}_i \rangle \equiv L_i \, q_i \,|\, s \rangle = q_i \, L_i \,|\, s \rangle = 0 \, .$$

c) $\displaystyle\sum_i \, [L_i \,,\, [L_i \,,\, q_j]] = -i\hbar \sum_{ik} \epsilon_{ijk} \, [L_i \,,\, q_k]$

$$= -\hbar^2 \sum_{ikl} \epsilon_{ijk} \, \epsilon_{ikl} \, q_l = 2\hbar^2 \, q_j$$

thanks to the identity $\sum_{ik} \epsilon_{jki} \, \epsilon_{lki} = +2\,\delta_{jl}$.

$$\sum_i \, [L_i \,,\, [L_i \,,\, q_j]] \,|\, s \rangle = \sum_i \left(L_i \, L_i \, q_j - 2 L_i \, q_j \, L_i + q_j \, L_i \, L_i \right) |\, s \rangle$$

$$= \vec{L}^2 \, q_j \,|\, s \rangle \quad \Rightarrow \quad \vec{L}^2 \, q_j \,|\, s \rangle = 2\hbar^2 \, q_j \,|\, s \rangle \, .$$

So all the states with wavefunction $x_i \, f(r)$ are eigenstates of $\vec{L}^2$ with the same eigenvalue $\hbar^2 \, l(l+1)$ with $l = 1$. As a consequence, also their linear combinations are such.

d) As above, from $[L_z \,,\, q_1 \pm i\, q_2] = \pm \hbar \, (q_1 \pm i\, q_2)$ it follows that $L_z \,|\, \mathcal{P}_\pm \rangle = \pm \hbar \,|\, \mathcal{P}_\pm \rangle$ and from $[L_z \,,\, q_3] = 0$ it follows $L_z \,|\, \mathcal{P}_0 \rangle = 0$.

Therefore L_z has $0, \pm \hbar$ as eigenvalues; they are nondegenerate because, for a given $f(r)$, the dimension of the space generated by the vectors $|\, \mathcal{P}_i \rangle$ is 3. As for L_x and L_y, the same conclusion applies: both have eigenvalues 0 and $\pm \hbar$ but, obviously, the eigenvectors are different.

e) The states with wavefunctions $\psi_1, \; \psi_2, \; \psi_3$ are orthogonal:

$$\int \psi_1^*(x, y, z) \, \psi_2(x, y, z) \, dV = \int |f(r)|^2 \, x\, y \, dV = 0$$

as the integrand is odd in x (or in y). Equivalently:

$$\int \psi_1^*(x,y,z)\,\psi_2(x,y,z)\,dV = \int |f(r)|^2\, r^2 \sin^2\theta\,\sin\phi\,\cos\phi\,r^2 dr\,d\Omega = 0$$

due to the integration on ϕ. Likewise for the other pairs. Furthermore:

$$\int |f(r)|^2 x^2\,dV = \int |f(r)|^2 y^2\,dV = \int |f(r)|^2 z^2\,dV = \frac{1}{3}\int |f(r)|^2 r^2\,dV$$

so the normalization coefficient is the same for the three states, that are normalized to 1 if $\int |f(r)|^2\, r^2\,dV = 3$. Since:

$$\int |\psi_\pm(x,y,z)|^2\,dV = \int |f(r)|^2\,(x^2+y^2)\,dV = 2\int |\psi_0(x,y,z)|^2\,dV$$

the following states are equally normalized:

$$|\,P_\pm\,\rangle \xrightarrow{\text{SR}} \frac{x\pm iy}{\sqrt2}\,f(r),\qquad |\,P_0\,\rangle \xrightarrow{\text{SR}} z\,f(r)$$

and their wavefunctions are proportional to the spherical harmonics $Y_{l=1,m}(\theta,\phi)$ with $m=\pm1,0$.

f) In the space of the states with $l=1$, the equally normalized eigenfunctions of L_x are $x\,f(r)$, $\dfrac{y\pm iz}{\sqrt2}\,f(r)$ and those of L_y are $y\,f(r)$, $\dfrac{z\pm ix}{\sqrt2}\,f(r)$.

8.2

a) From the commutation rules $[L_i,\,q_j]=i\hbar\,\epsilon_{ijk}\,q_k$ and with the same notation as in Problem 8.1, one has:

$$L_x\,|\,P_1\,\rangle=0,\quad L_x\,|\,P_2\,\rangle=L_x\,q_2\,|\,s\,\rangle=[L_x,\,q_2]\,|\,s\,\rangle=i\hbar\,q_3\,|\,s\,\rangle=i\hbar\,|\,P_3\,\rangle$$

so the only nonvanishing matrix elements of L_x are $\langle\,P_3\mid L_x\mid P_2\,\rangle=i\hbar$ and $\langle\,P_2\mid L_x\mid P_3\,\rangle=-i\hbar$, and they do not depend on $f(r)$. Likewise for L_y and L_z. So in the Cartesian basis (cb):

$$L_x\xrightarrow{\text{cb}}\hbar\begin{pmatrix}0&0&0\\0&0&-i\\0&i&0\end{pmatrix};\quad L_y\xrightarrow{\text{cb}}\hbar\begin{pmatrix}0&0&i\\0&0&0\\-i&0&0\end{pmatrix};\quad L_z\xrightarrow{\text{cb}}\hbar\begin{pmatrix}0&-i&0\\i&0&0\\0&0&0\end{pmatrix}.$$

It is worth noting (and it is not a mere mnemonic artifice!) that in this representation the matrices L_i are connected with the Levi-Civita tensor in the following way:

$$\big[L_x\big]_{jk}=-i\hbar\,\epsilon_{1jk},\qquad \big[L_y\big]_{jk}=-i\hbar\,\epsilon_{2jk},\qquad \big[L_z\big]_{jk}=-i\hbar\,\epsilon_{3jk}.$$

b) Let us calculate the representation of L_x: the only nonvanishing matrix elements are those between $|\,m=0\,\rangle$ and $|\,m=\pm1\,\rangle$, indeed:

$$L_x\,|\,m=\pm1\,\rangle=e^{i\alpha_\pm}L_x\frac{q_1\pm iq_2}{\sqrt2}\,|\,s\,\rangle=\pm\frac{ie^{i\alpha_\pm}}{\sqrt2}\,L_x\,q_2\,|\,s\,\rangle$$

$$=\mp e^{i\alpha_\pm}\frac{\hbar}{\sqrt2}\,q_3\,|\,s\,\rangle=\mp e^{i\alpha_\pm}\frac{\hbar}{\sqrt2}\,|\,m=0\,\rangle$$

and $\langle m = 0 \mid L_x \mid m = 0 \rangle = 0$. The representation of L_z is known (L_z is diagonal), and that of L_y can be found through the commutator of L_x with L_z. So in the spherical basis (sb) $\mid m = +1 \rangle$, $\mid m = 0 \rangle$, $\mid m = -1 \rangle$:

$$L_x \xrightarrow{\text{sb}} \frac{\hbar}{\sqrt{2}} \begin{pmatrix} 0 & -e^{-i\alpha_+} & 0 \\ -e^{i\alpha_+} & 0 & e^{i\alpha_-} \\ 0 & e^{-i\alpha_-} & 0 \end{pmatrix},$$

$$L_y \xrightarrow{\text{sb}} \frac{\hbar}{\sqrt{2}} \begin{pmatrix} 0 & ie^{-i\alpha_+} & 0 \\ -ie^{i\alpha_+} & 0 & -ie^{i\alpha_-} \\ 0 & ie^{-i\alpha_-} & 0 \end{pmatrix}, \qquad L_z \xrightarrow{\text{sb}} \hbar \begin{pmatrix} 1 & 0 & 0 \\ 0 & 0 & 0 \\ 0 & 0 & -1 \end{pmatrix}.$$

c) One has $L_- L_+ = L_x^2 + L_y^2 + i\,[L_x,\,L_y] = \vec{L}^2 - L_z^2 - \hbar L_z$, therefore

$$\langle l,\, m \mid L_- L_+ \mid l,\, m \rangle = \hbar^2 \big(l\,(l+1) - m\,(m+1)\big)$$

and likewise:

$$\langle l,\, m \mid L_+ L_- \mid l,\, m \rangle = \hbar^2 \big(l\,(l+1) - m\,(m-1)\big)$$

so, as $L_-^\dagger = L_+$, up to phase factors:

$$\begin{aligned} L_+ \mid l,\, m \rangle &= \hbar\sqrt{(l\,(l+1) - m\,(m+1)}\ \mid l,\, m+1 \rangle \\ L_- \mid l,\, m \rangle &= \hbar\sqrt{(l\,(l+1) - m\,(m-1)}\ \mid l,\, m-1 \rangle\,. \end{aligned} \tag{1}$$

As $L_x = \frac{1}{2}(L_+ + L_-)$, if the phase factors are put equal to 1 as in (1), the matrix elements of L_x: $\langle l, m \pm 1 \mid L_x \mid l, m \rangle = \frac{1}{2}\langle l, m \pm 1 \mid L_\pm \mid l, m \rangle$ are real and positive, and those of $L_y = -\frac{1}{2}i(L_+ - L_-)$ are imaginary. On the states with $l = 1$, with the above choice of the phases:

$$L_x \xrightarrow{\text{sb}} \frac{\hbar}{\sqrt{2}} \begin{pmatrix} 0 & 1 & 0 \\ 1 & 0 & 1 \\ 0 & 1 & 0 \end{pmatrix}, \quad L_y \xrightarrow{\text{sb}} \frac{\hbar}{\sqrt{2}} \begin{pmatrix} 0 & -i & 0 \\ i & 0 & -i \\ 0 & i & 0 \end{pmatrix}, \quad L_z \xrightarrow{\text{sb}} \hbar \begin{pmatrix} 1 & 0 & 0 \\ 0 & 0 & 0 \\ 0 & 0 & -1 \end{pmatrix}$$

that coincides with the result found in b) if $e^{i\alpha_\pm} = \mp 1$ is chosen.

d) Regardless of the representation, it does not exist: if M commutes with (the matrix that represents) L_z, the latter being nondegenerate, its eigenvectors $\mid m = 0, \pm 1 \rangle$ necessarily are eigenvectors of M; but M commutes also with L_+, so the states $\mid m = -1 \rangle$, $\mid m = 0 \rangle \propto L_+ \mid m = -1 \rangle$ and $\mid m = 1 \rangle \propto L_+ \mid m = 0 \rangle$ correspond to the same eigenvalue of M that, as a consequence, is a multiple of the identity.

Thanks to Schur's lemma, this means that the representation 3×3 of L_x, L_y, L_z is an irreducible representation. The same holds for the $(2l+1)$-dimensional representations on the states of angular momentum l.

If M commutes with L_x and L_y, then it commutes also with $[L_x,\, L_y]$ and, as a consequence, with L_z.

8.3

a) $U(\alpha) \mid A \rangle \xrightarrow{\text{SR}} e^{-\alpha\,\partial/\partial\phi} \psi_A(r,\, \theta,\, \phi) = \displaystyle\sum_n \frac{(-\alpha)^n}{n!} \Big(\frac{\partial}{\partial\phi}\Big)^n \psi_A(r,\, \theta,\, \phi)$

$= \psi_A(r,\, \theta,\, \phi - \alpha)\,.$

In Cartesian coordinates:

$$\begin{cases} x = r\sin\theta\cos\phi \xrightarrow{U(\alpha)} r\sin\theta\cos(\phi-\alpha) = x\cos\alpha + y\sin\alpha \\ y = r\sin\theta\sin\phi \xrightarrow{U(\alpha)} r\sin\theta\sin(\phi-\alpha) = y\cos\alpha - x\sin\alpha \end{cases} \Rightarrow$$

$$U(\alpha)\,\psi(x,y,z) = \psi(x\cos\alpha + y\sin\alpha,\ y\cos\alpha - x\sin\alpha,\ z)$$

therefore $U(\alpha)$ is the operator that rotates the state of the system counterclockwise by the angle α around the z axis.

b) i) In the Cartesian basis $|\mathcal{P}_i\rangle \xrightarrow{\text{SR}} x_i\,f(r)$ and, thanks to the above result, one has:

$$\begin{cases} U(\alpha)\,|\mathcal{P}_1\rangle \xrightarrow{\text{SR}} (x\cos\alpha + y\sin\alpha)\,f(r) = \cos\alpha\,|\mathcal{P}_1\rangle + \sin\alpha\,|\mathcal{P}_2\rangle \\ U(\alpha)\,|\mathcal{P}_2\rangle \xrightarrow{\text{SR}} (y\cos\alpha - x\sin\alpha)\,f(r) = \cos\alpha\,|\mathcal{P}_2\rangle - \sin\alpha\,|\mathcal{P}_1\rangle \\ U(\alpha)\,|\mathcal{P}_3\rangle \xrightarrow{\text{SR}} z\,f(r) = |\mathcal{P}_3\rangle \end{cases}$$

so that on the states with $l=1$:

$$U(\alpha) \xrightarrow{\text{cb}} \begin{pmatrix} \cos\alpha & -\sin\alpha & 0 \\ \sin\alpha & \cos\alpha & 0 \\ 0 & 0 & 1 \end{pmatrix}.$$

ii) From the expression of L_z in the basis $|\mathcal{P}_i\rangle$, found in Problem 8.2, one has:

$$\left(\frac{L_z}{\hbar}\right)^2 \xrightarrow{\text{cb}} \begin{pmatrix} 1 & 0 & 0 \\ 0 & 1 & 0 \\ 0 & 0 & 0 \end{pmatrix}, \quad \left(\frac{L_z}{\hbar}\right)^3 \xrightarrow{\text{cb}} \begin{pmatrix} 0 & -i & 0 \\ i & 0 & 0 \\ 0 & 0 & 0 \end{pmatrix}, \quad \left(\frac{L_z}{\hbar}\right)^4 \xrightarrow{\text{cb}} \begin{pmatrix} 1 & 0 & 0 \\ 0 & 1 & 0 \\ 0 & 0 & 0 \end{pmatrix}$$

so in the series expansion of $U(\alpha)$ the terms of even order contribute to the diagonal part $(\cos\alpha,\ \cos\alpha,\ 1)$, those of odd order to the sine.

iii) The matrix that transforms the Cartesian into the spherical basis is (its columns are the vectors of the 'arrival' basis: $M\,|\mathcal{P}_1\rangle = |\mathcal{P}_+\rangle$, $M\,|\mathcal{P}_2\rangle = |\mathcal{P}_0\rangle$, $M\,|\mathcal{P}_3\rangle = |\mathcal{P}_-\rangle$)

$$M = \frac{1}{\sqrt{2}} \begin{pmatrix} 1 & 0 & 1 \\ i & 0 & -i \\ 0 & \sqrt{2} & 0 \end{pmatrix}, \quad M^{-1} = M^\dagger = \frac{1}{\sqrt{2}} \begin{pmatrix} 1 & -i & 0 \\ 0 & 0 & \sqrt{2} \\ 1 & i & 0 \end{pmatrix}$$

and $U(\alpha)$ in the spherical basis is $U_{mm'} = \begin{pmatrix} e^{-i\alpha} & 0 & 0 \\ 0 & 1 & 0 \\ 0 & 0 & e^{i\alpha} \end{pmatrix}$; so in the Cartesian basis:

$$[U(\alpha)]_{ij} = \sum_{mm'} \langle i\,|\,m\rangle\, U_{mm'}\, \langle m'\,|\,j\rangle = [M\,U\,M^{-1}]_{ij}.$$

c) Let $|A\rangle \xrightarrow{\text{SR}} \psi_A(r,\theta,\phi)$ and let us put $(q_1+i q_2)\,|A\rangle \xrightarrow{\text{SR}} \tilde{\psi}_A(r,\theta,\phi)$. One has:

$$\tilde{\psi}_A(r,\theta,\phi) \equiv (x+iy)\,\psi_A(r,\theta,\phi) = r\sin\theta\,e^{i\phi}\,\psi_A(r,\theta,\phi).$$

$$U(\alpha)\,(q_1+i q_2)\,|A\rangle \xrightarrow{\text{SR}} U(\alpha)\,\tilde{\psi}_A(r,\theta,\phi) = \tilde{\psi}_A(r,\theta,\phi-\alpha)$$

$$= r\sin\theta\,e^{i(\phi-\alpha)}\,\psi_A(r,\theta,\phi-\alpha) = e^{-i\alpha}\,(x+iy)\,U(\alpha)\,\psi_A(r,\theta,\phi)$$

that is the Schrödinger representation of $e^{-i\alpha}(q_1 + i q_2) U(\alpha) | A \rangle$. Since $| A \rangle$ is arbitrary:

$$U(\alpha)(q_1 + i q_2) U^{-1}(\alpha) = e^{-i\alpha}(q_1 + i q_2)$$

$$U(\alpha)(q_1 - i q_2) U^{-1}(\alpha) = e^{+i\alpha}(q_1 - i q_2)$$

(the second obtains by taking the Hermitian conjugate of the first). So:

$$\begin{cases} U(\alpha) q_1 U^{-1}(\alpha) = q_1 \cos\alpha + q_2 \sin\alpha \\ U(\alpha) q_2 U^{-1}(\alpha) = q_2 \cos\alpha - q_1 \sin\alpha \\ U(\alpha) q_3 U^{-1}(\alpha) = q_3 . \end{cases}$$

d) From the commutation rules $[J_i , V_j] = i\hbar \epsilon_{ijk} V_k$ one has:

$$[J_3 , V_1 + i V_2] = \hbar(V_1 + i V_2) \quad \Rightarrow$$

$$J_3 (V_1 + i V_2) | m , \cdots \rangle = (V_1 + i V_2) J_3 | m , \cdots \rangle + \hbar(V_1 + i V_2) | m , \cdots \rangle$$

$$= (m + 1)\hbar(V_1 + i V_2) | m , \cdots \rangle .$$

e) $e^{-i J_3 \alpha/\hbar}(V_1 + i V_2) | m , \cdots \rangle = e^{-i(m+1)\alpha}(V_1 + i V_2) | m , \cdots \rangle$

$$= e^{-i\alpha}(V_1 + i V_2) e^{-i J_3 \alpha/\hbar} | m , \cdots \rangle$$

and, the set of the eigenvectors $| m , \cdots \rangle$ of J_3 being complete, one has:

$$U(\alpha)(V_1 + i V_2) U^{-1}(\alpha) = e^{-i\alpha}(V_1 + i V_2) .$$

8.4

a) Any component of the angular momentum, restricted to states of total angular momentum 1, has the eigenvalues $\pm 1, 0$: the displayed identity is the Hamilton–Cayley identity that can be immediately verified in the basis in which L_n is diagonal:

$$L_n (L_n + 1)(L_n - 1) | m_n = \pm 1, 0 \rangle = 0 .$$

b) We may consider – with no loss of generality – the case where the state is an eigenstate of L_z: indeed, the components of the angular momentum transforming under rotations as a vector, it is possible (by means of a rotation U in the plane containing both $\hat{n}$ and the z axis) to transform L_n into L_z: $U L_n U^{-1} = L_z$ and, correspondingly, $| A' \rangle \equiv U | A \rangle$ is an eigenvector of L_z. From the commutation rules:

$$0 = \langle A' | [L_z , L_x] | A' \rangle = i\hbar \langle A' | L_y | A' \rangle$$

likewise for L_x and, therefore, for all the linear combinations of L_x and L_y. Otherwise, without making use of rotations, if $\hat{m}$ is a unit vector orthogonal to $\hat{n}$ and $\hat{\rho} \equiv \hat{m} \wedge \hat{n}$, the statement follows directly from $[L_n , L_\rho] = i\hbar L_m$.

c) One can make the calculation either directly (the matrices representing L_x, L_y, L_z in the Cartesian basis have been calculated in Problem 8.2) or as a particular case of the answer to the following question. If instead one observes that α_1, α_2, α_3, being real, are the components of a vector $\vec{\alpha}$, putting $\vec{\alpha} = |\vec{\alpha}|\hat{n}$, one has $|A\rangle \xrightarrow{\text{SR}} \vec{r} \cdot \vec{\alpha} f(r)$ that is the eigenstate of L_n corresponding to the eigenvalue 0 and in addition, thanks to the result of b), also the orthogonal components of $\vec{L}$ have vanishing mean value.

d) By direct calculation (see Problem 8.2):

$$\bar{L}_x = \hbar \left(\alpha_1^* \; \alpha_2^* \; \alpha_3^* \right) \begin{pmatrix} 0 & 0 & 0 \\ 0 & 0 & -i \\ 0 & i & 0 \end{pmatrix} \begin{pmatrix} \alpha_1 \\ \alpha_2 \\ \alpha_3 \end{pmatrix} = 2\hbar\, \Im(\alpha_2^* \alpha_3)$$

and likewise $\bar{L}_y = 2\hbar\, \Im(\alpha_3^* \alpha_1)$, $\bar{L}_z = 2\hbar\, \Im(\alpha_1^* \alpha_2)$.

One has $i\left[\vec{\alpha} \wedge \vec{\alpha}^*\right]_1 = i(\alpha_2 \alpha_3^* - \alpha_3 \alpha_2^*) = 2\,\Im(\alpha_2^* \alpha_3)$ and similarly for the others components.

e) Let us put $\alpha_1 = |\alpha_1| e^{i\varphi_1}$, $\alpha_2 = |\alpha_2| e^{i\varphi_2}$, $\alpha_3 = |\alpha_3| e^{i\varphi_3}$. From the above result:

$$|\alpha_1 \alpha_2| \sin(\varphi_2 - \varphi_1) = |\alpha_1 \alpha_3| \sin(\varphi_1 - \varphi_3) = |\alpha_2 \alpha_3| \sin(\varphi_3 - \varphi_2) = 0 .$$

If two coefficients – for example α_1 and α_2 – vanish, the statement is true ($|A\rangle$ is an eigenstate of L_z); if one coefficient vanishes – for example α_1 – then $\varphi_3 = \varphi_2$ and, up to the common phase factor, α_1, α_2, α_3 are real and, thanks to c), the thesis follows; if all the coefficients are nonvanishing, one has $\varphi_1 = \varphi_2 = \varphi_3$ and the same conclusion applies. In any event, up to a phase factor, $\vec{\alpha}$ is a real vector and $\hat{n} \propto \vec{\alpha}$.

Otherwise: if $\vec{\alpha} \wedge \vec{\alpha}^* = 0$, it follows that $\vec{\alpha}$ and $\vec{\alpha}^*$ are parallel, therefore proportional to each other: $\vec{\alpha} = e^{i\varphi} \vec{\alpha}^*$, so, up to the common phase factor $e^{i\varphi/2}$, α_1, α_2, α_3 are real.

8.5

a) The operator $\vec{L} - L_n \hat{n}$ only has components orthogonal to $\hat{n}$, whence:

$$\langle A \,|\, \vec{L} - L_n \hat{n} \,|\, A \rangle = 0 \quad \Rightarrow \quad \langle A \,|\, \vec{L} \,|\, A \rangle = \langle A \,|\, L_n \,|\, A \rangle \hat{n} = \lambda \hat{n} .$$

b) If $\langle A \,|\, \vec{L} \,|\, A \rangle = \pm \hbar l \hat{n}$, then $\langle A \,|\, L_n \,|\, A \rangle = \pm \hbar l$. The mean value of any component L_n in the states of angular momentum l always is between the maximum eigenvalue $\hbar l$ and the minimum eigenvalue $-\hbar l$ and takes the maximum (minimum) value only if the state is the corresponding eigenstate of L_n:

$$|A\rangle = \sum_{mn} \alpha_{mn} \,|\, l, m_n \rangle \quad \Rightarrow$$

$$\bar{L}_n = \hbar \sum_{mn} |\alpha_{mn}|^2 \, m_n \leq \hbar\, m_n^{\max} \sum_{mn} |\alpha_{mn}|^2 = \hbar l$$

and similarly $\bar{L}_n \geq -\hbar l$; $\bar{L}_n = \pm \hbar l$ if and only if $|\alpha_{mn}| = \delta_{mn, \pm l}$.

c) From d) of Problem 8.4 one has that $\bar{L}_x = \bar{L}_y = 0$ and $\bar{L}_z = \hbar \sin 2\gamma$. For $\gamma = \pm\pi/4$, $\bar{L}_z = \pm\hbar$ and the state is, as a consequence, the eigenstate of

L_z belonging to the eigenvalue $\pm \hbar$; for $\gamma = 0$ the state is the eigenstate of L_x to the eigenvalue 0; for $\gamma = \pi/2$ it is an eigenstate of L_y. For no other value of γ it is an eigenstate of some component of $\vec{L}$: indeed $\sin 2\gamma \neq \pm 1,\, 0$.

d) If $L_n | A \rangle = \pm \hbar | A \rangle$, one has $\langle A \mid \vec{L} \mid A \rangle = \pm \hbar \hat{n} = \mathrm{i} \hbar \vec{\alpha} \wedge \vec{\alpha}^*$ (see Problem 8.4), therefore (sum over repeated indices is understood):

$$1 = n_i\, n_i = -\epsilon_{ijk}\, \alpha_j\, \alpha_k^*\, \epsilon_{ilm}\, \alpha_l\, \alpha_m^* = -(\delta_{jl}\delta_{km} - \delta_{jm}\delta_{kl})\, \alpha_j\, \alpha_k^*\, \alpha_l\, \alpha_m^*$$

$$= -(\vec{\alpha} \cdot \vec{\alpha})(\vec{\alpha}^* \cdot \vec{\alpha}^*) + (\vec{\alpha} \cdot \vec{\alpha}^*)^2 = -(\vec{\alpha} \cdot \vec{\alpha})(\vec{\alpha}^* \cdot \vec{\alpha}^*) + 1 \;\Rightarrow\; \vec{\alpha} \cdot \vec{\alpha} = 0 .$$

Vice versa, if $\vec{\alpha} \cdot \vec{\alpha} = 0$ then $(\mathrm{i}\vec{\alpha} \wedge \vec{\alpha}^*) \cdot (\mathrm{i}\vec{\alpha} \wedge \vec{\alpha}^*) = 1$ so, putting $\hat{n} = \mathrm{i}\vec{\alpha} \wedge \vec{\alpha}^*$, one has $\overline{L}_n = \hbar$.

8.6

a) No: the independent homogeneous polynomials of degree 2 are 6: $x_i\, x_j$ ($i,\, j = 1,\, 2,\, 3$), whereas the spherical harmonics with $l = 2$ are 5; indeed $r^{-2}(x^2 + y^2 + z^2)\, f(r) = f(r)$ is the wavefunction of a state with $l = 0$. Instead the polynomials $\alpha x + \beta y + \gamma z$ give rise to states with $l = 1$ (see Problem 8.1).

b) One has $z = r \cos \theta$, $(x + \mathrm{i}y) = r \sin \theta\, \mathrm{e}^{\mathrm{i}\phi}$ whence:

$$\cos^2 \theta\, \mathrm{e}^{2\mathrm{i}\phi} = \frac{1}{r^2}\left(z^2\, \frac{(x + \mathrm{i}y)^2}{x^2 + y^2} \right)$$

$$\neq \frac{1}{r^2} \times \text{(homogeneous polynomial of degree 2)} .$$

Likewise $\sin \theta \cos \theta\, \mathrm{e}^{2\mathrm{i}\phi} = r^{-2}\left(z\, (x + \mathrm{i}y)^2 / \sqrt{x^2 + y^2} \right)$. Instead:

$$\sin^2 \theta\, \mathrm{e}^{2\mathrm{i}\phi} = \frac{(x + \mathrm{i}y)^2}{r^2} ; \qquad \sin \theta \cos \theta\, \mathrm{e}^{\mathrm{i}\phi} = \frac{z\, (x + \mathrm{i}y)}{r^2} .$$

The degree of the polynomial says that in both cases $l = 2$, while the dependence on ϕ says that the first is proportional to $Y_{2,2}(\theta,\, \phi)$ and the second to $Y_{2,1}(\theta,\, \phi)$.

c) The polynomial one is after must be invariant under rotations around the z axis, so it has the form $a\,(x^2 + y^2) + b\, z^2$. The one that gives rise to $Y_{2,0}(\theta,\, \phi)$ must be orthogonal to $Y_{0,0}(\theta,\, \phi)$, which is a constant, so:

$$0 = \int \frac{a\,(x^2 + y^2) + b\, z^2}{r^2}\, \mathrm{d}\Omega = 2\pi \int_{-1}^{+1} (a \sin^2 \theta + b \cos^2 \theta)\, \mathrm{d} \cos \theta$$

$$= 2\pi\left(2\, a + \frac{2}{3}\,(b - a) \right) \;\Rightarrow\; b = -2\, a$$

$$\Rightarrow\; Y_{2,0}(\theta,\, \phi) \propto \frac{x^2 + y^2 - 2\, z^2}{r^2} = 1 - 3 \cos^2 \theta .$$

d) If I_y stands for the space inversion with respect to the plane $y = 0$: $x \to x$, $y \to -y$, $z \to z$, one has $I_y\, L_z\, I_y^{-1} = -L_z$, $I_y\, \vec{L}^2\, I_y^{-1} = \vec{L}^2 \Rightarrow$ $I_y\, |\, l,\, m\,\rangle = |\, l,\, -m\,\rangle$ therefore, as $y \to -y \Rightarrow \phi \to -\phi$, $Y_{l,-m}(\theta,\, \phi) = Y_{l,m}(\theta,\, -\phi)$, up to a phase factor that is usually chosen equal to ± 1: we shall put it equal to 1.

Due to b) and c) and to the above result (the calculation of the normalization factors requires the calculation of elementary integrals):

$$Y_{2,\pm 2}(\theta,\, \phi) \;=\; \sqrt{\frac{15}{32\,\pi}}\, \sin^2\theta\, e^{\pm 2i\phi} \;=\; \sqrt{\frac{15}{32\,\pi}}\, \frac{(x \pm iy)^2}{r^2}$$

$$Y_{2,\pm 1}(\theta,\, \phi) \;=\; \sqrt{\frac{15}{8\,\pi}}\, \sin\theta\cos\theta\, e^{\pm i\phi} \;=\; \sqrt{\frac{15}{8\,\pi}}\, \frac{z\,(x \pm iy)}{r^2}$$

$$Y_{2,0}(\theta,\, \phi) \;=\; \sqrt{\frac{5}{16\,\pi}}\, (1 - 3\cos^2\theta) \;=\; \sqrt{\frac{5}{16\,\pi}}\, \frac{r^2 - 3z^2}{r^2}\,.$$

8.7

a) Since rotations induce homogeneous linear transformations on the coordinates: $x_i' = R_{ij}x_j$ (see Problem 8.3), the degree of the polynomials does not change. Therefore $\mathcal{H}^{(n)}$ is also invariant under the L_i, that are the generators of the rotation group: for example, $L_z = -i\,\hbar(x\,\partial/\partial y - y\,\partial/\partial x)$ does not change the degree of polynomials.

$$\mathcal{H}^{(n)} \supset \frac{1}{r^n} \times (x^2 + y^2 + z^2) \times (\text{homog. polynomials of degree } n-2) \times f(r)$$

$$= \frac{1}{r^{n-2}} \times (\text{homog. polynomials of degree } n-2) \times f(r) = \mathcal{H}^{(n-2)}\,.$$

b) Among the polynomials of degree n there is $(x + iy)^n \propto e^{in\phi}$, therefore there is $m = n$ which obviously is the maximum value of m. It follows that $l_{\max} \geq n$, but since the polynomials have degree n (not higher), it follows that (see the text of Problem 8.6) $l_{\max} = n$. If n is even, $l_{\min} = 0$ since there is the polynomial $(x^2 + y^2 + z^2)^{n/2}$; if instead n is odd, $l_{\min} = 1$ because there are the polynomials $(\alpha\, x + \beta\, y + \gamma\, z) \times (x^2 + y^2 + z^2)^{(n-1)/2}$. As $\mathcal{H}^{(n)} \supset \mathcal{H}^{(n-2)} \supset \mathcal{H}^{(n-4)} \cdots$, $\vec{L}^2$ may only take the values $l(l+1)\hbar^2$ with $l = n$, $n-2$, $n-4$ $\cdots$ down to either $l = 0$ or 1.

c) Thanks to the above result:

$$Y_{n,n} \propto \frac{1}{r^n}\, (x + iy)^n = (\sin\theta)^n\, e^{in\phi}\,.$$

Thanks to the fact that the only polynomial of degree n that contains $(x + iy)^{n-1}$ is $z\,(x + iy)^{n-1}$, one has:

$$Y_{n,n-1} \propto \frac{1}{r^n}\, z\,(x + iy)^{n-1} = (\sin\theta)^{n-1}\cos\theta\, e^{i\,(n-1)\,\phi}\,.$$

d) There are two polynomials with $m = n - 2$: $(x + iy)^{n-2}(x^2 + y^2)$ and $(x + iy)^{n-2} z^2$. In $\mathcal{H}^{(n)}$ the wavefunctions with $m = n - 2$ are linear combinations only of wavefunctions with $l = n$ and $l = n-2$ $(n \geq l \geq m)$: indeed the sum of the two is $(x + iy)^{n-2} \times r^2$, that is proportional to $Y_{n-2,n-2}$, therefore the linear combination orthogonal (with respect to the integration over the angles) to the previous one is $Y_{n,n-2}$.

Since $\vec{L}^2$ has the eigenvalues $l = n, \; n-2, \; \cdots \;$ but not $l = n-1$, $n-3, \; \cdots \;$, there are always two polynomials with $m = n-3$: $(x + iy)^{n-3}(x^2 + y^2) z$ and $(x + iy)^{n-3} z^3$, the sum of which again is $r^2 Y_{n-2,n-3}$, so that the orthogonal linear combination provides $Y_{n,n-3}$. For $m = n-4$ things change: indeed, the polynomials of degree n with $m = n-4$ are three since they must give rise to $Y_{n,n-4}$, $Y_{n-2,n-4}$, $Y_{n-4,n-4}$; for example:

$$p_0 = (x+iy)^{n-4}(x^2+y^2)^2, \; p_1 = (x+iy)^{n-4}(x^2+y^2)z^2, \; p_2 = (x+iy)^{n-4}z^4.$$

The two linear combinations:

$$p_0 + p_1 = (x+iy)^{n-4}(x^2+y^2) \times r^2, \quad p_1 + p_2 = (x+iy)^{n-4}z^2 \times r^2$$

are, up to the factor r^2 that is irrelevant for angular momentum, homogeneous polynomials of degree $n-2$: from these it is possible to extract $Y_{n-4,n-4} \propto p_0 + 2p_1 + p_2 = (x+iy)^{n-4}$ and $Y_{n-2,n-4}$. The linear combination orthogonal to the two previous ones provides $Y_{n,n-4}$.

8.8

a) It is sufficient to recall that, in the Schrödinger representation, the eigenfunctions of L_z are proportional to $e^{im\phi}$ (ϕ is the azimuth angle): for $\xi_i = x$ one has:

$$x \propto \cos\phi \propto e^{i\phi} + e^{-i\phi} \; \Rightarrow \; q_1 \,|\, l'\,m'\rangle \xrightarrow{\text{SR}} f(r,\theta)\left(e^{i(m'+1)\phi} + e^{i(m'-1)\phi}\right)$$

so either $m'' = m'+1$ or $m'' = m'-1$ must be fulfilled, i.e. $\Delta m = \pm 1$. Likewise for y: $\Delta m = \pm 1$; whereas for z: $\Delta m = 0$.

$$x^2 \propto e^{2i\phi} + e^{-2i\phi} + 2 \; \Rightarrow \; \Delta m = \pm 2, 0; \quad y^2 \to \Delta m = \pm 2, 0;$$

$$z^2 \to \Delta m = 0; \quad xy \propto \sin 2\phi \to \Delta m = \pm 2;$$

$$x^2 - y^2 \to \Delta m = \pm 2; \quad x^2 + y^2 \to \Delta m = 0.$$

b) From Problem 8.7, denoting by $\mathcal{P}^{(l)}(x,y,z)$ a generic polynomial of degree l in x,y,z, one has:

$$q_1 \,|\, l\rangle \xrightarrow{\text{SR}} x \times \frac{\mathcal{P}^{(l)}(x,y,z)}{r^l} \times f(r) = \frac{\mathcal{P}^{(l+1)}(x,y,z)}{r^{l+1}} \times g(r), \quad g(r) = r\,f(r)$$

so $q_1 \,|\, l\,m\rangle \in \mathcal{H}^{(l+1)} \supset \mathcal{H}^{(l-1)} \supset \mathcal{H}^{(l-3)} \cdots$

whence the thesis follows. For y and z one may proceed in the same way.

c) From the above result, it follows that a necessary condition in order that $\langle l'' \,|\, q_i \,|\, l'\rangle \neq 0$, is either $l'' = l'+1$, or $l'' = l'-1$, or $l'' = l'-3$, $\cdots$, namely $l''-l' = 1, -1, -3, \cdots$; but in the same way one obtains that, in order that $\langle l' \,|\, q_i \,|\, l''\rangle \neq 0$, it is necessary that $l'-l'' = 1, -1, -3, \cdots$, therefore $\Delta l \equiv l''-l' = \pm 1$.

d) The procedure is the same for all the expressions that are quadratic in x, y, z: for example, if $\xi_i = x^2$, one has, as above,

$$q_i^2 \,|\,l'\,\rangle = \alpha\,|\,l'+2\,\rangle + \beta\,|\,l'\,\rangle + \gamma\,|\,l'-2\,\rangle + \delta\,|\,l'-4\,\rangle + \cdots$$

(obviously, the vector $|\,l'\,\rangle$ on the right hand side is different from the vector $|\,l'\,\rangle$ on the left hand side) whence $l''-l' = 2,\,0,\,-2,\,-4,\,\cdots$, but also $l'-l'' = 2,\,0,\,-2,\,-4,\,\cdots$ and, in conclusion, $\Delta l = \pm 2,\,0$. Of course, the above conditions are not sufficient to guarantee that $\langle\,l''\mid \xi_i \mid l'\,\rangle \neq 0$: for example, for the operator $\xi = x^2 + y^2 + z^2$ the selection rule is $\Delta l = 0$; the same reasoning applies for expressions of degree higher than 2: while, for example, for the operator $x^2\,y$ the selection rule is $\Delta l = \pm 3,\,\pm 1$, for the operator $x\,(x^2 + y^2 + z^2)$ it is $\Delta l = \pm 1$.

e) Selection rules on m: let (Λ is arbitrary)

$$U\,q_i\,U^{-1} = \Lambda\,p_i\,, \qquad U\,p_i\,U^{-1} = -\Lambda^{-1}q_i\,.$$

As $U\,L_i\,U^{-1} = L_i$, one has $U\,|\,m\,\rangle = |\,m\,\rangle'$ and, as a consequence,

$$\Lambda\,p_x\,|\,m\,\rangle = U^{-1}q_1\,U\,|\,m\,\rangle = U^{-1}q_1\,|\,m\,\rangle' = U^{-1}\big(\alpha\,|\,m+1\,\rangle + \beta\,|\,m-1\,\rangle\big)$$
$$= \alpha\,|\,m+1\,\rangle' + \beta\,|\,m-1\,\rangle'.$$

By direct calculation one obtains:

$$p_x\,|\,m\,\rangle \xrightarrow{\text{SR}} -\mathrm{i}\hbar\,\frac{\partial}{\partial x}\big(f(\rho,\,z)\,\mathrm{e}^{\mathrm{i}\,m\,\phi}\big) = -\mathrm{i}\hbar\Big(\frac{x}{\rho}\frac{\partial f}{\partial\rho} + \mathrm{i}\,m\,\frac{\partial\phi}{\partial x}\,f\Big)\mathrm{e}^{\mathrm{i}\,m\,\phi}$$

$$= -\mathrm{i}\hbar\Big(\frac{x}{\rho}\frac{\partial f}{\partial\rho} - \mathrm{i}\,m\,\frac{y}{\rho^2}\,f\Big)\mathrm{e}^{\mathrm{i}\,m\,\phi} = g(\rho,\,z)\,\mathrm{e}^{\mathrm{i}\,(m-1)\,\phi} + h(\rho,\,z)\,\mathrm{e}^{\mathrm{i}\,(m+1)\,\phi}$$

$$\Rightarrow \quad \Delta m = \pm 1$$

and the same for p_y, while for p_z one has $\Delta m = 0$ so that, for example,

$$p_y\,p_x\,|\,m\,\rangle = p_y\big(\alpha\,|\,m+1\,\rangle + \beta\,|\,m-1\,\rangle\big)$$
$$= \alpha'\,|\,m+2\,\rangle + \gamma\,|\,m\,\rangle + \beta'\,|\,m-2\,\rangle\,.$$

Selection rules on l: since $U\,|\,l\,\rangle = |\,l\,\rangle'$, exactly as above, one can show that $p_i\,|\,l\,\rangle = \alpha\,|\,l+1\,\rangle + \beta\,|\,l-1\,\rangle$; or by direct calculation:

$$p_x\,|\,l\,\rangle \xrightarrow{\text{SR}} -\mathrm{i}\hbar\,\frac{\partial}{\partial x}\Big(\frac{\mathcal{P}^{(l)}(x,\,y,\,z)}{r^l}\times f(r)\Big)$$

$$= -\mathrm{i}\hbar\Big(\frac{\mathcal{P}^{(l-1)}(x,\,y,\,z)}{r^{l-1}}\times\frac{f(r)}{r}\Big)$$

$$+\,\mathrm{i}\hbar\Big(\frac{\mathcal{P}^{(l+1)}(x,\,y,\,z)}{r^{l+1}}\times\big(l\,\frac{f(r)}{r} - f'(r)\big)\Big)\,.$$

8.9

a) The state with wavefunction ψ_1 is an eigenstate of L_z belonging to the eigenvalue 0 and is left unchanged by the measurement. The same is true for ψ_2 and ψ_4, since $x^2 + y^2$ is independent of ϕ.

$$\psi_3 \propto \cos^2\phi - \sin^2\phi = \frac{1}{2}\big(\mathrm{e}^{2\mathrm{i}\,\phi} + \mathrm{e}^{-2\mathrm{i}\,\phi}\big) \to L_z = \pm 2\,.$$

Since $\psi_3 \propto (x+iy)^2 f(r) + (x-iy)^2 f(r)$, if a measurement of L_z gives 2, then $\psi_{3'} = (x+iy)^2 f(r)$; if instead $L_z = -2$, $\psi_{3''} = (x-iy)^2 f(r)$.

b) $\psi_1 \rightarrow l = 0,\ 1$; $l = 0 \Rightarrow \psi_{1'} = a\, f(r)$; $l = 1 \Rightarrow \psi_{1''} = z\, f(r)$.

The wavefunction ψ_2 gives rise to both $l = 0$ and $l = 2$ ($x^2 + y^2$ is a homogeneous polynomial of degree 2, but it is not orthogonal to $Y_{0,0}$ for it has a nonvanishing angular mean); after the measurement, if $l = 0$, $\psi_{2'} = r^2 f(r)$; if $l = 2$, as ψ_2 is independent of ϕ, $\psi_{2''} = r^2 Y_{2,0}(\theta,\ \phi)\, f(r)$. Both ψ_3 and ψ_4 are eigenstates of $\vec{L}^2$ with $l = 2$: indeed they are homogeneous polynomials of degree 2 with vanishing angular mean.

c) One must write ψ_2 as $\left(\alpha\, Y_{2,0}(\theta,\ \phi) + \beta\, Y_{0,0}(\theta,\ \phi)\right) r^2 f(r)$ and the required probabilities are $|\alpha|^2/(|\alpha|^2 + |\beta|^2)$ and $|\beta|^2/(|\alpha|^2 + |\beta|^2)$. By using the expressions of the normalized spherical harmonics, found in Problem 8.6, and given that $Y_{0,0} = 1/\sqrt{4\pi}$, one has:

$$\frac{1}{r^2}(x^2 + y^2) = \sin^2\theta = 1 - \cos^2\theta = \frac{1}{\sqrt{4\pi}}\left(\alpha\sqrt{\frac{5}{4}}(1 - 3\cos^2\theta) + \beta\right)$$

$$\Rightarrow \quad \alpha \propto \sqrt{4/5},\quad \beta \propto 2$$

and therefore $P(l = 2) = 1/6$, $P(l = 0) = 5/6$.

8.10

a) One has $\overline{L_x^2} + \overline{L_y^2} = \overline{\vec{L}^2} - \overline{L_z^2} = \left(l(l+1) - m^2\right)\hbar^2$, furthermore $\overline{L_x^2} = \overline{L_y^2}$ because the eigenstates of L_z are invariant under rotations around the z axis and, by means of a rotation of $\pi/2$, $L_x \rightarrow L_y$, whence $\overline{L_x^2} = \frac{1}{2}\left(l(l+1) - m^2\right)\hbar^2$.

b) It is necessary to express the state $|l = 1,\ m = 1\rangle$, that in the Schrödinger representation is proportional to $(x+iy)/\sqrt{2}$, as a linear combination of the eigenstates of L_x, that in the Schrödinger representation are proportional (with the same proportionality factor) to x, $(y \pm iz)/\sqrt{2}$:

$$\frac{x+iy}{\sqrt{2}} = \frac{1}{\sqrt{2}}x + \frac{i}{2}\frac{y+iz}{\sqrt{2}} + \frac{i}{2}\frac{y-iz}{\sqrt{2}}$$

so the probabilities of finding the eigenvalues $0,\ 1,\ -1$ respectively are $\frac{1}{2}, \frac{1}{4}, \frac{1}{4}$ and $\overline{L_x^2} = \hbar^2/2$.

8.11

a) It is not an eigenstate of any of the three operators (see Problem 8.1):

$$L_x\left(x\,y\,f(r)\right) = x\,[L_x,\ y]\,f(r) = i\hbar\,x\,z\,f(r),\quad L_y\left(x\,y\,f(r)\right) = -i\hbar\,y\,z\,f(r)$$

$$L_z\left(x\,y\,f(r)\right) = [L_z,\ x\,y]\,f(r) = -i\hbar\,(x^2 - y^2)\,f(r)\ .$$

b) The wavefunctions of all the states are proportional to homogeneous polynomials of degree 2 with vanishing angular mean, then they have $l = 2$. They are mutually orthogonal, therefore independent: for example

$$\langle\, \mathcal{D}_{1,2} \mid \mathcal{D}_{2,3} \,\rangle = \int x\,z\,y^2\,|f(r)|^2\,dV = 0$$

because the integrand is odd in x (or in z).

c) They are all states with $l = 2$ (homogeneous polynomials of degree 2 with vanishing angular mean). They are not independent:

$$|\,\mathcal{D}_{1,1}\,\rangle + |\,\mathcal{D}_{2,2}\,\rangle + |\,\mathcal{D}_{3,3}\,\rangle = 0\,.$$

d) From Problem 8.3:

$$U(\alpha)\,\psi(x,y,z) = \psi(x\cos\alpha + y\sin\alpha,\ y\cos\alpha - x\sin\alpha,\ z)$$

so that, for example,

$$U(\alpha)\,x\,y\,f(r) = (x\cos\alpha + y\sin\alpha)\,(y\cos\alpha - x\sin\alpha)\,f(r)$$
$$= \big(-(x^2 - y^2)\sin\alpha\cos\alpha + x\,y\,(\cos^2\alpha - \sin^2\alpha)\big)\,f(r)$$

but $x^2 - y^2 = (x^2 - \tfrac{1}{3}r^2) - (y^2 - \tfrac{1}{3}r^2)$, so:

$$U(\alpha)\,|\,\mathcal{D}_{1,2}\,\rangle = -\sin\alpha\cos\alpha\big(|\,\mathcal{D}_{1,1}\,\rangle - |\,\mathcal{D}_{2,2}\,\rangle\big) + (\cos^2\alpha - \sin^2\alpha)\,|\,\mathcal{D}_{1,2}\,\rangle$$
$$= R_{11}R_{21}\,|\,\mathcal{D}_{1,1}\,\rangle + R_{11}R_{22}\,|\,\mathcal{D}_{1,2}\,\rangle + R_{12}R_{21}\,|\,\mathcal{D}_{2,1}\,\rangle + R_{12}R_{22}\,|\,\mathcal{D}_{2,2}\,\rangle\,.$$

In general:

$$\begin{pmatrix} x' \\ y' \\ z' \end{pmatrix} = \begin{pmatrix} \cos\alpha & \sin\alpha & 0 \\ -\sin\alpha & \cos\alpha & 0 \\ 0 & 0 & 1 \end{pmatrix} \begin{pmatrix} x \\ y \\ z \end{pmatrix} \quad \Longleftrightarrow \quad x_i' = R_{ik}\,x_k$$

and $x_i'\,x_j' = \sum_{kl} R_{ik}\,R_{jl}\,x_k\,x_l$; in addition, from $R\ {}^tR = \mathbb{1}$ (${}^tR$ is the transpose of R) it follows that $\delta_{ij} = \sum_{kl} R_{ik}\,R_{jl}\,\delta_{kl}$; so, if one considers the matrix $x_i\,x_j - \tfrac{1}{3}r^2\,\delta_{ij}$ (symmetrical traceless tensor) one has

$$|\,\mathcal{D}_{i,j}\,\rangle \ \xrightarrow{\ \text{SR}\ } \ \big(x_i\,x_j - \tfrac{1}{3}r^2\,\delta_{ij}\big)\,f(r) \quad \text{and:}$$

$$U(\alpha)\,|\,\mathcal{D}_{i,j}\,\rangle \ \xrightarrow{\ \text{SR}\ } \ \big(x_i'\,x_j' - \frac{1}{3}r^2\,\delta_{ij}\big)\,f(r)$$

$$= \sum_{kl} R_{ik}(\alpha)\,R_{jl}(\alpha)\,\big(x_k\,x_l - \frac{1}{3}r^2\,\delta_{kl}\big)\,f(r)$$

namely:

$$U(\alpha)\,|\,\mathcal{D}_{i,j}\,\rangle = \sum_{kl} R_{ik}(\alpha)\,R_{jl}(\alpha)\,|\,\mathcal{D}_{k,l}\,\rangle\,.$$

It should be clear that the result holds for whatever rotation: only the form of the orthogonal matrix R_{ij} changes.

9

Changes of Frame

Wigner's theorem; active and passive point of view; reference frame: translated, rotated; in uniform motion; in free fall, rotating.

Note. *Let K_1 and K_2 be two frames, i.e. two laboratories, endowed – by hypothesis – with the same observables, therefore all the states that can be realized in one frame can be realized also in the other. The transformation from K_1 to K_2 is implemented by a unitary operator V that associates to any state $|A\rangle$ prepared in K_1 the state $|A^{\mathrm{tr}}\rangle$ – viewed from K_1 – prepared in K_2 in the same way as $|A\rangle$ was prepared in K_1: $|A^{\mathrm{tr}}\rangle = V|A\rangle$. With reference to K_1 and K_2, $|A\rangle$ and $|A^{\mathrm{tr}}\rangle$ are named subjectively identical states. Similarly, two observables associated to identical apparatuses, one placed in K_1, the other in K_2, are said subjectively identical observables.*

9.1 It is known (Wigner's theorem) that if a correspondence between states is such that the probability transitions are preserved, then the correspondence can be promoted to a transformation between the vectors representative of the states, implemented by an either unitary or antiunitary operator, the latter occurring if the transformation entails the reversal of time, i.e. in formulae:

$$\mathbf{S}_A \to \mathbf{S}_{A'} , \quad \mathbf{S}_B \to \mathbf{S}_{B'} ; \quad |\langle B\,|\,A\rangle|^2 = |\langle B'\,|\,A'\rangle|^2 \qquad \Rightarrow$$

(if the transformation does not entail reversal of time)

$$|A'\rangle = V|A\rangle , \quad |B'\rangle = V|B\rangle ; \qquad V^\dagger V = V\,V^\dagger = \mathbb{1} .$$

a) Let K_1 and K_2 be two frames and assume that in the two frames time goes on in the same direction. Make use of the concept of subjectively identical states to show that the transformation from K_1 to K_2 preserves the transition probabilities (and therefore, thanks to Wigner's theorem, is implemented by a unitary operator V, as asserted in the note above).

9.2 Let K_1 and K_2 be two frames and, with the notation introduced above, V the unitary operator that implements the transformation from K_1 to K_2: $|A^{\mathrm{tr}}\rangle = V|A\rangle$. ("Active point of view": in the

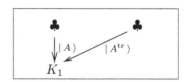

© Springer International Publishing AG 2017
E. d'Emilio and L.E. Picasso, *Problems in Quantum Mechanics*,
UNITEXT for Physics, DOI 10.1007/978-3-319-53267-7_9

same frame to any state and to any observable the transformed state and observable are associated).

a) Express the 'transformed observables' q', p', $[f(q, p)]'$ in terms of q, p (q', p' are subjectively identical to q, p).

b) Let $|A\rangle \xrightarrow{\text{SR}} \psi(x)$ be the wavefunction of some state in K_1. Write in K_2 the wavefunction of the subjectively identical state.

c) Let ξ_2 be in K_2 (the operator that represents) some observable. Write the operator η_1 that represents the *same* observable *seen* from K_1. Vice versa, given η_1 observable in K_1, write the operator ξ_2 that represents the same observable seen from K_2.

d) If a system in a given state $\mathbf{S}$ is observed from both frames and if $|A^{(1)}\rangle$ is the vector associated in K_1 to the state $\mathbf{S}$, which is the vector $|A^{(2)}\rangle$ associated in K_2 to the state $\mathbf{S}$? ("Passive point of view": the same state is observed from two frames).

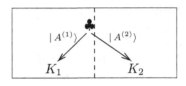

9.3   Let K_1 and K_2 be two frames and let $q' = V q V^\dagger$, $p' = V p V^\dagger$.

Assume that the time evolution operator $U_1(t)$ in K_1 is known.

a) Use the transformation law of states from the passive point of view (see Problem 9.2) and determine, in the general case when $V = V(t)$, the time evolution operator $U_2(t)$ in K_2.

b) In both cases when V is independent of time and when $V = V(t)$, find the Hamiltonian H_2 in the laboratory K_2 and verify that $H_2 = H_2^\dagger$.

c) Assume V is independent of time and that $|A, t\rangle$ is, in K_1, a solution of the Schrödinger equation:

$$i\hbar \frac{d}{dt} |A, t\rangle = H_1(q, p) |A, t\rangle .$$

Under what conditions is also $|A^{\text{tr}}, t\rangle$ a solution (always in K_1) of the Schrödinger equation?

9.4   Let K_1 and K_2 be two frames. Assume K_2 is obtained from K_1 by means of a translation of length a.

a) Write the transformation $(q, p) \rightarrow (q', p')$ of the canonical variables of a particle of mass m and the unitary operator that V implements it.

b) Let $|A\rangle \xrightarrow{\text{SR}} \psi_A(x) = e^{-\alpha x^2/2} e^{ikx}$ be (the vector that represents) some state in K_1. Write the wavefunction $\psi_{A^{\text{tr}}}(x)$ of the transformed state (active point of view: see Problem 9.2) and the wavefunction $\psi'_A(x)$

in the frame K_2 of the state $|A\rangle$ (i.e. the state $|A\rangle$ seen from K_2: passive point of view). Let $H_1(q, p)$ the be Hamiltonian in K_1: write the Hamiltonian $H_2(q, p)$ in K_2.

c) Let $|A, t\rangle$ solve in K_1 the Schrödinger equation:

$$i\hbar \frac{d}{dt}|A, t\rangle = H_1(q, p)|A, t\rangle .$$

Which condition must H_1 fulfill in order that also $|A^{tr}, t\rangle$ solves (still in K_1) the Schrödinger equation, for arbitrary a?

For a particle subject to the gravity the Hamiltonian is $H = p^2/2m + mgq$ (particle in free fall).

d) Show that the time evolution of the *states* (not of the vectors!) is invariant under translations, as in classical physics.

9.5 Assume the frame K_2 is obtained from K_1 through a counterclockwise rotation of angle ϕ around the z axis.

a) Write the transformation law $(q_i, p_i) \to (q_i', p_i')$ $\quad i = 1, 2, 3$, and the unitary operator V that implements it.

b) Let $|A\rangle \xrightarrow{\text{SR}} \psi_A(x, y, z) = e^{-\alpha(x^2+y^2+z^2)/2}e^{ikx}$ represent a state in K_1. Write the wavefunction $\psi_{A^{tr}}(x, y, z)$ of the transformed state and the wavefunction $\psi_A'(x, y, z)$ of the state $|A\rangle$ seen from K_2.
If $H_1(\vec{q}, \vec{p}) = \vec{p}^2/2m + V(\vec{q})$ is the Hamiltonian in K_1, write the Hamiltonian $H_2(\vec{q}, \vec{p})$ in K_2.

c) Let $|A, t\rangle$ solve, in K_1, the Schrödinger equation:

$$i\hbar \frac{d}{dt}|A, t\rangle = H_1(\vec{q}, \vec{p})|A, t\rangle .$$

Under what conditions does $|A^{tr}, t\rangle$ solve the Schrödinger equation in K_1 for arbitrary ϕ?

9.6 Assume the frame K_2 is in uniform rectilinear motion of velocity v with respect to the frame K_1.

a) Write the transformation law $(q, p) \to (q', p')$ of the canonical variables of a particle of mass m (in the Schrödinger picture for time evolution). Let, in K_1, $\bar{p}(t)$ and $\bar{q}(t)$ be the mean values of p and q in the state $|A, t\rangle$. Find the mean values of p and q in the transformed state $|A^{tr}, t\rangle$.

b) Find the (or a) unitary operator $G(v, t)$ that implements the transformation $(q, p) \to (q', p')$ (*Galilei transformation*) at time t:

$$q' = G(v, t)\, q\, G^\dagger(v, t), \qquad p' = G(v, t)\, p\, G^\dagger(v, t) .$$

c) Let $\psi_A(x)$ be the wavefunction of the state $|A\rangle$ at a given time $\bar{t}$. Write the wavefunction $\psi_{A^{tr}}(x)$ of the transformed state $|A^{tr}\rangle$.

d) If $H_1 = p^2/2m + V(q)$ is the Hamiltonian in K_1, write the Hamiltonian H_2 in K_2.

e) Assume that $H_1(q,\,p) = p^2/2m$ (free particle). If $|A,\,t\rangle$ is, in K_1, a solution of the Schrödinger equation:

$$i\hbar\,\frac{d}{dt}\,|A,\,t\rangle = H_1\,|A,\,t\rangle,$$

is $|A^{tr},\,t\rangle$ a solution (always in K_1) of the Schrödinger equation?

9.7   Consider a one-dimensional harmonic oscillator of mass m and angular frequency ω, whose center of oscillation moves with uniform velocity v.

a) Write the Hamiltonian of the system both in the laboratory frame and in the moving frame where the center of oscillation is at rest.

b) If, in the moving frame, the oscillator is in the ground state, which is its wavefunction in the frame of the laboratory?

c) Assume that the wavefunction in the laboratory $\psi(x,0)$ is known at time $t = 0$. Find the wavefunction $\psi(x,t)$ at time t. Explicitly verify that $\psi(x,t)$ satisfies the time dependent Schrödinger equation.

9.8   Let K_1 be an inertial frame and K_2 a frame that is in motion with respect to K_1 according to the time law $\xi(t)$.

a) Write the transformation law $(q,\,p) \to (q',\,p')$ of the canonical variables of a particle of mass m (in the Schrödinger picture for time evolution).

b) Let, in K_1, $\bar{q}(t)$ and $\bar{p}(t)$ be the mean values of q and p in the state $|A,\,t\rangle$. Find the mean values of q and p in the transformed state $|A^{tr},\,t\rangle$. If the Hamiltonian in K_1 is $H_1 = p^2/2m$ (free particle) and $\ddot{\xi}(t) \neq 0$, is $|A^{tr},\,t\rangle$ a solution, in K_1, of the Schrödinger equation? (It is not necessary to know the operator $V(t)$ that implements the transformation $(q,\,p) \to (q',\,p')$).

c) Find the (or a) unitary operator $V(t)$ that implements the transformation $(q,\,p) \to (q',\,p')$ at time t:

$$q' = V(t)\,q\,V^\dagger(t), \qquad p' = V(t)\,p\,V^\dagger(t)\,.$$

d) If $H_1 = p^2/2m + V(q)$ is the Hamiltonian in K_1, find the Hamiltonian H_2 in K_2, up to c–numerical constants.

9.9   The laboratory frame moves according to the time law $\xi(t) = \frac12 g\,t^2$ with respect to an inertial frame. At time $t = 0$ a particle of mass m, not subject to real forces, is (in the laboratory frame) in the state $|A,\,0\rangle$ of wavefunction
$$\psi_A(x,\,0) = \left(2\pi\,a^2\right)^{-1/4} e^{-x^2/4a^2}\,.$$

a) Find the mean values of q and p at time t and the uncertainties $\Delta q(t)$ and $\Delta p(t)$.

b) Use the results of Problem 7.7 and find $\psi_A(x, t)$.

9.10   The laboratory frame is rotating counterclockwise around the z axis with constant angular frequency ω with respect to an inertial frame. A particle of mass m (whose canonical variables are $q_i,\ p_i$), is subject to real time-independent forces described (in the laboratory) by the potential $\mathcal{V}(q_1, q_2, q_3)$.

a) Write the transformation law $(q_i,\ p_i) \to (q_i',\ p_i')$ of the canonical variables from the inertial frame to the rotating frame and the operator $V(t)$ that implements it:

$$q_i' = V(t)\, q_i\, V^\dagger(t), \qquad p_i' = V(t)\, p_i\, V^\dagger(t)\, .$$

b) Write the Hamiltonian of the system both in the inertial and in the rotating frame.

c) Verify that the Heisenberg equations in the rotating frame have the same form as the equations of the classical motion.

d) If in the rotating frame a particle is in an eigenstate of L_z belonging to the eigenvalue $\hbar l_z'$, show that also in the inertial frame the particle is in an eigenstate of L_z and find the eigenvalue. Explain the seemingly paradoxical result.

Solutions

9.1

a) Since the states represented by $|A^{\text{tr}}\rangle$ and $|B^{\text{tr}}\rangle$ are subjectively identical to (the states represented by) $|A\rangle$ and $|B\rangle$, one has $|\langle B | A\rangle|^2 = |\langle B^{\text{tr}} | A^{\text{tr}}\rangle|^2$: indeed the right hand side can be viewed as the probability transition from $|A\rangle$ to $|B\rangle$ measured in K_2 and the result of a measurement is an objective fact, independent of the laboratory in which the measurement is made.

9.2

a) Since $|A^{\text{tr}}\rangle$ and $[f(q, p)]'$ are subjectively identical to $|A\rangle$ and $f(q, p)$, one has $\langle A | f(q, p) | A\rangle = \langle A^{\text{tr}} | [f(q, p)]' | A^{\text{tr}}\rangle$: indeed the right hand side can be viewed as the mean value of the observable $f(q, p)$ in the state $|A\rangle$ in K_2 and, as in the previous problem, the result of a measurement is independent of the laboratory in which the measurement is made. Then:

$$\langle A | f(q, p) | A\rangle = \langle A^{\text{tr}} | [f(q, p)]' | A^{\text{tr}}\rangle = \langle A | V^\dagger [f(q, p)]' V | A\rangle \quad \Rightarrow$$
$$[f(q, p)]' = V f(q, p) V^\dagger = f(Vq V^\dagger, Vp V^\dagger) = f(q', p'),$$
$$q' = q'(q, p, t) = Vq V^\dagger, \quad p' = p'(q, p, t) = Vp V^\dagger.$$

b) In K_1 and in K_2 the wavefunctions of subjectively identical states are obviously the same: in general $\langle A | B\rangle$ in $K_1 = \langle A | B\rangle$ in K_2.

c) As we have seen in a), the transformation law $f(q', p') = V f(q, p) V^{-1}$ provides in K_1 the operator associated with any observable $f(q, p)$ of K_2, therefore $\eta_1 = V \xi_2 V^{-1}$, namely:

$$\eta_1(q, p) = V \xi_2(q, p) V^{-1} = \xi_2(Vq V^{-1}, Vp V^{-1}) = \xi_2(q', p') .$$

In order to find in K_2 the operator associated with η_1 it is sufficient to observe that K_1 is obtained from K_2 by means of the inverse transformation, so:

$$\xi_2 = V^{-1}\eta_1 V .$$

d) $|A^{(1)}\rangle$ is the transformed of the state (different from **S**) that in K_1 is represented by $|A^{(2)}\rangle$: $|A^{(1)}\rangle = V|A^{(2)}\rangle$ then:

$$|A^{(2)}\rangle = V^{-1}|A^{(1)}\rangle.$$

$|A\rangle$ and $V^{-1}|A\rangle$ represent the same state 'seen' from K_1 and from K_2; the above written relationship may also be interpreted by saying that $V^{-1}|A\rangle$ is the state that, in K_1, is obtained by effecting on $|A\rangle$

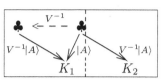

the active inverse transformation V^{-1}: so the transformation of the states is, from the passive point of view, the inverse of that of the active point of view.

9.3

a) As we know the time evolution operator in the frame K_1, it is sufficient to 'translate' in the frame K_2 the evolution of the states in K_1: also in classical physics the equations of motion and the time law of the motion for example in a noninertial frame, simply are the rewriting of the equations of motion and of the time law of the motion in an inertial frame in terms of the coordinates in the noninertial frame.

In K_1: $|A, t\rangle = U_1(t)|A, 0\rangle$; seen from K_2: $|A, t\rangle \to V^{-1}(t)|A, t\rangle \equiv |B, t\rangle$ (passive point of view), therefore:

$$|B, t\rangle = V^{-1}(t)\,U_1(t)\,|A, 0\rangle = V^{-1}(t)\,U_1(t)\,V(0)\,V^{-1}(0)\,|A, 0\rangle$$
$$= V^{-1}(t)\,U_1(t)\,V(0)\,|B, 0\rangle$$

so the time evolution operator in K_2 is $U_2(t) = V^{-1}(t)\,U_1(t)\,V(0)$.

b) If $U(t)$ is the time evolution operator and H the Hamiltonian, one has:

$$i\hbar\,\frac{dU(t)}{dt} = H\,U(t) \quad \Rightarrow \quad H = i\hbar\,\frac{dU(t)}{dt}\,U^\dagger(t).$$

If V does not depend on time:

$$H_2(q,\, p) = i\hbar\,\dot{U}_2\,U_2^\dagger = i\hbar\,V^{-1}\,\dot{U}_1\,U_1^\dagger\,V$$
$$= V^{-1}H_1(q,\, p)\,V = H_1(V^{-1}q\,V,\, V^{-1}p\,V).$$

We can rewrite the above relationship in the form:

$$H_1(q,\, p) = V\,H_2(q,\, p)\,V^{-1} = H_2(q',p')$$

that expresses the invariance *in value* of the Hamiltonian under canonical transformations independent of time $\big(\,H_2(q'(q,\, p),p'(q,\, p)) = H_1(q,\, p)\,\big)$.

If $V = V(t)$, one has:

$$H_2(q,\, p,\, t) = i\hbar\,\frac{dU_2(t)}{dt}\,U_2^\dagger(t) = i\hbar\,\Big(\frac{d}{dt}\,V^\dagger(t)\,U_1(t)\,V(0)\Big)U_2^\dagger(t)$$
$$= i\hbar\Big(\dot{V}^\dagger(t)\,U_1(t)\,V(0) + V^\dagger(t)\,\dot{U}_1(t)\,V(0)\Big)U_2^\dagger(t)$$

$$= i\hbar \dot{V}^\dagger(t)\, V(t)\, V^{-1}(t)\, U_1(t)\, V(0)\, U_2^\dagger(t) + V^{-1}(t)\, H_1\, U_1(t)\, V(0)\, U_2^\dagger(t)$$
$$= V^{-1}(t)\, H_1\, V(t) + i\hbar \dot{V}^\dagger(t)\, V(t)\ .$$

In order to verify that $H_2 = H_2^\dagger$, it is sufficient to show that:

$$\left(\dot{V}^\dagger(t)\, V(t)\right)^\dagger \equiv V^\dagger(t)\, \dot{V}(t) = -\dot{V}^\dagger(t)\, V(t)\ .$$

Indeed, taking the derivative of $V^\dagger(t)\, V(t) = \mathbb{1}$ with respect to t, one has:

$$\dot{V}^\dagger(t)\, V(t) + V^\dagger(t)\, \dot{V}(t) = 0 \quad \Rightarrow \quad V^\dagger(t)\, \dot{V}(t) = -\dot{V}^\dagger(t)\, V(t)\ .$$

c) $i\hbar \dfrac{d}{dt}\,|\,A^{\mathrm{tr}},\, t\,\rangle = i\hbar V \dfrac{d}{dt}\,|\,A,\, t\,\rangle = V\, H_1\,|\,A,\, t\,\rangle = V\, H_1\, V^{-1}\,|\,A^{\mathrm{tr}},\, t\,\rangle$

therefore $|\,A^{\mathrm{tr}},\, t\,\rangle$ solves the Schrödinger equation only if $H_1(q,\, p) = V\, H_1\, V^{-1} = H_1(q',p')$, i.e. only if H_1 is invariant *in form* under the transformation $(q,\, p) \to (q',p')$ or, equivalently, if V commutes with H_1. Or else: since $|\,A^{\mathrm{tr}}\,\rangle$ is subjectively identical to $|\,A\,\rangle$, in K_2 $|\,A\,\rangle$ satisfies the Schrödinger equation with H_1, not with H_2, as it should be, only if $H_1 = H_2$, i.e. $H_1(q,\, p) = V\, H_1\, V^{-1}$ as before.

9.4

a) One has:

$$q' = V(a)\, q\, V^{-1}(a) = q - a\, , \quad p' = p$$
$$V(a) = e^{-i\, p\, a/\hbar}.$$

b) $V(a)\,|\,x\,\rangle = |\,x + a\,\rangle \quad \Rightarrow \quad \langle\, x\,|\, V(a) = \langle\, x - a\,| \quad \Rightarrow$

$$\psi_{A^{\mathrm{tr}}}(x) = \langle\, x\,|\, A^{\mathrm{tr}}\,\rangle = \langle\, x\,|\, V\,|\, A\,\rangle = \langle\, x - a\,|\, A\,\rangle = \psi_A(x - a)$$
$$= e^{-\alpha\, (x-a)^2/2}\, e^{i\, k\, x}$$

(the phase factor $e^{-i\, k\, a}$ is inessential). The Gaussian relative to the transformed state is now centered at the point $x = a$: $\psi_{A^{\mathrm{tr}}}(x)$ is obtained by translating $\psi_A(x)$ by $+a$.

The state $|\,A\,\rangle$ is 'seen' from K_2 as the vector $V^{-1}\,|\,A\,\rangle$, so:

$$\psi'_A(x) = \langle\, x\,|\, V^{-1}\,|\, A\,\rangle = \langle\, x + a\,|\, A\,\rangle = \psi_A(x + a) = e^{-\alpha\, (x+a)^2/2}\, e^{i\, k\, x}.$$

To find the Hamiltonian in K_2, as the two frames do not move with respect to each other, due to the result of Problem 9.3 it is sufficient to write the Hamiltonian $H_1(q,\, p)$ 'seen' from K_2 (invariance in value):

$$H_2(q,\, p) = V^{-1} H_1(q,\, p)\, V = H_1(q + a, p)\ .$$

c) One must have (see Problem 9.3) $H_1(q,\, p) = H_1(q - a,\, p)$ for any a, so the particle must be free.

d) If $H = p^2/2m + m\, g\, q$, the Hamiltonian is not invariant in form:

$$V\, H(q,\, p)\, V^{-1} = H(q,\, p) - m\, g\, a$$

therefore the vector $|A^{\mathrm{tr}}, t\rangle$ evolves with the operator $\mathrm{e}^{img\,at/\hbar}\,\mathrm{e}^{-iHt/\hbar}$: it is however sufficient to redefine $|A^{\mathrm{tr'}}, t\rangle = \mathrm{e}^{-img\,at/\hbar}\,|A^{\mathrm{tr}}, t\rangle$ (since $\mathrm{e}^{-img\,at/\hbar}$ is a phase factor, the state is unchanged) and one obtains:

$$|A^{\mathrm{tr'}}, t\rangle = U(t)\,|A^{\mathrm{tr'}}, 0\rangle, \qquad U(t) \equiv \mathrm{e}^{-iHt/\hbar}\;.$$

As a matter of fact, also in classical physics it is well known that a massive body falls with the same time law, whatever the height it starts from: the Hamiltonian is not invariant (and therefore the momentum is not a constant of motion), but the equations of motion are invariant.

9.5

a) As already seen in Problem 8.3:

$$\begin{cases} q_1' = q_1 \cos\phi + q_2 \sin\phi \\ q_2' = q_2 \cos\phi - q_1 \sin\phi \\ q_3' = q_3 \end{cases} \qquad \begin{cases} p_1' = p_1 \cos\phi + p_2 \sin\phi \\ p_2' = p_2 \cos\phi - p_1 \sin\phi \\ p_3' = p_3 \end{cases}$$

$$V(\phi) = \mathrm{e}^{-iL_z\,\phi/\hbar}\,, \qquad L_z = q_1 p_2 - q_2 p_1\;.$$

b) In alternative to what has been found in Problem 8.3, one has:

$$q_1\,V^\dagger\,|x, y, z\rangle = V^\dagger V\,q_1\,V^\dagger\,|x, y, z\rangle = V^\dagger(q_1 \cos\phi + q_2 \sin\phi)\,|x, y, z\rangle$$
$$= (x\cos\phi + y\sin\phi)\,V^\dagger\,|x, y, z\rangle$$

and likewise:

$$q_2\,V^\dagger\,|x, y, z\rangle = (y\cos\phi - x\sin\phi)\,V^\dagger\,|x, y, z\rangle,$$
$$q_3\,V^\dagger\,|x, y, z\rangle = z\,V^\dagger\,|x, y, z\rangle$$

whence:

$$V^\dagger\,|x, y, z\rangle = |x\cos\phi + y\sin\phi,\ y\cos\phi - x\sin\phi,\ z\rangle \quad\Rightarrow$$
$$\psi_{A^{\mathrm{tr}}}(x, y, z) = \langle x, y, z\,|V|\,A\rangle = \langle x\cos\phi + y\sin\phi,\ y\cos\phi - x\sin\phi,\ z\,|\,A\rangle$$
$$= \psi_A(x\cos\phi + y\sin\phi,\ y\cos\phi - x\sin\phi,\ z)$$
$$= \mathrm{e}^{-\alpha\,(x^2+y^2+z^2)/2}\,\mathrm{e}^{i(k\,x\cos\phi + k\,y\sin\phi)}\;.$$

$$\psi_A'(x, y, z) = \langle x, y, z\,|V^{-1}|\,A\rangle = \psi_A(x\cos\phi - y\sin\phi,\ y\cos\phi + x\sin\phi,\ z)$$
$$= \mathrm{e}^{-\alpha\,(x^2+y^2+z^2)/2}\,\mathrm{e}^{i(k\,x\cos\phi - k\,y\sin\phi)}\;.$$

Since in this case K_2 is not in motion with respect to K_1 (see Problem 9.3):

$$H_2(\vec{q},\,\vec{p}) = V^{-1}H_1(\vec{q},\,\vec{p})\,V$$
$$= \frac{\vec{p}^{\,2}}{2m} + V(q_1 \cos\phi - q_2 \sin\phi,\ q_2 \cos\phi + q_1 \sin\phi,\ q_3)\;.$$

c) $H_1(\vec{q},\,\vec{p}) = H_2(\vec{q},\,\vec{p})$ if and only if the potential is invariant under rotations around the z axis and must, therefore, be a function of $q_1^2 + q_2^2$ and q_3.

9.6

a) $q' = q - vt, \qquad p' = p - mv$.

   As (see Problem 9.2) $\langle A, t \mid f(q, p) \mid A, t \rangle = \langle A^{\mathrm{tr}}, t \mid f(q',p') \mid A^{\mathrm{tr}}, t \rangle$, one has:

$$\langle A^{\mathrm{tr}}, t \mid q \mid A^{\mathrm{tr}}, t \rangle = \bar{q}(t) + vt, \qquad \langle A^{\mathrm{tr}}, t \mid p \mid A^{\mathrm{tr}}, t \rangle = \bar{p}(t) + mv.$$

b) Since the transformation $(q, p) \to (q',p')$ may be thought of as the composition of a translation of the q's and a translation of the p's, we may write either

$$G_1(v,t) = \mathrm{e}^{-i\,p\,v\,t/\hbar}\,\mathrm{e}^{i\,m\,q\,v/\hbar}, \qquad \text{or} \quad G_2(v,t) = \mathrm{e}^{i\,m\,q\,v/\hbar}\,\mathrm{e}^{-i\,p\,v\,t/\hbar}.$$

   Or, also (see Problem 5.8),

$$G_3(v,t) = \mathrm{e}^{-i\left(p\,v\,t - m\,q\,v\right)/\hbar} = \begin{cases} \mathrm{e}^{+i\,m v^2 t/2\hbar}\,\mathrm{e}^{-i\,p\,v\,t/\hbar}\,\mathrm{e}^{i\,m\,q\,v/\hbar} \\ \mathrm{e}^{-i\,m v^2 t/2\hbar}\,\mathrm{e}^{i\,m\,q\,v/\hbar}\,\mathrm{e}^{-i\,p\,v\,t/\hbar} \end{cases}.$$

   The three forms are equivalent as they differ by c–number factors. Let us put $G(t) \equiv G_3(v,t)$.

c) $\psi_{A^{\mathrm{tr}}}(x) = \langle x \mid A^{\mathrm{tr}} \rangle = \langle x \mid G(\bar{t}) \mid A \rangle$

$$= \mathrm{e}^{-i\,m v^2 \bar{t}/2\hbar}\langle x \mid \mathrm{e}^{i\,m v\,q/\hbar}\,\mathrm{e}^{-i\,v\,\bar{t}\,p/\hbar} \mid A \rangle$$

$$= \mathrm{e}^{-i\,m v^2 \bar{t}/2\hbar}\,\mathrm{e}^{i\,m v\,x/\hbar}\langle x \mid \mathrm{e}^{-i\,v\,\bar{t}\,p/\hbar} \mid A \rangle$$

$$= \mathrm{e}^{-i\,m v^2 \bar{t}/2\hbar}\,\mathrm{e}^{i\,m v\,x/\hbar}\,\psi_A(x - v\bar{t}).$$

d) As K_2 is in uniform rectilinear motion with respect to K_1, we expect that $H_2(q, p, t) = H_1(q + vt, p) = p^2/2m + V(q+vt)$. Indeed (see Problem 9.3)

$$H_2 = G^{-1}(t)\,H_1\,G(t) + i\hbar\,\dot{G}^\dagger(t)\,G(t).$$

   We have then to calculate $i\hbar\,\dot{G}^\dagger(t)\,G(t)$: one has $G^\dagger(t) = \mathrm{e}^{i\left(p\,v\,t - m\,q\,v\right)/\hbar}$ but note that $i\hbar\,\dot{G}^\dagger(t) \neq -v\,p\,G^\dagger(t)$ since the derivative of the exponent does not commute with the exponent itself. As a consequence, the factorized form of $G(t)$ is the most convenient:

$$G^\dagger(t) = \mathrm{e}^{i\,m v^2 t/2\hbar}\,\mathrm{e}^{i\,p\,v\,t/\hbar}\,\mathrm{e}^{-i\,m\,q\,v/\hbar} \quad \Rightarrow$$

$$i\hbar\,\dot{G}^\dagger(t) = -\left(\tfrac{1}{2}m v^2 + v\,p\right)G^\dagger(t) \quad \Rightarrow \quad i\hbar\,\dot{G}^\dagger(t)\,G(t) = -\tfrac{1}{2}m v^2 - v\,p$$

   therefore:

$$H_2(q, p, t) = G^{-1}(t)\,H_1(q, p)\,G(t) - v\,p - \tfrac{1}{2}m v^2$$

$$= \frac{(p + mv)^2}{2m} + V(q + vt) - v\,p - \tfrac{1}{2}m v^2 = \frac{p^2}{2m} + V(q + vt).$$

   If the phase factor $\mathrm{e}^{i\,m v^2 t/2\hbar}$ is omitted in the expression for $G^\dagger(t)$, H_2 changes by an additive constant: $H_2 \to H_2 + \tfrac{1}{2}m v^2$, that may be omitted as well.

e) Since the state $|A^{\mathrm{tr}}, t\rangle$ is the state $|A, t\rangle$ 'boosted in uniform rectilinear motion', and $V(q) = 0$, the answer should definitely be affirmative: indeed in this case $H_1(q, p) = H_2(q, p)$ and as a consequence also $U_1(t) = U_2(t)$.

9.7

a) Since in the moving frame the center of oscillation is at rest, the Hamiltonian is

$$\widetilde{H} = \frac{p^2}{2m} + \frac{1}{2}m\omega^2 q^2.$$

In the laboratory frame the abscissa of the center of oscillation is $x = vt$, therefore:

$$H = \frac{p^2}{2m} + \frac{1}{2}m\omega^2(q - vt)^2.$$

Note that (see Problem 9.6) $\widetilde{H} = G^\dagger(v,t)\, H\, G(v,t) + i\hbar\, \dot{G}^\dagger(v,t)\, G(v,t)$.

b) Any state $|A\rangle$ of the oscillator in the moving frame is seen from the laboratory as the state $|A^{\mathrm{tr}}\rangle$ obtained by transforming the state $|A\rangle$: so, as in the moving frame $|0\rangle \xrightarrow{\ \mathrm{SR}\ } \psi_0(x) = (m\omega/\pi\hbar)^{1/4}\, e^{-(m\omega/2\hbar)\,x^2}$, in the laboratory (see Problem 9.6):

$$|0^{\mathrm{tr}}\rangle \xrightarrow{\ \mathrm{SR}\ } \psi_{0^{\mathrm{tr}}}(x,t) = \left(\frac{m\omega}{\pi\hbar}\right)^{1/4} e^{-(m\omega/2\hbar)\,(x-vt)^2}\, e^{\,i\,mv\,x/\hbar}$$

(the phase factor $e^{-i\,mv^2 t/2\hbar}$ is inessential).

c) First we solve the problem in the frame where the center of the oscillator is at rest (moving frame), then we go back to the laboratory frame. In the moving frame the wavefunction $\widetilde\psi(x,0)$ at time $t = 0$ is

$$\widetilde\psi(x,0) = \psi(x,0)\, e^{-i\,mv\,x/\hbar} \equiv \sum_n a_n\, \psi_n(x)\,, \qquad a_n = \int \psi_n^*(x)\, \widetilde\psi(x,0)\,\mathrm{d}x$$

where $\psi_n(x)$ are the usual eigenfunctions of the Hamiltonian of the harmonic oscillator (i.e. of $\widetilde H$); then:

$$\widetilde\psi(x,t) = \sum_n a_n\, e^{-i(n+\frac{1}{2})\omega t}\, \psi_n(x) \quad \Rightarrow$$

$$\psi(x,t) = \sum_n a_n\, e^{-i(n+\frac{1}{2})\omega t}\, \psi_n(x - vt)\, e^{\,i\,mv\,x/\hbar}\,.$$

Let us verify that the single terms in the sum satisfy the Schrödinger equation (the phase factor $e^{-i\,mv^2 t/2\hbar}$ is reinserted):

$$\left(-\frac{\hbar^2}{2m}\frac{\mathrm{d}^2}{\mathrm{d}x^2} + \frac{1}{2}m\omega^2(x - vt)^2\right)\psi_n(x - vt)\, e^{\,i\,mv\,x/\hbar}\, e^{-i(n+\frac{1}{2})\omega t}\, e^{-i\,mv^2 t/2\hbar}$$

$$= \left(-\frac{\hbar^2}{2m}\psi_n''(x - vt) - i\hbar v\,\psi_n'(x - vt) + \frac{1}{2}mv^2\psi_n(x - vt)\right.$$

$$\left. + \frac{1}{2}m\omega^2(x - vt)^2\psi_n(x - vt)\right) \times e^{\,i\,mv\,x/\hbar}\, e^{-i(n+\frac{1}{2})\omega t}\, e^{-i\,mv^2 t/2\hbar}$$

$$= \left(\left(E_n + \frac{1}{2} m v^2 \right) \psi_n(x - v t) - i \hbar v \, \psi_n'(x - v t) \right)$$

$$\times e^{i m v x / \hbar} \, e^{-i(n + \frac{1}{2}) w t} \, e^{-i m v^2 t / 2 \hbar}$$

that is the same as $\; i \hbar \dfrac{\partial}{\partial t} \left(e^{-i(n + \frac{1}{2}) w t} \, e^{-i m v^2 t / 2 \hbar} \, \psi_n(x - v t) \right) e^{i m v x / \hbar}$.

9.8

a) $q' = q - \xi(t), \qquad p' = p - m \dot{\xi}(t)$.

b) As $\langle A, t \, | \, q \, | \, A t \rangle = \langle A^{tr}, t \, | \, q' \, | \, A^{tr}, t \rangle$ and likewise for p, one has:

$$\langle A^{tr}, t \, | \, q \, | \, A^{tr}, t \rangle = \bar{q}(t) + \xi(t), \qquad \langle A^{tr}, t \, | \, p \, | \, A^{tr}, t \rangle = \bar{p}(t) + m \dot{\xi}(t).$$

If $\ddot{\xi}(t) \neq 0$, the mean values of q and p in the state $| A^{tr}, t \rangle$ are not those of a free particle (in particular, for a free particle, p is a constant of motion), so $| A^{tr}, t \rangle$ is not a solution of the Schrödinger equation in K_1.

c) As in the case of the Galilei transformation (see Problem 9.6), we may equally write:

$$V_1(t) = e^{-i p \xi(t) / \hbar} \, e^{i m q \dot{\xi}(t) / \hbar}, \qquad V_2(t) = e^{i m q \dot{\xi}(t) / \hbar} \, e^{-i p \xi(t) / \hbar}$$

$$V_3(t) = e^{-i(p \xi(t) - m q \dot{\xi}(t)) / \hbar} = \begin{cases} e^{+i m \xi(t) \dot{\xi}(t) / 2\hbar} \, e^{-i p \xi(t) / \hbar} \, e^{i m q \dot{\xi}(t) / \hbar} \\ e^{-i m \xi(t) \dot{\xi}(t) / 2\hbar} \, e^{i m q \dot{\xi}(t) / \hbar} \, e^{-i p \xi(t) / \hbar} \end{cases}$$

and the three forms are equivalent.

d) In addition to the fact that the potential 'moves': $V(q + \xi(t))$, we expect to find in H_2 also the potential of the inertial force $-m \ddot{\xi}(t)$. Indeed:

$$H_2 = V^{-1}(t) H_1 V(t) + i \hbar \dot{V}^\dagger(t) V(t)$$

and since we will neglect, in the expression of H_2, c–numerical terms, we are allowed to use any of the three forms of $V(t)$: we shall use $V(t) \equiv V_1(t)$.

$$V^\dagger(t) = e^{-i m q \dot{\xi}(t) / \hbar} \, e^{i p \xi(t) / \hbar}$$

$$i \hbar \dot{V}^\dagger(t) V(t) = \left(m q \ddot{\xi}(t) V^\dagger(t) - e^{-i m q \dot{\xi}(t) / \hbar} \, p \, \dot{\xi}(t) \, e^{i p \xi(t) / \hbar} \right) V(t)$$

$$= m q \ddot{\xi}(t) - (p + m \dot{\xi}(t)) \, \dot{\xi}(t)$$

$$H_2(q, p, t) = \frac{(p + m \dot{\xi}(t))^2}{2m} + V(q + \xi(t)) + m q \ddot{\xi}(t) - (p + m \dot{\xi}(t)) \, \dot{\xi}(t)$$

$$= \frac{p^2}{2m} + V(q + \xi(t)) + m q \ddot{\xi}(t) \; -\frac{1}{2} m \dot{\xi}(t)^2 .$$

9.9

a) In the inertial frame the particle is free and the mean values of q and p are vanishing for $t = 0$, then for any t; therefore, in the laboratory, with respect to which the inertial frame has the time dependent position $-\frac{1}{2} g t^2$,

$$\bar{q}(t) = -\frac{1}{2} g t^2, \qquad \bar{p}(t) = -m g t.$$

If $|B, t\rangle$ stands for the vector that represents the state of the particle in the inertial frame, in the laboratory the same state is represented by the $|A, t\rangle = V^{-1}(t)|B, t\rangle$, where (see Problem 9.8):

$$V^{-1}(t) = e^{-iq\,mgt/\hbar}\,e^{ipg\,t^2/2\hbar}$$

so that at any time $|A, t\rangle$ is obtained from $|B, t\rangle$ by means of a translation in the space of coordinates and a translation in the space of momenta: it should be clear that both the width Δq in coordinate space and the width Δp in momentum space are the same in both frames, whence (see Problem 7.6):

$$\left(\Delta q(t)\right)^2 = (\Delta q)^2 + \frac{(\Delta p)^2}{m^2}\,t^2 = a^2 + \frac{\hbar^2}{4m^2a^2}\,t^2, \qquad \Delta p(t) = \Delta p(0) = \frac{\hbar}{2a}\,.$$

We have found again, from a different point of view, the results of Problem 7.8 (up to the substitution $mg \to -\gamma$), for in the laboratory frame the Hamiltonian is $H = p^2/2m + mg\,q$.

b) At time $t = 0$ the inertial frame coincides with the laboratory frame and is at rest with respect to the latter; so the wavefunction of the particle at $t = 0$ is the same in the two frames: $\psi_B(x, 0) = \psi_A(x, 0)$ $(V(0) = \mathbb{1})$. From Problem 7.7 we know the time evolution in the inertial frame of the state represented at $t = 0$ by $\psi_B(x, 0) = (2\pi a^2)^{-1/4}\,e^{-x^2/4a^2}$:

$$\psi_B(x, t) = \left(\frac{8\,a^2}{\pi}\right)^{1/4}\frac{e^{-x^2/(4a^2+2i\,\hbar t/m)}}{\sqrt{4a^2 + 2i\,\hbar t/m}}\,, \qquad \psi_B(x, 0) = \psi_A(x, 0)\,.$$

The wavefunction $\psi_A(x, t)$ in the laboratory is obtained by applying the operator $V^{-1}(t)$ to $\psi_B(x, t)$:

$$\psi_A(x, t) = \langle x\,|\,e^{-iq\,mgt/\hbar}\,e^{ipg\,t^2/2\hbar}\,|\,B, t\rangle = e^{-imgtx/\hbar}\,\psi_B(x + \tfrac{1}{2}g\,t^2, t)$$

$$= \left(\frac{8\,a^2}{\pi}\right)^{1/4}e^{-imgt\,x/\hbar}\,\frac{e^{-(x+\frac{1}{2}g\,t^2)^2/(4a^2+2i\,\hbar t/m)}}{\sqrt{4a^2 + 2i\,\hbar t/m}}\,.$$

9.10

a) One has (see Problem 9.5):

$$\begin{cases} q_1' = q_1\cos\omega t + q_2\sin\omega t \\ q_2' = q_2\cos\omega t - q_1\sin\omega t \\ q_3' = q_3 \end{cases} \qquad \begin{cases} p_1' = p_1\cos\omega t + p_2\sin\omega t \\ p_2' = p_2\cos\omega t - p_1\sin\omega t \\ p_3' = p_3 \end{cases}$$

$$V(t) = e^{-iL_z\,\omega t/\hbar}\,, \qquad L_z = q_1p_2 - q_2p_1\,.$$

b) Let H_I denote the Hamiltonian in the inertial frame and H that in the laboratory, namely in the rotating frame: one has (see Problem 9.3):

$$H_I(\vec{q}, \vec{p}, t) = \frac{\vec{p}^{\,2}}{2m} + V(q_1', q_2', q_3')$$

$$= \frac{\vec{p}^{\,2}}{2m} + V(q_1\cos\omega t + q_2\sin\omega t, q_2\cos\omega t - q_1\sin\omega t, q_3)$$

$$H(\vec{q}, \vec{p}) = V^{-1}(t)\, H_1\, V(t) + i\, \hbar\, \dot{V}^\dagger(t)\, V(t) = \frac{\vec{p}^2}{2m} + \mathcal{V}(q_1, q_2, q_3) - \omega\, L_z\ .$$

c) The Heisenberg equations for the variables q_1, q_2; p_1, p_2 are:

$$\begin{cases} \dot{q}_1 = \dfrac{i}{\hbar}[H\,,\,q_1] = \dfrac{1}{m}p_1 + \omega\, q_2 \quad &\Rightarrow \quad p_1 = m\,\dot{q}_1 - m\omega\, q_2 \\[2mm] \dot{q}_2 = \dfrac{i}{\hbar}[H\,,\,q_2] = \dfrac{1}{m}p_2 - \omega\, q_1 \quad &\Rightarrow \quad p_2 = m\,\dot{q}_2 + m\omega\, q_1 \\[2mm] \dot{p}_1 = \dfrac{i}{\hbar}[H\,,\,p_1] = \quad \omega\, p_2 - \dfrac{\partial \mathcal{V}}{\partial q_1} \\[2mm] \dot{p}_2 = \dfrac{i}{\hbar}[H\,,\,p_2] = -\omega\, p_1 - \dfrac{\partial \mathcal{V}}{\partial q_2} \end{cases}$$

therefore:

$$\begin{cases} m\,\ddot{q}_1 = \dot{p}_1 + m\omega\,\dot{q}_2 = -\dfrac{\partial \mathcal{V}}{\partial q_1} + m\omega^2 q_1 + 2m\omega\,\dot{q}_2 \\[3mm] m\,\ddot{q}_2 = \dot{p}_2 - m\omega\,\dot{q}_1 = -\dfrac{\partial \mathcal{V}}{\partial q_2} + m\omega^2 q_2 - 2m\omega\,\dot{q}_1\ . \end{cases}$$

In the right hand side one can recognize respectively the real force, the centrifugal force and the Coriolis force (see Problem 3.5).

d) Let $|\,l'_z\,\rangle$ be state of the particle in the rotating frame. From the inertial frame it is seen as $V(t)\,|\,l'_z\,\rangle$ (see Problem 9.2). One has:

$$L_z\, V(t)\,|\,l'_z\,\rangle \equiv L_z\, e^{-i\, L_z\, \omega\, t/\hbar}\,|\,l'_z\,\rangle = \hbar\, l'_z\, V(t)\,|\,l'_z\,\rangle$$

so, also in the inertial frame it is a state of definite angular momentum $L_z = \hbar\, l'_z$, not $\hbar\, l'_z + m\omega\, r^2$ – this is true also in classical physics. The result is not absurd; indeed, in the rotating frame the particle is subject to velocity dependent forces (the Coriolis force) so the canonical angular momentum $\vec{q} \wedge \vec{p}$ is not the same as the kinetic angular momentum $\vec{q} \wedge (m\,\dot{\vec{q}})$: indeed, as seen in c), $\vec{p} \neq m\,\dot{\vec{q}}$. Actually, while the kinetic angular momenta in the two frames are different, the canonical angular momentum in the laboratory coincides with the angular momentum in the inertial frame: $q'_1\, p'_2 - q'_2\, p'_1 = q_1\, p_2 - q_2\, p_1\ .$

10

Two and Three-Dimensional Systems

Separation of variables; degeneracy theorem; group of invariance of the two-dimensional isotropic oscillator.

10.1 Consider the Hamiltonian of a two-dimensional anisotropic harmonic oscillator:

$$H = \left(\frac{p_1^2}{2m} + \frac{1}{2}m\omega_1^2\, q_1^2 \right) + \left(\frac{p_2^2}{2m} + \frac{1}{2}m\omega_2^2\, q_2^2 \right); \qquad \omega_1 \neq \omega_2 \,.$$

a) Exploit the fact that the Schrödinger eigenvalue equation can be solved by separating the variables and find a complete set of eigenfunctions of H and the corresponding eigenvalues.

b) Assume that $\omega_1/\omega_2 = 3/4$. Find the first two degenerate energy levels. What can one say about the degeneracy of energy levels when the ratio between ω_1 and ω_2 is not a rational number?

c) Write the eigenfunctions of the Hamiltonian in the case $\omega_2 = 0$.

Consider now a particle of mass m in two dimensions subject to the potential:

$$V(q_1, q_2) = m\omega^2\left(q_1^2 - q_1 q_2 + q_2^2 \right) \,.$$

d) Say whether the problem of finding the eigenvalues of the Hamiltonian $H = (p_1^2 + p_2^2)/2m + V(q_1, q_2)$ can be solved by the method of separation of variables.

10.2 A particle of mass m in two dimensions is constrained inside a square whose edge is $2a$: $|x| \leq a$, $|y| \leq a$.

a) Write the Schrödinger equation, separate the variables and find a complete set of eigenfunctions of the Hamiltonian.

b) Find the energy levels of the system and say whether there is degeneracy.

c) Say whether there exist operators (i.e. transformations) that commute with the Hamiltonian but do not commute among themselves. In the affirmative case, give one or more examples.

© Springer International Publishing AG 2017

E. d'Emilio and L.E. Picasso, *Problems in Quantum Mechanics*,

UNITEXT for Physics, DOI 10.1007/978-3-319-53267-7_10

Assume now that within the square the potential:

$$V_a(x,y) = V_{0a} \cos(\pi x/2a) \cos(\pi y/2a)$$

is present.

d) Is it still possible to separate the variables in the Schrödinger equation? Do degenerate energy levels exist?

e) Say whether it is possible to guarantee the existence of degenerate energy levels if, instead, the potential is

$$V_b(x,y) = V_{0b} \sin(\pi x/a) \sin(\pi y/a) .$$

Is any relationship between the eigenfunctions of the Hamiltonian $\psi_E(x,y)$ and $\psi_E(y,x)$ expected? (Namely, are they equal? are they different? ...)

10.3   A particle of mass m in two dimensions is constrained inside the triangle whose vertices have the coordinates $(x = 0, y = 0)$; $(x = a, y = 0)$; $(x = a, y = a)$ (a half of the square with edge a).

a) Find eigenvectors and eigenvalues of the energy.

b) For the same system and exploiting the results of the previous question, find a complete set of eigenvectors of the operator that implements the reflection through the straight line $x + y = a$ (the dotted line in the figure).

10.4   A particle of mass m in three dimensions is confined within an infinite rectilinear guide with a cross section that is a square of edge a .

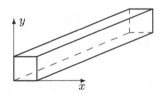

a) Find eigenfunctions and eigenvalues of the Hamiltonian. What is the minimum energy (threshold energy) the particle must have in order to propagate along the guide?

Consider the wavefunctions:

$$\psi_1(x,y,\dot{z}) = A \sin(2\pi x/a) \sin(\pi y/a) e^{i k_1 z}$$
$$\psi_2(x,y,z) = B \sin(\pi x/a) \sin(\pi y/a) e^{i k_2 z} .$$

b) Determine the normalization coefficients A and B in such a way that the integral of the densities $\rho_{1,2}$ over a slice of the guide of unit volume equals 1 ("one particle per unit volume" normalization).

c) Calculate the probability current densities:

$$\vec{j}_{1,2}(x,y,z) = \frac{\hbar}{m} \Im m \left(\psi_{1,2}^*(x,y,z) \nabla \psi_{1,2}(x,y,z) \right)$$

for the states represented by the wavefunctions ψ_1 and ψ_2 normalized as above, and verify that $\operatorname{div} \vec{j}_{1,2}(x, y, z) = 0$.

d) Say for which values of k_2 the probability current associated to the state represented by $\psi(x, y, z, t)$ with $\psi(x, y, z, 0) = \psi_1(x, y, z) + \psi_2(x, y, z)$, is divergenceless.

10.5 A particle is subject to the potential $V = V(q_1^2 + q_2^2, q_3)$.

a) Show that the Hamiltonian $H_0 = \vec{p}^2/2m + V$ commutes with the angular momentum operator $L_z = q_1 p_2 - q_2 p_1$.

b) Use the degeneracy theorem to show that there exist degenerate energy levels.

c) Say whether and how the degeneracy is removed if the system is on a platform rotating around the z axis with constant angular velocity ω.

10.6 The Hamiltonian of a two-dimensional isotropic harmonic oscillator of mass m and angular frequency ω is

$$H = \frac{1}{2m}(p_1^2 + p_2^2) + \frac{1}{2}m\omega^2(q_1^2 + q_2^2) = H_1(q_1, p_1) + H_2(q_2, p_2).$$

a) Exploit the separation of variables ($H = H_1 + H_2$) and find the eigenvalues of H and their degeneracies.

b) Write the eigenfunctions of the Hamiltonian in the Schrödinger representation, in the basis in which both H_1 and H_2 are diagonal.

c) Is the degeneracy found in a) in agreement with the result established in Problem 10.5? Find the maximum and the minimum of the eigenvalues m' of $L_3 = q_1 p_2 - q_2 p_1$ within each energy level. Do all its possible values ranging between m'_{max} and m'_{min} occur?

d) For each of the first three energy levels, say which eigenvalues of L_3 do occur and explicitly write the wavefunctions relative to the states $|E, m'\rangle$ (simultaneous eigenstates of H and L_3).

10.7 This problem is devoted to establish a priori the degeneracies of the two-dimensional isotropic harmonic oscillator found in Problem 10.6. Set

$$\eta_a = \frac{1}{\sqrt{2m\omega\hbar}}(p_a - \mathrm{i}\,m\omega\,q_a), \qquad a = 1,\, 2.$$

a) Write the Hamiltonian H of the two-dimensional oscillator in terms of the operators η_a and $\eta_a^\dagger$ and the commutation rules $[\eta_a, \eta_b^\dagger]$, $a, b = 1, 2$.

b) Show that the four operators $\eta_a^\dagger \eta_b$ commute with the Hamiltonian H.

Consider the operators:

$$j_1 = \frac{1}{2}(\eta_1^\dagger \eta_2 + \eta_2^\dagger \eta_1), \quad j_2 = \frac{1}{2i}(\eta_1^\dagger \eta_2 - \eta_2^\dagger \eta_1), \quad j_3 = \frac{1}{2}(\eta_1^\dagger \eta_1 - \eta_2^\dagger \eta_2).$$

c) Show that the operators j_a have the same commutation rules as the angular momentum (divided by $\hbar$). Write j_2 and j_3 in terms of the q's and p's and show that the 'angular momentum operators' j_a have both integer and half-integer eigenvalues.

Setting $h_0 = H/\hbar\omega - 1$, the identity

$$j^2 \equiv j_1^2 + j_2^2 + j_3^2 = \frac{h_0}{2} \times \left(\frac{h_0}{2} + 1\right)$$

holds (it may be verified using the commutation rules).

d) Exploit the theory of angular momentum (all the properties of the angular momentum follow uniquely from the commutation relations) and the above identity to find the eigenvalues of H and the relative degeneracies. Say which eigenvalues of L_3 do occur in each energy level.

10.8   In Problem 10.7 the energy levels of a two-dimensional isotropic harmonic oscillator and their degeneracies have been found starting from the commutation rules of the three constants of motion j_1, j_2, j_3 'given from the outside'. We now want to establish both the existence and the form of such constants of motion starting from the invariance group of the Hamiltonian. Adopting the notation of Problem 10.7 one has:

$$H = \hbar\omega \left(\sum_{a=1}^{2} \eta_a^\dagger \eta_a + 1\right); \quad [\eta_a, \eta_b] = 0, \quad [\eta_a, \eta_b^\dagger] = \delta_{ab}; \quad a, b = 1, 2.$$

Consider the linear transformation:

$$\tilde{\eta}_a = \sum_b u_{ab}\, \eta_b\,. \tag{1}$$

a) Show that (1) is an invariance transformation both for the Hamiltonian and for the commutation rules if and only if u is a unitary 2×2 matrix.

We shall consider only the transformations that fulfill $\det u = 1$.

b) Show that all the unitary 2×2 matrices, whose determinant is 1, may be written as:

$$\begin{pmatrix} z_1 & z_2 \\ -z_2^* & z_1^* \end{pmatrix}, \quad |z_1|^2 + |z_2|^2 = 1, \quad z_1, z_2 \in \mathbb{C}.$$

They, therefore, form a continuous, 3 parameter group – the group $SU(2)$.

The transformation (1) in a neighborhood of the identity takes the form:

$$\tilde{\eta}_a = \eta_a + i\epsilon \sum_b g_{ab}\, \eta_b, \quad u \approx \mathbb{1} + i\epsilon\, g, \quad \epsilon \ll 1. \tag{2}$$

c) Show that the matrix g is Hermitian and traceless.

Thanks to the von Neumann theorem, for any transformation (1) there exists a unitary operator that implements it:

$$\tilde{\eta}_a = U(u)\,\eta_a\,U^{-1}(u)\,.$$

d) Let $U(g,\,\epsilon) = e^{-i\epsilon G_g}$ be the unitary operator that implements the infinitesimal transformation (2) (the operators $G_g = G_g^\dagger$ are the generators of the group). Compare $\tilde{\eta}_a = U(g,\,\epsilon)\,\eta_a\,U^{-1}(g,\,\epsilon)$, expanded to the first order in ϵ, and (2) and show that $[G_g,\,\eta_a] = -\sum_b g_{ab}\,\eta_b$. Find the expression for G_g and show that $[G_g,\,H] = 0$.

e) Show that any traceless Hermitian 2×2 matrix g may be written in the form (the factor $\frac{1}{2}$ is there only for the sake of convenience):

$$g = a_1\,\tfrac{1}{2}\sigma_1 + a_2\,\tfrac{1}{2}\sigma_2 + a_3\,\tfrac{1}{2}\sigma_3\,, \qquad a_i \in \mathbb{R}$$

where σ_i are the Pauli matrices

$$\sigma_1 = \begin{pmatrix} 0 & 1 \\ 1 & 0 \end{pmatrix}; \quad \sigma_2 = \begin{pmatrix} 0 & -i \\ i & 0 \end{pmatrix}; \quad \sigma_3 = \begin{pmatrix} 1 & 0 \\ 0 & -1 \end{pmatrix}.$$

f) Write the expressions for G_g in the three particular cases when only one of the a_i equals 1 and the other two are vanishing: compare the generators G_1, G_2, G_3 so obtained with the operators j_1, j_2, j_3 of Problem 10.7. Show that $[G_{g'},\,G_{g''}] = \sum_{ab} \eta_a^\dagger\,[g',\,g'']_{ab}\,\eta_b$ and make use of the commutation relations of the Pauli matrices:

$$[\tfrac{1}{2}\sigma_a,\,\tfrac{1}{2}\sigma_b] = i\,\epsilon_{abc}\,\tfrac{1}{2}\sigma_c$$

to find the commutation rules of the generators: $[G_a,\,G_b] = i\,\epsilon_{abc}\,G_c$.

Solutions

10.1

a) The Schrödinger equation is

$$-\frac{\hbar^2}{2m}\frac{\partial^2\psi(x,y)}{\partial x^2} + \frac{m}{2}\,\omega_1^2\,x^2\,\psi(x,y) - \frac{\hbar^2}{2m}\frac{\partial^2\psi(x,y)}{\partial x^2} + \frac{m}{2}\,\omega_2^2\,y^2\,\psi(x,y)$$
$$= E\,\psi(x,y)$$

and one is after separate variables solutions $\psi(x,y) = \psi_1(x) \times \psi_2(y)$:

$$\left[-\frac{\hbar^2}{2m}\psi_1''(x) + \frac{m}{2}\,\omega_1^2\,x^2\,\psi_1(x)\right]\psi_2(y)$$

$$+\left[-\frac{\hbar^2}{2m}\psi_2''(y) + \frac{m}{2}\,\omega_2^2\,y^2\,\psi_2(y)\right]\psi_1(x) = E\,\psi_1(x) \times \psi_2(y)\ .$$

Dividing both sides of the equation by $\psi_1(x) \times \psi_2(y)$, one has:

$$\left(-\frac{\hbar^2}{2m}\frac{\psi_1''(x)}{\psi_1(x)} + \frac{m}{2}\,\omega_1^2\,x^2\right) + \left(-\frac{\hbar^2}{2m}\frac{\psi_2''(y)}{\psi_2(y)} + \frac{m}{2}\,\omega_2^2\,y^2\right) = E\ .$$

The left hand side is the sum of a term depending only on x and a term depending only on y: in order that their sum be a constant, each of them must be a constant:

$$-\frac{\hbar^2}{2m}\frac{\psi_1''(x)}{\psi_1(x)} + \frac{m}{2}\,\omega_1^2\,x^2 = E_1\,,\qquad -\frac{\hbar^2}{2m}\frac{\psi_2''(y)}{\psi_2(y)} + \frac{m}{2}\,\omega_2^2\,y^2 = E_2$$

$$E_1 + E_2 = E\ .$$

Multiplying the two equations respectively by $\psi_1(x)$ and $\psi_2(x)$ one obtains the eigenvalue equations relative to two independent one-dimensional harmonic oscillators. Therefore:

$$E_1 \equiv E_{n_1} = \hbar\omega_1\left(n_1 + \tfrac{1}{2}\right),\qquad E_2 \equiv E_{n_2} = \hbar\omega_2\left(n_2 + \tfrac{1}{2}\right)\quad \Rightarrow$$

$$E \equiv E_{n_1 n_2} = \hbar\omega_1\left(n_1 + \tfrac{1}{2}\right) + \hbar\omega_2\left(n_2 + \tfrac{1}{2}\right)$$

$$\psi_{n_1 n_2}(x,y) = \psi_{n_1}^{(1)}(x) \times \psi_{n_2}^{(2)}(y)$$

where $\psi_{n_1}^{(1)}(x)$ and $\psi_{n_2}^{(2)}(y)$ are the eigenfunctions of the Hamiltonians of one-dimensional oscillators of angular frequencies ω_1 and ω_2. As both $\psi_{n_1}^{(1)}$ and $\psi_{n_2}^{(2)}$ are a complete set in $L^2(\mathbb{R})$, $\psi_{n_1 n_2}(x, y) = \psi_{n_1}^{(1)}(x) \times \psi_{n_2}^{(2)}(y)$ is a complete set in the space of functions $L^2(x, y)$ and, as a consequence, the eigenvalues $E_{n_1 n_2}$ are all the eigenvalues of H. The eigenvectors of H corresponding to the wavefunctions $\psi_{n_1}^{(1)}(x) \times \psi_{n_2}^{(2)}(y)$ are usually denoted by $|\, n_1, n_2 \,\rangle$ or also $|\, n_1 \,\rangle |\, n_2 \,\rangle$.

b) Putting $\omega_1 = 3\omega$, $\omega_2 = 4\omega$, the energy levels write:

$E_{n_1 n_2} = \hbar\omega \left(3n_1 + 4n_2 + \frac{7}{2}\right) = \hbar\omega \left(N + \frac{7}{2}\right)$.

The first degenerate level is that with $N = 12$: $n_1 = 0$, $n_2 = 3$; $n_1 = 4$, $n_2 = 0$ and the following is that with $N = 15$: $n_1 = 1$, $n_2 = 3$; $n_1 = 5$, $n_2 = 0$.

If the ratio of ω_1 and ω_2 is not rational, all the levels are nondegenerate: it appears that the commensurability of the frequencies is a necessary and sufficient condition for the occurrence of degeneracy, much as in classical mechanics so it is for the closure of trajectories (Lissajous curves).

c) If $\omega_2 = 0$, the particle is a free particle along the y axis and the eigenfunctions of the Hamiltonian are:

$\psi_{n,k}(x, y) = \psi_n^{(1)}(x)\, e^{i k y}$.

d) The potential is a positive definite quadratic form:

$$V(q_1, q_2) = \frac{1}{2} m\omega^2 \begin{pmatrix} q_1 & q_2 \end{pmatrix} \begin{pmatrix} 2 & -1 \\ -1 & 2 \end{pmatrix} \begin{pmatrix} q_1 \\ q_2 \end{pmatrix}$$

and it is known (see Problem 1.8) that it can be brought to canonical form by means of a real orthogonal transformation, namely by introducing the normal coordinates:

$$\tilde{q}_1 = \frac{1}{\sqrt{2}}(q_1 + q_2), \qquad \tilde{q}_2 = \frac{1}{\sqrt{2}}(q_1 - q_2)$$

and correspondingly:

$$\tilde{p}_1 = \frac{1}{\sqrt{2}}(p_1 + p_2), \qquad \tilde{p}_2 = \frac{1}{\sqrt{2}}(p_1 - p_2).$$

One has:

$$H = \frac{\tilde{p}_1^2}{2m} + \frac{1}{2}m\omega_1^2\,\tilde{q}_1^2 + \frac{\tilde{p}_2^2}{2m} + \frac{1}{2}m\omega_2^2\,\tilde{q}_2^2\,; \qquad \omega_1^2 = \omega^2, \quad \omega_2^2 = 3\omega^2$$

and, since the transformation $q \to \tilde{q}$, $p \to \tilde{p}$ is canonical, H expressed in terms of the variables $\tilde{q}, \tilde{p}$ is the Hamiltonian of a two-dimensional oscillator.

10.2

a) The Schrödinger eigenvalue equation:

$$-\frac{\hbar^2}{2m}\left(\frac{\partial^2}{\partial x^2} + \frac{\partial^2}{\partial y^2}\right)\psi_E(x, y) = E\,\psi_E(x, y), \qquad\qquad |x| \le a,\ |y| \le a$$

can be solved by the method of separation of variables because both the following conditions are fulfilled: *i*) $H = H_1(q_1, p_1) + H_2(q_2, p_2)$ and, as a consequence, the Schrödinger equation possesses solutions with the form $\psi_E(x, y) = \psi_{E_1}(x)\,\psi_{E_2}(y)$, $E_1 + E_2 = E$; *ii*) the boundary conditions $(\psi_E(x, y) = 0$ on the edges of the square) give rise to separate boundary conditions for $\psi_{E_1}(x)$ and $\psi_{E_2}(y)$: $\psi_{E_1}(a) = \psi_{E_1}(-a) = 0$; $\psi_{E_2}(a) = \psi_{E_2}(-a) = 0$ (see Problems 6.9 and 6.10). The equations for $\psi_{E_1}(x)$ and $\psi_{E_2}(y)$ are those for a particle in the segments $|x| \le a$ and $|y| \le a$:

$$-\frac{\hbar^2}{2m}\psi_{E_1}''(x) = E_1\,\psi_{E_1}(x)\,, \qquad -\frac{\hbar^2}{2m}\psi_{E_2}''(y) = E_2\,\psi_{E_2}(y)$$

that, with the given boundary conditions, have the (nonnormalized) solutions:

$$\psi_{n_1}(x) = \begin{cases} \cos(n_1\pi\,x/2a) & n_1 > 0 \text{ odd} \\ \sin(n_1\pi\,x/2a) & n_1 > 0 \text{ even} \end{cases}$$

$$\psi_{n_2}(y) = \begin{cases} \cos(n_2\pi\,y/2a) & n_2 > 0 \text{ odd} \\ \sin(n_2\pi\,y/2a) & n_2 > 0 \text{ even} \end{cases}$$

and $\psi_{n_1,n_2}(x, y) = \psi_{n_1}(x)\,\psi_{n_2}(y)$ for any pair of positive integers n_1, n_2 gives rise to a complete set of eigenfunctions of the Hamiltonian H.

b) The energy levels are:

$$E_{n_1,n_2} = E_{n_2} + E_{n_2} = \frac{\hbar^2\pi^2}{8ma^2}\left(n_1^2 + n_2^2\right)$$

and, when $n_1 \ne n_2$, they are twice degenerate.

c) As there are degenerate energy levels, there must exist operators that commute with the Hamiltonian H, but do not commute with one another. Indeed H (with the given boundary conditions) exhibits all the symmetries of the square, therefore it commutes with the rotations by an angle that is an integer multiple of $\pi/2$, with the inversions $x \to -x$, $y \to y$ and $x \to x$, $y \to -y$ and therefore also with the exchange $x \leftrightarrow y$, that can be obtained as the product of the rotation by $\pi/2$ and the inversion of the x axis: not all of these transformations, and therefore the operators associated with them, commute with one another (the group of the square is non-Abelian), and this fact guarantees the existence of degenerate levels. All the $\psi_{n_1,n_2}(x, y)$ are simultaneous eigenfunctions of H and of the inversions; if $n_1 = n_2$ they also are eigenfunctions of the operator that exchanges x with y, and therefore of all the other transformations. On the contrary, if $n_1 \ne n_2$, $\psi_{n_1,n_2}(y, x) \ne \psi_{n_1,n_2}(x, y)$ but they have the same energy: in fact $\psi_{n_1,n_2}(y, x) = \psi_{n_2,n_1}(x, y)$.

d) The Schrödinger equation is no longer a separable differential equation: it is still true that, by means of elementary trigonometry formulae, $V_a(x, y)$ can be expressed as the sum of two terms respectively dependent on $x + y$ and $x - y$, but the boundary conditions cannot be expressed in terms of the latter variables.

The potential $V_a(x, y)$ still has all the symmetries of the square: $V_a(x, y) = V_a(-x, y) = V_a(y, x)$, so there exists degenerate levels, even if it is not possible to relate the degeneracy of levels in presence of the potential with that in the absence of $V_a(x, y)$ (for example, it is no longer possible to state that the second and the fourth level have degeneracy 2).

e) The potential $V_b(x, y)$ is no longer invariant under inversions, while still it is so under the exchange $x \leftrightarrow y$ and the rotation by π that commute with each other: as a consequence, it is no longer possible to establish the existence of degenerate levels. In the absence of degeneracy, the eigenstates of H must be eigenstates of the exchange operator $x \leftrightarrow y$ as well, so $\psi_E(x, y) = \pm \psi_E(y, x)$. For any degenerate level that might exist it is still possible to find simultaneous eigenstates of the energy and the exchange operator.

10.3

a) The eigenfunctions of the energy are those of the particle in the square of edge a: $0 \le x \le a$, $0 \le y \le a$, vanishing at $x = y$ (see Problems 6.9, 6.10) or, equivalently, that are odd under the exchange $x \leftrightarrow y$: the eigenfunctions of the energy for a particle in the square of edge a are:

$$\psi_{n_1, n_2}(x, y) = \alpha \sin(n_1\pi\, x/a) \sin(n_2\pi\, y/a) + \beta \sin(n_2\pi\, x/a) \sin(n_1\pi\, y/a)$$

with $n_1 = 1, 2, \cdots,\ n_2 = 1, 2, \cdots$ and those that vanish at $x = y$ are:

$$\psi_{n_1, n_2}(x, y) = \sin(n_1\pi\, x/a) \sin(n_2\pi\, y/a) - \sin(n_2\pi\, x/a) \sin(n_1\pi\, y/a)$$

with $n_1 \ne n_2$. The eigenvalues of the energy are:

$$E_{n_1, n_2} = \frac{\hbar^2\pi^2}{2ma^2} \left(n_1^2 + n_2^2 \right), \qquad n_1 \ne n_2 .$$

b) The Hamiltonian (as well as the boundary conditions) is invariant under the reflection with respect to the straight line $x + y = a$ and, possessing only nondegenerate eigenvalues, its eigenfunctions also are eigenfunctions of this reflection: indeed, if $x \to a - y$, $y \to a - x$, $\psi_{n_1, n_2}(x, y) \to (-)^{n_1 + n_2} \psi_{n_1, n_2}(x, y)$.

10.4

a) With the boundary condition $\psi_E(x, y, z) = 0$ on the surface of the wave guide, the Schrödinger equation is a separable variables one; taken the Cartesian axes as in the text, the eigenfunctions of the Hamiltonian are:

$$\psi_{n_1, n_2; k}(x, y, z) = \sin\left(n_1\pi\, \frac{x}{a} \right) \sin\left(n_2\pi\, \frac{y}{a} \right) e^{i\,k\,z}, \qquad n_1, n_2 = 1, 2, \cdots$$

and the eigenvalues:

$$E_{n_1, n_2}(k) = \frac{\hbar^2\pi^2}{2ma^2} \left(n_1^2 + n_2^2 \right) + \frac{\hbar^2 k^2}{2m} \qquad (k \in \mathbb{R}) .$$

As k is real, $E > E_t = E_{n_1=1, n_2=1}(k = 0) = \hbar^2 \pi^2/ma^2$; E_t is, for particles, the analogue of the cutoff frequency for electromagnetic waves within a wave guide, namely the minimum frequency that can give rise to propagation along the guide.

b) A piece of the guide with unitary volume has length $1/a^2$, therefore, as $|\psi_1|^2$ and $|\psi_2|^2$ do not depend on z,

$$1 = \frac{1}{a^2} A^2 \int_0^a \int_0^a \sin^2(2\pi\, x/a) \sin^2(\pi\, y/a)\, dx\, dy = \frac{A^2}{4} \quad \Rightarrow \quad A^2 = 4$$

and likewise $B = 2$.

c) In both cases the components x and y of $\vec{j}$ vanish, whereas:

$$j_{1z} = 4\,\frac{\hbar k_1}{m}\, \sin^2(2\pi\, x/a) \sin^2(\pi\, y/a)\,; \quad j_{2z} = 4\,\frac{\hbar k_2}{m}\, \sin^2(\pi\, x/a) \sin^2(\pi y/a)$$

and obviously in both cases $\text{div}\, \vec{j} = \partial j_z/\partial z = 0$.

d) The probability current is divergenceless if the state of wavefunction $\psi(x, y, z, t)$ is stationary: indeed, from the continuity equation:

$$\text{div}\, \vec{j} = -\frac{\partial \rho(x, y, z, t)}{\partial t} = -\frac{\partial}{\partial t}\, \left|\psi_1\, \mathrm{e}^{-\mathrm{i}\, E_1 t/\hbar} + \psi_2\, \mathrm{e}^{-\mathrm{i}\, E_2 t/\hbar}\right|^2$$

$$= -\mathrm{i}\frac{E_1 - E_2}{\hbar}\, \left(\psi_1^*\, \psi_2\, \mathrm{e}^{\mathrm{i}\,(E_1 - E_2)t} - \psi_2^*\, \psi_1\, \mathrm{e}^{-\mathrm{i}\,(E_1 - E_2)t}\right)$$

and, since $\psi_1^*\, \psi_2\, \mathrm{e}^{\mathrm{i}\,(E_1 - E_2)t} - \psi_2^*\, \psi_1\, \mathrm{e}^{-\mathrm{i}\,(E_1 - E_2)t}$ is not identically zero, $\text{div}\, \vec{j} = 0 \Leftrightarrow E_1 = E_2$, namely the state of wavefunction $\psi(x, y, z, t)$ must be an eigenstate of the energy. As a consequence $E_{2,1}(k_1) = E_{1,1}(k_2)$, i.e.

$$5 \times \frac{\hbar^2 \pi^2}{2ma^2} + \frac{\hbar^2 k_1^2}{2m} = 2 \times \frac{\hbar^2 \pi^2}{2ma^2} + \frac{\hbar^2 k_2^2}{2m} \quad \Rightarrow \quad k_2 = \pm\sqrt{k_1^2 + 3\,\frac{\pi^2}{a^2}}\;.$$

10.5

a) Both the kinetic energy and the potential are invariant under rotations around the z axis, therefore they commute with L_z.

b) The Hamiltonian H commutes also with the operator I_x, the inversion with respect to the plane $x = 0$ (as a matter of fact, thanks to the invariance under rotations, H commutes also with the inversion with respect to any plane containing the z axis), but I_x and L_z do not commute, so there must exist degenerate levels. Since $I_x\, L_z\, I_x^{-1} = -L_z$, if one considers the simultaneous eigenstates of H and L_z: $|E, m'\rangle$, one has $I_x\, |E, m'\rangle \propto |E, -m'\rangle$, and, as a consequence, all the energy levels with $m' \neq 0$ are at least twice degenerate. This result holds true whatever the potential, provided it is invariant under rotations around some axis and depends only on the q's: the invariance under reflections follows from these assumptions.

c) In the rotating frame the Hamiltonian is (see Problem 9.10):

$$H = \frac{\vec{p}^2}{2m} + V(q_1^2 + q_2^2, q_3) - \omega L_z \equiv H_0 - \omega L_z$$

that still commutes with L_z, but does no longer commute with the inversions (in the present case $V - \omega L_z$ no longer depends only on the q's), so the existence of degenerate levels cannot be guaranteed: indeed the states $|E_0, m'\rangle$ and $|E_0, -m'\rangle$ (eigenstates of H_0 and L_z, therefore of H and L_z) respectively have energies $E_0 \mp m' \hbar\omega$.

10.6

a) As in Problem 10.1 the eigenvalue equation can be split up into the two equations:

$$H_1 | E_1 \rangle = E_1 | E_1 \rangle, \qquad H_2 | E_2 \rangle = E_2 | E_2 \rangle; \qquad E = E_1 + E_2$$

that are the eigenvalue equations for two one-dimensional independent oscillators. Then:

$$E_1 = \left(n_1 + \tfrac{1}{2}\right) \hbar\omega, \quad E_2 = \left(n_2 + \tfrac{1}{2}\right) \hbar\omega$$
$$E = E_1 + E_2 \equiv E_{n_1, n_2} = (n_1 + n_2 + 1) \hbar\omega.$$

The degeneracy of the n-th level ($n \equiv n_1 + n_2 = 0, 1, \cdots$) is the number of ways in which $n_1 + n_2 = n$, namely $n + 1$.

b) In the basis in which H_1 and H_2 are diagonal, the eigenfunctions of H are the product of the eigenfunctions of H_1 and H_2, given in the text of Problem 5.14:

$$|E_{n_1, n_2}\rangle \xrightarrow{\text{SR}} \sqrt{\frac{m\omega/\pi\hbar}{2^n \, n_1! \, n_2!}} \, H_{n_1}\left(\sqrt{m\omega/\hbar}\, x\right) H_{n_2}\left(\sqrt{m\omega/\hbar}\, y\right) e^{-(m\omega/2\hbar)\,(x^2 + y^2)}$$

where $H_n(\xi)$ are the Hermite polynomials.

c) The degeneracy found in a) obviously does not disagree with the results of Problem 10.5, but, from the third energy level on ($n \geq 2$), it is greater than that imposed by rotation and reflection invariance. Evidently there must exist further operators that commute with H but do not commute with one another: in Problem 10.7 we shall find these operators and in Problem 10.8 we shall see that their existence and form is determined by the invariance properties of the Hamiltonian.
As $H_k(\xi)$ is a polynomial of degree k, $H_{n_1}(\sqrt{m\omega/\hbar}\, x) \times H_{n_2}(\sqrt{m\omega/\hbar}\, y)$ contains $x^{n_1} y^{n_2} = (x^2 + y^2)^{n/2} (\cos\phi)^{n_1} (\sin\phi)^{n_2}$ and therefore $e^{\pm i n \phi}$, so that $m'_{\max} = n$, $m'_{\min} = -n$. Certainly not all the values of m' ranging from n to $-n$ are possible: they would be $2n + 1 > n + 1$; only those with a definite parity, the same parity of n, are allowed: indeed $H_{n_1} \times H_{n_2}$ has parity $(-1)^{(n_1 + n_2)} = (-1)^n$. Note that the number of integers between n and $-n$ with the same parity as n exactly is $n + 1$, i.e. the degeneracy of the level: actually, in the next problem we will see that the $n + 1$ states of the n-th level are precisely those with $L_3 = n, n - 2, \cdots, -n$.

d) The ground state ($n_1 = n_2 = 0$) is nondegenerate, so it must have $m' = 0$; the nonnormalized wavefunction is

$$| E_0 = \hbar\omega,\, m' = 0 \rangle \xrightarrow{\text{SR}} e^{-(x^2+y^2)/2a^2}, \qquad a = \sqrt{\hbar/m\omega}$$

that indeed does not depend on the angle ϕ. The first excited level has degeneracy 2, so, thanks to what has been found in c), $m' = \pm 1$; as

$$| n_1 = 1,\, n_2 = 0 \rangle \xrightarrow{\text{SR}} x\, e^{-(x^2+y^2)/2a^2}$$

$$| n_1 = 0,\, n_2 = 1 \rangle \xrightarrow{\text{SR}} y\, e^{-(x^2+y^2)/2a^2}$$

one has:

$$| E_1 = 2\hbar\omega,\, m' = \pm 1 \rangle \xrightarrow{\text{SR}} (x \pm iy)\, e^{-(x^2+y^2)/2a^2}.$$

The eigenspace corresponding to the second excited level has dimension 3: thanks to what has been found in c), in it L_3 must have the eigenvalues $m' = \pm 2$ both nondegenerate, otherwise – due to (see Problem 10.5) $I_x | E, m' \rangle = | E, -m' \rangle$ – the degeneracy of the level would be 4. Then the third eigenvalue must be $m' = 0$. One has (the states are equally normalized):

$$| n_1 = 2,\, n_2 = 0 \rangle \xrightarrow{\text{SR}} (x^2 - \tfrac{1}{2}a^2)\, e^{-(x^2+y^2)/2a^2}$$

$$| n_1 = 1,\, n_2 = 1 \rangle \xrightarrow{\text{SR}} \sqrt{2}\, x\, y\, e^{-(x^2+y^2)/2a^2}$$

$$| n_1 = 0,\, n_2 = 2 \rangle \xrightarrow{\text{SR}} (y^2 - \tfrac{1}{2}a^2)\, e^{-(x^2+y^2)/2a^2}$$

and, since the states with $m' = \pm 2$ have wavefunctions proportional to $(x \pm iy)^2 = x^2 - y^2 \pm 2\,i\,x\,y$ (see Problem 8.6), one has:

$$| E_2 = 3\hbar\omega,\, m' = \pm 2 \rangle \xrightarrow{\text{SR}} (x \pm iy)^2\, e^{-(x^2+y^2)/2a^2}$$

namely (we shall omit $n_1 =,\ n_2 =$)

$$| E_2,\, m' = \pm 2 \rangle = | 2,\, 0 \rangle - | 0,\, 2 \rangle \pm \sqrt{2}\, i\, | 1,\, 1 \rangle$$

and therefore, by orthogonality:

$$| E_2,\, m' = 0 \rangle = | 2,\, 0 \rangle + | 0,\, 2 \rangle \xrightarrow{\text{SR}} (x^2 + y^2 - a^2)\, e^{-(x^2+y^2)/2a^2}.$$

10.7

a) The operators $\eta^\dagger_{1,2}$, $\eta_{1,2}$ respectively are the "raising and lowering operators" for the independent one-dimensional oscillators 1 and 2. Therefore:

$$H = \hbar\omega\big(\eta^\dagger_1\,\eta_1 + \eta^\dagger_2\,\eta_2 + 1\big); \qquad [\eta_a,\, \eta_b] = 0, \qquad [\eta_a,\, \eta^\dagger_b] = \delta_{ab}.$$

b) By direct use of the commutation rules one can verify that the operators $\eta^\dagger_a\,\eta_b$ commute with the Hamiltonian H. Otherwise observe that (notation as in Problem 10.6) $\eta^\dagger_1\,\eta_2 | n_1,\, n_2 \rangle \propto | n_1 + 1,\, n_2 - 1 \rangle$ and that $E_{n_1,n_2} = E_{n_1+1,n_2-1}$.

c) $[j_1 , j_2] = -\dfrac{1}{4\mathrm{i}} [\eta_1^\dagger \eta_2 , \eta_2^\dagger \eta_1] + \dfrac{1}{4\mathrm{i}} [\eta_2^\dagger \eta_1 , \eta_1^\dagger \eta_2] = \dfrac{1}{2\mathrm{i}} [\eta_2^\dagger \eta_1 , \eta_1^\dagger \eta_2]$

$\qquad = \dfrac{1}{2\mathrm{i}} \Big(\eta_2^\dagger [\eta_1 , \eta_1^\dagger \eta_2] + [\eta_2^\dagger , \eta_1^\dagger \eta_2] \eta_1 \Big) = \dfrac{1}{2\mathrm{i}} \big(\eta_2^\dagger \eta_2 - \eta_1^\dagger \eta_1 \big) = \mathrm{i} j_3$

and likewise for the other commutators. In conclusion $[j_a , j_b] = \mathrm{i}\,\epsilon_{abc} j_c$. With the notation of Problem 10.6 one has:

$$j_2 = \dfrac{1}{2\hbar} L_3 = \dfrac{1}{2\hbar} (q_1 p_2 - q_2 p_1), \qquad j_3 = \dfrac{1}{2\hbar\omega} (H_1 - H_2).$$

Since (see Problem 10.6) L_3 has both even and odd eigenvalues, j_2 (and therefore any j_a) has both integer and half-integer eigenvalues: the occurrence of both kinds of eigenvalues is not forbidden as it is instead, for a given system, in the case of the angular momentum.

d) From the theory of angular momentum we know that j^2 has the eigenvalues $j(j+1)$ with $j = \frac{1}{2}, 1, \frac{3}{2}, \cdots$ therefore, thanks to the identity given in the text, H has the eigenvalues $(2j+1)\hbar\omega = (n+1)\hbar\omega$ with $n \equiv 2j = 0, 1, 2, \cdots$. Furthermore, always from the theory of angular momentum, one has that the number of independent states with a given j is $2j+1 = n+1$ and the eigenvalues taken by $L_3/\hbar = 2j_2$ are those between $-2j$ and $+2j$, namely the integers between $-n$ and $+n$ with the same parity as n.

$\qquad$ The classification of the states by means of j^2 and j_2 is the same as the classifications in terms of n and m (the eigenvalue of L_3), whereas the classification by means of j^2 and j_3 is the same as the classifications in terms of $n = n_1 + n_2$ and $n_1 - n_2$, namely in terms of n_1, n_2.

10.8

a) By η with no index we shall denote the pair (η_1, η_2), so $H = \hbar\omega\,(\eta^\dagger \eta + 1)$. One has:

$$\tilde{\eta}_a^\dagger = \sum_b u_{ab}^* \, \eta_b^\dagger = \sum_b \eta_b^\dagger \, u_{ba}^\dagger; \qquad \tilde{\eta} = u\,\eta, \qquad \tilde{\eta}^\dagger = \eta^\dagger \, u^\dagger$$

$$\tilde{\eta}^\dagger \, \tilde{\eta} = \eta^\dagger \, u^\dagger \, u\,\eta = \eta^\dagger \, \eta \quad \Leftrightarrow \quad u^\dagger u = \mathbb{1}\,.$$

$$[\tilde{\eta}_a , \tilde{\eta}_b^\dagger] = \sum_{cd} u_{ac} \, [\eta_c , \eta_d^\dagger] \, u_{db}^\dagger = \sum_c u_{ac} \, u_{cb}^\dagger = \delta_{ab}\,.$$

b) Any matrix of the given form obviously is unitary and its determinant equals 1. Conversely:

$$\begin{pmatrix} z_1 & z_2 \\ z_3 & z_4 \end{pmatrix} \begin{pmatrix} z_1^* & z_3^* \\ z_2^* & z_4^* \end{pmatrix} = \mathbb{1} \quad \Leftrightarrow$$

$$\begin{cases} |z_1|^2 + |z_2|^2 = 1 \\ z_1 z_3^* = -z_2 z_4^* \\ |z_3|^2 + |z_4|^2 = 1 \end{cases} \quad \Rightarrow \quad \begin{matrix} z_3 = -\alpha\,z_2^*, \quad z_4 = \alpha\,z_1^* \\ |\alpha| = 1 \end{matrix}$$

and, if the determinant must equal 1, then $\alpha = 1$. Unitary 2×2 matrices form a group – the $U(2)$ group – and those with determinant equal to 1 form a subgroup, the group $SU(2)$ (special unitary group).

c) The condition that, to the first order in ϵ, the matrix $\mathbb{1} + i\epsilon g$ be unitary is

$$(\mathbb{1} + i\epsilon g)(\mathbb{1} - i\epsilon g^\dagger) = \mathbb{1} + O(\epsilon^2) \quad \Rightarrow \quad g = g^\dagger$$

$$\det(\mathbb{1} + i\epsilon g) = 1 + i\epsilon(g_{11} + g_{22}) + O(\epsilon^2) \quad \Rightarrow \quad \text{Tr } g = 0 .$$

d) To the first order in ϵ:

$$U(g, \epsilon)\, \eta_a\, U^{-1}(g, \epsilon) \approx (\mathbb{1} - i\epsilon G_g)\, \eta_a\, (\mathbb{1} + i\epsilon G_g) \approx \eta_a - i\epsilon\, [G_g\, , \eta_a]$$

and by comparison with (2) in the text, it follows that $[G_g\, , \eta] = -g\,\eta$. Therefore, by analogy with $[\eta^\dagger \eta, \eta] = -\eta$, one has:

$$G_g = \eta^\dagger g\, \eta : \quad \sum_{ab} [\eta_a^\dagger\, g_{ab}\, \eta_b\, , \eta_c] = \sum_{ab} [\eta_a^\dagger\, , \eta_c]\, g_{ab}\, \eta_b = -\sum_b g_{cb}\, \eta_b .$$

Since the operators G_g are generators of invariance transformations for the Hamiltonian H: $U(u)\, H\, U^{-1}(u) = H$, they commute with H itself.

e) Any 2×2 matrix can be written as a linear combination of the three Pauli matrices and of the identity:

$$\begin{pmatrix} \alpha & \beta \\ \gamma & \delta \end{pmatrix} = \frac{1}{2}(\alpha + \delta)\, \mathbb{1} + \frac{1}{2}(\beta + \gamma)\, \sigma_1 + \frac{i}{2}(\beta - \gamma)\, \sigma_2 + \frac{1}{2}(\alpha - \delta)\, \sigma_3 .$$

The Hermitian matrices are linear combinations with *real* coefficients (the Pauli matrices are Hermitian), and those with vanishing trace ($\alpha + \delta = 0$) are combinations only of the three Pauli matrices.

f) If $a_1 = 1$, $a_2 = a_3 = 0$ one has $g = \frac{1}{2}\sigma_1$; therefore:

$$G_1 = \frac{1}{2}\eta^\dagger \sigma_1 \eta = \frac{1}{2}(\eta_1^\dagger \eta_2 + \eta_2^\dagger \eta_1) \equiv j_1$$

and likewise:

$$G_2 = \frac{1}{2}\eta^\dagger \sigma_2 \eta = \frac{1}{2i}(\eta_1^\dagger \eta_2 - \eta_2^\dagger \eta_1) \equiv j_2 ,$$

$$G_3 = \frac{1}{2}\eta^\dagger \sigma_3 \eta = \frac{1}{2}(\eta_1^\dagger \eta_1 - \eta_2^\dagger \eta_2) \equiv j_3 .$$

$$[G_{g'}\, , G_{g''}] = \sum_{abcd} [\eta_a^\dagger\, g'_{ab}\, \eta_b\, , \eta_c^\dagger\, g''_{cd}\, \eta_d]$$

$$= \sum_{abcd} \left(\eta_a^\dagger\, g'_{ab}\, [\eta_b\, , \eta_c^\dagger]\, g''_{cd}\, \eta_d + \eta_c^\dagger\, g''_{cd}\, [\eta_a^\dagger\, , \eta_d]\, g'_{ab}\, \eta_b \right)$$

$$= \sum_{ab} \eta_a^\dagger\, [g'\, , g'']_{ab}\, \eta_b$$

therefore (and likewise for the others):

$$[G_1\, , G_2] = \eta^\dagger\, [\tfrac{1}{2}\sigma_1\, , \tfrac{1}{2}\sigma_2]\, \eta = \frac{i}{2}\eta^\dagger \sigma_3 \eta = i\, G_3$$

then, thanks to the identity of the G_a's with the j_a's, we have established again the commutation rules between the j_a's found in Problem 10.7.

11

Particle in Central Field

Schrödinger equation with radial potentials in two and three dimensions; vibrational and rotational energy levels of diatomic molecules.

11.1 Consider a particle of mass m in three dimensions. In polar spherical coordinates one has $|A\rangle \xrightarrow{\text{SR}} \psi_A(r, \theta, \phi)$ and the volume element is $\mathrm{d}V = r^2 \, \mathrm{d}r \, \mathrm{d}\Omega$.

a) Use the Schrödinger representation and show that the operator $p_r \xrightarrow{\text{SR}} -\mathrm{i}\,\hbar\,\partial/\partial r$ is not Hermitian. Find the Schrödinger representation of $p_r^\dagger$ (the specification of its domain is not required).

b) Show that $\tilde{p}_r = r^{-1} p_r\, r$ is Hermitian and that:
$$\tilde{p}_r^2 = p_r^\dagger \, p_r \xrightarrow{\text{SR}} -\frac{\hbar^2}{r}\frac{\partial^2}{\partial r^2}\, r \,.$$

c) Using the Schrödinger representation verify the identity:
$$\frac{\vec{p}^{\,2}}{2m} = \frac{1}{2m}\, p_r^\dagger \, p_r + \frac{\vec{L}^{\,2}}{2mr^2}$$
on the states with angular momentum $l = 0$.

d) For the free particle, find the wavefunctions of the eigenstates $|E, l = 0\rangle$ of the energy with angular momentum $l = 0$.

11.2 Consider a particle of mass m constrained inside a sphere of radius a.

a) Find the energy levels and the corresponding eigenfunctions for the states that have angular momentum $l = 0$.

Now, instead, assume that the particle is subject to the "spherical potential well":
$$V(r) = \begin{cases} -V_0 & r < a, \\ 0 & r > a. \end{cases} \qquad V_0 > 0$$

b) Find for which values of V_0 does a bound state with $l = 0$ exist. If no bound state with $l = 0$ exists, may a bound state with $l > 0$ exist?

© Springer International Publishing AG 2017
E. d'Emilio and L.E. Picasso, *Problems in Quantum Mechanics*,
UNITEXT for Physics, DOI 10.1007/978-3-319-53267-7_11

11.3   Consider a particle of mass μ in two dimensions. In polar coordinates:

$$|A\rangle \xrightarrow{\text{SR}} \psi_A(\rho,\,\phi)\,, \qquad \rho = \sqrt{x^2 + y^2}\,.$$

a) Say whether the operator $p_\rho \xrightarrow{\text{SR}} -i\,\hbar\,\partial/\partial\rho$ is Hermitian and, if not, find the Schrödinger representation of $p_\rho^\dagger$ and of $p_\rho^\dagger\,p_\rho$.

b) Use the Schrödinger representation and prove the identity:

$$\frac{\vec{p}^{\,2}}{2\mu} = \frac{1}{2\mu}\,p_\rho^\dagger\,p_\rho + \frac{L_3^2}{2\mu\rho^2}\,, \qquad L_3 = q_1 p_2 - q_2 p_1.$$

c) Demonstrate that if the particle is subject to a radial potential $V(\rho)$, the Schrödinger equation admits factorized solutions: $\psi_E(\rho,\,\phi) = R_E(\rho)\,\Phi(\phi)$. Write the equation for the function $\Phi(\phi)$ and find its solutions. Write the equation for the radial function $R_E(\rho)$ and find k in such a way that the equation for the *reduced radial function* $u_E(\rho) \equiv \rho^k\, R_E(\rho)$ has the same form as the one-dimensional Schrödinger equation.

11.4   Consider a two-dimensional isotropic harmonic oscillator of mass μ.

a) Exploit the results of Problem 10.6 and find the reduced radial functions $u_0(\rho)$ and $u_1(\rho)$ relative to the first two energy levels of the oscillator; write the radial equations they solve. How many radial functions belong to the second excited energy level?

Consider now a three-dimensional isotropic harmonic oscillator.

b) Find the wavefunctions relative to the simultaneous eigenstates $|E,l,m\rangle$ of H, $\vec{L}^2$, L_z, for the first two energy levels of the oscillator.

11.5   The Hamiltonian of a particle in a central field is $H = \vec{p}^{\,2}/2m + V(r)$. We shall assume that the spectrum of H is bounded from below (see Problem 11.10).

a) If the states of minimum energy with angular momentum l and $l-1$ are bound states with energies E_l^0 and E_{l-1}^0, show that $E_{l-1}^0 < E_l^0$.

b) If the ground state is a bound state, which is its angular momentum?

c) Assume that $\lim_{r\to\infty} V(r) = 0$. Prove that, if a bound state with angular momentum l exists, then also a bound state with angular momentum $l-1$ exists.

11.6   Consider a free particle of mass μ in two dimensions. The simultaneous eigenfunctions of H and L_z have the form $\psi_{E,m}(\rho,\,\phi) \equiv R_{E,m}(\rho)\,\Phi(\phi)$ (see Problem 11.3). For $\rho \approx 0$, $R_{E,m}(\rho)$ stays finite and $R_{E,m}(\rho) \approx \rho^{s_m}$.

a) Find s_m.

b) Exploit the asymptotic form of the equation for the reduced radial function $u_{E,m} \equiv \sqrt{\rho}\, R_{E,m}(\rho)$ and show that $R_{E,m}(\rho)$, for $\rho \to \infty$, has an oscillating behaviour with infinite zeroes.

The second order differential equation

$$x\, y''(x) + y'(x) + x\, y(x) = 0$$

is known as *Bessel equation*.

c) Show that, of the two independent solutions, at most one can be (and actually is) regular at the origin.

d) Make the change of variable $x \equiv k\rho$ ($k = \sqrt{2\mu E}/\hbar$) and prove that the knowledge of the regular solution $J_0(x)$ of the Bessel equation allows one to find all the radial functions $R_{E,m=0}(\rho)$ relative to the states with $m = 0$ of the free particle.

11.7 The Bessel function $J_0(x)$ defined in Problem 11.6 is plotted in the figure. The first zero of $J_0(x)$ is at $x_1 \simeq 2.40$.

Consider a particle of mass μ in two dimensions constrained inside a circle of radius a.

a) Find the energy of the ground state of the particle.

Consider now a particle of mass μ in an infinite rectilinear guide (see Problem 10.4) whose cross section is, in the present case, a circle of radius a.

b) Write the equation for the stationary states of the particle and find the threshold energy of such a guide.

c) Compare the threshold energy of the guide with circular section with that (calculated in Problem 10.4) of the guide with square section of edge $2a$. Is it possible to establish a priori which of the two is greater?

d) Compare the threshold energy of a circular guide with that of a squared guide, in the case the two cross sections have the same area.

11.8 A particle of mass m is subject to a central potential $V(r)$ such that $\lim_{r \to 0} r^2\, V(r) = 0$. Let $\psi_{E,l,m}(r, \theta, \phi) \equiv R_{E,l}(r)\, Y_{l,m}(\theta, \phi)$ be the wavefunctions of the simultaneous eigenstates of H, $\vec{L}^2$, L_z. For $r \approx 0$ $R_{E,l}(r)$ stays finite and $R_{E,l}(r) \approx r^{s_l}$.

a) Determine s_l.

Let $V(r) = -\lambda/r$, $\lambda > 0$.

b) Find the mean value of the Hamiltonian in the states represented by the trial functions $\psi_1(r\,;\,a) = (\pi\,a^2)^{-3/4}\,e^{-r^2/2a^2}$ and use the result to establish an upper bound for the energy E_0 of the ground state of the system. Compare the obtained result with the exact eigenvalue E_0.

c) Find an upper bound for the energy $E^0_{l=1}$ of the lowest energy level with $l = 1$ making use of the normalized trial functions

$$\psi_2(\vec{r}\,;\,a) = \left(\frac{8}{3\sqrt{\pi a^3}}\right)^{1/2}(r/a)\,e^{-r^2/2a^2}\,Y_{l=1,m}(\theta,\,\phi)\,.$$

Compare the obtained result with the exact result.

11.9   The following are six radial functions $R_{n,l}(\rho)$ relative to some eigenstates of the Hamiltonian of the hydrogen atom:

$$R_a = \frac{8}{27\sqrt{6}}\,\rho(1-\rho/6)\,e^{-\alpha_a\rho}; \quad R_b = \frac{1}{2\sqrt{2}}\,(2-\rho)\,e^{-\alpha_b\rho}; \quad R_c = 2\,e^{-\alpha_c\rho};$$

$$R_d = \frac{1}{2\sqrt{6}}\,\rho e^{-\alpha_d\rho}; \quad R_e = \frac{4}{81\sqrt{30}}\,\rho^2 e^{-\alpha_e\rho}\,;$$

$$R_f = \frac{2}{81\sqrt{3}}\,(27-18\rho+2\rho^2)\,e^{-\alpha_f\rho}$$

where $\alpha_a\cdots\alpha_f$ are suitable coefficients and $\rho = r/a_{\mathrm{B}}$.

a) Assign the correct quantum numbers n, l and the value of the corresponding coefficient α to each of the above radial functions. Which couples of the above functions are orthogonal to each other ($\int_0^\infty R_\mu(r)\,R_\nu(r)\,r^2 dr = 0$)?

b) Prove that the Hamiltonian H_Z of a hydrogen-like ion with nuclear charge $Z\,e$ is unitarily equivalent to $Z^2 H_{Z=1}$, where $H_{Z=1}$ is the Hamiltonian of the hydrogen atom. Use the result and find the eigenvalues of H_Z, the energy levels of the hydrogen atom being known.

The functions $R_a \cdots R_f$ are normalized with respect to the dimensionless variable ρ:

$$\int_0^\infty R^2_{n,l}(\rho)\,\rho^2\,d\rho = 1\,.$$

c) Rewrite $R_a\cdots R_f$ as functions of r for a hydrogen-like ion with nuclear charge $Z\,e$, so that they are normalized with respect to the measure $r^2\,dr$.

11.10   Consider a particle of mass m in three dimensions subject to the central potential $V(r) = -\lambda/r^s$, with $\lambda > 0$.

a) Find how the normalization factor N depends on a for the states represented by the trial functions $\psi(r\,;\,a)$ depending on r and a only through the ratio r/a: $\psi(r\,;\,a) = N\,f(r/a)$, $a > 0$. Do the same for the mean values of the kinetic energy and of the potential energy.

b) Use the above result and prove that, if $s > 2$, the spectrum of the Hamiltonian is not bounded from below.

c) Use the method of the 'inscribed well' exposed in Problem 6.4 (see also Problem 6.6) and prove that, if $0 < s < 2$, the system admits an infinite number of bound states. In the latter case, where is it relevant that the potential has the asymptotic behaviour r^{-s} – for $r \to 0$ or, instead, for $r \to \infty$?

11.11   A particle of mass m, subject to a central potential $V(r)$, is in a laboratory rotating with constant angular velocity $\vec{\omega}_0$ (see Problem 9.10).

a) Give the quantum numbers that characterize the energy levels of the system and say which are their (minimum) degeneracies.

Assume now that the particle has a charge q and that the system, instead of being in a rotating laboratory, is in a constant uniform electric field $\vec{\mathcal{E}}$.

b) Give the quantum numbers that characterize the energy levels of the system and say which are their degeneracies.

11.12   The structure of the energy levels of diatomic molecules is complicated: however, at least in a first approximation, a classification is possible in terms of electronic energy levels, vibrational levels and rotational levels. The first are the energy levels of the electrons in the field of the nuclei assumed in fixed positions; the second are due to the small oscillations of the distance between the nuclei and the latter are due to the rotation of the molecule considered as a rigid body. In the above approximation, the energy levels are given by:

$$E_{n,v,l} = E_n + E_v + E_l = E_n + A_n\, v + B_n\, l(l+1) \tag{1}$$

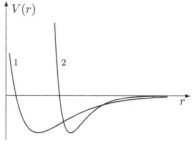

where E_n are the electronic levels, v and l (respectively the vibrational and the rotational quantum numbers) are nonnegative integers; the constants A_n, B_n depend on the electronic state. The approximation turns out to be acceptable only for values of the quantum numbers n, v, l not too large. In order to find the vibrational and rotational energy levels, we may consider the two atoms as two material points that interact through a potential $V(r)$ that, for a given electronic state, only depends on their distance and has the behaviour of the two plots reported in the figure.

a) Using polar spherical coordinates, write the Schrödinger equation for the two atoms in the center-of-mass frame.

The vibrational and rotational energy levels are determined by approximating $V(r)$ with a harmonic potential around its minimum:

$$V(r) + \frac{\hbar^2 l(l+1)}{2\mu r^2} \rightarrow V(r_0) + \frac{1}{2} V''(r_0)(r-r_0)^2 + \frac{\hbar^2 l(l+1)}{2\mu r_0^2}$$

(μ is the reduced mass of the two atoms); the centrifugal potential has been approximated by a constant: in this way the rotational motion has been decoupled from the vibrational one.

b) Say for which of the potentials reported in the figure the approximation of the centrifugal potential by a constant is more tenable.

c) Derive (1) that, the electronic state being assigned, takes the form $E_{v,l} = E_0 + A v + B l(l+1)$ where E_0 is a constant. To this end, it is necessary to extend the domain of the variable r down to $-\infty$: show that this is legitimate for values of the vibrational quantum number v not exceedingly bigger than 1 and if the inequality

$$\sqrt{\frac{\hbar^2}{\mu V''(r_0)}} \ll r_0^2 \tag{2}$$

is satisfied. Express (2) in terms of the constants A and B. What is the degeneracy of the levels $E_{v,l}$?

11.13   The experimental data relative to the vibrational and rotational energy levels

$$E_{v,l} = E_0 + A v + B l(l+1)$$

(see Problem 11.12) of the heteropolar molecules HF, HCl, HBr are the following (the energies are given in cm^{-1}: $E[eV] = hc \times E[cm^{-1}] = 1.24 \times 10^{-4} \times E[cm^{-1}]$):

HF :	$A = 4003$	$B = 41.1$
HCl:	$A = 2907$	$B = 20.8$
HBr:	$A = 2575$	$B = 16.7$.

a) The atomic masses of F, Cl and Br respectively are 19, 35 and 79 a.m.u. Use the results of Problem 11.12 to estimate both the dimensions of the three molecules and the amplitude of the small oscillations. Is it possible to estimate the dissociation energy of the molecules from the given data?

b) Say, for the three molecules, in which region of the electromagnetic spectrum do the first purely rotational absorption line $E_{0,0} \rightarrow E_{0,1}$ and the first roto-vibrational line $E_{0,0} \rightarrow E_{1,1}$ fall.

c) Calculate the constants A and B for the molecule DCl, where D is deuterium (isotope of hydrogen with atomic weight 2).

Solutions

11.1

a) From the definition $\langle A \mid \xi^\dagger \mid B \rangle = \langle B \mid \xi \mid A \rangle^*$ and by a partial integration one has:

$$\left(-i\hbar \int \psi_B^*(r, \theta, \phi) \frac{\partial \psi_A(r, \theta, \phi)}{\partial r} r^2 \, dr \, d\Omega \right)^* = -i\hbar \int \psi_A^* \frac{\partial (r^2 \psi_B)}{\partial r} \, dr \, d\Omega$$

$$= -i\hbar \int \psi_A^* \left(\frac{\partial}{\partial r} + \frac{2}{r}\right) \psi_B \, r^2 \, dr \, d\Omega \quad \Rightarrow \quad p_r^\dagger \xrightarrow{\text{SR}} -i\hbar\left(\frac{\partial}{\partial r} + \frac{2}{r}\right).$$

b) If $f(r)$ is an arbitrary function, one has

$$-i\hbar \frac{1}{r} \frac{\partial}{\partial r} (r f(r)) = -i\hbar\left(\frac{\partial}{\partial r} + \frac{1}{r}\right) f(r) \quad \Rightarrow \quad \tilde{p}_r = \frac{1}{2}(p_r + p_r^\dagger) = \tilde{p}_r^\dagger.$$

$$\tilde{p}_r^2 = \frac{1}{r} p_r^2 r \xrightarrow{\text{SR}} -\frac{\hbar^2}{r} \frac{\partial^2}{\partial r^2} r$$

$$p_r^\dagger p_r \xrightarrow{\text{SR}} -\hbar^2\left(\frac{\partial}{\partial r} + \frac{2}{r}\right)\frac{\partial}{\partial r} = -\hbar^2\left(\frac{\partial^2}{\partial r^2} + \frac{2}{r}\frac{\partial}{\partial r}\right)$$

and

$$\frac{1}{r} \frac{\partial^2}{\partial r^2} (r f(r)) = \frac{1}{r} \frac{\partial}{\partial r}\left(f(r) + r \frac{\partial f(r)}{\partial r}\right) = \left(\frac{\partial^2}{\partial r^2} + \frac{2}{r}\frac{\partial}{\partial r}\right) f(r).$$

c) In the Schrödinger representation, the wavefunctions of the states with $l = 0$ do not depend on the angles θ, ϕ, therefore

$$\vec{p}^{\,2} \mid l = 0 \rangle \xrightarrow{\text{SR}} -\hbar^2 \Delta f(r) = -\hbar^2 \sum_i \frac{\partial}{\partial x_i}\left(\frac{x_i}{r}\frac{\partial f}{\partial r}\right)$$

$$= -\hbar^2\left(\frac{\partial^2}{\partial r^2} + \frac{2}{r}\frac{\partial}{\partial r}\right) f(r).$$

d) The Schrödinger equation for the states with $l = 0$ relative to a free particle is

$$-\frac{\hbar^2}{2m} \frac{1}{r} \frac{\partial^2}{\partial r^2} (r R_E(r)) = E R_E(r)$$

and putting $u_E(r) = r\, R_E(r)$ ($u_E(r)$ is the *reduced radial function*), one has:

$$\begin{cases} u_E''(r) = -k^2\, u_E(r) \\ u_E(0) = 0 \end{cases} \qquad \Rightarrow \qquad R_E(r) = \frac{\sin k r}{r}, \qquad k^2 \equiv \frac{2mE}{\hbar^2}.$$

11.2

a) The Schrödinger equation for the reduced radial function $u_E(r) \equiv r R_E(r)$:

$$-\frac{\hbar^2}{2m}\, u_E''(r) = E\, u_E(r), \qquad u_E(0) = u_E(a) = 0$$

is the same as for a particle in one dimension constrained in the segment $0 \le x \le a$, therefore:

$$u_E(r) = \sin k r, \qquad k = \sqrt{\frac{2mE}{\hbar^2}}, \qquad ka = n\pi \quad \Rightarrow \quad E_n = \frac{\hbar^2 \pi^2}{2ma^2}\, n^2.$$

b) Thanks to the condition $u_E(0) = 0$, the present problem coincides with Problem 6.9: $V(x) = -V_0$ for $0 \le x \le a$, $V(x) = 0$ for $x > a$ and we know that the bound state exists if and only if $V_0 > \pi^2 \hbar^2 / 8ma^2$. Therefore in three dimensions, contrary to the one-dimensional case, not necessarily an attractive potential possesses bound states. In a central potential, the bound state of minimum energy, if it exists, must have angular momentum $l = 0$: indeed, the centrifugal potential $\hbar^2 l(l+1)/2mr^2$ has the effect of raising the eigenvalues of the energy (see also Problems 6.5 and 11.5).

11.3

a) As $dx\, dy = \rho\, d\rho\, d\phi$, proceeding as in Problem 11.1, one finds:

$$p_\rho^\dagger \xrightarrow{\text{SR}} -i\hbar\left(\frac{\partial}{\partial\rho} + \frac{1}{\rho}\right); \qquad p_\rho^\dagger p_\rho \xrightarrow{\text{SR}} -\hbar^2\left(\frac{\partial^2}{\partial\rho^2} + \frac{1}{\rho}\frac{\partial}{\partial\rho}\right).$$

b) $\vec{p}^{\,2} \xrightarrow{\text{SR}} -\hbar^2 \Delta_2, \qquad \Delta_2 = \frac{\partial^2}{\partial x^2} + \frac{\partial^2}{\partial y^2}.$

One has:

$$\frac{\partial}{\partial x} = \frac{\partial\rho}{\partial x}\frac{\partial}{\partial\rho} + \frac{\partial\phi}{\partial x}\frac{\partial}{\partial\phi} = \frac{x}{\rho}\frac{\partial}{\partial\rho} - \frac{y}{\rho^2}\frac{\partial}{\partial\phi} = \cos\phi\,\frac{\partial}{\partial\rho} - \frac{\sin\phi}{\rho}\frac{\partial}{\partial\phi}$$

$$\frac{\partial}{\partial y} = \frac{\partial\rho}{\partial y}\frac{\partial}{\partial\rho} + \frac{\partial\phi}{\partial y}\frac{\partial}{\partial\phi} = \frac{y}{\rho}\frac{\partial}{\partial\rho} + \frac{x}{\rho^2}\frac{\partial}{\partial\phi} = \sin\phi\,\frac{\partial}{\partial\rho} + \frac{\cos\phi}{\rho}\frac{\partial}{\partial\phi}$$

whence:

$$\Delta_2 = \frac{\partial^2}{\partial\rho^2} + \frac{1}{\rho}\frac{\partial}{\partial\rho} + \frac{1}{\rho^2}\frac{\partial^2}{\partial\phi^2} \quad \Rightarrow \quad \vec{p}^{\,2} = p_\rho^\dagger p_\rho + \frac{L_3^2}{\rho^2}.$$

c) The Schrödinger equation for factorized solutions $\psi_E(\rho, \phi) = R_E(\rho)\,\Phi(\phi)$ is

$$\left[-\frac{\hbar^2}{2\mu}\left(\frac{\partial^2}{\partial\rho^2} + \frac{1}{\rho}\frac{\partial}{\partial\rho}\right) - \frac{\hbar^2}{2\mu\rho^2}\frac{\partial^2}{\partial\phi^2} + V(\rho)\right] R_E(\rho)\,\Phi(\phi) = E\, R_E(\rho)\,\Phi(\phi).$$

By the same technique one uses for the separation of variables (see e.g. Problem 10.1) one obtains the equations for $\Phi(\phi)$ and for $R_E(\rho)$:

$$\frac{d^2}{d\phi^2}\Phi(\phi) = -m^2\,\Phi(\phi) \;\Rightarrow\; \Phi(\phi) = e^{\pm i\, m\phi}, \quad \Phi(\phi+2\pi) = \Phi(\phi) \;\Rightarrow\; m\in\mathbb{N}$$

$$\left[-\frac{\hbar^2}{2\mu}\left(\frac{d^2}{d\rho^2}+\frac{1}{\rho}\frac{d}{d\rho}\right)+\frac{\hbar^2 m^2}{2\mu\rho^2}+V(\rho)\right]R_E(\rho) = E\,R_E(\rho)\,.$$

From the identity:

$$\frac{1}{x^k}\frac{d^2}{dx^2}x^k f(x) = \frac{1}{x^k}\frac{d}{dx}\left(k\,x^{k-1}+x^k\,\frac{d}{dx}\right)f(x)$$

$$= \left(\frac{k(k-1)}{x^2}+\frac{2k}{x}\frac{d}{dx}+\frac{d^2}{dx^2}\right)f(x)$$

with $k=1$ one finds again the identity established in Problem 11.1:

$$\frac{1}{r}\frac{d^2}{dr^2}r = \frac{d^2}{dr^2}+\frac{2}{r}\frac{d}{dr}$$

whereas in two dimensions one must choose $k=1/2$ and the radial equation with $u_E(\rho) \equiv \sqrt{\rho}\,R_E(\rho)$ becomes:

$$\left[-\frac{\hbar^2}{2\mu}\frac{d^2}{d\rho^2}+\frac{\hbar^2(m^2-1/4)}{2\mu\rho^2}+V(\rho)\right]u_E(\rho) = E\,u_E(\rho)\,.$$

The effective potential therefore is $V_{\text{eff}}(\rho) = V(\rho)+\hbar^2(m^2-1/4)/2\mu\rho^2$.

11.4

a) The wavefunction of the ground state is (see Problem 10.6):

$$\psi_0(\rho) = e^{-(\mu\omega/2\hbar)\rho^2} \quad\Rightarrow\quad u_0(\rho) = \sqrt{\rho}\,e^{-(\mu\omega/2\hbar)\rho^2}$$

$$\left[-\frac{\hbar^2}{2\mu}\frac{d^2}{d\rho^2}-\frac{\hbar^2}{8\mu\rho^2}+\frac{1}{2}\mu\omega^2\rho^2\right]u_0(\rho) = \hbar\omega\,u_0(\rho)\,.$$

The wavefunctions with well defined L_z, relative to first excited level, are (see Problem 10.6):

$$\psi_{1,\pm1}(\rho,\phi) = \rho\,e^{-(\mu\omega/2\hbar)\rho^2}\,e^{\pm i\phi} \quad\Rightarrow\quad u_1(\rho) = \rho^{3/2}\,e^{-(\mu\omega/2\hbar)\rho^2}$$

$$\left[-\frac{\hbar^2}{2\mu}\frac{d^2}{d\rho^2}+\frac{3\hbar^2}{8\mu\rho^2}+\frac{1}{2}\mu\omega^2\rho^2\right]u_1(\rho) = 2\hbar\omega\,u_1(\rho)\,.$$

The second excited level has degeneracy 3: two states with $m = \pm2$, that have the same radial function: $m^2 = 4$, $R_{2,\pm2}(\rho) = \rho^2\,e^{-(\mu\omega/2\hbar)\rho^2}$; and one state with $m=0$, with radial function: $m^2 = 0$, $R_{2,0}(\rho) = (\rho^2-\hbar/\mu\omega)\,e^{-(\mu\omega/2\hbar)\rho^2}$.

b) With the notation of Problem 10.6, the ground state of the three-dimensional oscillator is

$$|\,n_1=0,\,n_2=0,\,n_3=0\,\rangle \;\xrightarrow{\text{SR}}\; e^{-(\mu\omega/2\hbar)r^2}$$

and has, therefore, angular momentum $l = 0$, whereas the first excited level has degeneracy 3: $n_1 + n_2 + n_2 = 1$ and all the states have $l = 1$:

$$| E, l = 1, m = \pm 1, 0 \rangle \xrightarrow{\text{SR}} r \, e^{-(\mu\omega/2\hbar) \, r^2} Y_{l=1,m}(\theta, \phi) .$$

11.5

a) The Hamiltonian for the states with angular momentum l may be written as (see Problem 11.1):

$$H_l = \frac{1}{2m} \, p_r^\dagger \, p_r + \frac{\hbar^2 \, l(l+1)}{2mr^2} + V(r) \quad \Rightarrow \quad H_l = H_{l-1} + \frac{\hbar^2 \, l}{m \, r^2} .$$

As each H_l gives rise to a one-dimensional Schrödinger equation for the reduced radial function:

$$-\frac{\hbar^2}{2m} \, u_{E,l}''(r) + \frac{\hbar^2 \, l(l+1)}{2mr^2} \, u_{E,l}(r) + V(r) \, u_{E,l}(r) = E \, u_{E,l}(r)$$

the present problem is similar to Problem 6.5: let $| E_l^0 \rangle$ and $| E_{l-1}^0 \rangle$ stand for the normalized eigenvectors relative to the minimum energies respectively of H_l and H_{l-1}; one has:

$$E_l^0 \equiv \langle E_l^0 \mid H_l \mid E_l^0 \rangle = \langle E_l^0 \mid H_{l-1} \mid E_l^0 \rangle + \langle E_l^0 \mid \frac{\hbar^2 \, l}{m \, r^2} \mid E_l^0 \rangle$$

$$> \langle E_l^0 \mid H_{l-1} \mid E_l^0 \rangle$$

so the spectrum of H_{l-1} extends below E_l^0: as a consequence, if – as assumed – the state of minimum energy of H_{l-1} is a bound state, one has $E_{l-1}^0 < E_l^0$.

b) It follows from the above result that, if the ground state is a bound state, it must have $l = 0$.

c) As the "effective potential" $V(r) + \hbar^2 l(l+1)/2mr^2 \to 0$ for $r \to \infty$, for any l the continuous spectrum is $E_l \geq 0$ and the possible bound states always have negative energies. It follows from what has been seen above that, if H_l has a bound state $E_l^0 < 0$, also H_{l-1} has a bound state, since its spectrum extends below E_l^0.

11.6

a) Let us impose that in the limit $\rho \to 0$ the following equation be satisfied:

$$\left[-\frac{\hbar^2}{2\mu} \left(\frac{d^2}{d\rho^2} + \frac{1}{\rho} \frac{d}{d\rho} \right) + \frac{\hbar^2 m^2}{2\mu\rho^2} \right] R_{E,m}(\rho) = E \, R_{E,m}(\rho) \qquad (1)$$

with $R_{E,m}(\rho) = \rho^{s_m}$:

$$-\frac{\hbar^2}{2\mu} \, s_m(s_m - 1) \, \rho^{s_m - 2} - \frac{\hbar^2}{2\mu} \, (s_m - m^2) \, \rho^{s_m - 2} - E \, \rho^{s_m} = 0 .$$

Dividing by ρ^{sm-2} and taking the limit $\rho \to 0$ one finds $s_m^2 = m^2$: only the solution with $s_m = |m|$ is finite at the origin.

For $m = 0$ the second solution $S_{E,m=0}(\rho)$ of equation (1) has, for $\rho \to 0$, the asymptotic behaviour $S_{E,m=0}(\rho) \approx \log \rho$ (see also point c).

b) The asymptotic form of the equation for the reduced radial function is obtained by neglecting the centrifugal term $\propto \rho^{-2}$; putting $k \equiv \sqrt{2\mu E}/\hbar$, one has:

$$-\frac{\hbar^2}{2\mu} \frac{d^2}{d\rho^2} u_{E,m}(\rho) \approx E\, u_{E,m}(\rho) \quad \Rightarrow \quad u_{E,m}(\rho) \approx \alpha \sin k\rho + \beta \cos k\rho \quad \Rightarrow$$

$$R_{E,m}(\rho) \approx \frac{\alpha \sin k\rho + \beta \cos k\rho}{\sqrt{\rho}}\,.$$

c) If the solution is regular at the origin, $x\, y''(x)$ and $x\, y(x)$ vanish at $x = 0$, so $y'(0) = 0$ and, the equation being homogeneous and of the second order, the solution is determined up to a multiplicative factor.

The second solution, usually denoted by $N_0(x)$ (named Bessel function of the second kind), can be found by the method of reduction of the degree (see Problem 6.22) and diverges at the origin as $\log x$.

d) Putting $x \equiv k\rho$ and having in mind that $d/d\rho = k\, d/dx$, the equation for the radial function $y_E(x) \equiv R_{E,m=0}(x/k)$ writes:

$$y_E''(x) + \frac{1}{x} y_E'(x) + y_E(x) = 0$$

that (after being multiplied by x) is the Bessel equation; the only solution that is regular at the origin is therefore $y_E(x) = J_0(x)$, whence $R_{E,m=0}(\rho) = J_0(k\rho)$.

11.7

a) The ground state must have $L_3 = 0$: indeed, in analogy with the three-dimensional case (Problem 11.5), one has $H_{|m|} < H_{|m|+1}$. The equation for the states with $m = 0$ is

$$-\frac{\hbar^2}{2\mu} \left(\frac{d^2}{d\rho^2} + \frac{1}{\rho} \frac{d}{d\rho} \right) R_{E,m=0}(\rho) = E\, R_{E,m=0}(\rho)\,, \qquad R_{E,m=0}(a) = 0\,.$$

As (see Problem 11.6) $R_{E,m=0}(\rho) = J_0(k\rho)$, one must have $J_0(ka) = 0$ and, as a consequence, for the ground state, $ka = x_1$ where $x_1 \simeq 2.40$ is the first zero of the Bessel function. Therefore:

$$E_0 = \frac{\hbar^2 k^2}{2\mu} = \frac{\hbar^2}{2\mu a^2} x_1^2 \simeq 5.8 \frac{\hbar^2}{2\mu a^2}\,.$$

b) Since the cross section of the guide is circular, it is convenient to write the Schrödinger equation for the stationary states of the particle in cylindrical coordinates ρ, ϕ, z. The factorized solutions $\psi_E(\rho, \phi, z) = R(\rho)\, \Phi(\phi)\, Z(z)$ satisfy the equations:

$$Z''(z) + k_z^2\, Z(z) = 0\,, \qquad\qquad k_z \in R$$

$$\Phi''(\phi) + m^2\, \Phi(\phi) = 0\,, \qquad\qquad m \in R$$

$$-\frac{\hbar^2}{2\mu}\Big(\frac{d^2}{d\rho^2} + \frac{1}{\rho}\frac{d}{d\rho} - \frac{m^2}{\rho^2}\Big) R_{n,m}(\rho) = E_n^{(2)}\, R_{n,m}(\rho)\,, \qquad R_{n,m}(a) = 0$$

where $E_n^{(2)}$ are the energy levels for a particle constrained within a circle of radius a. As $E = E_n^{(2)} + \hbar^2 k_z^2/2\mu$ and the threshold energy E_t is the minimum of E, one has $k_z = 0$ and:

$$E_t = \frac{\hbar^2}{2\mu a^2}\, x_1^2 \simeq 5.8\,\frac{\hbar^2}{2\mu a^2}\,.$$

c) The threshold energy for a guide with square cross section of edge $2a$ is (see Problem 10.4):

$$E_t^S = \frac{\hbar^2}{2\mu a^2}\,\frac{\pi^2}{4}\,(1^2 + 1^2) \simeq 4.9\,\frac{\hbar^2}{2\mu a^2} < E_t^C \simeq 5.8\,\frac{\hbar^2}{2\mu a^2}\,.$$

The inequality $E_t^S < E_t^C$ follows from the observation that E_t^S and E_t^C are the lowest eigenvalues of the Hamiltonians H_S and H_C for a free particle constrained respectively in a square and in the circle inscribed in the square, and from the fact that $H_S < H_C$: indeed H_C can be obtained by first adding to H_S the potential that is vanishing within the circle and equals $V_0 > 0$ in the region bounded by circle and square (see the figure), then letting $V_0 \to \infty$.

d) A guide of circular cross section, whose area is $4a^2$, has the radius $b = a\sqrt{4/\pi}$ and therefore the threshold energy:

$$E_t^{C'} = \frac{\hbar^2}{2\mu a^2}\,\frac{\pi}{4}\,x_1^2 \simeq 4.5\,\frac{\hbar^2}{2\mu a^2} < E_t^S\,.$$

11.8

a) Let us impose that in the limit $r \to 0$ the following equation be satisfied:

$$-\frac{\hbar^2}{2m}\frac{1}{r}\frac{d^2}{dr^2}\big(r\,R_{E,l}(r)\big) + \frac{\hbar^2 l(l+1)}{2mr^2}\,R_{E,l}(r) + V(r)\,R_{E,l}(r) = E\,R_{E,l}(r)$$

with $R_{E,l}(r) = r^{s_l}$:

$$-\frac{\hbar^2}{2m}\,s_l(s_l+1)\,r^{s_l-2} + \frac{\hbar^2}{2m}\,l(l+1)\,r^{s_l-2} + \big(V(r) - E\big)\,r^{s_l} = 0\,.$$

Dividing by r^{s_l-2} and taking the limit $r \to 0$ one obtains:

$$s_l(s_l+1) = l(l+1) \quad \Rightarrow \quad s_l = l\,, \quad s_l = -(l+1)$$

where only the first solution is acceptable, for $R_{E,l}(r)$ must be finite at the origin.

b) The wavefunction $\psi_1(r)$ is the product of three one-dimensional normalized Gaussian functions, so it is normalized and the mean value of the kinetic energy is three times that for a particle in one dimension with Gaussian wavefunction: from Problem 5.17 one has:

$$\frac{\overline{p}^2}{2m} = \frac{3\hbar^2}{4ma^2} .$$

Equivalently: $\psi_1(r)$ is the wavefunction of the ground state for a isotropic three-dimensional harmonic oscillator with angular frequency $w = \hbar/ma^2$, therefore the mean value of kinetic energy is a half of the total energy $E_0 = \frac{3}{2}\hbar w$.

The mean value of the potential energy is

$$\overline{V} = -\frac{\lambda}{(\sqrt{\pi}\,a)^3} \int e^{-r^2/a^2}\frac{1}{r}r^2\,dr\,d\Omega = -\frac{4\lambda}{\sqrt{\pi}\,a} \int_0^\infty e^{-x^2}x\,dx = -\frac{2\lambda}{\sqrt{\pi}\,a} \quad \Rightarrow$$

$$\overline{H} \equiv h(a) = \frac{3\hbar^2}{4ma^2} - \frac{2\lambda}{\sqrt{\pi}\,a} .$$

$h(a)$ has a minimum at $a^{-1} = 4\lambda\,m/(3\sqrt{\pi}\,\hbar^2)$ with value $-4\lambda^2 m/(3\pi\,\hbar^2)$ that is, therefore, an upper bound for E_0. The exact value can be obtained from the energy of the ground state of the hydrogen atom, provided e^2 is replaced by λ: $E_0 = -\lambda^2 m/2\hbar^2$; so the upper bound found above equals $(8/3\pi)\,E_0 \simeq 0.85\,E_0$, i.e. 15% higher than the exact result.

c) $\psi_2(\vec{r})$ is a wavefunction belonging to the first excited level (energy $\frac{5}{2}\hbar w$) of the three-dimensional harmonic oscillator with $w = \hbar/ma^2$, so the mean value of the kinetic energy is $5\hbar^2/4ma^2$. The mean value of the potential energy is

$$\overline{V} = -\lambda\,N^2 a^2 \int_0^\infty x^2 e^{-x^2}x\,dx = -\frac{4\lambda}{3\sqrt{\pi}\,a} \quad \Rightarrow$$

$$\overline{H} \equiv h(a) = \frac{5\hbar^2}{4ma^2} - \frac{4\lambda}{3\sqrt{\pi}\,a} \quad \Rightarrow \quad E^0_{l=1} < -\frac{16\lambda^2 m}{45\pi\,\hbar^2} .$$

As $E^0_{l=1} = -\lambda^2 m/8\hbar^2$ (the first excited level in the Coulomb field), the upper bound found above equals $0.91\,E^0_{l=1}$.

11.9

a) The asymptotic behaviour at the origin determines the value of l (see Problem 11.8), whereas the number of zeroes for $\rho > 0$, given by $n-l-1$, determines the value of n; in addition, the coefficient α is $1/n$. Therefore:

$$R_a = R_{3,1},\ \alpha_a = 1/3;\quad R_b = R_{2,0},\ \alpha_b = 1/2;\quad R_c = R_{1,0},\ \alpha_c = 1;$$

$$R_d = R_{2,1},\ \alpha_d = 1/2;\quad R_e = R_{3,2},\ \alpha_e = 1/3;\quad R_f = R_{3,0},\ \alpha_f = 1/3 .$$

Those with the same l and different values of n are orthogonal: indeed, the corresponding reduced radial functions $u_{n,l}(r) \equiv r\,R_{n,l}(r)$ satisfy the same Schrödinger equation, but with different eigenvalues. Otherwise: while $\psi_{n,l,m}$ with different values of l are orthogonal thanks to the orthogonality of the spherical harmonics, those with the same value of l are orthogonal only if such are the radial functions.

b) The transformation $\tilde{q}_i = Z^{-1}q_i$, $\tilde{p}_i = Z\,p_i$ is a canonical transformation (see Problem 5.6), therefore $Z^{-1}q_i = U\,q_i\,U^{-1}$, $Z\,p_i = U\,p_i\,U^{-1}$ whence:

$$Z^2 H_{Z=1} \equiv \frac{(Z\,\vec{p})^2}{2m} - \frac{Z\,e^2}{r/Z} = U\left(\frac{\vec{p}^2}{2m} - \frac{Z\,e^2}{r}\right)U^{-1} \;\Rightarrow\; E_n^{(Z)} = Z^2 E_n^{(Z=1)}\,.$$

c) To pass from the normalization with respect to the measure $\rho^2\,\mathrm{d}\rho$ to that with respect to the measure $r^2\,\mathrm{d}r \equiv a_{\rm B}^3\,\rho^2\,\mathrm{d}\rho$, the wavefunctions must be multiplied by $a_{\rm B}^{-3/2}$, regardless of the value of Z.
One has $|E_n^{(Z)}\rangle = U^{-1}|E_n^{(Z=1)}\rangle$, therefore:

$$\psi_n^{(Z)}(x,y,x) \equiv \langle x,y,z\,|\,E_n^{(Z)}\rangle = \langle x,y,z\,|\,U^\dagger\,|\,E_n^{(Z=1)}\rangle$$

and, since $U\,|\,x,y,z\rangle = Z^{3/2}\,|\,Zx,Zy,Zz\rangle$ (the factor $Z^{3/2}$ comes from imposing the Dirac delta normalization: see Problem 5.14), one has $\psi_n^{(Z)}(x,y,z) = Z^{3/2}\,\psi_n^{(Z=1)}(Zx,Zy,Zz)$.   So:

$$R_{1,0}(r) = \left(\frac{Z}{a_{\rm B}}\right)^{3/2} 2\,e^{-Z\,r/a_{\rm B}}\;;\quad R_{2,0}(r) = \left(\frac{Z}{a_{\rm B}}\right)^{3/2}\frac{1}{2\sqrt{2}}\left(2 - \frac{Z\,r}{a_{\rm B}}\right)e^{-Z\,r/2a_{\rm B}}\;;$$

$$R_{2,1}(r) = \left(\frac{Z}{a_{\rm B}}\right)^{3/2}\frac{Z\,r}{2\sqrt{6}\,a_{\rm B}}\,e^{-Z\,r/2a_{\rm B}}\;;$$

etc.

11.10

a) $1 = N^2\displaystyle\int |\psi(r\,;a)|^2\,r^2\,\mathrm{d}r\,\mathrm{d}\Omega = N^2 a^3\int |f(\xi)|^2\,\xi^2\,\mathrm{d}\xi\,\mathrm{d}\Omega \;\Rightarrow\; N \propto a^{-3/2}.$

Having in mind that $\mathrm{d}^2 f(\xi)/\mathrm{d}r^2 = a^{-2}\mathrm{d}^2 f(\xi)/\mathrm{d}\xi^2$, one has:

$$\overline{\frac{\vec{p}^2}{2m}} = -\frac{\hbar^2}{2m}\,N^2\int f^*(\xi)\,\xi^{-1}\frac{1}{a^2}\frac{\mathrm{d}^2\big(\xi\,f(\xi)\big)}{\mathrm{d}\xi^2}\,a^3\,\xi^2\,\mathrm{d}\xi\,\mathrm{d}\Omega = \frac{c_1}{a^2}\,,\qquad c_1 > 0\,.$$

Likewise:

$$\overline{V(r)} = -\lambda N^2\int |f(\xi)|^2\,a^{-s}\frac{1}{\xi^s}\,a^3\,\xi^2\,\mathrm{d}\xi\,\mathrm{d}\Omega = -\frac{c_2}{a^s}\,,\qquad c_2 > 0\,.$$

b) The mean value of the Hamiltonian in the states of wavefunction $\psi(r\,;a)$ is

$$\overline{H} \equiv h(a) = \frac{c_1}{a^2} - \frac{c_2}{a^s}\,,\qquad c_1,\,c_2 > 0$$

and, when $s > 2$, for a sufficiently small, $h(a)$ takes arbitrarily large negative values. This means that the spectrum of the operator H when $s > 2$ extends to $-\infty$: this situation, lacking of any physical significance, is described as "fall of the particle in the center" (see also Problem 12.11).

c) To start, let us limit ourselves to the case of the energy levels with $l = 0$: the Schrödinger equation for the reduced radial function $u(r)$ is that of a particle in one dimension subject to the potential $V(x) = -\lambda/x^s$ for $x > 0$ and $V(x) = \infty$ for $x \leq 0$ and, thanks to the condition $u(0) = 0$, the levels are the odd levels of a particle subject to the potential $V(x) = -\lambda/|x|^s$ for $-\infty < x < \infty$. Since $V(x) < 0$ and $V(x) \to 0$ for $x \to \infty$, for $E < 0$ the Hamiltonian may only have discrete eigenvalues; furthermore, the inscribed rectangular well of width $2a$ has depth $V_0 = \lambda/a^s$ and, owing to the fact that $V_0 a^2 = \lambda a^{2-s}$ grows indefinitely with the growing of a, the number of bound states is infinite for any $\lambda > 0$, $0 < s < 2$. What is relevant for the above result is the growing of λa^{2-s} with a, therefore the behaviour of the potential for $r \to \infty$: so the same result obtains for the states with $l > 0$, because for $s < 2$ the centrifugal potential does not change the asymptotic behaviour of $V(r)$.

11.11

a) The Hamiltonian is (see Problem 9.10):

$$H = \frac{\vec{p}^{\,2}}{2m} + V(r) - \vec{L} \cdot \vec{\omega}_0$$

and, taking the z axis parallel to $\vec{\omega}_0$, it commutes with $\vec{L}^2$, L_z and with the space inversion operator I that, in the case of one single particle, is a function of $\vec{L}^2$. As a consequence, the energy levels are characterized as $|E, l, m'\rangle$ and, for a generic potential $V(r)$, are nondegenerate.

b) Taking the z axis parallel to $\vec{\mathcal{E}}$, the Hamiltonian is

$$H = \frac{\vec{p}^{\,2}}{2m} + V(r) - q E z$$

and commutes with L_z and with the space inversions with respect to the planes that contain the z axis, e.g. the inversion I_x with respect to the plane $x = 0$. As $I_x L_z I_x^{-1} = -L_z$, the energy levels are characterized as $|E, m'\rangle$ and, due to $I_x|E, m'\rangle \propto |E, -m'\rangle$, the degeneracy of all the energy levels with $m' \neq 0$ is (at least) 2 (see Problem 10.5).

11.12

a) If $\vec{r}_1$, $\vec{r}_2$ denote the coordinates of the two nuclei, $\vec{r} = \vec{r}_2 - \vec{r}_1$ and one has (μ is the reduced mass of the two atoms):

$$\psi_{E,l,m}(\vec{r}) = R_{E,l}(r)\, Y_{l,m}(\theta, \phi), \qquad u_{E,l}(r) = r\, R_{E,l}(r)$$

$$-\frac{\hbar^2}{2\mu} u_{E,l}''(r) + \left[V(r) + \frac{\hbar^2 l(l+1)}{2\mu r^2}\right] u_{E,l}(r) = E\, u_{E,l}(r), \qquad u_{E,l}(0) = 0\,.$$

b) In the case of graph 2, the region of the 'small oscillations' is narrower and more distant from the origin than in the case of graph 1, therefore in that region the variation of the centrifugal potential is smaller.

c) If the domain of r is extended down to $-\infty$, the Schrödinger equation for the reduced radial function $u_{E,l}(r)$ is, up to the additive constant $V(r_0) + \hbar^2 l(l+1)/2\mu r_0^2$, the equation for a one-dimensional harmonic oscillator with the coordinate $x = r - r_0$, therefore the energy levels are:

$$E_{v,l} = V(r_0) + \hbar \sqrt{\frac{V''(r_0)}{\mu}} \left(v + \tfrac{1}{2} \right) + \frac{\hbar^2}{2\mu r_0^2} l(l+1) \qquad v, l \in \mathbb{N}$$

and the degeneracy is $2l+1$: $-l \le m \le l$. The extension of the domain of r down to $-\infty$ is equivalent to ignore the condition $u_{E,l}(0) = 0$: putting $\xi_0^2 = \hbar/\mu\omega = \sqrt{\hbar^2/\mu V''(r_0)}$, for the stationary states $\psi_v(x)$ of the oscillator with v not much bigger than 1 one has $\psi_v(-r_0) \propto e^{-r_0^2/2\xi_0^2}$ that, if the condition (2) given in the text is satisfied, is practically vanishing. As:

$$A = E_{1,l} - E_{0,l} = \hbar \sqrt{V''(r_0)/\mu}\,, \qquad B = \frac{1}{2}(E_{v,1} - E_{v,0}) = \hbar^2/2\mu r_0^2$$

condition (2) is equivalent to $A \gg B$.

11.13   We shall use the notation of Problem 11.12.

a) The dimensions of the molecules are given by r_0, i.e. the distance between the nuclei at equilibrium. One has:

$$r_0 \simeq \frac{\hbar}{\sqrt{2\mu B}} = \frac{1}{2\pi} \frac{hc}{\sqrt{2(\mu c^2) B}} \simeq 0.66,\, 0.92,\, 1.02 \text{ Å}$$

respectively.
The amplitude of the small oscillations is of the order of $\sqrt{v}\,\xi_0$ with, as in Problem 11.12, $\xi_0 = \sqrt{\hbar/\mu\omega}$:

$$\xi_0 = \frac{1}{2\pi} \frac{hc}{\sqrt{\mu c^2 A}} = r_0 \sqrt{\frac{2B}{A}} \simeq 0.09,\, 0.11,\, 0.12 \text{ Å}\,.$$

It is not possible to give an estimate of the dissociation energy, as we do not know $V(r_0)$.

b) $\Delta E_{\text{rot}}[\text{cm}^{-1}] = 2B \;\Rightarrow\; \lambda_{\text{rot}}[\text{cm}] = 1/\Delta E_{\text{rot}}[\text{cm}^{-1}] = 0.012,\, 0.024,\, 0.030$

and likewise:

$$\Delta E_{\text{rot vib}}[\text{cm}^{-1}] = A + 2B \;\Rightarrow$$

$\lambda_{\text{rot vib}}[\text{cm}] = 1/\Delta E_{\text{rot vib}} = 2.45 \times 10^{-4},\quad 3.39 \times 10^{-4},\quad 3.83 \times 10^{-4}\,.$

Since the purely rotational lines fall in the far infrared, they are of more difficult observation than the roto-vibrational ones, that fall in the nearer infrared. For the latter, indeed, the available spectroscopic precision is such that $E_{1,1} - E_{0,0} = A + 2B$ can be distinguished from $E_{1,2} - E_{0,1} = A + 4B$: as a matter of fact, the rotational structure is observed only inasmuch as superimposed to the purely vibrational one.

c) Since r_0 is much greater than the range of the nuclear forces, the potential $V(r)$ between the atoms is purely electrostatic in nature, so it does not appreciably vary when the hydrogen atom is replaced by a deuterium atom. As a consequence, the most relevant correction to the energy levels comes from the variation of the reduced mass: as (see Problem 11.12) $A \propto 1/\sqrt{\mu}$ and $B \propto 1/\mu$, one has:

$$A_D = A_H \sqrt{\frac{\mu_H}{\mu_D}} = 2907 \times \sqrt{\frac{35 \times 1/(35+1)}{35 \times 2/(35+2)}} = 2907 \times \sqrt{\frac{37}{72}} = 2084$$

$$B_D = B_H \frac{\mu_H}{\mu_D} = 20.8 \times \frac{37}{72} = 10.7 \ .$$

Then it appears that, going from light to heavy isotopes, the distance among roto-vibrational lines, that is proportional to B, decreases.

12

Perturbations to Energy Levels

Perturbations in one-dimensional systems; Bender–Wu method for the anharmonic oscillator; Feynman–Hellmann and virial theorems; "no-crossing theorem"; external and internal perturbations in hydrogen-like ions.

12.1 Consider a particle of mass m constrained in the segment $-a \le x \le a$ and subject to the repulsive potential $V(x) = \lambda\,\delta(x)\,,\quad \lambda > 0$.

a) Consider $V(x)$ as a perturbation and calculate the first order corrections $\Delta E_0^{(1)}$, $\Delta E_1^{(1)}$ to the energies of the ground state and of the first excited state.

b) Show that the result entails the existence of a value of the coupling constant λ such that the first two energy levels are degenerate. Say whether this is acceptable.

c) Find the exact energy of the first excited level for any value of λ and show that the equation that determines the exact energy of the ground state is

$$\tan ka = -\frac{\hbar^2}{m\lambda a}\,ka\,, \qquad k = \sqrt{\frac{2mE}{\hbar^2}}\,.$$

The solution of the above equation is given by the intersection of the curve $y = \tan ka$ with the straight line $y = -(\hbar^2/m\lambda a)\,ka$. Exploit this fact to illustrate, with the help of a graph, the behaviour of the lowest energy level $E_0(\lambda)$ for λ from 0 to $+\infty$.

d) Let $\psi_0(x, \lambda)$ be the eigenfunction corresponding to the ground state. Prove that $\lim_{\lambda \to \infty} \psi_0(0, \lambda) = 0$ Find the eigenfunctions of the Hamiltonian corresponding to $E_0(\lambda = \infty)$ and $E_1(\lambda = \infty)$.

12.2 Consider a particle of mass $m = 0.51\,\mathrm{MeV}/c^2$ (i.e. the mass of the electron), in a potential well:

$$V(x) = \begin{cases} -V_0 & |x| \le a\,, \\ 0 & |x| > a\,. \end{cases} \qquad V_0 = 6\,\mathrm{eV}, \quad a = 1.4\,\text{Å}$$

a) Say how many bound states the Hamiltonian admits and numerically calculate their energies.

© Springer International Publishing AG 2017
E. d'Emilio and L.E. Picasso, *Problems in Quantum Mechanics*,
UNITEXT for Physics, DOI 10.1007/978-3-319-53267-7_12

On such a particle the following positive definite perturbation acts:

$$V'(x) = \lambda \, \delta(x - b), \qquad \lambda > 0, \qquad -a < b < a .$$

b) Say for which value of b the first order correction $\Delta E_0^{(1)}$ to the energy of the ground state is a maximum and calculate it numerically for $\lambda = 1 \, \text{eV} \, \text{Å}$. For which values of λ is the result acceptable?

c) Say if there is a value of λ above which $H_0 + V'$ has no more bound states.

12.3   Consider two noninteracting particles, both of mass $m = 0.51 \, \text{MeV}/c^2$ and coordinates x_1, x_2, in a potential well:

$$V(x) = \begin{cases} -V_0 & |x| \le a, \\ 0 & |x| > a. \end{cases} \qquad V_0 = 6 \, \text{eV}, \quad a = 1.4 \, \text{Å}$$

It is known (see Problem 12.2) that, for the single particle, $V(x)$ gives rise to two bound states with energies E_0 and E_1.

a) Write the Hamiltonian H_0 for the system consisting of the two particles and find the bound states, the eigenvalues they belong to and their degeneracies.

Consider the "exchange operator" Π defined, up to a phase, by:

$$\Pi \, x_1 \, \Pi^{-1} = x_2, \quad \Pi \, p_1 \, \Pi^{-1} = p_2, \quad \Pi \, x_2 \, \Pi^{-1} = x_1, \quad \Pi \, p_2 \, \Pi^{-1} = p_1 .$$

b) Show that H_0 commutes with Π and that it is possible to choose Π in such a way that its eigenvalues are $+1$ and -1.

c) Find one or more observables that commute with H_0, but not with Π. Use this result and explain the degeneracies found in a).

Assume now that the particles interact through the repulsive potential:

$$V'(x_1, x_2) = \lambda \, \delta(x_1 - x_2), \qquad \lambda > 0 .$$

d) Prove that $H \equiv H_0 + V'$ commutes with Π. Is it still possible to guarantee the degeneracies found in a)?

e) Treating V' as a perturbation to the first order on H_0, prove that the discrete eigenvalues of $H = H_0 + V'$ are nondegenerate. May one say that the found sequence of the energy levels is the same (to the first perturbative order) for any short range potential? ($V'(|x_1 - x_2|) \ne 0$ for $|x_1 - x_2| \simeq 0$).

12.4   The Hamiltonian of a particle in one dimension is

$$H = \frac{p^2}{2m} + \frac{1}{2} k \, q^2 + \frac{1}{2} k' q^2 \equiv H_0 + \frac{1}{2} k' q^2, \qquad k > 0$$

and consider the term $\frac{1}{2} k' q^2$ (for suitable values of k', see below) as a perturbation to H_0.

a) Calculate the first order corrections $\Delta E_n^{(1)}$ to the eigenvalues $E_n^{(0)}$ of H_0.

b) Calculate the second order correction $\Delta E_0^{(2)}$ to the energy of the ground state.

c) Compare the obtained results with the exact eigenvalues.

d) Say in which conditions the first order result may be considered satisfactory.

The perturbative series is (most of the times only formally) the expansion of the eigenvalues in a power series of the parameter that defines the strength of the perturbation.

e) Find the radius of convergence of the perturbative series in the complex variable $z = k'/k$.

12.5 The Hamiltonian of a one-dimensional harmonic oscillator subject to an external constant force F is

$$H = \frac{p^2}{2m} + \frac{1}{2}m\omega^2 q^2 - Fq \equiv H_0 - Fq \, .$$

a) Find the eigenvalues and the eigenfunctions of the Hamiltonian H.

Let us now consider the term $-Fq$ as a perturbation. We want to calculate its effect on the unperturbed levels of the oscillator by means of perturbation theory.

b) Say at which orders of perturbation theory does the generic energy level receive a contribution from the perturbation and calculate them.

c) At which orders does perturbation theory give a contribution to the wavefunctions $\psi_n(x)$ of the exact eigenstates?

The expressions of the corrections of the third and fourth order to a nondegenerate energy level E_n due to the perturbation V are, if $\Delta E_n^{(1)} = 0$,

$$\Delta E_n^{(3)} = \sum_{a,b\neq n} \frac{V_{na} V_{ab} V_{bn}}{(E_a - E_n)(E_b - E_n)}$$

$$\Delta E_n^{(4)} = -\sum_{a,b,c\neq n} \frac{V_{na} V_{ab} V_{bc} V_{cn}}{(E_a - E_n)(E_b - E_n)(E_c - E_n)} - \Delta E_n^{(2)} \times \sum_{a\neq n} \frac{|V_{na}|^2}{(E_a - E_n)^2} \, .$$

d) Verify that the third and fourth order corrections to the energy of the ground state, due to the perturbation $-Fq$, are vanishing.

12.6 Consider the one-dimensional harmonic oscillator whose unperturbed Hamiltonian is H_0 and:

$$H = \frac{p^2}{2m} + \frac{1}{2}m\omega^2 q^2 + g\,q^4 = H_0 + \gamma\,\frac{m^2\omega^3}{\hbar}\,q^4 \equiv H_0 + H' \, .$$

a) Find the dimensions of the coupling constants g and γ.

b) Give the selection rules for the matrix elements of the perturbation H' between the eigenstates of H_0: i.e. establish *necessary* conditions on n' and n'' such that $\langle n'' \mid q^4 \mid n' \rangle$ may be nonvanishing.

Consider H' as a perturbation on H_0.

c) Find the first and second order effect of the perturbation H' on the ground state. Is the result meaningful when $g < 0$?

d) For $\gamma = 0.1$ and in units of $\hbar\omega$, calculate the energy of the ground state corrected up to the second order; compare it with the value $E_0 = 0.559\,\hbar\omega$, obtained by numerically integrating the Schrödinger equation.

12.7 Consider the one-dimensional harmonic oscillator of mass m and angular frequency ω perturbed by the potential:

$$V(x) = g\,x^4\,e^{-x^2/b^2}$$

where $b \gg a \equiv \sqrt{\hbar/m\omega}$. It is intended to describe the situation in which the perturbation is $g\,x^4$ in a region much larger than the characteristic length a of the oscillator, and rapidly tends to 0 for $|x| \to \infty$.

a) Calculate, to the first order, the effect of the perturbation on the ground state of the oscillator. Say whether, within the limits of applicability of perturbation theory ($|g| \ll \cdots$), the result is meaningful both for $g > 0$ and for $g < 0$.

b) Show that for $b/a \to \infty$ the result relative to the perturbation $g\,x^4$ of Problem 12.6 is recovered. If $b = 10\,a$, what is the fractional difference with respect to the case with $b = \infty$?

Certainly it is more meaningful to describe the situations when the perturbation is $g\,x^4$ in a region $|x| \le b \gg a$, by means of the potential:

$$V(x) = g\,x^4\,f(x/b)\,, \qquad f(\xi) = \begin{cases} 1 & |\xi| \le 1 \\ 1/\xi^4 & |\xi| \ge 1 \end{cases}$$

so that $V(x)$ is constant for $|x| \ge b$ (absence of forces).

c) How is the spectrum of the Hamiltonian $H = H_0 + V(q)$ (discrete and/or continuous)?

d) Show that for $b \to \infty$ the first order correction to the energy of the ground state of the oscillator converges to that relative to the perturbation $g\,x^4$.

12.8 Consider the one-dimensional harmonic oscillator whose unperturbed Hamiltonian is H_0 and:

$$H = \frac{p^2}{2m} + \frac{m\omega^2}{2}\,q^2 + g\,q^3 \equiv H_0 + H'\,.$$

a) Find the dimensions of the coupling constant g and write H' (as in Problem 12.6) in the form $H' = \gamma \cdots q^3$ (without useless numerical factors) so that the coupling constant γ is dimensionless.

b) Find the first order effect of the perturbation H' on the unperturbed energy levels.

c) Calculate the second order corrections $\Delta E_0^{(2)}$ to the energy of the ground state.

d) Say which is the spectrum of the Hamiltonian and if the use of perturbation theory is, in the present case, legitimate. What is the meaning to be attributed to the results found in b) and c)?

12.9 Problem 12.6 has emphasized that presumably, in the case of the perturbation $g\,q^4$ to the Hamiltonian of the harmonic oscillator, the perturbative series has a vanishing radius of convergence and, therefore, only a formal meaning. However we have seen that the second order result is in good agreement with the result obtained by numerical integration of the Schrödinger equation. The problem requires to understand whether, pushing the perturbative calculation to higher orders improves the agreement with the exact result, or not. C. Bender and T.T. Wu (1975) proposed a method for the calculation of higher orders that only requires the solution of algebraic equations and that therefore, possibly using a computer, may be pushed to very high orders.

Consider the Schrödinger equation for an oscillator with anharmonic term $g\,x^4$, in the dimensionless form found in Problem 6.3:

$$-\frac{1}{2}\,f''(\xi;\gamma) + \left(\frac{1}{2}\,\xi^2 + \gamma\,\xi^4\right) f(\xi;\gamma) = \epsilon(\gamma)\,f(\xi;\gamma) \qquad \gamma \ge 0 \, .$$

a) Set $f(\xi;\gamma) = h(\xi;\gamma)\,e^{-\xi^2/2}$ and show that $h(\xi;\gamma)$ satisfies the equation:

$$h'' - 2\xi\,h' - 2\gamma\,\xi^4 h + 2\big(\epsilon(\gamma) - \tfrac{1}{2}\big)\,h = 0 \, .$$

Show that, for $\gamma \neq 0$, $h(\xi;\gamma)$ cannot be a polynomial in ξ.

To find the ground state $f_0(\xi;\gamma) = f_0(-\xi;\gamma)$, $f_0(0,\gamma) = 1$, one puts:

$$\delta_0(\gamma) \equiv \epsilon_0(\gamma) - \tfrac{1}{2} = \sum_{n=1}^{\infty} a_n\,\gamma^n, \qquad h_0(\xi;\gamma) = 1 + \sum_{n=1}^{\infty} B_n(\xi)\,\gamma^n$$

where $B_n(\xi)$ are even polynomials of unknown degree $\big(B_n(0) = 0\big)$, and:

$$h_0'' - 2\xi\,h_0' - 2\gamma\,\xi^4 h_0 + 2\delta_0(\gamma)\,h_0 \equiv \sum_{n=1}^{\infty} C_n(\xi)\,\gamma^n = 0 \, .$$

With the above positions one has $\big(n \ge 1, \quad B_0(\xi) \equiv 1\big)$:

$$C_n \equiv B_n'' - 2\xi\,B_n' - 2\xi^4 B_{n-1} + 2\Big(a_n + \sum_{k=1}^{n-1} a_{n-k} B_k\Big) \, .$$

b) Demonstrate (by induction) that from the equation $C_n(\xi) = 0$ it follows that the degree of the polynomial $B_n(\xi)$ is $4n$.

c) Determine $B_1(\xi)$ and a_1 and show that $f_0(\xi; \gamma) \approx e^{-\xi^2/2}(1 + \gamma B_1(\xi))$ has two zeroes for any value of $\gamma > 0$; say whether $\epsilon_0(\gamma) \approx \frac{1}{2} + a_1\gamma$ either overestimates or underestimates the exact value.

d) Demonstrate that the knowledge of $B_i(\xi)$ and of a_i for $i = 1 \cdots n - 1$ determines $B_n(\xi)$ and a_n.

Limiting oneself to the first 9 (Bender and Wu calculated the first 75) orders, one finds:

$$\epsilon_0(\gamma) = \frac{1}{2} + \frac{3}{4}\gamma - \frac{21}{8}\gamma^2 + \frac{333}{16}\gamma^3 - \frac{30\,885}{128}\gamma^4 + \frac{916\,731}{256}\gamma^5$$
$$- \frac{65\,518\,401}{1024}\gamma^6 + \frac{2\,723\,294\,673}{2048}\gamma^7 - \frac{1\,030\,495\,099\,053}{32\,768}\gamma^8$$
$$+ \frac{54\,626\,982\,511\,455}{65\,536}\gamma^9 - \cdots.$$

e) For $\gamma = 0.1$, report in a graph the values of $\epsilon_0(\gamma)$ to the several orders $n = 0, \cdots, 9$ and compare them with the 'exact' eigenvalue $\epsilon_0 \simeq 0.559$. Explain why pushing the calculation to higher orders is meaningless.

12.10 Let the Hamiltonian $H(\lambda)$ of a system depend on a parameter λ in a continuous and differentiable way. Assume that $E(\lambda_0)$ is a discrete nondegenerate eigenvalue of $H(\lambda_0)$.

a) Demonstrate (Feynman–Hellmann theorem) that:

$$\left.\frac{dE(\lambda)}{d\lambda}\right|_{\lambda=\lambda_0} = \left.\langle E(\lambda_0) \mid \frac{dH(\lambda)}{d\lambda}\right|_{\lambda=\lambda_0} \mid E(\lambda_0)\rangle .$$

(Exploit $E(\lambda) = \langle E(\lambda) \mid H(\lambda) \mid E(\lambda)\rangle$ and $\langle E(\lambda) \mid E(\lambda)\rangle = 1$).
Show that, if $H(\lambda) = H_0 + \lambda V$, the Feynman–Hellmann theorem provides the well known expression for the first order correction to the eigenvalues of H_0 due to the perturbation λV.

b) Show that, if $E(\lambda_0)$ is a degenerate eigenvalue, the above result is, in general, false. Reformulate the theorem for the case when $E(\lambda_0)$ is a degenerate eigenvalue.

c) Let $H(Z)$ be the Hamiltonian of a hydrogen-like ion and consider Z as a continuous parameter. Exploit the preceding result and calculate the mean value of the potential energy $-Z\,e^2/r$ in the stationary states $|n, l, m\rangle$.

12.11 Let $H(q_i, p_i) = \sum_i p_i^2/2m_i + V(q)$ be the Hamiltonian of one or more particles and consider the canonical transformation (see Problem 5.6):

$$\tilde{q}_i = \lambda q_i, \qquad \tilde{p}_i = \lambda^{-1}p_i; \qquad H(q_i, p_i) \to H(\lambda q_i, \lambda^{-1}p_i).$$

a) Use the Feynman–Hellmann theorem (see Problem 12.10) to demonstrate that the mean value $\overline{T}$ of the kinetic energy in any normalizable eigenstate

$|E\rangle$ of H equals $\frac{1}{2}\langle E \mid \sum_i q_i \left(\partial V/\partial q_i\right) \mid E\rangle$ (virial theorem). If V is a homogeneous function of the coordinates of degree k (i.e. $\sum_i q_i \, \partial V/\partial q_i = kV$), find the relationships among the energy E and the mean values of the kinetic and potential energy.

b) Use the preceding result and find the mean values of the kinetic and potential energy in the states $|n,l,m\rangle$ of a hydrogen-like ion.

c) Demonstrate that in no case the attractive potential $V(r) = -g/r^2$ with $g > 0$ admits bound states with energy $E < 0$.

12.12   Let $H(\lambda) = H_0 + \lambda V$ be a Hamiltonian and let E_1, E_2 two different nondegenerate eigenvalues of H_0. Assume that the distance $|E_1 - E_2|$ is small with respect to the distance from the other energy levels of the system.

a) Make use of perturbation theory for quasi-degenerate levels and show that, if $V_{12} \equiv \langle E_1 \mid V \mid E_2\rangle \neq 0$, for no value of λ the two levels $E_1(\lambda)$, $E_2(\lambda)$ can cross each other: $E_1(\lambda) \neq E_2(\lambda)$.

b) Let $|E_1\rangle$ and $|E_2\rangle$ be eigenvectors relative to different eigenvalues of some observable ξ that commutes with $H(\lambda)$ for any value of λ. Is it possible that the two levels $E_1(\lambda)$, $E_2(\lambda)$ cross each other?

12.13   Consider a hydrogen atom in a static uniform electric field $\vec{\mathcal{E}}$.

a) Say which, among the following operators, are constants of motion: $\vec{L}^2$, one or more components of the angular momentum $\vec{L}$, the space inversion operator I $(x \to -x,\ y \to -y,\ z \to -z$ and the analogue for $p_x, p_y, p_z)$, space inversion operators with respect to suitable planes.

If the interaction between the atom and the electric field is considered as a first order perturbation, it is known that the level $n = 2$ splits into three levels, one of them being degenerate.

b) Could the residual degeneracy of the $n = 2$ level be completely removed by the perturbation at the orders higher than the first?

Consider now the level $n = 3$ of the hydrogen atom.

c) Say whether for a field of intensity $10^4 \, \mathrm{V/cm}$ the use of perturbation theory limited to the first order is reliable.

d) Use the results of a) and establish a priori what is the maximum number of energy sublevels the $n = 3$ level may split into, due to the perturbation.

e) Write the matrix representing the perturbation for the level $n = 3$ in the basis $|n,l,m\rangle$, after having ordered the elements of the basis in such a way that the resulting matrix is a diagonal block matrix, with blocks of dimensions respectively 3, 2, 2, 1, 1: denote by A, B, $C \cdots$ the matrix elements that are not vanishing because of some selection rule and show that the 3×3 block has a vanishing eigenvalue; show that the corresponding eigenvector is an eigenvector of the space inversion I.

f) Calculate the nonvanishing matrix elements of the perturbation, making use of the radial wavefunctions given in Problem 11.9 and of the spherical harmonics given in the solution of Problem 8.6. Eventually find the first order eigenvalues originating from the unperturbed level $n = 3$.

12.14 A hydrogen atom, whose unperturbed Hamiltonian is H_0, is subject to the perturbation (x, y, z are the electron–proton relative coordinates):

$$V_1 = \frac{V_0}{a_B^2} \, x \, y \, ; \qquad V_0 \simeq 10^{-2} \, \text{eV} \, , \qquad a_B = \frac{\hbar^2}{m_e e^2} \, .$$

a) Write V_1 in polar coordinates and say which conditions must $\Delta m \equiv m' - m''$ satisfy in order that the matrix elements $\langle \, m' \mid V_1 \mid m'' \, \rangle$ are nonvanishing (selection rule on L_z: see Problem 8.8).

We want to study the first order effect of the perturbation on the first two energy levels of the atom ($n = 1, 2$).

b) Show that the matrix elements of the perturbation on the levels $n = 1, 2$ that fulfill the selection rule on L_z are indeed nonvanishing. Find, for the first two energy levels of the atom, the eigenvalues to the first order and the corresponding approximate eigenvectors: take advantage of the following result:

$$\langle \, n, l, m = 0 \mid r^2 \mid n, l, m = 0 \, \rangle = \frac{n^2}{2} \left[5n^2 + 1 - 3l(l+1) \right] a_B^2 \, .$$

12.15 When a hydrogen atom occupies the center of a cubic cell of a crystal consisting of alternatively positive and negative ions, it is subject to the perturbation:

$$V = \frac{V_0}{a_B^3} \, x \, y \, z \, , \qquad (x, \, y, \, z) \equiv \vec{r} = \vec{r}_e - \vec{r}_p \, ; \qquad a_B = \frac{\hbar^2}{m_e e^2} \, .$$

a) Exploit suitable selection rules for V to show that the effect of the perturbation on the levels $n = 1$ and $n = 2$ is vanishing to the first order.

b) Exploit the invariance of the trace of a matrix under a change of basis and show that the sum of the eigenvalues of the (first order) perturbation on the states with $n = 3$ is vanishing (each eigenvalue is counted as many times as its degeneracy).

In order to study the effect of the perturbation on the level $n = 3$, the use of the Cartesian basis (see Problems 8.1, 8.2, 8.6 and 8.11) turns out to be convenient: the first six vectors of this basis are the following:

$$|\mathcal{P}_1\rangle \xrightarrow{\text{SR}} \sqrt{\frac{3}{4\pi}}\frac{x}{r}R_{3,1}(r), \quad |\mathcal{P}_2\rangle \xrightarrow{\text{SR}} \sqrt{\frac{3}{4\pi}}\frac{y}{r}R_{3,1}(r),$$

$$|\mathcal{P}_3\rangle \xrightarrow{\text{SR}} \sqrt{\frac{3}{4\pi}}\frac{z}{r}R_{3,1}(r),$$

$$|\mathcal{D}_{1,2}\rangle \xrightarrow{\text{SR}} \sqrt{\frac{15}{4\pi}}\frac{xy}{r^2}R_{3,2}(r), \quad |\mathcal{D}_{2,3}\rangle \xrightarrow{\text{SR}} \sqrt{\frac{15}{4\pi}}\frac{yz}{r^2}R_{3,2}(r),$$

$$|\mathcal{D}_{3,1}\rangle \xrightarrow{\text{SR}} \sqrt{\frac{15}{4\pi}}\frac{zx}{r^2}R_{3,2}(r).$$

c) Find three more vectors that complete the basis.

Since for each of the vectors of the basis there is at most only one vector such that the matrix element of the perturbation between them is nonvanishing, it is possible to arrange the vectors of the basis so that the matrix that represents the perturbation consists of three identical 2×2 blocks and of an identically vanishing 3×3 block.

d) In how many sublevels does the level $n = 3$ split into and which are the relative degeneracies?

12.16 A hydrogen atom is at the center of the cell of a crystal consisting of equal atoms forming a rectangular parallelepiped lattice with edges a, b, c parallel to the axes $\hat{x}$, $\hat{y}$, $\hat{z}$.

a) Exploit the symmetry of the crystal lattice and write the expansion of the electrostatic potential energy $V(x, y, z)$ generated by the lattice on the atom, up to the second order in $(x, y, z) \equiv \vec{r} = \vec{r}_e - \vec{r}_p$.

Consider the three cases: $i)$ $a = b = c$, $ii)$ $a = b \neq c$, $iii)$ $a \neq b \neq c \neq a$. Approximate $V(x, y, z)$ with its expansion $V^{(2)}$ to the second order in x, y, z (being the potential energy $V(x, y, z)$ electrostatic, one has $V(0) = 0$).

b) For each of the above cases say which, among the following observables, are constants of motion: L_x, L_y, L_z, $\vec{L}^2$, I_x (the space inversion with respect to the plane $x = 0$), I_y, I_z and what can be concluded, as a consequence, on the degeneracy of the energy levels of the hydrogen atom.

Consider now $V^{(2)}(x, y, z)$ as a perturbation.

c) Calculate its first order effect on the levels $n = 1$ and $n = 2$ when a, b, c are all different (case iii): for the states with $n = 2$, $l = 1$ it is convenient the use of the Cartesian basis (see Problem 8.2) or, which is the same thing, the basis of the simultaneous eigenvectors of I_x, I_y, I_z. Use the identity given in the text of Problem 12.14 and:

$$\langle 2,1,0 \mid z^2 \mid 2,1,0 \rangle = \tfrac{3}{5} \langle 2,1,0 \mid r^2 \mid 2,1,0 \rangle .$$

Let now $a = b = c$ (cubic crystal), and do not approximate $V(x,y,z)$ by $V^{(2)}$ any more.

d) Is it possible that $V(x,y,z)$ completely removes the degeneracies of the energy levels of the hydrogen atom?

12.17   Let us assume that the potential felt by the electron in the hydrogen atom is not exactly Coulombic:

$$-\frac{e^2}{r} \rightarrow -\frac{e^2}{r} + V(r) .$$

a) Let $V(r)$ be of constant sign for $r \le a \ll a_{\rm B}$ and vanishing for $r > a$. Consider $V(r)$ as a first order perturbation. Say on which states, among the ones with the same n, does the perturbation give rise to the more sizable effect and, in the two cases $V(r) \gtrless 0$, which is their sequence in order of growing energy.

Let us now assume that the proton charge does not exactly equal (in absolute value) the electron charge, but is $e\,(1+\epsilon)$.

b) Calculate both exactly and to the first order in $V(r)$ the energy levels of the hydrogen atom and verify the compatibility of the two results.

Let, finally, $V(r) = \epsilon \times \hbar^2/(2m_e\,r^2)$.

c) Calculate exactly the energy levels of the hydrogen atom and derive, from their expression, the correction of order ϵ. Use the found result to calculate the matrix elements $\langle n,l,m \mid r^{-2} \mid n,l,m \rangle$, where $\mid n,l,m \rangle$ are the unperturbed eigenstates of the hydrogen atom.

12.18   Consider the ion $C^{\rm VI}$ consisting of the nucleus of the carbon atom ($Z = 6$) and of only one electron ($C^{\rm II}$, $C^{\rm III}$, $\cdots$ respectively are the carbon atoms that have been ionized once, twice, $\cdots$). Assume that the nucleus is a uniformly charged sphere of radius $R \simeq 2.5 \times 10^{-13}$ cm (finite nuclear extension).

a) Draw a graph of the potential energy $U(r)$ of the electron and write the Hamiltonian of the system in the form $H = H_0 + V(r)$, where $V(r)$ is the difference between $U(r)$ and the potential energy of the electron in the field of the nucleus assumed pointlike.

b) Considering $V(r)$ as a perturbation, calculate the first order corrections $\Delta E_{1s}^{(1)}$, $\Delta E_{2s}^{(1)}$ and $\Delta E_{2p}^{(1)}$: it is sufficient to keep the lowest nonvanishing order in $R/a_{\rm B} \simeq 4.7 \times 10^{-5}$ (for the needed radial functions see Problem 11.9). What is the fractional correction to the ionization energy of $C^{\rm VI}$?

c) Taking into account that, presently, the spectroscopic measurements may arrive at a relative precision of one part per 10^{14}, say if and how the degeneracy of the levels with $n \geq 3$ may be experimentally resolved.

Let us assume that the nuclear charge is uniformly distributed on the surface of the sphere of radius R.

d) Establish a priori if, in this case, the corrections $\Delta E^{(1)}$ are greater or smaller with respect to the previous case and calculate those relative to the states $1s$, $2s$ and $2p$.

12.19 An α particle captures a μ^- meson and gives rise to the mesic ion $He^{++}-\mu^-$. The mass m_μ of the meson μ^- is about 207 times the mass of the electron. Let μ be the reduced mass of the system, which is in its ground state.

a) Write the wavefunction $\psi_{1,0,0}(r)$ of the μ^- meson.

b) Calculate the probability $P(r \leq r_0)$ of finding the μ^- meson inside the sphere of radius $r_0 = 10^{-2}a_B$, where $a_B \equiv \hbar^2/m_e e^2$ is the electron Bohr radius.

c) Calculate the potential $\varphi(r)$ generated by the charge distribution $\rho(r) = -e\,|\psi_{1,0,0}(r)|^2$ associated to the probability distribution for the position of the μ^- meson. Find, in particular, the asymptotic behaviour of the potential $\varphi(r)$ for $r \gg \hbar^2/\mu e^2$ and numerically evaluate the fractional difference between $\varphi(10^{-2}a_B)$ and the potential generated by a single negative pointlike charge at the same distance.

The mesic ion $He^{++}-\mu^-$ captures an electron in the ground state and forms a neutral mesic atom $He^{++}-\mu^--e^-$.

d) Calculate, in the approximation in which the finite extension of the meson charge is neglected, the probability of finding the electron inside the sphere of radius $10^{-2}a_B$.

e) Calculate the first order correction to the electron energy levels $n = 1$ and $n = 2$, due to the finite extension of the charge distribution of the μ^- meson, assuming that the latter is not perturbed by the electron.

12.20 There are several perturbations to the energy levels of the hydrogen atom due to relativistic effects. One of these is the different form of the kinetic energy:

$$\sqrt{m_e^2 c^4 + \vec{p}^{\,2}c^2} = m_e\,c^2 + \frac{\vec{p}^{\,2}}{2m_e} - \frac{(\vec{p}^{\,2})^2}{8m_e^3 c^2} + O(v^6/c^6)\,.$$

a) Calculate $v/c \stackrel{\text{def}}{=} \left(\overline{\vec{p}^{\,2}}/m_e^2 c^2\right)^{1/2}$ in the stationary states of the (nonrelativistic) hydrogen atom.

Consider the term $-(\vec{p}^{\,2})^2/8m_e^3 c^2$ as a perturbation and set:

$$H = \frac{\vec{p}^{\,2}}{2m_e} - \frac{e^2}{r} - \frac{(\vec{p}^{\,2})^2}{8m_e^3 c^2} \equiv H_0 - \frac{(\vec{p}^{\,2})^2}{8m_e^3 c^2} \; .$$

b) Estimate the order of magnitude of the fractional corrections the perturbation causes on the levels E_n .

c) Calculate the first order effect of the perturbation on the energy levels of the hydrogen atom: the identity $\vec{p}^{\,2} = 2m_e \, (H_0 + e^2/r)$ and (see Problem 12.17):

$$\langle \, n, l, m \mid r^{-2} \mid n, l, m \, \rangle = \frac{1}{n^3(l + \frac{1}{2}) \, a_B^2}$$

are useful.

d) Calculate the separation between the levels $2s$ and $2p$ both in eV and in cm^{-1} (see Problem 11.13).

Solutions

12.1

a) The normalized wavefunction of the ground state is

$$\psi_0(x) = \frac{1}{\sqrt{a}} \cos \frac{\pi x}{2a} \quad \Rightarrow \quad \Delta E_0^{(1)} = \int_{-a}^{+a} V(x) |\psi_0(x)|^2 \, dx = \frac{\lambda}{a} .$$

The wavefunction $\psi_1(x)$ of the first excited state vanishes at the origin, whence $\Delta E_1^{(1)} = 0$.

b) The lowest level crosses the first excited one when $\lambda/a = E_1^{(0)} - E_0^{(0)} = 3\hbar^2\pi^2/8ma^2$. The result is not acceptable for two reasons: first of all, the discrete energy levels of one-dimensional systems always are nondegenerate; secondly, the crossing occurs for $\Delta E_0^{(1)} = E_1^{(0)} - E_0^{(0)}$ and the validity of the first order perturbative calculation is not guaranteed in this condition.

c) The first excited level is not displaced: this holds true not only at first order ($\Delta E_1^{(1)} = 0$), but exactly: with a δ–potential as a perturbation this is a general result that applies to any eigenfunction $\psi_E(x)$ of a Hamiltonian H_0, such that $\psi_E(0) = 0$. Indeed, $\lambda \, \delta(x) \, \psi_E(x) = 0$ so that:

$$\Big(H_0 + \lambda \, \delta(x) \Big) \, \psi_E(x) = H_0 \, \psi_E(x) = E \, \psi_E(x) .$$

As for the ground state, that is an even state, let us put:

$$\psi_0(x, \lambda) = \begin{cases} \sin k(x+a) & -a \le x \le 0 \\ -\sin k(x-a) & 0 \le x \le a \end{cases} \tag{1}$$

and let us impose that the discontinuity of the logarithmic derivative be $2m\lambda/\hbar^2$ (see Problem 6.18; note that in Problem 6.18 the potential is $-\lambda \, \delta(x)$); one obtains the equation:

$$\tan ka = -\frac{\hbar^2}{m\lambda a} ka .$$

As k is determined by the intersection point between the tangent and a straight line with negative slope, proportional to $1/\lambda$, the solution lies between $ka = \pi/2$ (for $\lambda = 0$) and $ka = \pi$ (for $\lambda = \infty$), therefore the energy of the ground state starts, for $\lambda = 0$, from $\hbar^2\pi^2/8ma^2$, is a monotonically increasing function of λ and asymptotically ($ka \to \pi$) tends to $E_1 = \hbar^2\pi^2/2ma^2$: therefore it does never cross the first excited level, as it must be.

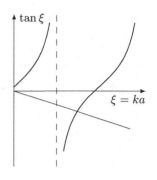

d) Apart from the normalization coefficient that, from (1), is seen to be bounded for any λ, $\psi_0(x, \lambda)$ takes the value $\sin ka$ at the origin and for $ka \to \pi$ tends to zero. Otherwise: as the logarithmic derivative diverges for $\lambda \to \infty$ and the derivative of (1) remains finite, $\psi_0(0, \lambda)$ must tend to zero. The wavefunctions relative to the energy level, degenerate in the limit $\lambda = \infty$, $E_0(\lambda = \infty) = E_1$ are:

$$\psi_0(x) = \frac{1}{\sqrt{a}}\left| \sin\frac{\pi x}{a} \right|, \qquad \psi_1(x) = \frac{1}{\sqrt{a}}\sin\frac{\pi x}{a}$$

and their linear combinations: in particular $\psi_r(x) = \frac{1}{\sqrt{2}}(\psi_0(x) + \psi_1(x))$, that is nonvanishing only for $x > 0$, and $\psi_1(x) = \frac{1}{\sqrt{2}}(\psi_0(x) - \psi_1(x))$, that is nonvanishing only for $x < 0$. As a consequence, in the limit $\lambda \to \infty$, $V(x) = \lambda\,\delta(x)$ behaves as an impenetrable barrier of potential.

12.2

a) It is known (see Problem 6.6) that a rectangular potential well of depth V_0 and width $2a$ possesses a number of bound states equal to the lowest integer greater or equal to $\sqrt{2mV_0a^2/\hbar^2}/(\pi/2) \simeq \sqrt{3.1}/(\pi/2) = 1.1$ $\left(\sqrt{2mV_0a^2/\hbar^2} = \sqrt{(a/a_B)^2 \times V_0/(e^2/2a_B)} \right)$. So the system has two bound states whose energies are obtained by numerically solving the equations, respectively for the ground and the first excited state (see Problem 6.11 and 6.9):

$$\xi\tan\xi = \sqrt{(2mV_0a^2/\hbar^2) - \xi^2}, \qquad -\xi/\tan\xi = \sqrt{(2mV_0a^2/\hbar^2) - \xi^2}$$

where $\xi = ka = \sqrt{2ma^2(V_0 - |E|)/\hbar^2}$.   One finds $\xi_0 = 0.97$ and $\xi_1 = 1.71$ whence:

$$E_0 = -(\xi_0\tan\xi_0)^2\,\frac{\hbar^2}{2ma^2} = -(\xi_0\tan\xi_0)^2 \times \left(\frac{a_B}{a}\right)^2\frac{e^2}{2a_B} \simeq -4\,\text{eV};$$

$$E_1 = -(\xi_1/\tan\xi_1)^2\,\frac{\hbar^2}{2ma^2} \simeq -0.11\,\text{eV}.$$

b) If $\psi_0(x)$ is the normalized wavefunction of the unperturbed ground state, one has $\Delta E_0^{(1)} = \lambda |\psi_0(b)|^2$, that is a maximum for $b = 0$. One has:

$$\psi_0(x) = \psi_0(-x) = A \begin{cases} \cos(k_0 x) & 0 \le x \le a \\ \cos(k_0 a)\, e^{-\kappa_0(x-a)} & x \ge a \end{cases}$$

where $k_0 a = \xi_0$ and $\xi_0^2 + (\kappa_0 a)^2 = 2m V_0 a^2/\hbar^2$, whence:

$$|A|^2 = \left[a\left(1 + \frac{\cos^2(k_0 a)}{\kappa_0 a} + \frac{\sin(2k_0 a)}{2k_0 a}\right) \right]^{-1} \simeq 0.42\,\text{Å}^{-1} \quad \Rightarrow$$

$$\Delta E_0^{(1)} = \lambda |A|^2 \simeq 0.42\,\text{eV} .$$

The first order correction provides an acceptable result until $\Delta E_0^{(1)} \ll E_1 - E_0 \simeq 3.9\,\text{eV} \Rightarrow \lambda \ll 9.3\,\text{eV Å}$.

c) No: for any (finite) value of λ, E_1 is a discrete eigenvalue of $H_0 + V'$ and since (see Problem 12.1) $E_0(\lambda) < E_1$, there always exist two bound states.

12.3

a) One has:

$$H_0 = \frac{p_1^2}{2m} + V(x_1) + \frac{p_2^2}{2m} + V(x_2) \equiv H_1 + H_2$$

that is a separate variable Hamiltonian, with the following normalizable eigenvectors: $|E_0\rangle_1 |E_0\rangle_2 \equiv |E_0, E_0\rangle$ belonging to the nondegenerate eigenvalue $2E_0$; $|E_0, E_1\rangle$ and $|E_1, E_0\rangle$, both belonging to the eigenvalue $E_0 + E_1$; $|E_1, E_1\rangle$ belonging to the nondegenerate eigenvalue $2E_1$. In the remaining stationary states at least one particle must be in a non-normalizable eigenstate belonging to the continuous spectrum.

b) From the definition of Π one has:

$$\Pi H_1 \Pi^{-1} = H_2, \qquad \Pi H_2 \Pi^{-1} = H_1 \quad \Rightarrow \quad \Pi H_0 \Pi^{-1} = H_0 .$$

Thanks to the von Neumann theorem, the operator Π is a unitary operator defined up to a phase factor; as Π^2 is a multiple of the identity: $\Pi^2 = e^{i\varphi}\,\mathbb{1}$, by means of the redefinition $\Pi \to e^{-i\varphi/2}\,\Pi$ one has $\Pi^2 = \mathbb{1}$ whence its eigenvalues are ± 1.

c) The operators H_1 and H_2 (obviously) commute with H_0, but they do not commute with Π. This explains the existence of twice degenerate eigenvalues of H_0: indeed, from $\Pi H_{1,2} \Pi^{-1} = H_{2,1}$ it follows that $\Pi |E_a, E_b\rangle = |E_b, E_a\rangle$, therefore, if $E_a \ne E_b$, the level $E_a + E_b$ is degenerate (exchange degeneracy).

d) One has:

$$\Pi\, \delta(x_1 - x_2)\, \Pi^{-1} = \delta(x_2 - x_1) = \delta(x_1 - x_2) \quad \Rightarrow \quad \Pi H \Pi^{-1} = H .$$

The argument used above to explain the exchange degeneracy does not apply in the present case: H_1 and H_2 no longer commute with H. However, and this is a limit of the degeneracy theorem, one cannot exclude the existence of further constants of motion that enable one to conclude that the degeneracy is not removed by the potential V'. In several cases perturbation theory provides a solution to the problem: if there exist reasons for the persistence of the degeneracy, the persistence occurs also in the framework of a perturbative calculation.

e) The only degenerate unperturbed level is that with energy $E_0 + E_1$. To the first order, one must diagonalize the 2×2 matrix of the restriction of the perturbation to the subspace generated by the vectors $|E_0, E_1\rangle$ and $|E_1, E_0\rangle$. Since V' and Π commute, it is convenient to take the (normalized) vectors:

$$|E_\pm\rangle = \frac{1}{\sqrt{2}}\left(|E_0, E_1\rangle \pm |E_1, E_0\rangle\right) \xrightarrow{\text{SR}}$$

$$\frac{1}{\sqrt{2}}\left(\psi_0(x_1)\psi_1(x_2) \pm \psi_1(x_1)\psi_0(x_2)\right)$$

as basis vectors, for they are eigenvectors of Π belonging to different eigenvalues ($+1$ and -1), so that the matrix of the perturbation is already diagonal and its eigenvalues are:

$$\Delta E_\pm = \langle E_\pm | V' | E_\pm \rangle$$

$$= \frac{\lambda}{2}\int \left|\psi_0(x_1)\psi_1(x_2) \pm \psi_1(x_1)\psi_0(x_2)\right|^2 \delta(x_1 - x_2)\,dx_1 dx_2$$

whence:

$$\Delta E_- = 0, \qquad \Delta E_+ = 2\lambda \int |\psi_0(x)|^2 |\psi_1(x)|^2\,dx > 0.$$

The state $|E_-\rangle$ is an exact eigenstate of H $(V'|E_-\rangle = 0)$.
In general, if $V'(|x_1 - x_2|)$ is the interaction potential, the difference in energy between the states $|E_+\rangle$ and $|E_-\rangle$ is given, to the first order in V', by the *exchange integral*:

$$E_+ - E_- = 2\int \left[\psi_0^*(x_1)\psi_1(x_1)\right] V'(|x_1 - x_2|)\left[\psi_0(x_2)\psi_1^*(x_2)\right] dx_1 dx_2.$$

If V' is a short range potential, the integrand has, in the region where it is appreciably different from zero, the same sign as V'. Therefore, to the first order, $E_+ \gtrless E_-$ according to the sign of V'.

12.4

a) One has $\Delta E_n^{(1)} = \frac{1}{2}k'\langle n | q^2 | n\rangle$. The term $\langle n | q^2 | n\rangle$ can be calculated in several ways, for instance recalling that (see Problems 4.15 and 5.7):

$$\langle n | \frac{1}{2}k\,q^2 | n\rangle = \frac{1}{2}E_n^{(0)} \quad \Rightarrow \quad \Delta E_n^{(1)} = \frac{1}{2}\left(\frac{k'}{k}\right)E_n^{(0)}.$$

b) The expression for the second order correction is $\left(w \equiv \sqrt{k/m}\right)$

$$\Delta E_0^{(2)} = -\sum_{s>0} \frac{\left|\langle 0 \mid \frac{1}{2} k' q^2 \mid s \rangle\right|^2}{E_s^{(0)} - E_0^{(0)}} = -\frac{k'^2}{4} \frac{\left|\langle 0 \mid q^2 \mid 2 \rangle\right|^2}{2\hbar\omega}$$

$$= -\frac{k'^2}{8\hbar\omega} \left(\frac{\hbar}{2m\omega}\right)^2 \left|\langle 0 \mid (\eta - \eta^\dagger)^2 \mid 2 \rangle\right|^2$$

$$= -\frac{1}{32}\left(\frac{k'}{k}\right)^2 \hbar\omega \left|\langle 0 \mid \eta^2 \mid 2 \rangle\right|^2 = -\frac{1}{8}\left(\frac{k'}{k}\right)^2 E_0^{(0)}.$$

c) The exact energy levels are:

$$E_n = \left(n + \tfrac{1}{2}\right)\hbar\sqrt{\frac{k + k'}{m}} = \left(n + \tfrac{1}{2}\right)\hbar\omega\sqrt{1 + \left(\frac{k'}{k}\right)}$$

$$= E_n^{(0)}\left(1 + \tfrac{1}{2}(k'/k) - \tfrac{1}{8}(k'/k)^2 + O\big((k'/k)^3\big)\right)$$

in agreement with the previous results.

d) In the present case, due to the knowledge of the second order corrections, we can require that $|\Delta E_n^{(2)}| \ll |\Delta E_n^{(1)}|$, namely:

$$\frac{1}{8}\left(\frac{k'}{k}\right)^2 \ll \frac{1}{2}\frac{|k'|}{k} \quad \Rightarrow \quad |k'| \ll k .$$

Usually, when the second order correction is not available, one limits oneself by requiring that $\Delta E_n^{(1)}$ be much smaller than the distance of the unperturbed level $E_n^{(0)}$ from the nearest one which, in this case, equals $\hbar\omega$ and leads to a too restrictive condition: $\left(n + \tfrac{1}{2}\right)|k'/k| \ll 1$.

e) As $E_n \propto \sqrt{1 + z}$, the radius of convergence of the perturbative series is 1; the singularity at $z = -1$ is due to the fact that for $k' \leq -k$ the system has no more bound states: H, to which no physical meaning can be any longer attached, should have a continuous spectrum from $-\infty$ to $+\infty$.

12.5

a) The eigenvalues of H, calculated in Problem 5.11, are:

$$E_n = \left(n + \tfrac{1}{2}\right)\hbar\omega - \frac{1}{2}\frac{F^2}{m\omega^2}.$$

The eigenfunctions of H are obtained by translating those of H_0 by $F/m\omega^2$: $x \to x - F/m\omega^2$.

b) Since the correction to the unperturbed eigenvalues is proportional to F^2, the only contribution must come from the second order: indeed, one has:

$$\Delta E_n^{(2)} = -F^2 \sum_{s \neq n} \frac{\left|\langle n \mid q \mid s \rangle\right|^2}{E_s - E_n}$$

$$= -\frac{F^2}{\hbar\omega}\left(\left|\langle n \mid q \mid n+1 \rangle\right|^2 - \left|\langle n \mid q \mid n-1 \rangle\right|^2\right)$$

and, as $q_{n,n+1} = i\sqrt{(n+1)\,\hbar/2m\omega}$, $q_{n,n-1} = -i\sqrt{n\,\hbar/2m\omega}$, one finds again the exact result $\Delta E_n^{(2)} = -F^2/2m\omega^2$.

c) Since the exact eigenfunctions of H are obtained by translating those of H_0 by $F/m\omega^2$ (for example $\psi_0(x) \propto e^{-(m\omega/2\hbar)(x-F/m\omega^2)^2}$), then their power series expansion contains terms of any order.

d) $\Delta E_0^{(1)} = 0$. As the only nonvanishing matrix elements of q are $q_{n,n\pm1}$, $\Delta E_0^{(3)}$ is vanishing (to go from 0 to 0 by steps of ±1, an even number of steps is necessary).

$$\Delta E_0^{(4)} = -F^4\left(\frac{|q_{01}\,q_{12}|^2}{2(\hbar\omega)^3} - \frac{|q_{01}|^2}{\hbar\omega} \times \frac{|q_{01}|^2}{(\hbar\omega)^2}\right) = 0 .$$

12.6

a) The dimensions of the coupling constant g are energy/(length)^4, whereas γ is dimensionless and is, therefore, the parameter of the perturbative expansion.

b) Recalling that the eigenvectors $|n\rangle$ of H_0 have parity $(-1)^n$ under space inversion and that q^4 is even, the first selection rule is that n' and n'' must have the same parity (parity selection rule: see Problem 6.1). The second rule, that applies only to the harmonic oscillator, comes from the observation that, being $q^4 \propto (\eta^\dagger - \eta)^4$, n' and n'' must differ by 0, ±2, ±4:

$$\Delta n \equiv n'' - n' = 0,\ \pm2,\ \pm4 .$$

c) To the first order: $\Delta E_0^{(1)} = g\langle 0\,|\,q^4\,|\,0\rangle$. It is convenient to calculate $\langle 0\,|\,q^4\,|\,0\rangle$ as the squared norm of the vector $q^2\,|\,0\rangle$:

$$q = -i\sqrt{\frac{\hbar}{2m\omega}}\,(\eta^\dagger - \eta); \quad (\eta^\dagger - \eta)^2\,|\,0\rangle = \left((\eta^\dagger)^2 - \eta\,\eta^\dagger\right)|\,0\rangle = \sqrt{2!}\,|\,2\rangle - |\,0\rangle$$

$$(\eta^\dagger\eta\,|\,0\rangle = 0 = \eta^2\,|\,0\rangle), \text{ so:}$$

$$\Delta E_0^{(1)} = g\langle 0\,|\,q^4\,|\,0\rangle = 3g\left(\frac{\hbar}{2m\omega}\right)^2 = \frac{3}{4}\gamma\hbar\omega .$$

To the second order:

$$\Delta E_0^{(2)} = -g^2\left(\frac{|\langle 0\,|\,q^4\,|\,2\rangle|^2}{2\hbar\omega} + \frac{|\langle 0\,|\,q^4\,|\,4\rangle|^2}{4\hbar\omega}\right)$$

$$\langle 0\,|\,q^4\,|\,4\rangle = \left(\frac{\hbar}{2m\omega}\right)^2\langle 0\,|\,\eta^4\,|\,4\rangle = \sqrt{4!}\left(\frac{\hbar}{2m\omega}\right)^2$$

$$\langle 0\,|\,q^4\,|\,2\rangle = \left(\frac{\hbar}{2m\omega}\right)^2\langle 0\,|\,(\eta^\dagger - \eta)^4\,|\,2\rangle .$$

In expanding $(\eta^\dagger - \eta)^4$ we only keep the terms that give rise to a nonvanishing contribution:

$$\langle 0 \mid (\eta^\dagger - \eta)^4 \mid 2 \rangle = -\langle 0 \mid \eta\,\eta^\dagger\eta^2 + \eta^2\eta^\dagger\eta + \eta^3\eta^\dagger \mid 2 \rangle$$
$$= -\langle 0 \mid [\eta, \eta^\dagger]\eta^2 + [\eta^2, \eta^\dagger]\eta + [\eta^3, \eta^\dagger] \mid 2 \rangle$$
$$= -\langle 0 \mid \eta^2 + 2\eta^2 + 3\eta^2 \mid 2 \rangle = -6\sqrt{2}$$

whence $\langle 0 \mid q^4 \mid 2 \rangle = -6\sqrt{2}\,(\hbar/2m\omega)^2$, and:

$$\Delta E_0^{(2)} = -\frac{21}{8}\gamma^2\,\hbar\omega\;.$$

If $g < 0$ the potential tends to $-\infty$ for $x \to \pm\infty$, therefore the system has no longer bound states: actually, if $g < 0$ the system would have a continuous spectrum from $-\infty$ to $+\infty$, so the perturbative calculation is meaningless.

d) $\dfrac{1}{2}\hbar\omega + \left(\dfrac{3}{4}\gamma - \dfrac{21}{8}\gamma^2\right)\hbar\omega = 0.55\,\hbar\omega\;.$

12.7

a) Let us use the Schrödinger representation:

$$\Delta E_0^{(1)} = \frac{g}{\sqrt{\pi}\,a}\int_{-\infty}^{+\infty} x^4\,e^{-x^2(1/a^2 + 1/b^2)}\,\mathrm{d}x = \frac{3g}{4a}\left(\frac{1}{a^2} + \frac{1}{b^2}\right)^{-5/2}$$
$$= \frac{3}{4}g\,\frac{a^4 b^5}{(a^2 + b^2)^{5/2}}$$

(for the calculation of the integral, see Problem 5.17). In this case, for $|g| \ll m^2\omega^3/\hbar$, the result is meaningful: indeed, independently of the sign of g, the Hamiltonian – even in presence of the perturbation – only has bound states, so it is presumable (and it is true) that the perturbative series has a nonvanishing radius of convergence.

b) $\displaystyle\lim_{b\to\infty}\frac{b^5}{(a^2 + b^2)^{5/2}} = 1 \quad\Rightarrow\quad \Delta E_0^{(1)} \to \frac{3}{4}g\,a^4.$

$$\frac{b^5}{(a^2 + b^2)^{5/2}} = \left(1 + (a/b)^2\right)^{-5/2} \simeq 1 - \frac{5}{2}\frac{a^2}{b^2}$$

so for $b/a = 10$ the relative correction is as large as $2.5\,\%$.

c) At large distances the overall potential is – up to an additive constant – that of a harmonic oscillator, so the system only has bound states, also in the present case independently of the sign of g.

d) One has:

$$\Delta E_0^{(1)} = 2\,\frac{g}{\sqrt{\pi}\,a}\int_0^\infty x^4\,e^{-x^2/a^2}\,f(x/b)\,\mathrm{d}x$$

whence, since $f(x/b) \le 1$ entails that the limit $b \to \infty$ can be taken before the integration, the thesis follows. It is also possible to give an estimate of the rate of convergence:

$$\left| \frac{2g}{\sqrt{\pi}\,a} \int_b^\infty e^{-x^2/a^2} x^4 \left(1 - (b/x)^4 \right) dx \right| < \frac{2\,|g|}{\sqrt{\pi}\,a} \int_b^\infty e^{-x^2/a^2}\, x^4 \left(1 - (b/x)^4 \right) \frac{2x}{2b}\, dx$$

$$= \frac{|g|}{\sqrt{\pi}\,a\,b} \int_{b^2}^\infty (y^2 - b^4)\, e^{-y/a^2}\, dy = \frac{|g|\,a^4}{\sqrt{\pi}} \frac{a^2 + b^2}{a\,b}\, e^{-b^2/a^2}$$

that for $b = 10\,a$ approximatively equals $(|g|\,a^4/\sqrt{\pi}) \times 3.8 \times 10^{-43}$.

12.8

a) The dimensions of the coupling constant g are energy/(length)^3. As $\sqrt{\hbar/m\omega}$ is the characteristic length of the harmonic oscillator, putting $g = \gamma\,\hbar\omega/(\hbar/m\omega)^{3/2}$, γ is dimensionless and:

$$H' = \gamma\,m\omega^2 \sqrt{\frac{m\omega}{\hbar}}\, q^3 \,.$$

b) To the first order $\Delta E_n^{(1)} \propto \langle n\,|\,q^3\,|\,n\rangle = 0$, owing to the selection rule on parity (q^3 is an odd operator).

c) To the second order:

$$\Delta E_0^{(2)} = -g^2 \sum_{s>0} \frac{|\langle 0\,|\,q^3\,|\,s\rangle|^2}{E_s^{(0)} - E_0^{(0)}} = -\frac{g^2}{\hbar\omega} \left(|\langle 0\,|\,q^3\,|\,1\rangle|^2 + \frac{|\langle 0\,|\,q^3\,|\,3\rangle|^2}{3} \right).$$

The matrix elements $\langle 0\,|\,q^3\,|\,1\rangle$ and $\langle 0\,|\,q^3\,|\,3\rangle$ can be calculated in several ways: for example, from Problem 5.12 one has:

$$q^3\,|\,0\rangle = (-i)^3 \left(\frac{\hbar}{2m\omega} \right)^{3/2} \left(\sqrt{3!}\,|\,3\rangle - 3\,|\,1\rangle \right)$$

whence:

$$\Delta E_0^{(2)} = -\frac{11}{8} \frac{g^2}{\hbar\omega} \left(\frac{\hbar}{2m\omega} \right)^3 = -\frac{11}{8}\,\gamma^2\hbar\omega \,.$$

d) The potential is of the type $\alpha\,x^2 + \beta\,x^3$, therefore H only has the continuous spectrum, consisting of nondegenerate eigenvalues, from $-\infty$ to $+\infty$: it follows that the use of perturbation theory is not legitimate. However, if (as in Problem 12.7) the perturbation is $g\,x^3$ only in a region of space bounded, but large with respect to the characteristic length of the oscillator, and outside this region it stays limited, the result of the perturbative calculation with $g\,x^3$ for the lowest energy levels provides in a simpler way and to a good approximation the correction to the energy levels of the system subject to the 'physical' potential.

12.9

a) If $h(\xi;\gamma)$ were a polynomial, no term in the equation for $h(\xi;\gamma)$ could cancel $\xi^4 h(\xi;\gamma)$. Obviously this argument does not apply if $\gamma = 0$.

b) The degree of $\xi\,B_n'$, namely of B_n, must equal that of $\xi^4 B_{n-1}$, i.e. it must be equal to $4(n-1) + 4 = 4n$.

c) Let us put $B_1(\xi) = a\,\xi^2 + b\,\xi^4$ $(B_n(0) = 0)$. The equation $C_1(\xi) = 0$ writes:

$$B_1'' - 2\xi\,B_1' - 2\xi^4 + 2a_1 = 0 \;\Rightarrow\; 4a\,\xi^2 + 8b\,\xi^4 - 2a - 12b\,\xi^2 + 2\xi^4 - 2a_1 = 0$$

then:

$$a = -\frac{3}{4}, \quad b = -\frac{1}{4}; \quad a_1 = \frac{3}{4}; \qquad f_0(\xi;\gamma) \approx \frac{1}{4}\,e^{-\xi^2/2}\left(4 - 3\gamma\,\xi^2 - \gamma\,\xi^4\right).$$

The polynomial in ξ^2 in parentheses has, for $\gamma > 0$, a positive root; therefore, to the first order, $f_0(\xi;\gamma)$ has two (symmetrical) zeroes, whence (see Problem 6.3) $\frac{1}{2} + a_1\gamma = \frac{1}{2} + \frac{3}{4}\gamma$ is greater that the exact eigenvalue.

d) Let us write the equation $C_n = 0$ in the form:

$$B_n'' - 2\xi\,B_n' + 2a_n = 2\xi^4\,B_{n-1} - 2\left(a_{n-1}B_1 + \cdots + a_1B_{n-1}\right).$$

As B_n is an even polynomial of degree $2n$ in ξ^2 and $B_n(0) = 0$, the unknowns are the $2n$ coefficients of the polynomial and a_n; in the right hand side we have a known polynomial of degree $2n$: one obtains a system of $2n+1$ linear equations (the right hand side also has the term ξ^0) that determine $B_n(\xi)$ and a_n.

e) $\epsilon^{(0)} = 0.5$
    $\epsilon^{(1)} = 0.575$
    $\epsilon^{(2)} = 0.549$
    $\epsilon^{(3)} = 0.570$
    $\epsilon^{(4)} = 0.545$
    $\epsilon^{(5)} = 0.581$
    $\epsilon^{(6)} = 0.517$
    $\epsilon^{(7)} = 0.650$
    $\epsilon^{(8)} = 0.336$
    $\epsilon^{(9)} = 1.169$

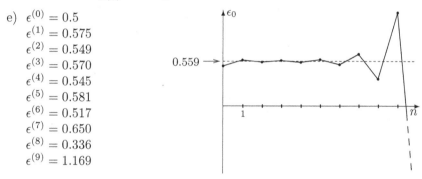

(the next terms are $\epsilon^{(10)} = -1.279$, $\epsilon^{(11)} = 6.615$). It is evident that the perturbative series is a series with alternating signs and rapidly divergent coefficients (the asymptotic estimate given by the Authors is $a_n \approx (-1)^{n+1}n!\,3^n$), so the partial sums, no matter how small the value of γ, give rise to large oscillations: it is therefore meaningless to push the calculation to high orders.

There exists, in the present case, a way to find in a unique way the function $\epsilon_0(\gamma)$ starting from the divergent perturbative series: this way is known with the name of Borel resummation of the series, but it goes beyond the scope of our treatment.

12.10

a) Put $\left|\,\mathrm{d}_\lambda E(\lambda_0)\right\rangle \equiv \dfrac{\mathrm{d}}{\mathrm{d}\lambda}\left|\,E(\lambda)\right\rangle\Big|_{\lambda_0}$. One has:

$$\frac{dE(\lambda)}{d\lambda}\Big|_{\lambda_0} = \frac{d}{d\lambda}\langle E(\lambda) \mid H(\lambda) \mid E(\lambda)\rangle\Big|_{\lambda_0} = \langle E(\lambda_0) \mid \frac{dH(\lambda)}{d\lambda}\Big|_{\lambda_0} \mid E(\lambda_0)\rangle$$

$$+ \langle d_\lambda E(\lambda_0) \mid H(\lambda_0) \mid E(\lambda_0)\rangle + \langle E(\lambda_0) \mid H(\lambda_0) \mid d_\lambda E(\lambda_0)\rangle .$$

The last two terms give vanishing contribution thanks to:

$$\langle d_\lambda E(\lambda_0) \mid H(\lambda_0) \mid E(\lambda_0)\rangle = E_0 \langle d_\lambda E(\lambda_0) \mid E(\lambda_0)\rangle$$

and:

$$\frac{d}{d\lambda}\langle E(\lambda) \mid E(\lambda)\rangle = 0 .$$

If $H(\lambda) = H_0 + \lambda V$, one has $(dH(\lambda)/d\lambda)_{\lambda=0} = V$, therefore:

$$\Delta E^{(1)} = \langle E^0 \mid \lambda V \mid E^0\rangle = \lambda \langle E^0 \mid \frac{dH(\lambda)}{d\lambda}\Big|_{\lambda=0} \mid E^0\rangle = \lambda \frac{dE(\lambda)}{d\lambda}\Big|_{\lambda=0} .$$

b) If $E(\lambda_0)$ is a degenerate eigenvalue and $\mid E(\lambda_0)\rangle$ is a generic eigenvector of $H(\lambda_0)$, in general $\lim_{\lambda\to\lambda_0} \mid E(\lambda)\rangle \neq \mid E(\lambda_0)\rangle$ and, in the latter case, $\mid d_\lambda E(\lambda_0)\rangle$ does not exist.

The theorem is still valid for the eigenvectors $\mid E_\nu(\lambda_0)\rangle$ of $H(\lambda_0)$, corresponding to the degenerate eigenvalue $E(\lambda_0)$, that are the vectors to which $\mid E_\nu(\lambda)\rangle$ tend for $\lambda \to \lambda_0$: $\mid E_\nu(\lambda_0)\rangle = \lim_{\lambda\to\lambda_0} \mid E_\nu(\lambda)\rangle$ (the approximate eigenvectors of $H(\lambda)$ in the framework of the perturbative theory of degenerate energy levels).

c) One has $H(Z) = p^2/2m_e - Z e^2/r$, $E_n(Z) = -Z^2 e^2/(2n^2 a_B)$ and, considering Z as a continuous parameter,

$$\langle n,l,m \mid \frac{-e^2}{r} \mid n,l,m\rangle = -\frac{2Z e^2}{2n^2 a_B} \;\Rightarrow\; \langle n,l,m \mid \frac{-Z e^2}{r} \mid n,l,m\rangle = 2E_n(Z) .$$

12.11

a) Since the transformation is a canonical one, H and $H(\lambda) \equiv H(\lambda q_i, \lambda^{-1} p_i)$ have the same spectrum:

$$H \mid E\rangle = E \mid E\rangle \;\Rightarrow\; U(\lambda) H U^{-1}(\lambda) \left(U(\lambda) \mid E\rangle\right) = E \left(U(\lambda) \mid E\rangle\right)$$

and, in particular, have the same discrete eigenvalues, that – as a consequence – are independent of λ. So:

$$0 = \langle E \mid \frac{dH(\lambda)}{d\lambda}\Big|_{\lambda=1} \mid E\rangle = \langle E \mid \left(-\sum_i \frac{p_i^2}{m_i} + \sum_i q_i \frac{\partial V}{\partial q_i}\right) \mid E\rangle$$

from which the thesis follows.

If V is a homogeneous function of degree k, one has $\sum_i q_i (\partial V/\partial q_i) = k V$ whence $2\overline{T} = k\overline{V}$. Since $\overline{T} + \overline{V} = E$, one has:

$$\overline{T} = \frac{k}{2+k} E, \qquad \overline{V} = \frac{2}{2+k} E; \qquad k \neq -2 .$$

b) For a hydrogen-like ion $k = -1$, so:

$$\overline{T} \equiv \langle\, n,l,m \mid \frac{\vec{p}^{\,2}}{2m_e} \mid n,l,m \,\rangle = -E_n = \frac{Z^2\,e^2}{2a_B\,n^2}$$

$$\overline{V} \equiv \langle\, n,l,m \mid \frac{-Z\,e^2}{r} \mid n,l,m \,\rangle = 2E_n = -\frac{Z^2\,e^2}{a_B\,n^2}\;.$$

c) If a normalizable eigenvector of the Hamiltonian existed, from the virial theorem $2\overline{T} = k\,\overline{V}$ with $k = -2$ one would have:

$$2\overline{T} = -2\overline{V} \quad \Rightarrow \quad E = \overline{T} + \overline{V} = 0\;.$$

Alternatively:

$$H(\lambda) \equiv H(\lambda\,q_i,\ \lambda^{-1}p_i) = \lambda^{-2}H(q_i,\ p_i)$$

so, as H and $H(\lambda)$ have the same spectrum, if E is an (either proper or improper) eigenvalue, also $\lambda^2 E$ is an eigenvalue for any real λ. As a consequence H has only the continuous spectrum, but – possibly – for the proper eigenvalue $E = 0$. The result holds for any homogeneous potential of degree -2, provided H be well defined as a self-adjoint operator – this does not happen, for example, for $V(r) = -g/r^2$ with $g \geq \hbar^2/8m$.

12.12

a) Perturbation theory for quasi-degenerate levels consists in the neglect of all the matrix elements of $H(\lambda)$ among the states belonging to the levels of interest and the states relative to other energy levels; so, in conclusion, the Hamiltonian is restricted to the space generated by the quasi-degenerate states. In the present case it is necessary to diagonalize the 2×2 matrix:

$$H \rightarrow \begin{pmatrix} E_1 + \lambda\,V_{11} & \lambda\,V_{12} \\ \lambda\,V_{21} & E_2 + \lambda\,V_{22} \end{pmatrix}$$

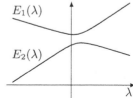

that for no value of λ is a multiple of the identity if $V_{12} \neq 0$. So $E_1(\lambda) \neq E_2(\lambda)$.

b) As ξ commutes with $H(\lambda)$ for any value of λ, it commutes also with V, whence $V_{12} = 0$ and, provided $V_{11} - V_{22} \neq 0$, the two levels intersect each other for $\lambda = (E_1 - E_2)/(V_{22} - V_{11})$.

12.13

a) Let us take the z axis parallel to $\vec{\mathcal{E}}$. The Hamiltonian of the system is

$$H = H_0 + e\mathcal{E}\,z, \qquad (x,\,y,\,z) \equiv \vec{r} = \vec{r}_e - \vec{r}_p$$

where H_0 is the Hamiltonian of the unperturbed hydrogen atom. Neither $\vec{L}^2$ nor I commute with z, whereas both L_z and all the reflections with respect to the planes containing the z axis do commute with H.

b) To the first order, the two states with $n = 2$, $l = 1$, $m = \pm 1$ are degenerate. This degeneration is due to the fact that L_z and the reflection with respect to any plane containing the z axis, e.g. the inversion I_x with respect to the plane $x = 0$, commute with H, but $I_x L_z = -L_z I_x$: as a consequence $I_x \, | \, E, m \, \rangle \propto \, | \, E, -m \, \rangle$ and all the states with opposite $m \neq 0$ remain degenerate, giving rise to energy levels at least twice degenerate in any order of perturbation theory (see Problems 10.5 and 11.11).

c) The order of magnitude of the effect of the perturbation is $n^2 a_{\rm B} \times e\mathcal{E}$, that for $n = 3$ equals $4.5 \times 10^{-4}\,{\rm eV}$. This must be compared with the distance between the unperturbed levels $n = 3$ and $n = 4$: $E_4 - E_3 \simeq 13.6 \times (1/9 - 1/16) \simeq 0.66\,{\rm eV}$, so for $n = 3$ the perturbative calculation may be considered acceptable.

d) The level $n = 3$ consists of three states with $m = 0$, two states with $m = 1$, two states with $m = -1$ and, finally, one state with $m = 2$ and one state with $m = -2$. Having in mind that any state with $m \neq 0$ is degenerate with a state with $-m$, the possible maximum number of energy levels is $3 + 2 + 1 = 6$. However we shall see that, to the first order, there are only five energy levels.

e) As the perturbation commutes with L_z (and therefore $\langle m' \, | \, z \, | \, m'' \rangle = 0$ if $m' \neq m''$), we may order the basis vectors grouping those with the same m:

$$m = 0: \; l = 0, 1, 2 \quad \Rightarrow \quad e\mathcal{E}\, a_{\rm B} \begin{pmatrix} 0 & A & 0 \\ A^* & 0 & B \\ 0 & B^* & 0 \end{pmatrix}$$

where the selection rule on the space inversion has been used:

$$\langle l \, | \, z \, | \, l \rangle = 0, \qquad \langle l = 0 \, | \, z \, | \, l = 2 \rangle = 0$$

$$m = \pm 1: \; l = 1, 2 \quad \Rightarrow \quad e\mathcal{E}\, a_{\rm B} \begin{pmatrix} 0 & C \\ C^* & 0 \end{pmatrix};$$

$$m = \pm 2: \; \langle m = \pm 2 \, | \, e\mathcal{E}\, z \, | \, m = \pm 2 \rangle = 0\,.$$

The 3×3 block has the first and the last row proportional to each other, so it has one vanishing eigenvalue (the other two are $\pm\sqrt{|A|^2 + |B|^2}$): therefore the energy levels are five.

The (nonnormalized) state corresponding to the vanishing eigenvalue of the 3×3 block is $B \, | \, 3, 0, 0 \rangle - A^* \, | \, 3, 2, 0 \rangle$, so it is an eigenvector of the space inversion corresponding to the eigenvalue $+1$.

f) One has:

$$A = \int_0^\infty R_{3,0}(r)\frac{r}{a_{\rm B}} R_{3,1}(r)\, r^2 {\rm d}r \int Y_{0,0}^*(\theta, \phi) \cos\theta\, Y_{1,0}(\theta, \phi)\, {\rm d}\Omega$$

$$= -9\sqrt{2}\,\frac{1}{\sqrt{3}} = -\frac{3}{2}\sqrt{6}$$

$$B = \int_0^\infty R_{3,1}(r) \frac{r}{a_B} R_{3,2}(r)\, r^2 dr \int Y_{1,0}^*(\theta,\phi) \cos\theta\, Y_{2,0}(\theta,\phi)\, d\Omega$$

$$= -\frac{9\sqrt{5}}{2} \frac{2}{\sqrt{15}} = -3\sqrt{3}$$

$$C = \int_0^\infty R_{3,1}(r) \frac{r}{a_B} R_{3,2}(r)\, r^2 dr \int Y_{1,1}^*(\theta,\phi) \cos\theta\, Y_{2,1}(\theta,\phi)\, d\Omega$$

$$= -\frac{9\sqrt{5}}{2} \frac{1}{\sqrt{5}} = -\frac{9}{2}.$$

Therefore the energy levels up to the first order are:

$m = 0:$ $E_3,$ $E_3 \pm \dfrac{9\sqrt{2}}{2} e\mathcal{E}\, a_B;$

$m = \pm 1:$ $E_3 \pm \dfrac{9}{2} e\mathcal{E}\, a_B$ (twice degenerate)

$m = \pm 2:$ $E_3.$

————— $m=0$
————— $m=\pm1$

————— $m=0,\ m=\pm2$

————— $m=\pm1$
————— $m=0$

12.14

a) $x\, y = \dfrac{1}{2} r^2 \sin^2\theta \sin 2\phi = -\dfrac{i}{4} r^2 \sin^2\theta \left(e^{2i\phi} - e^{-2i\phi}\right)$ $\Rightarrow$ $\Delta m = \pm 2.$

b) One has $\Delta E_1^{(1)} = \langle 1,0,0 \mid V_1 \mid 1,0,0 \rangle = 0$ ($\Delta m = 0$). The only matrix elements with $\Delta m = \pm 2$ from among the states of the level $n = 2$ are $\langle 2,1,1 \mid V_1 \mid 2,1,-1 \rangle$ and its complex conjugate:

$$\langle 2,1,1 \mid V_1 \mid 2,1,-1 \rangle = -i\frac{V_0}{4} \int_0^\infty R_{21}^2(r)\left(\frac{r}{a_B}\right)^2 r^2 dr$$

$$\times \int Y_{11}^*(\theta,\phi) \sin^2\theta\, e^{2i\phi}\, Y_{1\,-1}(\theta,\phi)\, d\Omega$$

that is indeed different from zero as both the radial and the angular integrals are positive $\left(Y_{1,\pm1} = \sqrt{3/8\pi} \sin\theta\, e^{\pm i\phi}\right)$. The expression given in the text coincides with the radial integral and equals $30\, a_B^2$; the angular integration gives $4/5$, so $\langle 2,1,1 \mid V_1 \mid 2,1,-1 \rangle = -6\,i\,V_0$ and in conclusion, for the level $n = 2$ the eigenvalues corrected up to the first order and the corresponding approximate eigenvectors are:

$E_1:$ $\mid 2,0,0 \rangle,\ \mid 2,1,0 \rangle;$ $E_1 \pm 6\,V_0:$ $\dfrac{1}{\sqrt{2}}\left(\mid 2,1,1 \rangle \pm i \mid 2,1,-1 \rangle\right).$

12.15

a) For the potential V the selection rule $\Delta m = \pm 2$ applies and, among the states from the levels $n = 1$ and $n = 2$, the only matrix elements with $\Delta m = \pm 2$ are $\langle 2,1,1 \mid V \mid 2,1,-1 \rangle$ and its complex conjugate; but V is an odd operator, therefore also $\langle 2,1,1 \mid V \mid 2,1,-1 \rangle = 0.$

b) The trace of a matrix equals the sum of its eigenvalues; in the basis $|n, l, m\rangle$ all the diagonal matrix elements of the perturbation are vanishing, due to space inversion.

c) The first three vectors given in the text correspond to $l = 1$, the other three to $l = 2$, so two vectors with $l = 2$ and that with $l = 0$ are missing: the latter is, in the basis $|n, l, m\rangle$,

$$|3, 0, 0\rangle \xrightarrow{\text{SR}} \frac{1}{\sqrt{4\pi}} R_{3,0}(r) .$$

The two missing vectors can be found by comparing with the spherical harmonics expressed in terms of Cartesian coordinates (see Problem 8.6) and are, for example, that with $m = 0$ and the one proportional to $x^2 - y^2$ (as well as their linear combinations):

$$|3, 2, 0\rangle \xrightarrow{\text{SR}} \sqrt{\frac{5}{16\pi}} \frac{r^2 - 3z^2}{r^2} R_{3,2}(r) ,$$

$$\frac{1}{\sqrt{2}} \left(|3, 2, 2\rangle + |3, 2, -2\rangle \right) \xrightarrow{\text{SR}} \sqrt{\frac{15}{16\pi}} \frac{x^2 - y^2}{r^2} R_{3,2}(r) .$$

These three vectors generate a space that we shall denote by $\mathcal{H}^{(3)}$.

d) Since all the integrals, whose integrands are odd in at least one of the coordinates, do vanish, the only nonvanishing matrix elements are:

$$\langle \mathcal{P}_1 | V | \mathcal{D}_{2,3} \rangle, \quad \langle \mathcal{P}_2 | V | \mathcal{D}_{3,1} \rangle, \quad \langle \mathcal{P}_3 | V | \mathcal{D}_{1,2} \rangle .$$

Therefore, after ordering the vectors in the following way:

$$|\mathcal{P}_1\rangle, \ |\mathcal{D}_{2,3}\rangle; \quad |\mathcal{P}_2\rangle, \ |\mathcal{D}_{3,1}\rangle; \quad |\mathcal{P}_3\rangle, \ |\mathcal{D}_{1,2}\rangle; \qquad \mathcal{H}^{(3)}$$

the matrix relative to the perturbation consists of three 2×2 blocks and of one 3×3 identically vanishing block. As:

$$A \equiv \langle \mathcal{P}_1 | V | \mathcal{D}_{2,3} \rangle = \langle \mathcal{P}_2 | V | \mathcal{D}_{3,1} \rangle = \langle \mathcal{P}_3 | V | \mathcal{D}_{1,2} \rangle$$

$$= \frac{3V_0\sqrt{5}}{4\pi\, a_{\text{B}}^3} \int \frac{x^2 y^2 z^2}{r^3} R_{3,1}(r)\, R_{3,2}(r)\, dV$$

the 2×2 blocks, that have the form $\begin{pmatrix} 0 & A \\ A & 0 \end{pmatrix}$, $A \in \mathbb{R}$, are identical.

So the perturbation has the 0 eigenvalue three times degenerate (whose eigenspace is $\mathcal{H}^{(3)}$) and the eigenvalues $\pm A$, they too three times degenerate.

12.16

a) The lattice is invariant under the inversion of the single axes, so:

$$V^{(2)}(x, y, z) = \alpha\, x^2 + \beta\, y^2 + \gamma\, z^2 .$$

If $a = b$, the lattice is invariant under rotations by $90°$ around the z axis, therefore $\alpha = \beta$; if $a = b = c$, then $\alpha = \beta = \gamma$.

b) If $\alpha = \beta = \gamma$ all the observables given in the text are constants of motion: in the latter case $V^{(2)}$ is a central potential and the eigenvectors of the Hamiltonian can be classified as $|E, l, m\rangle$. The degeneracy on l of the hydrogen atom is removed and only the degeneracy $2l + 1$ on m is guaranteed.

If $\alpha = \beta \neq \gamma$, the observables L_z, I_x, I_y, I_z are constants of motion; the eigenstates of the Hamiltonian can be classified as $|E, m\rangle$, degenerate with $|E, -m\rangle$ ($I_x L_z = -L_z I_x$).

If $\alpha \neq \beta \neq \gamma$, only I_x, I_y, I_z are constants of motion commuting with one another, so the eigenstates of the Hamiltonian can be classified by means of the eigenvalues w_x, w_y, w_z of these operators ($w_i = \pm 1$) and one must expect that all the energy levels are nondegenerate.

c) In the s (i.e. $l = 0$) states: $\overline{x^2} = \overline{y^2} = \overline{z^2} = \frac{1}{3}\overline{r^2}$, so that:

$$\Delta E_{1s}^{(1)} = \langle 1,0,0 \, | \, V^{(2)} \, | \, 1,0,0 \rangle = \frac{1}{3}(\alpha + \beta + \gamma)\langle 1,0,0 \, | \, r^2 \, | \, 1,0,0 \rangle$$

$$= (\alpha + \beta + \gamma) \, a_{\mathrm{B}}^2 \, .$$

Level $n = 2$: $V^{(2)}$ has no nonvanishing matrix element between the state $2s$ ($|2,0,0\rangle$) and the states $2p$ ($|2,1,m\rangle$), that have opposite parity, so:

$$\Delta E_{2s}^{(1)} = \frac{1}{3}(\alpha + \beta + \gamma)\langle 2,0,0 \, | \, r^2 \, | \, 2,0,0 \rangle = 14 \, (\alpha + \beta + \gamma) \, a_{\mathrm{B}}^2 \, .$$

The states $2p$, classified by w_x, w_y, w_z, are:

$$|+,+,-\rangle = |n = 2, l = 1, m_z = 0\rangle,$$
$$|+,-,+\rangle = |n = 2, l = 1, m_y = 0\rangle,$$
$$|-,+,+\rangle = |n = 2, l = 1, m_x = 0\rangle$$

and in this basis the perturbation is diagonal (the terns w_x, w_y, w_z are all different). In the state with $m_z = 0$ one has $\overline{x^2} = \overline{y^2}$, so:

$$\Delta E_{2p, m_z=0}^{(1)} = \alpha \, \overline{x^2} + \beta \, \overline{y^2} + \gamma \, \overline{z^2} = \frac{1}{2}(\alpha + \beta) \left(\overline{x^2} + \overline{y^2} \right) + \gamma \, \overline{z^2}$$

$$= \frac{1}{2}(\alpha + \beta) \left(\overline{r^2} - \overline{z^2} \right) + \gamma \, \overline{z^2} = \left[6(\alpha + \beta) + 18\gamma \right] a_{\mathrm{B}}^2 \, ;$$

likewise:

$$\Delta E_{2p, m_y=0}^{(1)} = \left[6(\alpha + \gamma) + 18\beta \right] a_{\mathrm{B}}^2 \, ; \quad \Delta E_{2p, m_x=0}^{(1)} = \left[6(\beta + \gamma) + 18\alpha \right] a_{\mathrm{B}}^2 \, .$$

d) The Hamiltonian $H_0 + V$ commutes with all the transformations that leave the cube invariant (the cube group): rotations by $90°$ around each of the three axes orthogonal to the faces, rotations by $120°$ around the diagonals of the cube, inversions etc.: as the group of the cube is noncommutative (indeed it contains the group of the square: see Problem 10.2), due to the degeneracy theorem there must exist degenerate energy levels.

12.17

a) The potential $V(r)$ removes the degeneracy on l. One has:

$$\Delta E_{n,l}^{(1)} = \int_0^a R_{n,l}^2(r)\,V(r)\,r^2\,dr$$

and, since for $r \ll a_B$ one has $R_{n,l}^2(r) \propto r^{2l}$, the smaller l, the greater $|\Delta E_{n,l}^{(1)}|$. Therefore the order of the levels with the same n is ($E_n^{(0)}$ is the unperturbed level):

$$V(r) < 0: \quad E_{n,l=0} < E_{n,l=1} < \cdots < E_{n,l=n-1} < E_n^{(0)}$$
$$V(r) > 0: \quad E_n^{(0)} < E_{n,l=n-1} < E_{n,l=n-2} < \cdots < E_{n,l=0}\,.$$

b) The exact energy levels are those of the hydrogen-like ion with atomic number $Z = 1 + \epsilon$:

$$E_n(\epsilon) = -(1+\epsilon)^2\,\frac{e^2}{2\,n^2 a_B} \approx -\frac{e^2}{2\,n^2 a_B} - 2\epsilon\,\frac{e^2}{2\,n^2 a_B}\,.$$

The first order correction due to the potential $V(r) = -\epsilon\,\dfrac{e^2}{r}$ is (see Problem 12.11):

$$\Delta E_n^{(1)} = \langle\, n,l,m\mid -\epsilon\,\frac{e^2}{r}\mid n,l,m\,\rangle = -\epsilon\,\frac{e^2}{n^2 a_B}$$

in agreement with the preceding result. In this case the degeneracy on l is not removed.

c) Since $V(r)$ has the same form as the centrifugal potential $\hbar^2 l(l+1)/2m_e r^2$, in the Schrödinger equation we put:

$$\frac{\hbar^2\big(l(l+1)+\epsilon\big)}{2m_e r^2} = \frac{\hbar^2 l'(l'+1)}{2m_e r^2}\,, \qquad l' = \frac{1}{2}\Big(-1 + \sqrt{1 + 4l(l+1) + 4\epsilon}\,\Big)\,.$$

The well known calculation of the energy levels of the (unperturbed) hydrogen atom leads one to define the principal quantum number n as $n \equiv \bar{\imath} + l + 1$, where $\bar{\imath}$ is an integer number (the degree of the polynomial that in the radial function multiplies $r^l\,e^{-r/na_B}$); in the latter case it is sufficient to substitute l with $l' \equiv l + \Delta l$ and, as a consequence, $n \to n + \Delta l$; one obtains:

$$E_{n,l} = -\frac{e^2}{2a_B\,(n + \Delta l)^2}\,.$$

Since $\Delta l = l' - l \approx \epsilon/(2l+1)$, by expanding $E_{n,l}$ one has:

$$E_{n,l} \approx -\frac{e^2}{2a_B\,n^2} + \epsilon\,\frac{e^2}{2a_B\,n^3(l + \frac{1}{2})}$$

whence:

$$\epsilon\,\frac{e^2}{2a_B\,n^3(l + \frac{1}{2})} = \Delta E_{n,l}^{(1)} = \epsilon\,\frac{\hbar^2}{2m_e}\langle\, n,l,m\mid r^{-2}\mid n,l,m\,\rangle \quad\Rightarrow$$

$$\langle\, n,l,m\mid r^{-2}\mid n,l,m\,\rangle = \big(n^3(l + \tfrac{1}{2})\,a_B^2\big)^{-1}\,.$$

12.18

a) The potential generated by a charge uniformly distributed within a sphere, for $r \leq R$, is given in Problem 1.1; for $r \geq R$ the potential energy is $-Z\,e^2/r$. One has:

$$H = H_0 + V(r) ; \qquad H_0 = \frac{\vec{p}^2}{2m_e} - \frac{Z\,e^2}{r}$$

$$V(r) = \begin{cases} \dfrac{Z\,e^2}{r} + \dfrac{Z\,e^2}{R}\left[\dfrac{1}{2}\dfrac{r^2}{R^2} - \dfrac{3}{2}\right] & r \leq R \\ 0 & r \geq R . \end{cases}$$

b)
$$\Delta E_{1s}^{(1)} = \int_0^R R_{1,0}^2(r)\left[\frac{Z\,e^2}{r} + \frac{Z\,e^2}{R}\left(\frac{1}{2}\frac{r^2}{R^2} - \frac{3}{2}\right)\right] r^2 dr$$

$$\simeq \int_0^R 4\frac{Z^3}{a_B^3}\left[\frac{Z\,e^2}{r} + \frac{Z\,e^2}{R}\left(\frac{1}{2}\frac{r^2}{R^2} - \frac{3}{2}\right)\right] r^2 dr = \frac{2Z^4}{5}\frac{e^2}{a_B}\left(\frac{R}{a_B}\right)^2$$

$$= 3.1 \times 10^{-5}\,\text{eV} .$$

The ionization energy of C^{VI} is $Z^2 \cdot 13.6 \simeq 490\,\text{eV}$, so $\Delta E/E \simeq 6 \times 10^{-8}$.

$$\Delta E_{2s}^{(1)} \simeq \int_0^R \frac{1}{2}\frac{Z^3}{a_B^3}\left[\frac{Z\,e^2}{r} + \frac{Z\,e^2}{R}\left(\frac{1}{2}\frac{r^2}{R^2} - \frac{3}{2}\right)\right] r^2 dr = \frac{Z^4}{20}\frac{e^2}{a_B}\left(\frac{R}{a_B}\right)^2$$

$$= 3.9 \times 10^{-6}\,\text{eV} .$$

$$\Delta E_{2p}^{(1)} \simeq \int_0^R \frac{1}{24}\frac{Z^5}{a_B^5}r^2\left[\frac{Z\,e^2}{r} + \frac{Z\,e^2}{R}\left(\frac{1}{2}\frac{r^2}{R^2} - \frac{3}{2}\right)\right] r^2 dr = \frac{Z^6}{1120}\frac{e^2}{a_B}\left(\frac{R}{a_B}\right)^4$$

$$= 5.6 \times 10^{-15}\,\text{eV} .$$

c) The corrections $\Delta E^{(1)}$ we have found must be compared with the distance $E_2 - E_1 \simeq 367\,\text{eV}$ between the unperturbed levels: whereas for the levels $1s$ and $2s$ $\Delta E/E \simeq 10^{-7} \div 10^{-8} \gg 10^{-14}$, in the case of the level $2p$ $\Delta E/E \simeq 10^{-17}$ therefore (and the same conclusion applies a fortiori for the levels with $n \geq 3$) it is at most possible to observe the (positive) shift of the s energy levels with respect to the others that remain unresolved.

d) In the present case the potential energy $U(r)$ of the electron is $-Z\,e^2/R$ for $r \leq R$: the latter is greater than the energy of the electron in the field of the uniformly charged sphere. As a consequence, also the energy shift of the levels is greater. One has:

$$V(r) = \begin{cases} \dfrac{Z\,e^2}{r} - \dfrac{Z\,e^2}{R} & r \leq R \\ 0 & r \geq R \end{cases} \Rightarrow$$

$$\Delta E_{1s}^{(1)} \simeq \int_0^R 4\frac{Z^3}{a_B^3}\left(\frac{Z\,e^2}{r} - \frac{Z\,e^2}{R}\right) r^2 dr = \frac{2Z^4}{3}\frac{e^2}{a_B}\left(\frac{R}{a_B}\right)^2 = 5.2 \times 10^{-5}\,\text{eV} .$$

$$\Delta E_{2s}^{(1)} \simeq \int_0^R \frac{1}{2}\frac{Z^3}{a_B^3}\left(\frac{Z\,e^2}{r} - \frac{Z\,e^2}{R}\right)r^2 \mathrm{d}r = \frac{Z^4}{12}\frac{e^2}{a_B}\left(\frac{R}{a_B}\right)^2 = 6.5 \times 10^{-6}\,\mathrm{eV}\,.$$

$$\Delta E_{2p}^{(1)} \simeq \int_0^R \frac{1}{24}\frac{Z^5}{a_B^5}r^2\left(\frac{Z\,e^2}{r} - \frac{Z\,e^2}{R}\right)r^2 \mathrm{d}r = \frac{Z^6}{480}\frac{e^2}{a_B}\left(\frac{R}{a_B}\right)^4$$

$$= 1.3 \times 10^{-14}\,\mathrm{eV}\,.$$

Although increased, these values do not alter the conclusions of point c).

12.19

a) Let $a_\mu \equiv \hbar^2/\mu e^2$, where $\mu \simeq 200\,m_e$ is the reduced mass of the system $He^{++}-\mu^-$.

$$\psi_{1,0,0}(r) = R_{1,0}(r)\,Y_{0,0}(\theta,\phi) = \frac{1}{\sqrt{4\pi}}\left(\frac{Z}{a_\mu}\right)^{3/2}2\,\mathrm{e}^{-Z\,r/a_\mu}, \qquad Z = 2\,.$$

b) $P(r \le r_0) = \left(\dfrac{Z}{a_\mu}\right)^3 4\displaystyle\int_0^{r_0}\mathrm{e}^{-2Z\,r/a_\mu}\,r^2 \mathrm{d}r$

$$= 1 - \mathrm{e}^{-2Z\,r_0/a_\mu}\left(1 + 2Z\,r/a_\mu + 2(Z\,r/a_\mu)^2\right)$$

and letting $Z = 2$, $r_0/a_\mu = 10^{-2}\,\mu/m_e \simeq 2$, one has $P = 0.986 \simeq 1$.

c) Due to Gauss' theorem, one has:

$$E(r) = \frac{1}{r^2}\int_0^r \rho(r')\,\mathrm{d}V' = -\frac{e}{r^2}\left(1 - \mathrm{e}^{-4r/a_\mu}\left(1 + 4r/a_\mu + 8(r/a_\mu)^2\right)\right) \;\Rightarrow$$

$$\varphi(r) = -\frac{e}{r} + e\left(\frac{1}{r} + \frac{2}{a_\mu}\right)\mathrm{e}^{-4r/a_\mu}; \qquad \varphi(r) \approx -\frac{e}{r}\quad \text{for } r \gg a_\mu\,.$$

$$\left.\frac{\varphi(r) + e/r}{e/r}\right|_{r=a_B/100} \simeq 1.7 \times 10^{-3}\,.$$

d) The probability of finding the electron within the sphere of radius $r_0 = 10^{-2}a_B$ can be obtained from the answer to question b), by replacing a_μ by a_B and putting $Z = 1$:

$$P(r \le r_0) = 1.3 \times 10^{-6} \simeq 0\,.$$

e) In the approximation in which the distribution of the μ^- meson is 'frozen', it is legitimate to consider the Hamiltonian relative to the electron alone:

$$H = \frac{\vec{p}^{\,2}}{2m_e} - \frac{2e^2}{r} - e\,\varphi(r) = \frac{\vec{p}^{\,2}}{2m_e} - \frac{e^2}{r} - e^2\left(\frac{1}{r} + \frac{2}{a_\mu}\right)\mathrm{e}^{-4r/a_\mu}\,.$$

Owing to the results above, the main effect of the μ^- meson is the screening of one unit of the nuclear charge, so we put:

$$H = H_0 + V(r)\,; \qquad H_0 = \frac{\vec{p}^{\,2}}{2m_e} - \frac{e^2}{r}\,, \qquad V(r) = -e^2\left(\frac{1}{r} + \frac{2}{a_\mu}\right)\mathrm{e}^{-4r/a_\mu}$$

and consider $V(r)$ as a perturbation. Limiting the calculation to the lowest nonvanishing order in a_μ/a_B (see Problem 12.18), one has:

$$\Delta E_{1,0}^{(1)} \simeq -e^2 \int_0^\infty \frac{4}{a_B^3} \times \left(\frac{1}{r} + \frac{2}{a_\mu}\right) e^{-4r/a_\mu} r^2 dr = -\frac{e^2}{2\,a_B}\left(\frac{a_\mu}{a_B}\right)^2$$

$$\simeq -3.4 \times 10^{-4}\,\text{eV}$$

$$\Delta E_{2,0}^{(1)} \simeq -e^2 \int_0^\infty \frac{1}{2\,a_B^3} \times \left(\frac{1}{r} + \frac{2}{a_\mu}\right) e^{-4r/a_\mu} r^2 dr = -\frac{e^2}{16\,a_B}\left(\frac{a_\mu}{a_B}\right)^2$$

$$\simeq -4.3 \times 10^{-5}\,\text{eV}$$

whereas, for the p states, corrections smaller by a factor $(a_\mu/a_B)^2 \simeq 2.5 \times 10^{-5}$ are expected (see Problem 12.18).

12.20

a) Use of the virial theorem is convenient: $\overline{\vec{p}^2}/2m_e = -E_n$ (see Problem 12.11),

$$\langle n \mid \frac{\vec{p}^2}{m_e^2 c^2} \mid n \rangle = \frac{2}{m_e c^2}\langle n \mid \frac{\vec{p}^2}{2m_e} \mid n \rangle = -\frac{2E_n}{m_e c^2} = \frac{\alpha^2}{n^2}$$

where $\alpha \equiv e^2/\hbar c \simeq 1/137$ is the fine structure constant (see Problem 2.2). Therefore $v/c \simeq \alpha/n$.

b) If one takes $\overline{(\vec{p}^2)^2} \simeq \left(\overline{\vec{p}^2}\right)^2$, one has $\Delta E_n \simeq (v/c)^2 E_n$ whence $\Delta E_n/E_n \simeq \alpha^2/n^2$.

c) The perturbation is rotationally invariant, so the selection rule $\Delta l = 0$ applies and the degeneracy on m is not removed. One has:

$$\Delta E_{n,l}^{(1)} = \langle n, l, m \mid \frac{-(\vec{p}^2)^2}{8m_e^3 c^2} \mid n, l, m \rangle$$

$$= -\frac{1}{2m_e c^2}\langle n, l, m \mid \left(H_0 + \frac{e^2}{r}\right)^2 \mid n, l, m \rangle$$

$$= -\frac{1}{2m_e c^2}\left[E_n^2 + 2E_n\langle n, l, m \mid \frac{e^2}{r} \mid n, l, m \rangle + \langle n, l, m \mid \frac{e^4}{r^2} \mid n, l, m \rangle\right]$$

$$= -\frac{E_n^2}{2m_e c^2}\left[1 + 2\times(-2) + \frac{(2n^2)^2}{n^3(l+1/2)}\right] = E_n \frac{\alpha^2}{n^2}\left(\frac{n}{l+1/2} - \frac{3}{4}\right).$$

d) $E_{2p} - E_{2s} = (2\alpha^2/3)\,E_2 = 1.2 \times 10^{-4}\,\text{eV} \simeq 1\,\text{cm}^{-1}$.

Since this is not the only relativistic contribution and, in addition, it is essential to take into account that the electron has an intrinsic magnetic moment, a comparison with the experimental data would be meaningless: in any event the number just found is about three times the maximum separation within the multiplet $n = 2$.

13

Spin and Magnetic Field

Spin $\frac{1}{2}$; Stern and Gerlach apparatus; spin rotations; minimal interaction; Landau levels; Aharonov–Bohm effect.

13.1 Consider a particle of spin $\frac{1}{2}$ in the spin state:

$$|s\rangle = \alpha|+\rangle + \beta|-\rangle\,; \qquad |\alpha|^2 + |\beta|^2 = 1\,, \qquad \sigma_z|\pm\rangle = \pm|\pm\rangle\,.$$

a) Calculate the mean value $\langle\vec\sigma\,\rangle \equiv \langle\,s\,|\,\vec\sigma\,|\,s\,\rangle$ of the Pauli matrices $\vec\sigma$ in the state $|\,s\,\rangle$ and show that $\langle\vec\sigma\,\rangle$ is a unit vector.

b) Show that $|\,s\,\rangle$ is the eigenstate, belonging to the eigenvalue $+1$, of a suitable component $\sigma_n \equiv \vec\sigma\cdot\hat n$ of the spin: find $\hat n$. Say whether, also for particles of spin 1, any state is eigenstate of some spin component.

c) Show that the mean value of $\vec\sigma\cdot\hat m$ in the state $|\,s\,\rangle$ ($\hat m$ is an arbitrary unit vector) is given by $\hat n\cdot\hat m$.

d) Using the $\vec\sigma$ matrices, write the operator $\mathcal{P}_+$ that projects onto the state $|+\rangle$ and the operator $\mathcal{P}_n$ that projects onto the state $|\,s\,\rangle$.

13.2 Consider a particle of spin $\frac{1}{2}$. We shall denote by $|\pm\rangle$ the eigenstates of σ_z: $\sigma_z|\pm\rangle = \pm|\pm\rangle$ and by $|\pm\hat n\rangle$ the eigenstates of $\sigma_n \equiv \vec\sigma\cdot\hat n$: $\sigma_n|\pm\hat n\rangle = \pm|\pm\hat n\rangle$. The operator that effects, on spin states, a counterclockwise rotation by an angle ϕ around the axis $\hat\varrho$ is (see Problems 8.3 and 7.13) $U(\hat\varrho,\,\phi) = \mathrm{e}^{-\mathrm{i}\phi\,\vec s\cdot\hat\varrho} = \mathrm{e}^{-\frac12\mathrm{i}\phi\,\vec\sigma\cdot\hat\varrho} = \mathbb{1}\cos(\phi/2) - \mathrm{i}\vec\sigma\cdot\hat\varrho\,\mathrm{sen}(\phi/2)$.

a) In the particular case $\hat\varrho$ parallel to the z axis, calculate:

$$\sigma_i' = U(\hat z,\,\phi)\,\sigma_i\,U^{-1}(\hat z,\,\phi) \qquad i = 1,\,2,\,3\,.$$

If $|\,s\,\rangle$ is an arbitrary spin state, find $U(\hat\varrho,\,2\pi)\,|\,s\,\rangle$.

b) Given $|+\hat n\rangle$, $|+\hat m\rangle$ and $\hat\varrho$ orthogonal to both $\hat n$ and $\hat m$, determine ϕ such that $U(\hat\varrho,\,\phi)\,|+\hat n\rangle = |+\hat m\rangle$. Calculate the scalar product $\langle+\hat n\,|+\hat m\rangle$, in particular when $\hat n\perp\hat m$.

c) Write $|+\hat n\rangle$ in terms of the eigenvectors of σ_z.

© Springer International Publishing AG 2017
E. d'Emilio and L.E. Picasso, *Problems in Quantum Mechanics*,
UNITEXT for Physics, DOI 10.1007/978-3-319-53267-7_13

13.3 The Stern–Gerlach apparatus is an experimental set up in which a beam of atoms endowed with magnetic moment $\vec{\mu} = -g_J \mu_B \vec{J}/\hbar$ (where $\mu_B = e\hbar/2m_e c \simeq 6 \times 10^{-9}\,\mathrm{eV/G}$ is the *Bohr magneton* and g_J a c–number) enters a region where there

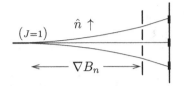

is a magnetic field with a strong gradient in a given direction $\hat{n}$. The force $F_n = \left[\nabla(\vec{\mu} \cdot \vec{B})\right]_n$ exerted on the atoms in the direction $\hat{n}$ spatially separates the atoms that have the different values of the component J_n of the angular momentum. If the different components of the beam are intercepted, for example by either a photographic plate or by counters, the Stern–Gerlach apparatus is an instrument that measures J_n and emphasizes its quantization. The experiment performed by Stern and Gerlach in 1922 made use of a beam of silver atoms that, in the ground state, have an angular momentum $J = \frac{1}{2}\hbar$ and $g_J = 2$.

a) Estimate the separation s between the beams after the atoms have gone a distance $d = 20\,\mathrm{cm}$, having exited an oven at the temperature of boiling Ag ($T = 2485\,\mathrm{K}$) with a kinetic energy $E_c \simeq k_B T$ and $|\nabla B| = 10^5\,\mathrm{G/cm}$.

b) Say how many spots are seen on the photographic plate in the case when the Ag atoms that enter the region in which there is the magnetic field:
   i) are totally unpolarized: statistical mixture $\{|\,J_z = \frac{1}{2}\,\rangle, \frac{1}{2}; \ |\,J_z = -\frac{1}{2}\,\rangle, \frac{1}{2}\}$;
   ii) all are in an eigenstate of a component of $\vec{J}$ orthogonal to $\hat{n}$.

In the figure on the right the poles of the magnet and the position of the beam travelling in direction of the x axis (orthogonal to the sheet) are drawn. If instead of atoms one uses a beam of electrons, they are subject also to the Lorentz force $\vec{F}^L$: as a consequence the force in the direction of the

field gradient exhibits an uncertainty $\Delta F_z^L \simeq e(v_x/c)\,(\partial B_y/\partial y)\,\Delta y$ due to the spreading Δy of the wave packet in the direction orthogonal to the motion and to the direction of the field in the region occupied by the beam (the z axis).

c) Let $\vec{B} = (0, B_y, B_z)$ ($\mathrm{div}\,\vec{B} = 0$: $\partial B_y/\partial y = -\partial B_z/\partial z$). The condition necessary to be able to observe the spatial separation of the beams is $\Delta F_z^L/|F_z| \ll 1$. Show that it requires $\Delta y \ll \lambda_x$ ($\lambda_x = h/p_x \simeq h/m_e v_x$), and that, even if this condition is met before entering the apparatus, certainly is not fulfilled when the electrons leave it, having gone a distance $d \gg \lambda_x$: in order to determine $\Delta y(t)$, the approximation of considering the electrons as free particles is acceptable. [The above argument was proposed by Bohr and Pauli.]

13.4 Atoms of Ag ($J = \frac{1}{2}\hbar$) are polarized (namely they are all in the same unknown spin state) and are sent in a Stern–Gerlach apparatus. A first measurement provides the result that 50% of the atoms are deflected in the direction of the field gradient (the z axis) and, obviously, 50% in the opposite direction.

a) Calculate $\langle \sigma_z \rangle$. What can be said about the spin state $|s\rangle$ of the atoms?

A second measurement is performed, always on atoms in the same initial state of polarization, but with the field rotated by 90° around the direction of the beam. It is observed that 75% of the atoms are deflected in the direction of the gradient of the field (y axis).

b) Is now the polarization state of the atoms completely determined? If not, which states are compatible with the results of the measurements?

A third measurement is effected with the gradient of the field at 45° in the y-z plane.

c) What is the result of the third measurement? Is any further piece of information obtained on the polarization state of the atoms?

13.5   Free particles of spin $\frac{1}{2}$ are prepared in the state:

$$|A\rangle = \alpha\,|r,t\rangle\,|+\rangle + \beta\,|l,t\rangle\,|-\rangle, \qquad \langle r,t\,|\,r,t\rangle = \langle l,t\,|\,l,t\rangle = 1$$

where $|r,t\rangle$, $|l,t\rangle$ respectively are states of the particle travelling toward the right and the left and the corresponding wavefunctions $\psi_r(x,t)$ and $\psi_l(x,t)$ have, for any $t > 0$, disjoint supports $x > 0$ and $x < 0$ (this is an approximation); $|+\rangle$, $|-\rangle$ are the normalized eigenstates of σ_z. By means of a Stern–Gerlach apparatus, able to detect the particles travelling towards the right, measurements are made, for $t > 0$, of the component $\vec{\sigma} \cdot \hat{n}$ of the spin.

a) If N is the number of the produced particles, what are the numbers N_+ and N_- of the particles respectively detected in the spin states $|+\,\hat{n}\rangle$ and $|-\,\hat{n}\rangle$? What is the mean value of such measurements, i.e. relative only to the particles revealed by the apparatus?

b) If we conventionally assign the value 0 to the result of a measurement in which the particle is not seen by the apparatus, what is the mean value of the results of such measurements? Which is the operator $\xi^{\text{right}} = f(q, p, \vec{\sigma})$ corresponding to such an observable?

Now a second Stern–Gerlach apparatus, able to measure the component $\hat{n}$ of the spin of particles travelling towards the left, is available. Let ξ^{left} be the operator corresponding to this observable.

c) What is the mean value of the measurements performed by the instrument consisting of both apparatuses ($\xi = \xi^{\text{left}} + \xi^{\text{right}}$)?

13.6 Neutrons (spin $\frac{1}{2}$, magnetic moment $\vec{\mu} = \mu_n \vec{\sigma}$, $\mu_n = -1.9 \, e\hbar/2m_n c$ $= -1.9 \cdot 5 \times 10^{-24}$ erg/G) travelling in the direction of the x axis with energy E enter the region $0 \le x \le L$ where a uniform magnetic field $\vec{B}$ parallel to the z axis is present. The Hamiltonian of the neutron is

$$H = \frac{\vec{p}^2}{2m_n} - \vec{\mu} \cdot \vec{B} = \frac{\vec{p}^2}{2m_n} - \chi(x)\,\mu_n B\,\sigma_z$$

where $\chi(x)$ equals 1 in the region where $\vec{B} \ne 0$, and 0 elsewhere. Assume that $E \gg \mu_n B$, so that (see Problems 6.12 and 6.13) the wave reflected by the region where $\vec{B} \ne 0$ can be neglected (in such a case only the continuity of the wavefunction must be required, not that of its derivative).

a) Find the eigenfunctions $\psi_+(x)$ and $\psi_-(x)$ of the Hamiltonian H belonging to the eigenvalue E in the two cases in which the neutron spin is either parallel or antiparallel to the magnetic field.

Let $|\,a\,\rangle$ be the (generic) spin state of the incident neutrons.

b) Find the spin state $|\,b\,\rangle$ at the exit of the region where $\vec{B} \ne 0$ and show that $|\,b\,\rangle$ is obtained from $|\,a\,\rangle$ through a rotation by a suitable angle ϕ around the direction of $\vec{B}$. For which values of ϕ and, correspondingly, of B, is the final spin state equal to the initial one?

The neutrons are now sent in a Bonse–Hart interferometer in which the magnetic field is present only in one of the two paths. The interferometer is tuned in such a way that, when the magnetic field is switched off, all the neutrons arrive at counter C_1: the phase difference between the two components of the wavefunctions is, therefore, due only to the effect of the magnetic field on the spin state. The semi-transparent mirrors have equal reflection and transmission coefficients. Let N be the number of incident neutrons.

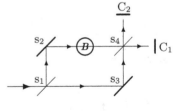

c) Write the component of the wavefunction with support in the region between s_4 and C_1 (see figure) and calculate the number N_1 of the neutrons that arrive at counter C_1.

d) For which values of B do all the neutrons arrive at the counter C_1? Make the numerical calculation with the data of the first experiment (Rauch, Zeilinger et al. 1975): $\lambda_n = 1.8 \, \text{Å}$, $L = 1 \, \text{cm}$.

13.7 Let the Hamiltonian of a particle in one dimension be

$$H = \frac{1}{2m}\bigl(p - f(x)\bigr)^2.$$

a) For the *classical* particle of energy E in motion in the direction of the x-axis, calculate the canonical momentum $p(x)$ as a function of the position of the particle and the kinetic momentum $m\,\dot{q}$. For the *quantum* particle, find the eigenfunction $\psi_E(x)$ of the Hamiltonian corresponding to the eigenvalue E.

Consider now a particle in three dimensions, whose Hamiltonian is

$$H = \frac{1}{2m}\left(\vec{p} - \hbar\,\nabla\Phi(x,y,z)\right)^2 .$$

b) Find the relationship between the canonical momentum $\vec{p}$ and the kinetic momentum $m\,\dot{\vec{q}}$.

c) Prove that H is unitarily equivalent to the Hamiltonian $H_0 = \vec{p}^{\,2}/2m$ of the free particle: $U\,H_0\,U^{-1} = H$. Find U:

$$U q_i U^{-1} = q_i, \quad U p_i U^{-1} = p_i - \hbar\,\partial_i \Phi(x,y,z) .$$

d) The eigenfunctions $\psi_E^0(\vec{r}) = e^{\,i\,\vec{k}\cdot\vec{r}}$ of H_0 are known. Find the eigenfunctions $\psi_E(\vec{r})$ of the Hamiltonian H.

13.8 A particle of mass m and charge $-e$ subject to the potential $V(\vec{r})$, is also subject to a magnetic field $\vec{B}(\vec{r})$. Let $\vec{A}(\vec{r})$ be a possible choice for the vector potential: $\vec{B} = \mathrm{curl}\,\vec{A}$. It is known that the Hamiltonian H is obtained by replacing $\vec{p}$ with $\vec{p} + (e/c)\,\vec{A}$ in the Hamiltonian in absence of magnetic field (*minimal substitution*).

a) Find the velocity operators $\dot{q}_i(q,\,p)$ and calculate the commutators $[\dot{q}_i,\,\dot{q}_j]$ and $[\dot{q}_i,\,q_j]$.

b) Demonstrate that the Hamiltonians $H^{(1)}$ and $H^{(2)}$ relative to the two choices $\vec{A}_1(\vec{r})$, $\vec{A}_2(\vec{r})$ of the vector potential, differing by a *gauge transformation* $\vec{A}_2(\vec{r}) = \vec{A}_1(\vec{r}) + \nabla\Lambda(\vec{r})$, are unitarily equivalent.

c) Assume the solutions of the Schrödinger equation $i\hbar\,\partial\psi_1(\vec{r},t)/\partial t = H^{(1)}\psi_1(\vec{r},t)$ are known. Find the solutions $\psi_2(\vec{r},t)$ of the Schrödinger equation with $H^{(2)}$.

d) Verify that the probability current given by:

$$\vec{j}(\vec{r},t) = -i\,\frac{\hbar}{2m}\left[\psi^*(\vec{r},t)\left(\nabla + i\,\frac{e}{\hbar c}\vec{A}\right)\psi(\vec{r},t) - \psi(\vec{r},t)\left(\nabla - i\,\frac{e}{\hbar c}\vec{A}\right)\psi^*(\vec{r},t)\right]$$

obeys the continuity equation:

$$\mathrm{div}\,\vec{j}(\vec{r},t) + \partial\rho\,(\vec{r},t)/\partial t = 0$$

and is "gauge-invariant", namely independent of the choice of the vector potential.

13.9 Consider an electron in a uniform magnetic field $\vec{B}$ parallel to the z axis.

a) Show that the following vector potentials:

$$\vec{A}_1(\vec{r}) = \frac{B}{2}\left(-y,\, x,\, 0\right), \quad \vec{A}_2(\vec{r}) = \frac{B}{2}\left(-y - a_2,\, x + a_1,\, 0\right),$$

$$\vec{A}_3(\vec{r}) = B\left(0,\, x,\, 0\right)$$

differ by a gauge transformation and that, for any choice among the three potentials $\vec{A}_1$, $\vec{A}_2$, $\vec{A}_3$, the z coordinate is separable in the Hamiltonian.

From now on the motion along the z axis will be ignored and the system will be considered two-dimensional.

b) Show that, if the magnetic moment associated to the spin is ignored, the Hamiltonian in terms of the vector potential $\vec{A}_1(\vec{r}) = \frac{1}{2}\vec{B}\wedge\vec{r}$ takes the form:

$$H = \frac{p_x^2}{2m_e} + \frac{p_y^2}{2m_e} + \frac{1}{2}m_e\,\omega_{\mathrm{L}}^2(x^2 + y^2) + \omega_{\mathrm{L}}L_z \equiv H_0 + \omega_{\mathrm{L}}L_z$$

where $\omega_{\mathrm{L}}/2\pi$ is Larmor frequency ($\omega_{\mathrm{L}} = e\,B/2m_e\,c$) and $L_z = x\,p_y - y\,p_x$ is the canonical (see Problem 9.10) angular momentum. Use the results of Problem 10.6 and find the eigenvalues of H (Landau levels: see Problem 2.9) starting from the energy levels of the two-dimensional oscillator and show that each eigenvalue has infinite degeneracy. Which is the separation between two adjacent Landau levels?

Let us now consider also the interaction $-\vec{\mu}\cdot\vec{B}$ between the magnetic field and the intrinsic magnetic moment of the electron: $\vec{\mu} = -g\,(e/2m_e\,c)\,\vec{s}$, $\vec{s} = \frac{1}{2}\hbar\vec{\sigma}$, where $g \simeq 2$ is the gyromagnetic factor of the electron (actually $g \simeq 2.002$).

c) Write the Hamiltonian and find its eigenvalues first assuming $g = 2$, then $g > 2$.

13.10   Consider an electron subject to a uniform magnetic field $\vec{B}$ parallel to the z axis. As in Problem 13.9, the motion along the z axis will be ignored. The purpose of this problem is to derive the Landau levels and their degeneracies in a way that is independent from the choice of the vector potential. We shall ignore the electron spin (that can be taken into account as in Problem 13.9).

a) Let x, y; v_x, v_y respectively be the position and the velocity of the electron at a certain instant, the electron being considered as a *classical* particle. Find the coordinates x_c, y_c of the center of the orbit described by the electron.

b) Calculate the (quantum) commutator $[x_c,\, y_c]$ and show that v_x, v_y; x_c, y_c constitute, up to a multiplicative factor, two couples of canonical variables P, Q; p, q.

c) Write the Hamiltonian in terms of the canonical variables defined above and show that x_c and y_c are constants of motion. Find the Landau levels and their degeneracies.

d) Calculate the *kinetic* angular momentum of the electron with respect to the center of the orbit $L_z^{kin} \equiv m_e (x - x_c) v_y - m_e (y - y_c) v_x$ and find its eigenvalues.

To study the quantum Hall effect, suitable devices (heterostructures) are realized in which the electron is confined, in the direction of the z axis, in the ground state of a very deep potential well $V(z)$ so that the motion of the electron is two-dimensional in the x-y plane.

e) Let A be the area in the x-y plane of the device. Calculate the volume (the area, in the present case) of the (classical) phase space associated to the canonical variables p and q defined above and use the result to estimate (see Problem 2.6) the finite degeneracy of each Landau level for $A = 1\,\text{cm}^2$, $B = 10^4\,\text{G}$.

13.11 Consider a particle of spin $\frac{1}{2}$, mass m, charge $-e$ and magnetic moment $\vec{\mu} = -g\,(e/2mc)\,\vec{s}$, $\vec{s} = \frac{1}{2}\hbar\vec{\sigma}$, in a uniform magnetic field $\vec{B}$.

a) Take the z axis parallel to $\vec{B}$ and calculate the angular velocity by which the spin precedes around the z axis.

b) Show that, if $g = 2$, the *helicity* operator $\vec{\sigma} \cdot \vec{v}$ ($\vec{v} \equiv \dot{\vec{q}}$) is a constant of motion.

Assume that the particle (now $g \neq 2$) enters the region where $\vec{B} \neq 0$ with velocity $\vec{v}$ orthogonal to $\vec{B}$ and with the spin parallel to the velocity $\vec{v}$ (state with helicity $h = \vec{\sigma} \cdot \hat{v} = +1$).

c) Calculate the angular velocity of the particle and determine the angle $\phi(t)$, as a function of time, between the direction of the spin and the direction $\hat{v}$ of the velocity.

In some experiments aimed at determining the "magnetic anomaly" $a_\mu \equiv \frac{1}{2}(g - 2)$ of the μ^- meson (mass $m_\mu = 207\,m_e$), the μ^- mesons, produced in the decay of the π^- mesons, enter a magnetic field $B = 1.45 \times 10^4\,\text{G}$ and $w_a \equiv \dot{\phi}(t)$ is measured. It is necessary to keep into account that the μ^- is relativistic: $\gamma(v) = 29.3$, and therefore both the angular velocity and the spin precession velocity are reduced by the factor $\gamma(v)$ (and the lifetime is dilated by the same factor).

d) Knowing that $T_a^{(\text{exp})} \equiv 2\pi/w_a^{(\text{exp})} \simeq 1.23 \times 10^{-4}\,\text{s}$, find a_μ.

13.12 Consider the system constituted by two particles of masses m_1, m_2 and charges $e_1 = -e$, $e_2 = e$, interacting through a potential $V(q)$, $q \equiv |\vec{q}_1 - \vec{q}_2|$. The system is subject to a uniform magnetic field $\vec{B}$ parallel the z axis.

a) After having effected the minimal substitutions $\vec{p}_i \to \vec{p}_i - (e_i/c)\vec{A}(\vec{r}_i)$ with $\vec{A}(\vec{r}) = \frac{1}{2}\vec{B} \wedge \vec{r} = \frac{1}{2}B(-y, x, 0)$, write the complete Hamiltonian H in terms of the variables of the center of mass $\vec{Q}$, $\vec{P}$ and of the relative variables $\vec{q}$, $\vec{p}$:

$$\vec{Q} = \frac{m_1\vec{q}_1 + m_2\vec{q}_2}{m_1 + m_2}, \quad \vec{P} = \vec{p}_1 + \vec{p}_2; \quad \vec{q} = \vec{q}_1 - \vec{q}_2, \quad \vec{p} = \frac{m_2\vec{p}_1 - m_1\vec{p}_2}{m_1 + m_2}.$$

Say whether H is separable: $H \overset{?}{=} H_{CM}(\vec{Q}, \vec{P}) + H_{\text{rel}}(\vec{q}, \vec{p})$.

b) Take the z axis parallel to $\vec{B}$ and let $U = e^{-ieB(q_x Q_y - q_y Q_x)/2\hbar c}$. Write the canonical transformation induced by U on the variables of the system ($\tilde{Q}_i = UQ_iU^\dagger$, $\cdots$). Show that $\tilde{H} = UHU^\dagger$ is given by (M is the total mass and μ the reduced mass):

$$\tilde{H} = \frac{\vec{P}^2}{2M} + \frac{eB}{Mc}(\vec{q} \wedge \vec{P})_z$$
$$+ \left[\frac{\vec{p}^2}{2\mu} + V(q) + \frac{eB}{2c}\frac{m_2 - m_1}{m_1 m_2}(\vec{q} \wedge \vec{p})_z + \frac{e^2 B^2}{8\mu c^2}(q_x^2 + q_y^2) \right]$$

and say whether the momentum $\vec{P}$ and/or the velocity $\dot{\vec{Q}}$ of the center of mass are constants of motion.

Assume now that the system is the hydrogen atom: $V(q) = -e^2/q$ (the spins of electron and proton are ignored). Let $m_1 = m_e$, $m_2 = m_p$.

c) Calculate the effect of the term $(\vec{q} \wedge \vec{p})_z$ on the levels of $\vec{p}^2/2\mu - e^2/q$ for $B = 10^4\,\text{G}$ and say whether the neglect of the term quadratic in B on the first energy levels of the hydrogen atom is legitimate.

d) Estimate the order of magnitude of the term $(eB/Mc)(q_x P_y - q_y P_x)$ for hydrogen atoms whose center of mass has the room temperature thermal velocity.

13.13 It is possible to realize an interferometer for electrons similar to that of Bonse–Hart for neutrons (see Problem 3.4). At the center of the interferometer a long solenoid, of radius a and with the axis orthogonal to the plane containing the trajectories of the electrons, is present. The interferometer is tuned in such a way that, when the magnetic field inside the solenoid is vanishing, all the electrons arrive at the counter C_1. Let I be the intensity of the beam of electrons (of energy E), $\vec{B}$ the magnetic field inside the solenoid, $\vec{A}(\vec{r}) = \frac{1}{2}\vec{B} \wedge \vec{r}$ the vector potential inside the solenoid.

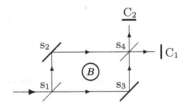

a) Show that the vector potential outside the solenoid (where $\vec{B} = 0$) is

$$\vec{A}(\vec{r}) = \frac{a^2}{2(x^2 + y^2)} \vec{B} \wedge \vec{r} = \frac{a^2 B}{2(x^2 + y^2)} (-y,\, x,\, 0), \qquad \hat{z} \parallel \vec{B}.$$

Calculate the line integral of $\vec{A}(\vec{r})$ along the closed circuit $s_1 \to s_3 \to s_4 \to s_2 \to s_1$.

b) Show that the wavefunction along each of the two paths $\gamma_1 = s_1 \to s_3 \to s_4$ and $\gamma_2 = s_1 \to s_2 \to s_4$ is given by:

$$\psi(x, y, z) = \exp\left(i \int \left(\vec{k} + (e/\hbar c)\, \vec{A} \right) \cdot d\vec{l} \right), \qquad k = \sqrt{2 m_e E}/\hbar$$

where the integral is taken from the point where the electrons enter the interferometer up to the point (x, y, z), of the path it belongs to.

c) Calculate the difference of phase φ between the two components of the electron wavefunction that arrive at s_4 from the paths γ_1 and γ_2; calculate (see Problem 13.6) the intensities I_1, I_2 of the electrons detected by the counters C_1 and C_2.

The above effect has been predicted by Aharonov and Bohm in 1959.

Solutions

13.1

a) $\langle \sigma_x \rangle = 2\,\Re e(\alpha^*\beta)$, $\quad \langle \sigma_y \rangle = 2\,\Im m(\alpha^*\beta)$, $\quad \langle \sigma_z \rangle = |\alpha|^2 - |\beta|^2$.

$\langle \sigma_x \rangle^2 + \langle \sigma_y \rangle^2 + \langle \sigma_z \rangle^2 = 4|\alpha^*\beta|^2 + (|\alpha|^2 - |\beta|^2)^2 = (|\alpha|^2 + |\beta|^2)^2 = 1$.

b) Several ways of proceeding are possible. Thanks to the above results, the simplest is the following: putting $\hat{n} = \langle \vec{\sigma} \rangle$, one has $\langle s \mid \vec{\sigma} \cdot \hat{n} \mid s \rangle = \hat{n}^2 = 1 \Rightarrow \vec{\sigma} \cdot \hat{n} \mid s \rangle = \mid s \rangle$ because the maximum eigenvalue of $\vec{\sigma} \cdot \hat{n}$ is 1 ($\sigma_n^2 = 1$). For particles with spin higher than $\frac{1}{2}$ not all the states are eigenstates of some component of the spin (see Problem 8.5).

c) $\langle s \mid \vec{\sigma} \cdot \hat{m} \mid s \rangle = \langle \vec{\sigma} \rangle \cdot \hat{m} = \hat{n} \cdot \hat{m} \equiv \cos\theta_{nm}$.

d) $\mathcal{P}_+ = \frac{1}{2}(\mathbb{1} + \sigma_z)$: $\frac{1}{2}(\mathbb{1} + \sigma_z)\mid + \rangle = \mid + \rangle$, $\frac{1}{2}(\mathbb{1} + \sigma_z)\mid - \rangle = 0$;

$\mathcal{P}_n = \frac{1}{2}(\mathbb{1} + \sigma_n)$.

13.2

a) $\sigma_z' = \sigma_z$;

$\sigma_x' = \big(\mathbb{1}\cos(\phi/2) - i\sigma_z\,\sin(\phi/2)\big)\sigma_x\big(\mathbb{1}\cos(\phi/2) + i\sigma_z\,\sin(\phi/2)\big)$

$= \big(\cos^2(\phi/2) - \sin^2(\phi/2)\big)\,\sigma_x + 2\sin(\phi/2)\cos(\phi/2)\,\sigma_y$

$= \sigma_x\,\cos\phi + \sigma_y\,\sin\phi$

and analogously:

$\sigma_y' = -\sigma_x\,\sin\phi + \sigma_y\,\cos\phi$.

$U(\hat{\varrho},\,2\pi) = -\mathbb{1} \quad \Rightarrow \quad U(\hat{\varrho},\,2\pi)\mid s \rangle = -\mid s \rangle$. Instead $U(\hat{\varrho},\,4\pi) = \mathbb{1}$.

b) Since $\mid + \hat{n} \rangle$ and $\mid + \hat{m} \rangle$ are respectively eigenvectors of σ_n and σ_m with the same eigenvalue, then $\sigma_m = U(\hat{\varrho},\,\phi)\,\sigma_n\,U^{-1}(\hat{\varrho},\,\phi)$. Since $\hat{\varrho}$ is orthogonal to both $\hat{n}$ and $\hat{m}$, from the previous question with $\hat{z} = \hat{\varrho}$, it follows that ϕ is the angle between $\hat{n}$ and $\hat{m}$, in particular $\hat{n} \cdot \hat{m} = \cos\phi$. Instead, $\langle +\hat{n} \mid +\hat{m} \rangle = \cos(\phi/2)$:

$$\langle +\hat{n} \,|\, +\hat{m} \rangle = \langle +\hat{n} \,|\, U(\hat{\varrho}, \phi) \,|\, +\hat{n} \rangle = \cos(\phi/2) - \mathrm{i}\, \sin(\phi/2) \langle +\hat{n} \,|\, \sigma_\varrho \,|\, +\hat{n} \rangle = \cos(\phi/2)$$

$$= \cos(\phi/2) - \mathrm{i}\, \sin(\phi/2) \langle +\hat{n} \,|\, \sigma_n \sigma_\varrho \sigma_n \,|\, +\hat{n} \rangle = \cos(\phi/2)$$

since, as $\hat{\varrho} \perp \hat{n}$, σ_ϱ and σ_n anticommute: $\sigma_n \, \sigma_\varrho \, \sigma_n = -\sigma_\varrho$.

Contrary to the case of photons (for which, if $\hat{e}_\alpha \perp \hat{e}_\beta$ then $\langle e_\alpha \,|\, e_\beta \rangle = 0$), if $\hat{n} \perp \hat{m}$ the states $|+\hat{n} \rangle$ and $|+\hat{m} \rangle$ are not orthogonal to each other, but $\langle +\hat{n} \,|\, +\hat{m} \rangle = \sqrt{2}/2$.

c) Several ways of proceeding are possible. With $|+\hat{n} \rangle = \alpha |+\rangle + \beta |-\rangle$:

$$\sigma_n |+\hat{n} \rangle = |+\hat{n} \rangle \quad \Rightarrow \quad \begin{pmatrix} n_z & n_x - \mathrm{i}\, n_y \\ n_x + \mathrm{i}\, n_y & n_z \end{pmatrix} \begin{pmatrix} \alpha \\ \beta \end{pmatrix} = \begin{pmatrix} \alpha \\ \beta \end{pmatrix} \quad \Rightarrow$$

$$\alpha = \sqrt{\frac{1 + n_z}{2}} \; ; \quad \beta = \frac{n_x + \mathrm{i}\, n_y}{\sqrt{2(1 + n_z)}} \; .$$

Alternatively, with $n_z = \cos\theta$, $n_x + \mathrm{i}\, n_y = \sin\theta\, \mathrm{e}^{\mathrm{i}\phi}$, $|+\hat{n} \rangle = U(\hat{\varrho}, \theta) |+\rangle$ where $\hat{\varrho} \parallel (\hat{z} \wedge \hat{n}) = (-\sin\phi, \cos\phi, 0)$, then:

$$|+\hat{n} \rangle = \big(\cos(\theta/2) - \mathrm{i}\, \vec{\sigma} \cdot \hat{\varrho}\, \sin(\theta/2) \big) |+\rangle \rightarrow$$

$$\begin{pmatrix} \cos(\theta/2) & -\mathrm{e}^{-\mathrm{i}\phi} \sin(\theta/2) \\ \mathrm{e}^{\mathrm{i}\phi} \sin(\theta/2) & \cos(\theta/2) \end{pmatrix} \begin{pmatrix} 1 \\ 0 \end{pmatrix} = \begin{pmatrix} \cos(\theta/2) \\ \mathrm{e}^{\mathrm{i}\phi} \sin(\theta/2) \end{pmatrix} \quad \Rightarrow$$

$$|+\hat{n} \rangle = \cos(\theta/2) |+\rangle + \mathrm{e}^{\mathrm{i}\phi} \sin(\theta/2) |-\rangle \; .$$

13.3

a) $\dfrac{s}{2} \simeq \dfrac{F\, t^2}{2 m_{\mathrm{Ag}}} = \dfrac{F\, d^2}{2 m_{\mathrm{Ag}}\, v^2} = \dfrac{\mu_{\mathrm{B}}\, |\nabla B|\, d^2}{2 m_{\mathrm{Ag}}\, v^2} \simeq \dfrac{6 \times 10^{-9} \cdot 10^5 \cdot 4 \times 10^2}{0.8} \simeq 0.3\,\mathrm{cm} \; .$

b) In the case of the statistical mixture obviously two spots of equal intensity are observed; in the other case let us take the z axis parallel to $\hat{n}$ and the x axis in the direction of the polarization of the atoms. In the Schrödinger representation and with s_z diagonal, the state of the atoms just before entering the apparatus is

$$|A\rangle \rightarrow \frac{1}{\sqrt{2}}\, \psi(x, y, z) \begin{pmatrix} 1 \\ 1 \end{pmatrix}$$

whereas for the atoms leaving the apparatus the two components are spatially separated:

$$|B\rangle \rightarrow \frac{1}{\sqrt{2}} \begin{pmatrix} \psi_+(x, y, z) \\ \psi_-(x, y, z) \end{pmatrix} , \qquad |\psi_\pm(x, y, z)|^2 = |f(x, y, z \mp s/2)|^2$$

so still two spots of equal intensity are observed.

c) $F_z^{\mathrm{L}} = e\, \dfrac{v_x}{c}\, B_y \quad \Rightarrow \quad \Delta F_z^{\mathrm{L}} = e\, \dfrac{v_x}{c}\, \dfrac{\partial B_y}{\partial y}\, \Delta y = -e\, \dfrac{v_x}{c}\, \dfrac{\partial B_z}{\partial z}\, \Delta y \; .$ Then:

$$\frac{\Delta F_z^{\mathrm{L}}}{F_z} = \frac{e\, v_x\, \Delta y}{c\, \mu_{\mathrm{B}}} = 4\pi\, \frac{m_e\, v_x}{h}\, \Delta y \simeq 4\pi\, \frac{p_x}{h}\, \Delta y = 4\pi\, \frac{\Delta y}{\lambda_x} \; .$$

$$\Delta y(t) \simeq \frac{\Delta p_y}{m_e} t, \quad t = \frac{d}{v_x} \simeq \frac{m_e d}{p_x} \quad \Rightarrow \quad \Delta y(t) \simeq \frac{\Delta p_y}{p_x} d \quad \Rightarrow$$

$$\frac{\Delta y(t)}{\lambda_x} \simeq \frac{\Delta p_y}{h} d \simeq \frac{d}{\Delta y(0)} \gg \frac{\lambda_x}{\Delta y(0)} \gg 1 .$$

Indeed the condition $\Delta y \ll \lambda_x$ means diffraction by a hole of dimensions much smaller than the wavelength, i.e. spreading of the beam $\theta \gg 1$.

13.4

a) $\langle \sigma_z \rangle = \frac{1}{2} \times 1 + \frac{1}{2} \times (-1) = 0$. So (see Problem 13.1) the atoms are polarized along some direction of the x-y plane: $|s\rangle = \frac{1}{\sqrt{2}} (|+\rangle + e^{i\varphi} |-\rangle)$ with φ unknown.

b) $\langle \sigma_y \rangle = \frac{1}{2}$. Therefore (see Problem 13.1) $\langle \sigma_x \rangle^2 = 1 - \langle \sigma_y \rangle^2 - \langle \sigma_z \rangle^2 = \frac{3}{4}$. The result is compatible with two states of polarization: $\hat{n}_1 = (\frac{\sqrt{3}}{2}, \frac{1}{2}, 0)$, $\hat{n}_2 = (-\frac{\sqrt{3}}{2}, \frac{1}{2}, 0)$. Alternatively: $\langle \sigma_y \rangle = \frac{1}{2} = \sin\varphi \Rightarrow \varphi = 90° \pm 60°$.

c) The mean value of $\frac{1}{\sqrt{2}}(\sigma_y + \sigma_z)$ is $\sqrt{2}/4$, so the fraction $(4+\sqrt{2})/8 \simeq 68\%$ is deviated in the direction of the field gradient. Since σ_x is not involved in the measurement, no additional information is acquired.

13.5

a) $N_+ = N |\alpha|^2 |\langle +\hat{n} | +\rangle|^2 = N |\alpha|^2 |\cos^2(\theta/2)$, $N_- = N |\alpha|^2 \sin^2(\theta/2)$

$$\overline{\vec{\sigma} \cdot \hat{n}} = \frac{N_+ - N_-}{N_+ + N_-} = \cos\theta .$$

b) $\langle \vec{\sigma} \cdot \hat{n} \rangle = \dfrac{N_+ - N_-}{N} = |\alpha|^2 \cos\theta = \langle A \,|\, E_{x>0} \, \vec{\sigma} \cdot \hat{n} \,|\, A \rangle$

where $E_{x>0}$ is the operator that projects onto the states whose wavefunction has support for $x > 0$. The operator $\xi^{\text{right}} = E_{x>0} \, \vec{\sigma} \cdot \hat{n}$ has the eigenvalues $\pm 1, 0$.

c) $\langle A \,|\, \xi \,|\, A \rangle = \langle A \,|\, E_{x>0} \, \vec{\sigma} \cdot \hat{n} \,|\, A \rangle + \langle A \,|\, E_{x<0} \, \vec{\sigma} \cdot \hat{n} \,|\, A \rangle = \langle A \,|\, \vec{\sigma} \cdot \hat{n} \,|\, A \rangle$
   $= (|\alpha|^2 - |\beta|^2) \cos\theta .$

13.6

a) In the representation with s_z diagonal the Schrödinger equations for the neutrons with the spin parallel to the magnetic field is

$$-\frac{\hbar^2}{2m_n} \begin{pmatrix} \psi_+''(x) \\ 0 \end{pmatrix} - \chi(x) \, \mu_n B \begin{pmatrix} \psi_+(x) \\ 0 \end{pmatrix} = E \begin{pmatrix} \psi_+(x) \\ 0 \end{pmatrix}$$

that is the Schrödinger equation for a particle subject to a rectangular potential barrier of width L and height $V_0 = -\mu_n B$ ($\mu_n < 0$). Neglecting the reflected wave and putting $k = \sqrt{2m_n E}/\hbar$ one has, respectively in the three regions $x < 0$, $0 < x < L$, $x > L$,

$$\psi_+(x) = e^{ikx}, \quad e^{ik_+x}, \quad e^{i(kx+\varphi_+)}$$

$$k_+ = \hbar^{-1}\sqrt{2m_n(E + \mu_n B)} \simeq k + \frac{\mu_n B}{2E}k, \quad \varphi_+ = (k_+ - k)L \simeq \frac{\mu_n B}{2E}kL.$$

Analogously:

$$-\frac{\hbar^2}{2m_n}\psi_-''(x) + \chi(x)\,\mu_n B\,\psi_-(x) = E\,\psi_-(x)$$

is the equation for a particle subject to a rectangular potential well of depth V_0 and one has:

$$\psi_-(x) = e^{ikx}, \quad e^{ik_-x}, \quad e^{i(kx+\varphi_-)}$$

$$k_- = \sqrt{2m_n(E - \mu_n B)}\,/\,\hbar \simeq k - \frac{\mu_n B}{2E}k, \quad \varphi_- = (k_- - k)L \simeq -\frac{\mu_n B}{2E}kL.$$

b) The Schrödinger equation is

$$-\frac{\hbar^2}{2m_n}\begin{pmatrix}\psi_+''(x)\\ \psi_-''(x)\end{pmatrix} - \chi(x)\,\mu_n B\begin{pmatrix}\psi_+(x)\\ -\psi_-(x)\end{pmatrix} = E\begin{pmatrix}\psi_+(x)\\ \psi_-(x)\end{pmatrix}$$

and still one has two separate equations for $\psi_+(x)$ and $\psi_-(x)$. Putting $|a\rangle = \alpha\,|+\rangle + \beta\,|-\rangle$ one has, in the three regions $x < 0$, $0 < x < L$, $x > L$:

$$\begin{pmatrix}\psi_+(x)\\ \psi_-(x)\end{pmatrix} = e^{ikx}\begin{pmatrix}\alpha\\ \beta\end{pmatrix}, \quad \begin{pmatrix}\alpha\,e^{ik_+x}\\ \beta\,e^{ik_-x}\end{pmatrix}, \quad \begin{pmatrix}\alpha\,e^{i(kx+\varphi_+)}\\ \beta\,e^{i(kx+\varphi_-)}\end{pmatrix} = e^{ikx}\begin{pmatrix}\alpha\,e^{i\varphi_+}\\ \beta\,e^{i\varphi_-}\end{pmatrix}$$

$$\Rightarrow \quad |b\rangle = \alpha\,e^{i\varphi_+}\,|+\rangle + \beta\,e^{i\varphi_-}\,|-\rangle\,.$$

The result can be interpreted as a rotation of the spin by an angle $\phi = -2\varphi_+ = -\mu_n B\,kL/E$ around the z axis:

$$U(\phi)\,|a\rangle = e^{-\frac{1}{2}i\phi\,\sigma_z}\,|a\rangle = \begin{pmatrix}e^{-\frac{1}{2}i\phi} & 0\\ 0 & e^{+\frac{1}{2}i\phi}\end{pmatrix}\begin{pmatrix}\alpha\\ \beta\end{pmatrix} = \begin{pmatrix}\alpha\,e^{-\frac{1}{2}i\phi}\\ \beta\,e^{+\frac{1}{2}i\phi}\end{pmatrix}.$$

The final spin state equals the initial one when $e^{-\frac{1}{2}i\phi} = e^{+\frac{1}{2}i\phi}$, namely for $\phi = 2\pi n$ (if n is odd, $|b\rangle = -|a\rangle$), i.e. for $B = 2\pi n\,E/\mu_n kL$.

c) After s_4, the component of the wavefunction towards the counter C_1 is (the factor $\frac{1}{2}$ is due to the fact that the amplitude of the wavefunction either transmitted or reflected by a semi-transparent beam splitter is $1/\sqrt{2}$ times the amplitude of the incident wavefunction):

$$\frac{1}{2}e^{ikx}\left[\begin{pmatrix}\alpha\,e^{i\varphi_+}\\ \beta\,e^{i\varphi_-}\end{pmatrix} + \begin{pmatrix}\alpha\\ \beta\end{pmatrix}\right] = \frac{1}{2}e^{ikx}\begin{pmatrix}\alpha\,(1 + e^{i\varphi_+})\\ \beta\,(1 + e^{i\varphi_-})\end{pmatrix} \quad \Rightarrow$$

$$N_1 = \frac{1}{4}N\Big(2|\alpha|^2(1 + \cos\varphi_+) + 2|\beta|^2(1 + \cos\varphi_-)\Big) = \frac{N}{2}(1 + \cos\tfrac{1}{2}\phi)\,.$$

d) The period of $N_1(\phi)$ is 4π, not 2π: after a rotation by 2π the spin state of the neutron is the same, but the vector – as seen above – changes its sign; therefore the component of the wavefunction that goes through the region with the magnetic field and the other one, that follows the alternative path, interfere destructively with each other: $N_1(2\pi) = 0$.

$$B = \frac{4\pi\, E}{\mu_n k L} = \frac{4\pi\, h^2}{2m_n \mu_n \lambda^2} \times \frac{\lambda}{2\pi\, L} = \frac{h^2}{m_n \mu_n \lambda L} \simeq 149\,\mathrm{G}\,.$$

13.7

a) $H = E \;\Rightarrow\; p(x) = \sqrt{2mE} + f(x)\,; \qquad \dot{q} = \dfrac{\partial H}{\partial p} = \dfrac{1}{m}\big(p - f(x)\big) \quad\Rightarrow$

$m\dot{q} = \sqrt{2mE}\,.$

$-\mathrm{i}\,\hbar\, \psi_E'(x) = \big(\sqrt{2mE} + f(x)\big)\, \psi_E(x) \quad\Rightarrow$

$\psi_E(x) = \mathrm{e}^{\,\mathrm{i}\int_0^x \left(\sqrt{2mE}+f(x')\right)\,\mathrm{d}x'/\hbar} = \mathrm{e}^{\,\mathrm{i}\int_0^x p(x')\,\mathrm{d}x'/\hbar}\,.$

So the wave number $k(x)$ is proportional to the canonical momentum, not to the velocity: $k(x) = p(x)/\hbar$. This is particularly important in interference problems, where the phase of the wavefunction plays a crucial role (see Problems 3.5 and 13.13).

b) $\dot{\vec{q}} = \dfrac{\mathrm{i}}{\hbar}\,[H\,,\,\vec{q}\,] = \dfrac{1}{m}\big(\vec{p} - \hbar\,\nabla\Phi(x,y,z)\big)\,.$

c) The transformation $\tilde{q}_i = q_i$, $\tilde{p}_i = p_i - \hbar\,\partial_i\Phi(x,y,z)$ (gauge transformation) is canonical: $[\tilde{p}_i\,,\,\tilde{p}_j] = 0$ thanks to $\partial^2\Phi(x,y,z)/\partial x_i\partial x_j = \partial^2\Phi(x,y,z)/\partial x_j\partial x_i$. Therefore, thanks to the von Neumann theorem, U exists and depends only on the q_i: $U = \mathrm{e}^{\,\mathrm{i}\,\Phi(x,y,z)}\,.$

d) $\psi_E(\vec{r}) = \mathrm{e}^{\,\mathrm{i}\,\Phi(x,y,z)}\,\psi_E^0(\vec{r}) = \mathrm{e}^{\,\mathrm{i}\left(\vec{k}\cdot\vec{r}+\Phi(x,y,z)\right)}\,, \qquad \hbar k = \sqrt{2mE} = mv\,.$

13.8

a) One has:

$$H = \frac{1}{2m}\sum_{i=1}^{3}\Big(p_i + \frac{e}{c}\,A_i(\vec{q}\,)\Big)^2 + V(\vec{q}\,) \quad\Rightarrow$$

$$\dot{q}_i = \frac{\mathrm{i}}{\hbar}\,[H\,,\,q_i] = \frac{1}{m}\Big(p_i + \frac{e}{c}\,A_i(\vec{q}\,)\Big)\,,$$

$$[\dot{q}_i\,,\,\dot{q}_j] = -\mathrm{i}\,\frac{e\,\hbar}{m^2 c}\Big(\frac{\partial A_j}{\partial x_i} - \frac{\partial A_i}{\partial x_j}\Big) = -\mathrm{i}\,\frac{e\,\hbar}{m^2 c}\,\epsilon_{ijk}\,B_k\,;$$

$$[\dot{q}_i\,,\,q_j] = -\mathrm{i}\,\frac{\hbar}{m}\,\delta_{ij}\,.$$

b) One has:

$$H^{(2)} = \frac{1}{2m}\Big(\vec{p} + \frac{e}{c}\,\vec{A}_2(\vec{q}\,)\Big)^2 + V(\vec{q}\,) = \frac{1}{2m}\Big(\vec{p} + \frac{e}{c}\,\nabla\Lambda + \frac{e}{c}\,\vec{A}_1(\vec{q}\,)\Big)^2 + V(\vec{q}\,)\,.$$

The transformation $\tilde{q}_i = q_i$, $\tilde{p}_i = p_i + f_i(\vec{q}\,)$ is canonical if and only if $f_i = \partial F/\partial q_i$ $\big([\tilde{p}_i\,,\,\tilde{p}_j] = 0\big)$. In this case $\tilde{p}_i = U\,p_i\,U^{-1}$ with $U = \mathrm{e}^{-\mathrm{i}\,F(\vec{q}\,)/\hbar}$. Whence:

$$H^{(2)} = \mathrm{e}^{-\mathrm{i}\,e\,\Lambda/\hbar c}\,H^{(1)}\,\mathrm{e}^{\,\mathrm{i}\,e\,\Lambda/\hbar c}\,.$$

c) Since Λ does not depend on t, $UH^{(1)}\psi_1 = i\hbar \partial U\psi_1/\partial t$, but
$$UH^{(1)}\psi_1 = UH^{(1)}U^{-1}U\psi_1 = H^{(2)}U\psi_1 \, .$$
Therefore $\psi_2(\vec{r}, t) = U\psi_1(\vec{r}, t) = e^{-i e \Lambda(\vec{r})/\hbar c}\psi_1(\vec{r}, t)$.

d) The verification can be made, starting from the time dependent Schrödinger equation, as in the case where $\vec{A} = 0$.
Let $\vec{j}_1$ and $\vec{j}_2$ be the probability currents relative to the vector potentials
$\vec{A}_1(\vec{r})$ and $\vec{A}_2(\vec{r}) = \vec{A}_1(\vec{r}) + \nabla\Lambda(\vec{r})$. One has:
$$\vec{j}_2 = -i\frac{\hbar}{2m}\left[\psi_2^*\left(\nabla + i\frac{e}{\hbar c}(\vec{A}_1 + \nabla\Lambda)\right)\psi_2 - \psi_2\left(\nabla - i\frac{e}{\hbar c}(\vec{A}_1 + \nabla\Lambda)\right)\psi_2^*\right]$$
where $\psi_2(\vec{r}, t) = e^{-i e \Lambda(\vec{r})/\hbar c}\psi_1(\vec{r}, t)$ with ψ_1 and ψ_2 solving the respec-
tive Schrödinger equations. As $\nabla\psi_2 = (-i e \nabla\Lambda/\hbar c)\psi_2 + \cdots$, one obtains
$\vec{j}_2 = \vec{j}_1$: indeed $(-i\hbar/m)(\nabla + (i e/\hbar c)\vec{A})$ is the velocity operator and is
gauge invariant.

13.9

a) $\vec{A}_2 = \vec{A}_1 + \dfrac{B}{2}\nabla(-a_2 x + a_1 y)$; $\vec{A}_3 = \vec{A}_1 + \dfrac{B}{2}\nabla(x y)$.

$\vec{A}_1$, $\vec{A}_2$ and $\vec{A}_3$ depend only on the variables x and y.

b) $H = \dfrac{1}{2m_e}\left[\left(p_x - \dfrac{e B}{2c}y\right)^2 + \left(p_y + \dfrac{e B}{2c}x\right)^2\right]$.

The Hamiltonian given in the text obtains by expanding the squares.
The eigenvalues of H_0 are $E_n^0 = (n + 1)\hbar\omega_{\rm L}$ with degeneracy $n + 1$
and the simultaneous eigenvectors of H^0 and L_z: $|n, m\rangle$, $m = -n$,
$-n + 2, \cdots n - 2, n$, also are eigenvectors of H belonging to the eigenvalues $E_N = (N + 1)\hbar\omega_{\rm L}$, $N = n + m$. Since m has the same parity
as n, N is even, so the distance between adjacent levels is $\Delta E = 2\hbar\omega_{\rm L}$
and all the states $|n, m\rangle$, for which $n + m$ is constant, are degenerate:
for example, the level with $N = 0$ is obtained from $n = 0$, $m = 0$; $n =$
1, $m = -1$; $n = 2$, $m = -2 \cdots$ and has, therefore, infinite degeneracy;
likewise for the other energy levels, as illustrated by the following table
where the numbers between parentheses are (n, m):

$$
\begin{array}{llllllll}
N = 0: & (0, 0); & (1, -1); & (2, -2); & (3, -3); & (4, -4); & (5, -5); & \cdots \\
N = 2: & (1, 1); & (2, 0); & (3, -1); & (4, -2); & (5, -3); & (6, -4); & \cdots \\
N = 4: & (2, 2); & (3, 1); & (4, 0); & (5, -1); & (6, -2); & (7, -3); & \cdots \\
N = 6: & (3, 3); & (4, 2); & (5, 1); & (6, 0); & (7, -1); & (8, -2); & \cdots \\
\vdots
\end{array}
$$

If the cyclotron frequency $\omega_{\rm c} = 2\omega_{\rm L} = e B/m_e c$ is introduced, the Landau
levels are given by $E_n = (n + \frac{1}{2})\hbar\omega_{\rm c}$, $n = 0, 1, \cdots$ (see Problem 2.9).

c) The Hamiltonian is $H = H_0 + \omega_{\rm L}(L_z + g s_z)$. The term $g\omega_{\rm L}s_z$ commutes with $H_0 + \omega_{\rm L}L_z$ therefore, as the eigenvalues of s_z are $\pm\frac{1}{2}\hbar$, the
eigenvalues of H are:

$$E_{n,s} = (n + \tfrac{1}{2} + \tfrac{1}{2} g\, s)\, \hbar\omega_c, \quad n = 0, 1, \cdots ; \; s = \pm\tfrac{1}{2} .$$

As $g \simeq 2.002 > 2$, the structure of the energy levels is that represented in the right part of the figure (not a scale drawing). In the approximation in which $g = 2$, the levels $E_{n,\,s=+\frac{1}{2}}$ and $E_{n+1,\,s=-\frac{1}{2}}$ coincide.

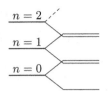

13.10

a) The radius of the orbit is v/ω_c ($\omega_c = eB/m_e c$ is the cyclotron frequency) and the vector $(x - x_c,\; y - y_c)$ is orthogonal to $\vec{v}$ so, since the electron moves counterclockwise:

$$x_c = x - \frac{v_y}{\omega_c} , \qquad y_c = y + \frac{v_x}{\omega_c} .$$

b) From Problem 13.8 $[v_x , v_y] = -\mathrm{i}\, e\, \hbar\, B/m_e^2 c = -\mathrm{i}\, \hbar\omega_c/m_e$, whence:

$$[x_c , y_c] = \mathrm{i}\, \frac{\hbar}{m_e \omega_c} .$$

In addition, always from Problem 13.8, $[x_c , v_x] = [x_c , v_y] = [y_c , v_x] = [y_c , v_y] = 0$, so we may put:

$$P = m_e v_x , \quad Q = \frac{1}{\omega_c} v_y \quad \Rightarrow \quad [Q , P] = \mathrm{i}\, \hbar$$

$$p = m_e \omega_c x_c , \quad q = y_c \quad \Rightarrow \quad [q , p] = \mathrm{i}\, \hbar .$$

c) $H = \dfrac{1}{2} m_e (v_x^2 + v_y^2) = \dfrac{P^2}{2m_e} + \dfrac{1}{2} m_e \omega_c^2 Q^2$

so the eigenvalues are those of a one-dimensional harmonic oscillator with angular frequency ω_c. In addition, p and q, therefore also x_c and y_c, obviously are constants of motion since they do not appear in H: for this reason each eigenvalue of H has as eigenspace the Hilbert space of a particle in one dimension: the eigenfunctions of H have the form $\Psi_E(X, x) = \psi_n(X) \times \phi(x)$, with $\psi_n(X)$ standing for the eigenfunctions of the harmonic oscillator and $\phi(x)$ an arbitrary function in $L^2(\mathbb{R})$.

d) $L_z^{\mathrm{kin}} = m_e (v_x^2 + v_y^2)/\omega_c = 2H/\omega_c$ whose eigenvalues are $(2n+1)\,\hbar$, where $n = 0, 1, \cdots$: note that they are non negative and spaced by $2\,\hbar$.

e) Since the point with coordinates x_c, y_c is classically constrained within the area A, the volume in the phase space of the variables p and q is $\mathcal{A} = m_e \omega_c A$. In order to estimate the number of quantum states associated with the variables p and q, one assumes (see Problem 2.6) that each state occupies, in the phase space, the volume h, so the degeneracy of each Landau level is $\mathcal{A}/h \simeq (e\, B/h\, c) \times A = 2.4 \times 10^{10}$.

13.11

a) The Hamiltonian is (see Problem 13.9) $H = \frac{1}{2} m \vec{v}^2 + \omega_L\, g\, s_z$, with $\omega_L = e\,B/2mc$. One has:

$$\dot{\sigma}_x = i\,\hbar^{-1}[H\,,\,\sigma_x] = i\,\hbar^{-1}\omega_L\, g\,[s_z\,,\,\sigma_x] = -g\,\omega_L\,\sigma_y\,, \qquad \dot{\sigma}_y = g\,\omega_L\,\sigma_x$$

so the angular velocity of the spin precession is $\omega_{\rm spin} = g\,\omega_L$.

b) Thanks to the commutation rules of $\dot{q}_i$ and $\dot{q}_j$ (see Problem 13.8) and $[s_i\,,\,\sigma_j] = i\,\hbar\,\epsilon_{ijk}\,\sigma_k$ one has:

$$[H\,,\,\sigma_j\,\dot{q}_j] = [\frac{1}{2} m\,\dot{q}_i\,\dot{q}_i + g\,\frac{e}{2mc}\,B_i\,s_i\,,\,\sigma_j\,\dot{q}_j]$$

$$= -i\,\frac{e\,\hbar}{mc}\,\epsilon_{ijk}\,\dot{q}_i\,\sigma_j\,B_k + i\,g\,\frac{e\,\hbar}{2mc}\,\epsilon_{ijk}\,B_i\,\dot{q}_j\,\sigma_k = i\,\hbar\,(g-2)\,\frac{e}{2mc}\,\vec{B}\cdot\vec{v}\wedge\vec{\sigma}\,.$$

c) $\dot{q}_x = i\,\hbar^{-1}[H\,,\,q_x] = -\dfrac{e\,B}{mc}\,q_y = -2\omega_L\,q_y\,, \qquad \dot{q}_y = 2\omega_L\,q_x$

so the angular velocity is $\omega_c \equiv 2\,\omega_L$, as expected. As a consequence:

$$\phi(t) = (\omega_{\rm spin} - \omega_c)\,t = (g-2)\,\omega_L\,t = \frac{g-2}{2}\,\omega_c\,t\,.$$

d) $\omega_c = \dfrac{e\,B}{2m_\mu\gamma(v)\,c} = 4.25\times10^7\,{\rm s}^{-1}\,, \qquad a_\mu = \dfrac{\omega_a}{\omega_c} = 1.2\times10^{-3}\,.$

The theoretical and the experimental values coincide at least up to the fifth significant digit: $a_\mu = 1.1659\times10^{-3}$.

13.12

a) As a result of the minimal substitution on the variables $\vec{p}_1$, $\vec{p}_2$ we have ($\gamma \equiv e\,B/2c$):

$$P_x \to P_x - \gamma\,q_y\,, \quad P_y \to P_y + \gamma\,q_x\,, \quad P_z \to P_z$$

$$p_x \to p_x - \gamma\left(Q_y + \frac{m_2 - m_1}{M}\,q_y\right)\,, \quad p_y \to p_y + \gamma\left(Q_x + \frac{m_2 - m_1}{M}\,q_x\right)\,,$$

$$p_z \to p_z\,.$$

The Hamiltonian in the absence of magnetic field is

$$H_0 = \frac{\vec{P}^{\,2}}{2M} + \frac{\vec{p}^{\,2}}{2\mu} + V(q)\,, \qquad M = m_1 + m_2\,, \quad \mu = \frac{m_1 m_2}{m_1 + m_2}$$

whence ($\delta m \equiv m_2 - m_1$):

$$H = \frac{1}{2M}\left[(P_x - \gamma\,q_y)^2 + (P_y + \gamma\,q_x)^2 + P_z^2\right]$$

$$+ \frac{1}{2\mu}\left[\left(p_x - \gamma\left(Q_y + \frac{\delta m}{M}\,q_y\right)\right)^2 + \left(p_y + \gamma\left(Q_x + \frac{\delta m}{M}\,q_x\right)\right)^2 + p_z^2\right] + V(q)\,.$$

Terms proportional to $\gamma\,(P_x q_y - P_y q_x)$ and $\gamma\,(p_x Q_y - p_y Q_x)$ appear: therefore only the motion parallel to $\vec{B}$ of the center of mass is separable.

b) $\widetilde{Q}_i = Q_i \, ; \qquad \widetilde{P}_x = P_x - \gamma \, q_y \, , \quad \widetilde{P}_y = P_y + \gamma \, q_x \, , \quad \widetilde{P}_z = P_z$

$\widetilde{q}_i = q_i \, ; \qquad \widetilde{p}_x = p_x + \gamma \, Q_y \, , \quad \widetilde{p}_y = p_y - \gamma \, Q_x \, , \quad \widetilde{p}_z = p_z$

whence $\widetilde{H}$ given in the text. Note that, in $\widetilde{H}$, the terms $\gamma \, p_i \, Q_j$ are absent, but the terms $\gamma \, P_i \, q_j$ are present, so not even $\widetilde{H}$ is separable. By using U^{-1} instead of U, it is possible to eliminate the second terms instead of the first, but it is not possible to dispose of both. One has:

$$\dot{Q}_x = \frac{\partial \widetilde{H}}{\partial P_x} = \frac{1}{M} (P_x - 2\gamma \, q_y), \quad \dot{Q}_y = \frac{\partial \widetilde{H}}{\partial P_y} = \frac{1}{M} (P_y + 2\gamma \, q_x), \quad \dot{Q}_z = \frac{1}{M} P_z$$

and since P_x, P_y, P_z (but not q_x, q_y) commute with $\widetilde{H}$, only P_x, P_y, P_z and $\dot{Q}_z$ are constants of the motion.

c) Up to terms of the order of $m_e/m_p \simeq 5.4 \times 10^{-4}$, the term we are interested in is $eB/(2m_e c) \, L_z$ that is diagonal in the basis $| \, n, l, m \, \rangle$, so any energy level E_n, $n = 1, 2, \cdots$, of the hydrogen atom splits up into a multiplet of levels (Zeeman multiplet):

$$E_n \to E_n + \frac{e \hbar B}{2 m_e c} \, m \, ; \qquad m = 0, \pm 1, \pm 2, \cdots, \pm (n-1)$$

respectively with degeneracies n, $(n-1)$, $\cdots$, 1.
The Bohr magneton $\mu_B \equiv e\hbar/2m_e c$ has the numerical value $\mu_B \simeq 0.93 \times 10^{-20} \mathrm{erg/G} \simeq 5.8 \times 10^{-9} \mathrm{eV/G}$ so, with $B = 10^4$ G, $\Delta E = \mu_B \, B \simeq 5.8 \times 10^{-5} \mathrm{eV}$. The term quadratic in B produces, on the first energy levels of the atom ($q_x^2 + q_y^2 \simeq a_B^2$) effects of the order of

$$\frac{e^2 B^2}{8 m c^2} \, a_B^2 = \frac{1}{2} \Big(\frac{e \hbar B}{2mc} \Big)^2 \frac{m e^2}{\hbar^2} \frac{a_B^2}{e^2} = \frac{1}{2} \frac{(\Delta E)^2}{e^2/a_B} \simeq 10^{-6} \Delta E \, .$$

d) The thermal motion velocity V of the center of mass is of the order of

$$V \simeq \sqrt{k_B T/M} = c \sqrt{k_B T/Mc^2} \quad \Rightarrow \quad \frac{V}{c} \simeq 5 \times 10^{-6}$$

and $2\gamma \, q/Mc \simeq 1.7 \times 10^{-11}$, so $P/M \simeq V$.
   So the term $(eB/Mc)(q_x P_y - q_y P_x)$ can be interpreted as the interaction $e \vec{q} \cdot \vec{E}$ of the atom with the electric field $\vec{E} = (\vec{V}/c) \wedge \vec{B}$, which is the electric field in the center-of-mass frame; as $VB/c \simeq 5 \times 10^{-2} \mathrm{esu} = 15 \, \mathrm{V/cm}$, the effect on the first energy levels of the hydrogen atom is of the order $10^{-7} \mathrm{eV}$ (see Problem 12.13).

13.13

a) $\vec{A}$ is continuous on the surface of the solenoid and, out of it, curl $\vec{A} = 0$. Thanks to Stokes theorem, the line integral of $\vec{A}$ is given by the flux of $\vec{B}$ through the surface, namely $\pi \, a^2 B$.

b) Along each of the two paths the problem is one-dimensional, with Hamiltonian $H = \big(p + (e/c) A_t \big)^2 / 2m$ (A_t is the component of $\vec{A}$ along the path), so it is identical with the problem discussed in question a) of Problem 13.7. Alternatively: since the region where $\operatorname{curl}\vec{A} = 0$ is not simply connected, out of the solenoid $\vec{A}(\vec{r})$ is the gradient of a multivalued function Φ ($\Phi = a^2 B\phi/2\sqrt{x^2 + y^2}$, where ϕ is the azimuth angle around the axis of the solenoid); however, limiting to simply connected regions, as the two single circuits γ_1 and γ_2, the function Φ is one valued (Φ_1 on γ_1, Φ_2 on γ_2) and, as a consequence, the problem is the same as the three-dimensional one discussed in 13.7.

c) The difference in phase φ is given by:

$$\varphi = \int_{\gamma_1} \left(\vec{k} + (e/\hbar c)\,\vec{A}\right) \cdot d\vec{l}_1 - \int_{\gamma_2} \left(\vec{k} + (e/\hbar c)\,\vec{A}\right) \cdot d\vec{l}_2$$

$$= \frac{e}{\hbar c} \oint \vec{A}(\vec{r}) \cdot d\vec{l} = \frac{\pi a^2 e B}{\hbar c}$$

(the line integral of $\vec{k}$ is vanishing because, by assumption, the interferometer is well calibrated). It is remarkable that the phase difference is proportional to the flux of $\vec{B}$, even if only regions where $\vec{B} = 0$ are accessible to the electrons: this fact, known as Aharonov–Bohm effect, has been experimentally verified. One has (see Problem 13.6):

$$I_1 = \frac{I}{2}(1 + \cos\varphi), \qquad I_2 = \frac{I}{2}(1 - \cos\varphi).$$

14

Electromagnetic Transitions

Coherent and incoherent radiation; photoelectric effect; transitions in dipole approximation; angular distribution and polarization of the emitted radiation; lifetimes.

14.1 An anisotropic harmonic oscillator, of Hamiltonian:

$$H_0 = \frac{\vec{p}^{\,2}}{2m} + \frac{1}{2}m\left(\omega_1^2\, x^2 + \omega_2^2\, y^2 + \omega_3^2\, z^2\right)$$

and charge $-e$ is subject, in the time interval $0 \le t \le \tau$, to a coherent electromagnetic wave generated by a laser, whose electric field is

$$\vec{\mathcal{E}}(t) = \left(\mathcal{E}_0 \sin(k\,y - \omega\,t), 0, 0\right).$$

For $t \le 0$ the oscillator is in the ground state $|\,0,0,0\,\rangle$ (notation $|\,n_1, n_2, n_3\,\rangle$: see Problems 10.6 and 11.4).

a) Assume $\omega \simeq \omega_1$. Calculate, in the dipole approximation and to the first order of time dependent perturbation theory, the probability $P_{1\leftarrow0}(t)$ to find the oscillator in the excited state $|\,1,0,0\,\rangle$, both for $0 < t < \tau$ and for $t \ge \tau$ (here and in the following, neglect the terms whose denominators contain $\omega + \omega_1$ with respect to those whose denominators contain $\omega - \omega_1$). Is it possible, in this case, to define a transition probability per unit time independent of t? What is the value, to the first order and for $t \ge \tau$, of the transition probability to the states $|\,n_1, 0, 0\,\rangle$, $n_1 > 1$?

b) Calculate numerically, for $t \ge \tau$, the probability $P_{1\leftarrow0}$ in the case the particle is an electron, the electromagnetic wave is generated by a He–Ne laser that emits the frequency $\nu = \omega/2\pi = 453 \times 10^{12}\,\mathrm{Hz}$, $\omega_1 = \omega$, $\tau = 10^{-4}\,\mathrm{s}$, in the two cases when the intensities are:

$$I_1 \equiv \frac{c}{8\pi}\,\mathcal{E}_0^2 = 10^{-2}\,\mathrm{erg/cm^2\,s}, \qquad I_2 = 10^{-3}\,\mathrm{erg/cm^2\,s}\,.$$

Say whether, with the above data, the dipole approximation is legitimate and whether such is the perturbative approximation to the first order.

Instead of sending on the oscillator the coherent radiation of a laser, assume now to send light with the same frequency and polarization, and intensity

© Springer International Publishing AG 2017   283
E. d'Emilio and L.E. Picasso, *Problems in Quantum Mechanics*,
UNITEXT for Physics, DOI 10.1007/978-3-319-53267-7_14

$I = I_2 = 10^{-3}\,\mathrm{erg/cm^2\,s}$, but generated by a lamp that emits incoherent radiation, namely wave packets (photons) each having a time duration $\tau_{\mathrm{ph}} \simeq 10^{-8}\,\mathrm{s}$. The spectral intensity $I(\omega_1)$ may be evaluated assuming that the total intensity I is uniformly distributed over a frequency interval $\Delta\omega \simeq 1/\tau_{\mathrm{ph}}$.

c) Find the expression for the first order transition probability per unit time $W_{1\leftarrow0}$ from the ground state to the state $|1,0,0\rangle$. Compare, at time $t = \tau$, the transition probability in the case of the incoherent radiation with the transition probability in the case of coherent radiation.

14.2 Consider, as in Problem 14.1, a charged harmonic oscillator (charge $-e$) of Hamiltonian:

$$H_0 = \frac{\vec{p}^2}{2m} + \frac{1}{2}m(\omega_1^2\,x^2 + \omega_2^2\,y^2 + \omega_3^2\,z^2)$$

subject, in the time interval $0 \le t \le \tau$, to a coherent electromagnetic wave whose electric field is

$$\vec{\mathcal{E}}(t) = (\mathcal{E}_0\sin(k\,y - \omega\,t), 0, 0)\,.$$

For $t \le 0$ the oscillator is in the ground state $|0,0,0\rangle$ (notation $|n_1, n_2, n_3\rangle$). In the dipole approximation, thanks to the results of Problem 7.16, it is possible to find exactly the state of the system at time t and to make a comparison with the perturbative results obtained in Problem 14.1.

a) Find the state of the oscillator for $t \ge \tau$.

b) Assume $\omega \simeq \omega_1$. Exploit the result found in a) to determine the probability $P_{1\leftarrow0}(t \ge \tau)$ to find the oscillator in the excited state $|1,0,0\rangle$. Make the numerical calculation with the same data of Problem 14.1: $m = m_e$, $\nu = \omega/2\pi = 453 \times 10^{12}\,\mathrm{Hz}$, $\omega_1 = \omega$, $\tau = 10^{-4}\,\mathrm{s}$, $I_1 = 10^{-2}\,\mathrm{erg/cm^2\,s}$, $I_2 = 10^{-3}\,\mathrm{erg/cm^2\,s}$. Compare the results with those obtained in the Problem 14.1 to the first order of perturbation theory. Calculate, in addition, the total probability (for $t \ge \tau$) to find the oscillator in any state $|n_1, 0, 0\rangle$, $n_1 > 1$.

c) Draw a graph of the probability $P_{1\leftarrow0}$ as a function of the intensity I of the radiation and find the maximum value $P_{1\leftarrow0}$ may have.

d) For $I = 10\,I_1$, for which, among the states of the oscillator, is the transition probability from the ground state a maximum?

14.3 A system has the ground state of energy E_0 and a continuum of states of energies $E_1 - \frac{1}{2}\Delta E \le E \le E_1 + \frac{1}{2}\Delta E$, nondegenerate and normalized according to $\langle E' | E''\rangle = \delta(E' - E'')$. The system is subject, for $t \ge 0$, to a coherent electromagnetic wave of frequency $\omega = (E_1 - E_0)/\hbar$ whose electric field is

$$\vec{\mathcal{E}}(t) = (\mathcal{E}_0\sin(k\,z - \omega\,t), 0, 0)\,.$$

For $t < 0$ the system is in its ground state $|E_0\rangle$.

a) Show that, to the first order, the total transition probability $P(t)$ to the states of energy $E_1 - \frac{1}{2}\Delta E \leq E \leq E_1 + \frac{1}{2}\Delta E$ is given by:

$$P(t) = \hbar^{-2} \int_{E_1-\frac{1}{2}\Delta E}^{E_1+\frac{1}{2}\Delta E} dE \left| \int_0^t \langle E \mid H'(t') \mid E_0 \rangle e^{i(E-E_0)t'/\hbar} dt' \right|^2$$

where $H'(t)$ is the Hamiltonian of interaction between the system and the electromagnetic field.

Assume that, in the given energy interval, the matrix elements $\langle E \mid D_x \mid E_0 \rangle$ of the dipole moment operator are independent of E.

b) Calculate $P(t)$ in the dipole approximation and show that for $t \gg \hbar/\Delta E$ it is possible to define the transition probability per unit time $W \equiv dP(t)/dt$, independent of t.

Assume now that the system is a hydrogen atom and that $\omega > E_I/\hbar$, where E_I is the ionization energy (photoelectric effect); take the states $\mid E, l, m \rangle$ ($E > 0$) as a basis for the states of the continuum. Also in this case it can be assumed that, in a small energy interval ΔE, the matrix elements $\langle E, l, m \mid D_x \mid E_0 \rangle$ of the dipole moment operator are independent of E (but not of l and of m).

c) Show that the transition probability per unit time to the states of the continuum with fixed l and m is given by:

$$W = \frac{\pi \mathcal{E}_0^2}{2\hbar} \left| \langle E_f, l, m \mid D_x \mid E_0 \rangle \right|^2.$$

Find the energy E_f of the emitted electron and say for which values of l, m the probability W is nonvanishing.

d) Find the angular distribution of the emitted electrons with respect to the direction of propagation of the incident wave (i.e. the dependence on the angles of the probability of detecting an electron within the solid angle $d\Omega$).

14.4 A three-dimensional isotropic harmonic oscillator (charge $-e$) is in the ground state. Electromagnetic radiation induces, in the dipole approximation, a transition to the first excited level E_1. Let $\hat{e}$ be the complex unit vector describing the polarization state of the radiation (denoted by $\hat{e}_{\vartheta\varphi}$ in Chapters 3 and 4).

a) Find, among the states of the level E_1, the state (notation $\mid n_1, n_2, n_3 \rangle$ as in Problems and 14.1 and 14.2):

$$\mid E_1, \hat{a} \rangle = \alpha_x \mid 1,0,0 \rangle + \alpha_y \mid 0,1,0 \rangle + \alpha_z \mid 0,0,1 \rangle, \quad |\alpha_x|^2 + |\alpha_y|^2 + |\alpha_z|^2 = 1$$

for which the transition probability is a maximum and demonstrate the vanishing of the transition probability to the states belonging to the same level E_1 and orthogonal to it.

b) Exploit the transversality of the electromagnetic waves and find the direction of propagation of the wave that induces the transition to the state $|E_1, \alpha\rangle$, in the general case in which $\hat{\alpha} = (\alpha_x, \alpha_y, \alpha_z)$ is a complex unit vector: $\hat{\alpha} = \hat{\alpha}_1 + i\hat{\alpha}_2$, with $\hat{\alpha}_1$ and $\hat{\alpha}_2$ real and not parallel to each other.

The radiation emitted in a given direction of observation $\hat{n}$ in the transition between two fixed states is polarized and the polarization vector $\hat{e}$ is the vector that maximizes the transition probability, subject to the transversality condition $\hat{e} \cdot \hat{n} = 0$.

c) Find the polarization of the emitted radiation (spontaneous emission) in the transition from the state $|E_1, \hat{\alpha}\rangle$ to the ground state, as a function of the direction of observation $\hat{n}$. Still assuming that $\hat{\alpha} = \hat{\alpha}_1 + i\hat{\alpha}_2$ with $\hat{\alpha}_1$ and $\hat{\alpha}_2$ not parallel to each other, show that there exist two directions of observation in which the radiation is linearly polarized.

d) Find the angular distribution of the emitted radiation. In which direction is the intensity $I(\hat{n})$ a maximum and how is, in this direction, the radiation polarized?

14.5   Consider a three-dimensional isotropic harmonic oscillator of mass m, angular frequency ω and charge $-e$, initially in a state $|E_1, \hat{\alpha}\rangle$ of the first excited level:

$$|E_1, \hat{\alpha}\rangle = \alpha_x\,|1,0,0\rangle + \alpha_y\,|0,1,0\rangle + \alpha_z\,|0,0,1\rangle, \quad |\alpha_x|^2 + |\alpha_y|^2 + |\alpha_z|^2 = 1.$$

The oscillator makes a transition to the ground state by spontaneous emission.

a) Show that the transition probability per unit time $w_{0\leftarrow1} = \int I(\hat{n})\,\mathrm{d}\Omega$ is independent of the initial state (i.e. does not depend on $\hat{\alpha} = (\alpha_x, \alpha_y, \alpha_z)$).

b) Calculate the lifetime (the reciprocal of the transition probability per unit time) of the first excited energy level. Make the numerical calculation in the case the particle is an electron and $\hbar\omega = 2\,\mathrm{eV}$.

Assume now, for the sake of simplicity, that the oscillator is one-dimensional (namely a three-dimensional anisotropic oscillator with $\omega_2, \omega_3 \gg \omega_1$).

c) Calculate the transition probability for the spontaneous emission from the level E_n to the level E_{n-1} and show that, if $n \gg 1$ (and the approximation of sums by integrals is therefore legitimate), the average time for the transition $n \to n-1 \to n-2 \cdots \to n/e$ coincides with the lifetime of the classical oscillator (see Problem 1.2) and with the lifetime of the transition $E_1 \to E_0$.

14.6   Inside a cavity atoms with two nondegenerate energy levels E_a and E_b, $E_a < E_b$, are in thermal equilibrium with the radiation (*black body radiation*).

a) At which temperature $\overline{T}$ of the cavity does the transition probability between the two states of energies E_a, E_b, induced by the radiation in the cavity, equal the probability of spontaneous emission $E_b \to E_a$? Make the numerical calculation in the case $E_b - E_a = 1\,\mathrm{eV}$.

The black body temperature typical of a lamp used in the laboratory to induce transitions in an atomic system is about $T_L = 3000\,\mathrm{K}$.

b) What is the value of the ratio between the spectral intensity $I(\omega_{ba}, T)$ $(\omega_{ba} = (E_b - E_a)/\hbar)$ of a black body at the temperature $\overline{T}$ determined in a) and the spectral intensity $I(\omega_{ba}, T_L)$ of a lamp at the temperature $T_L = 3000\,\mathrm{K}$?

14.7   Vapors of sodium in a bottle of volume $V = 50\,\mathrm{cm}^3$, at the pressure $10^{-1}\,\mathrm{Pa}$, are kept at the temperature $T = 3000\,\mathrm{K}$. The first excited energy level of the sodium atom is separated by $2.1\,\mathrm{eV}$ from the lowest energy level, its degeneracy is three times that of the lowest energy level, its lifetime is $\tau = 1.6 \times 10^{-8}\,\mathrm{s}$.

a) Find the average number of atoms in the first excited energy level.

b) Calculate the power emitted in the transitions from the first excited energy level to the lowest energy level.

14.8   A hydrogen atom is in a static uniform electric field $\vec{\mathcal{E}}$. It is known that, due to the perturbation, the level $n = 2$ splits up into three sublevels E_2, $E_2^{(+)} \equiv E_2 + \Delta E$, $E_2^{(-)} \equiv E_2 - \Delta E$ (see also Problem 12.13). Assume that the atom initially is in the ground state and that radiation in a direction orthogonal to the static field $\vec{\mathcal{E}}$ is sent on it.

a) Say how many absorption lines are observed in the dipole transitions $n = 1 \to n = 2$:   *i*) if the radiation is polarized in the direction parallel to the static field $\vec{\mathcal{E}}$,   *ii*) if the radiation is polarized in a direction orthogonal to the field $\vec{\mathcal{E}}$; which is, in this case, the final state of the atom that has undergone the transition?   *iii*) How many lines are observed if the radiation is circularly polarized?

Assume now that the radiation is sent in the direction parallel to the static field $\vec{\mathcal{E}}$ and that it is not polarized.

b) How many absorption lines $n = 1 \to n = 2$ are observed in this case?

c) Show that, if the radiation is sufficiently monochromatic, it is possible to induce the transition from the ground state to an arbitrary stationary state with $n = 2$ of the atom in the static field $\vec{\mathcal{E}}$. How small must the degree of monochromaticity $\Delta\nu/\nu$ of the incident radiation be?

Consider now the transitions $n = 1 \to n = 3$. From Problem 12.13 one knows that, due to the perturbation, the level $n = 3$ splits up into five sublevels.

d) Show that only four (not five) absorption lines can be observed and that it is possible to observe simultaneously all of them by means of polarized radiation.

14.9   A hydrogen atom is subject to the perturbation:

$$V = \frac{V_0}{a_B^3}\, x y z\,, \qquad (x,\, y,\, z) \equiv \vec{r} = \vec{r}_e - \vec{r}_p\,; \qquad a_B = \frac{\hbar^2}{m_e e^2}\,.$$

From Problem 12.15 one knows that the level $n = 3$ splits up into the three sublevels: E_3, $E_3 \pm \Delta E$ ($\Delta E \equiv A$), each of them three times degenerate.

a) Exploit the results of Problem 12.15 and write the states corresponding to these sublevels.

b) How many absorption lines are observed in the transitions $n = 1 \to n = 3$?

c) Radiation rectilinearly polarized along the z axis and with frequency corresponding to the transition $E_1 \to E_3 + \Delta E$ is sent on the atom in the ground state. Which is the state of the atom that has undergone the transition?

14.10   Atomic hydrogen in the gaseous state, in conditions of pressure and temperature such that the interactions among different atoms can be neglected, has been excited in such a way that the atoms in the level $n = 2$ are described, in the basis $|\,2, 0, 0\,\rangle$, $|\,2, 1, m_z\,\rangle$ $(m_z = 0, +1, -1)$, by the density matrix:

$$\varrho^{at} = \begin{pmatrix} \varrho_0 & 0 & 0 & 0 \\ 0 & \varrho_3 & 0 & 0 \\ 0 & 0 & \varrho_+ & 0 \\ 0 & 0 & 0 & \varrho_- \end{pmatrix}.$$

The radiation due to spontaneous emission is observed from the z direction.

a) Write the density matrix relative to state of polarization of the observed photons and find the degree of polarization (see Problem 5.2) of the radiation.

b) One now observes the radiation emitted in the direction of the z axis and that crosses a polaroid filter whose transmission axis is parallel to the x axis. If $I^{(z)}$ is the intensity of the radiation emitted in the direction of the z axis, find the intensity of the observed radiation.

c) The polaroid filter is now rotated by an angle α around the z axis. How does the observed intensity depend on the angle α?

Radiation emitted in the direction $\hat{n} = (\sin\theta\cos\phi,\ \sin\theta\sin\phi,\ \cos\theta)$ is observed (θ and ϕ polar angles with respect to the z-axis).

d) Find the angular distribution of the radiation, namely how the intensity $I(\theta, \phi)$ depends on the angles.

Let $\hat{e}_{n1} = (-\cos\theta\cos\phi, -\cos\theta\sin\phi, \sin\theta)$ be the unit vector orthogonal to $\hat{n}$ in the plane containing $\hat{n}$ and the z axis, and $\hat{e}_{n2} = \hat{n} \wedge \hat{e}_{n1} = (\sin\phi, -\cos\phi, 0)$.

e) Write the density matrix relative to the state of polarization of the photons emitted in the direction $\hat{n}$ and calculate the ratio between the intensities $I_1(\theta, \phi)$, $I_2(\theta, \phi)$ of the radiation that has crossed a polaroid filter with transmission axis parallel to $\hat{e}_{n1}$ and of the radiation that has crossed a polaroid filter with transmission axis parallel to $\hat{e}_{n2}$.

14.11   Cesium atoms ($Z = 55$, $A = 133$) exhibit a doublet of levels E_1 and E_2 separated from the lowest energy level E_0 respectively by $\lambda_1^{-1} = 11178\,\mathrm{cm}^{-1}$ and by $\lambda_2^{-1} = 11732\,\mathrm{cm}^{-1}$. Vapors of cesium at room temperature and pressure $10^{-4}\,\mathrm{Pa}$ are available in a container whose linear dimensions l are some centimeters. The system is irradiated with radiation of frequency $(E_2 - E_0)/h$ and spectral width $\Delta\nu/\nu \simeq 10^{-5}$, so that only the level E_2 is populated. The radiation scattered by the atoms (spontaneous emission to the lowest energy level) is observed.

a) Verify that for transitions in the optical region one has $\hbar\omega \lesssim m_e c^2\alpha^2$ (α is the fine structure constant) and that the lifetime of a level is estimated by $\tau_{\mathrm{rad}}^{-1} \simeq \omega\alpha^3$. Estimate the lifetime of the level E_2 and compare the result with the experimental value $\tau_{\mathrm{rad}}^{\mathrm{exp}} \simeq 30.5\,\mathrm{ns}$.

b) Recalling that the free mean path of an atom is $\bar{\lambda} = (n\sigma)^{-1}$, where n is the number of atoms per unit volume and σ the cross section of the atom, evaluate, in the given conditions, both the average collisional time τ_{coll} between two atoms, and the time τ_{walls} between two consecutive collisions with the walls and compare them with $\tau_{\mathrm{rad}}^{\mathrm{exp}} \simeq 30.5\,\mathrm{ns}$.

c) How many lines of spontaneous emission are seen in the described conditions?

Helium at the pressure $10^4\,\mathrm{Pa}$ is added to the cesium vapor. The presence of the noble gas has the unique effect of increasing the number of collisions of cesium atoms.

d) Calculate the average collisional time among cesium atoms and helium atoms and say how many lines of spontaneous emission are observed.

14.12   The lifetime of the $2p$ states of the hydrogen is $\tau_{\mathrm{H}} = 1.6 \times 10^{-9}\,\mathrm{s}$.

a) Calculate the lifetime of the $2p$ states of the hydrogen-like ions He^+, Li^{++}, C^{VI}.

14.13   An atom decays by spontaneous emission from a state $|E_{J=0}\rangle$, with angular momentum $J = 0$, to a level $E_{J=1}$ of angular momentum $J = 1$.

a) Within the dipole approximation, write the expression of the transition probability per unit time to the state $|E_{J=1},\ J_z = M\rangle$ (assume the matrix elements of the electric dipole operator $\vec{D}$ are known).

b) Write the expression for the transition probability per unit time from the level $E_{J=0}$ to the level $E_{J=1}$.

c) Find the angular distribution of the emitted radiation.

d) Say whether the atoms, that have undergone the transition, are in a pure state (namely all in one and the same state of the level $E_{J=1}$); in the affirmative case, find the state of the atoms; otherwise, find the density matrix relative to the final state of the atoms.

e) Say whether the photons emitted in a given direction $\hat{n}$ are polarized; in the affirmative case, find the state of polarization; otherwise, find the density matrix relative to the state of polarization of the photons.

Solutions

14.1

a) In the dipole approximation $\sin(k\,y - \omega t) \to -\sin\omega t$, therefore in this framework the problem is a separate variables one and boils down to that of a one-dimensional harmonic oscillator of angular frequency ω_1 subject to the electric field $\mathcal{E}(t) = -\mathcal{E}_0 \sin\omega t$. To the first order in perturbation theory:

$$P_{f\leftarrow i}(t) = \hbar^{-2} \left| \int_0^t \langle E_f \mid H'(t') \mid E_i \rangle \, e^{i\,(E_f - E_i)\,t'/\hbar}\, dt' \right|^2$$

where:

$$H'(t') = -e\,q\,\mathcal{E}_0 \sin\omega t' = -\frac{i}{2}\,e\,q\,\mathcal{E}_0 \left(e^{-i\omega t'} - e^{i\omega t'} \right), \qquad t' \leq \tau.$$

The term $\propto e^{i\omega t'}$ can be neglected: indeed, after integrating with respect to t', it gives rise to a term, whose denominator is $\omega + \omega_1$, which for $\omega \simeq \omega_1$ is important only in the emission processes. Therefore for $t \leq \tau$ one has:

$$P_{1\leftarrow 0}(t) = \frac{e^2 \mathcal{E}_0^2}{4\hbar^2} |\langle 1 \mid q \mid 0 \rangle|^2 \, \frac{\sin^2\left[\frac{1}{2}(\omega_1 - \omega)t\right]}{\left[\frac{1}{2}(\omega_1 - \omega)\right]^2}$$

$$= \frac{e^2 \mathcal{E}_0^2 \, t^2}{8\hbar\, m\omega_1} \, \frac{\sin^2\left[\frac{1}{2}(\omega_1 - \omega)t\right]}{\left[\frac{1}{2}(\omega_1 - \omega)t\right]^2}$$

and for $t \geq \tau$:

$$P_{1\leftarrow 0}(t) = \frac{e^2 \mathcal{E}_0^2 \, \tau^2}{8\hbar\, m\omega_1} \, \frac{\sin^2\left[\frac{1}{2}(\omega_1 - \omega)\tau\right]}{\left[\frac{1}{2}(\omega_1 - \omega)\tau\right]^2}.$$

For $|\omega - \omega_1| \ll 1/\tau$, $\quad \sin^2\left[\frac{1}{2}(\omega_1 - \omega)t\right] / \left[\frac{1}{2}(\omega_1 - \omega)t\right]^2 \simeq 1 \qquad 0 \leq t \leq \tau$.

In this case (transitions between discrete energy levels induced by coherent radiation) it is not possible to define a transition rate: for $t \geq \tau$, $P_{1\leftarrow 0}(t)$

is constant and the transition rate vanishes; for $t < \tau$, $P_{1\leftarrow 0}(t)$ is proportional to t^2 and the rate $dP_{1\leftarrow 0}(t)/dt$ is not constant. The transition probability to the states with $n_1 > 1$ is vanishing since $\langle n_1 \mid q \mid 0 \rangle = 0$; in any event, for systems for which the matrix element of the dipole operator is nonvanishing, for $\omega \simeq \omega_1$ the factor $\sin^2\left[\tfrac{1}{2}(\omega_{n_1} - \omega)\tau\right] / \left[\tfrac{1}{2}(\omega_{n_1} - \omega)\tau\right]^2$ is practically vanishing.

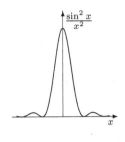

b) For $\omega = \omega_1$ and $t > \tau$, if I is the intensity of the radiation and $\alpha = e^2/\hbar c \simeq 1/137$ the fine structure constant:

$$P_{1\leftarrow 0} = \frac{\pi e^2 I \tau^2}{mc\,\hbar\omega_1} = \alpha\,\frac{\pi I \tau^2}{m\omega_1}\,.$$

With the data given in the text and $m = m_e$:

$$I_1 : P_{1\leftarrow 0}(t > \tau) \simeq 0.9;\qquad I_2 : P_{1\leftarrow 0}(t > \tau) \simeq 0.09\,.$$

The first order approximation of perturbation theory requires that the transition probability be $\ll 1$: this condition is not fulfilled in the first case ($I = I_1$), whereas it can be considered acceptable in the second. The dipole approximation requires that:

$$\lambda \equiv \frac{2\pi c}{\omega} \simeq 6.6 \times 10^{-5}\,\mathrm{cm} \gg \sqrt{\frac{\hbar}{2m_e\,\omega}} \simeq 1.4 \times 10^{-8}\,\mathrm{cm}$$

so it is legitimate ($\sqrt{\hbar/2m_e\,\omega}$ is the characteristic length of the oscillator).

c) In the case of incoherent light of low intensity it is possible to define the transition rate: it is given by

$$W_{1\leftarrow 0} = \frac{4\pi^2 e^2}{\hbar^2 c}\,I(\omega_1)\,|\langle 1 \mid q \mid 0 \rangle|^2 = \alpha\,\frac{2\pi^2 I(\omega_1)}{m\omega_1} \simeq \alpha\,\frac{2\pi^2 I\,\tau_{\mathrm{ph}}}{m\omega_1}$$

so $P_{1\leftarrow 0}(\tau)/W_{1\leftarrow 0} \times \tau = \tau/2\pi\tau_{\mathrm{ph}} = 1.6 \times 10^3$. Therefore, time and intensity being equal, the transition probability is much higher with the coherent radiation.

14.2

a) In the dipole approximation the problem boils down to that of a one-dimensional harmonic oscillator with angular frequency ω_1 subject to the external force $F(t) = e\,\mathcal{E}_0 \sin\omega t$. From Problem 7.16 we know that for $t \geq \tau$ the state of the oscillator is the coherent state $|\gamma(t)\rangle$ (we shall not use the letter α, as in Problem 7.16, to avoid confusion with the fine structure constant), with:

$$\gamma(t \geq \tau) = \frac{e^{-i\omega_1 t}}{\sqrt{2m\hbar\omega_1}} \int_0^\tau e^{i\omega_1 t'}\,e\,\mathcal{E}(t')\,dt' = \frac{e\,\mathcal{E}_0\,e^{-i\omega_1 t}}{\sqrt{2m\hbar\omega_1}} \int_0^\tau e^{i\omega_1 t'}\sin\omega\,t'\,dt'$$

$$= \frac{e\,\mathcal{E}_0}{2\sqrt{2m\hbar\omega_1}}\left(\frac{e^{i(\omega_1-\omega)\tau} - 1}{\omega_1 - \omega} - \frac{e^{i(\omega_1+\omega)\tau} - 1}{\omega_1 + \omega}\right)e^{-i\omega_1 t}\,. \qquad (1)$$

b) As $\omega \simeq \omega_1$, the neglect of the second term in (1) is legitimate. Putting $\gamma \equiv \gamma(t \geq \tau)$, one has (see Problem 5.9):

$$P_{0\leftarrow 0} = e^{-|\gamma|^2}; \qquad P_{1\leftarrow 0} = |\gamma|^2 e^{-|\gamma|^2}$$

$$|\gamma|^2 = \frac{e^2 \mathcal{E}_0^2}{8m\hbar\omega_1} \tau^2 \frac{\sin^2\left[\frac{1}{2}(\omega_1 - \omega)\tau\right]}{\left[\frac{1}{2}(\omega_1 - \omega)\tau\right]^2} \simeq \alpha \frac{\pi I \tau^2}{m\omega_1}$$

so (between parentheses the first order result):

$$I_1 \; : \; |\gamma|^2 = 0.89 \,, \quad \begin{cases} P_{0\leftarrow 0} = 0.4 \,; \qquad P_{1\leftarrow 0} = 0.366 \;\; (0.89) \\ \sum_{n_1 > 1} P_{n_1 \leftarrow 0} = 1 - (0.4 + 0.37) = 0.23 \,; \end{cases}$$

$$I_2 \; : \; |\gamma|^2 = 0.09 \,, \quad \begin{cases} P_{0\leftarrow 0} = 0.91 \,; \qquad P_{1\leftarrow 0} = 0.08 \;\; (0.09) \\ \sum_{n_1 > 1} P_{n_1 \leftarrow 0} \lesssim 0.01 \,. \end{cases}$$

This result confirms that, in the first case, the perturbative result is totally unreliable.

c) Putting $x = (\alpha \pi \tau^2/m\omega_1) I$, the plot of $P_{1\leftarrow 0}$ versus I is the graph of the function $x e^{-x}$ that has a maximum for $x = 1$, corresponding to the intensity $I = 1.12 \times 10^{-2} \,\mathrm{erg/cm^2\,s}$ and $P_{1\leftarrow 0}$ equal to $1/e = 0.368$.

d) $P_{n\leftarrow 0} = (|\gamma|^{2n}/n!) \, e^{-|\gamma|^2}$. Since $x^{n+1}/(n+1)! = (x^n/n!) \times (x/n+1)$ the probability grows with n until $n+1 < |\gamma|^2$, therefore it has a maximum when n equals the integer part of $|\gamma|^2$. For $I = 10\,I_1$, $|\gamma|^2 = 8.9$, so the transition probability is a maximum for $n = 8$ and equals 0.133 (for $n = 7$ its value is 0.120 for $n = 9$ its value is 0.132).

14.3

a) Since the states $|\,E\,\rangle$ are improper states and, as a consequence, the transition probability $|\,E_0\,\rangle \to |\,E\,\rangle$ is meaningless, we proceed ab initio. If $U(t)$ is the time evolution operator and $|\,n\,\rangle$ whatever orthonormal basis in the subspace $\mathcal{H}_{\Delta E}$ of the states with energy $E_1 - \frac{1}{2}\Delta E \leq E \leq E_1 + \frac{1}{2}\Delta E$, (the states $|\,n\,\rangle$ are not stationary states), owing to the von Neumann postulate ("sum over the final states") one has:

$$P(t) = \sum_n |\langle\, n \mid U(t) \mid E_0 \,\rangle|^2 = \langle\, E_0 \mid U^\dagger(t) \mathcal{P}_{\Delta E} \, U(t) \mid E_0 \,\rangle$$

where:

$$\mathcal{P}_{\Delta E} \equiv \sum_n |\, n\,\rangle\langle\, n\,| = \int_{E_1 - \frac{1}{2}\Delta E}^{E_1 + \frac{1}{2}\Delta E} |\, E\,\rangle \, \mathrm{d}E \, \langle\, E\,|$$

is the projection operator onto $\mathcal{H}_{\Delta E}$. As the time evolution operator $U_0(t) = e^{-\mathrm{i}\,H_0\, t/\hbar}$ of the free system commutes with $\mathcal{P}_{\Delta E}$, if $\widetilde{U}(t) \equiv U_0^\dagger(t)\, U(t)$ is the time evolution operator in the interaction picture (see Problem 7.11), one has:

$$P(t) = \int_{E_1 - \frac{1}{2}\Delta E}^{E_1 + \frac{1}{2}\Delta E} |\langle\, E \mid \widetilde{U}(t) \mid E_0 \,\rangle|^2 \, \mathrm{d}E \,.$$

To the first order:

$$\tilde{U}(t) \simeq \mathbb{1} - \frac{i}{\hbar} \int_0^t \tilde{H}'(t')\, dt', \qquad \tilde{H}'(t) = U_0^\dagger(t)\, H'(t)\, U_0(t)$$

whence the formula given in the text.

b) In the dipole approximation $\sin(k\,z - \omega t) \to -\sin \omega t$:

$$H'(t') = D_x\, \mathcal{E}_0 \sin \omega t' = \frac{i}{2} D_x\, \mathcal{E}_0 \left(e^{-i\omega t'} - e^{i\omega t'} \right)$$

so, neglecting the term $\propto e^{i\omega t'}$ that is relevant only in an emission process, and putting:

$$\omega' \equiv (E - E_0)/\hbar - \omega = (E - E_1)/\hbar, \quad x \equiv \tfrac{1}{2}\omega' t, \quad \Delta x \equiv \tfrac{1}{2} \Delta E \times t/\hbar$$

$$P(t) = \frac{\mathcal{E}_0^2}{4\hbar^2} \int_{E_1 - \frac{1}{2}\Delta E}^{E_1 + \frac{1}{2}\Delta E} \left| \langle\, E \mid D_x \mid E_0 \,\rangle \right|^2 \frac{\sin^2 \frac{1}{2}\omega' t}{(\frac{1}{2}\omega')^2}\, dE$$

$$= \frac{\left| \langle\, E_1 \mid D_x \mid E_0 \,\rangle \right|^2 \mathcal{E}_0^2\, t}{2\hbar} \int_{-\Delta x}^{+\Delta x} \frac{\sin^2 x}{x^2}\, dx$$

and since $t \gg \hbar/\Delta E \Rightarrow \Delta x \gg 1$, the integral is extended from $-\infty$ to $+\infty$ and equals π; therefore:

$$W = \frac{d\,P(t)}{dt} = \frac{\pi \mathcal{E}_0^2 \left| \langle\, E_1 \mid D_x \mid E_0 \,\rangle \right|^2}{2\hbar} = \frac{4\pi^2\, I \left| \langle\, E_1 \mid D_x \mid E_0 \,\rangle \right|^2}{c\,\hbar}$$

where $I \equiv c\mathcal{E}_0^2/8\pi$ is the intensity of the electromagnetic wave.

c) Comparing with the preceding case, it suffices to replace $\mid E \,\rangle$ by $\mid E, l, m \,\rangle$, with l and m fixed. In this way the expression given in the text obtains, with $E_f = \hbar\omega - E_I$. Since the ground state has $l = 0$, thanks to the selection rules for the operator D_x, the final state must have $l = 1$ and $m_x = 0$ (if instead m refers to the z component of the angular momentum, the final state is a superposition of states with $m = \pm 1$).

d) As the electron is emitted in a state with $l = 1$ and $m_x = 0$, the angular part of the electron wavefunction is $x/r = \sin\theta \cos\phi$, where the polar axis is the z axis, i.e. the direction of propagation of the incident wave. So the intensity $I(\theta, \phi)$ of the photoelectrons is proportional to $\sin^2 \theta \cos^2 \phi$, i.e. (in the dipole approximation) the photoelectrons are emitted mainly in the direction of the polarization of the wave.

14.4

a) In the dipole approximation the transition probability to the state $\mid E_1, \hat{a} \,\rangle$ is proportional to $\left| \langle\, E_1, \hat{a} \mid \vec{D}\cdot\hat{e} \mid E_0 \,\rangle \right|^2 \equiv |\vec{d}\cdot\hat{e}|^2$, $\vec{d} \equiv -e\langle\, E_1, \alpha \mid \vec{q} \mid E_0 \,\rangle$. One has:

$$\langle\, 0,0,0 \mid q_i \mid E_1, \hat{a} \,\rangle = i \sqrt{\frac{\hbar}{2m\omega}}\, \alpha_i \quad \Rightarrow \quad \vec{d} \propto (\alpha_x^*, \alpha_y^*, \alpha_z^*) \equiv \hat{a}^*$$

and $|\hat{a}^* \cdot \hat{e}|^2$ is a maximum (and equals 1) for $\hat{a} = \hat{e}$.

If $|\,E_1,\,\hat{\beta}\,\rangle \equiv \beta_x\,|\,1,0,0\,\rangle + \beta_y\,|\,0,1,0\,\rangle + \beta_z\,|\,0,0,1\,\rangle$ is orthogonal to $|\,E_1,\,\hat{\alpha}\,\rangle$, one has $\hat{\beta}^*\cdot\hat{\alpha} = 0$ and, as a consequence, $\langle\,E_1,\,\hat{\beta}\,|\,\vec{D}\cdot\hat{e}\,|\,E_0\,\rangle \propto \hat{\beta}^*\cdot\hat{e} = \hat{\beta}^*\cdot\hat{\alpha} = 0$.

b) If $\hat{k}$ stands for the direction of propagation of the wave, one must have $0 = \hat{k}\cdot\hat{e} = \hat{k}\cdot\hat{\alpha}$. As $\hat{\alpha} = \hat{\alpha}_1 + \mathrm{i}\hat{\alpha}_2$, one has $\hat{k}\cdot\hat{\alpha}_1 = \hat{k}\cdot\hat{\alpha}_2 = 0$ namely $\hat{k}\propto\hat{\alpha}_1\wedge\hat{\alpha}_2$.

c) The emission probability is proportional to:

$$|\langle\,E_0\,|\,\vec{D}\cdot\hat{e}^*\,|\,E_1,\,\hat{\alpha}\,\rangle|^2 \equiv |\vec{d}\cdot\hat{e}^*|^2, \quad \vec{d}\equiv -e\langle\,E_0\,|\,\vec{q}\,|\,E_1,\,\hat{\alpha}\,\rangle, \quad \hat{e}\cdot\hat{n} = 0\,.$$

Its maximum with the condition $\hat{e}\cdot\hat{n} = 0$ can be found by means of the Lagrange multipliers and one obtains (as in the classical theory of radiation, see Problem 1.1):

$$\hat{e} = \frac{\vec{d} - (\vec{d}\cdot\hat{n})\,\hat{n}}{\sqrt{|\vec{d}|^2 - |\vec{d}\cdot\hat{n}|^2}} = \frac{\hat{\alpha} - (\hat{\alpha}\cdot\hat{n})\,\hat{n}}{\sqrt{|\hat{\alpha}|^2 - |\hat{\alpha}\cdot\hat{n}|^2}}\,, \qquad \hat{e}^*\cdot\hat{e} = 1\,.$$

The radiation is linearly polarized if $\hat{e}$ is proportional to a real vector: this happens if $\hat{n}$ is parallel to either $\hat{\alpha}_1$ or $\hat{\alpha}_2$:

$$\hat{n}\,\|\,\hat{\alpha}_1 \quad\Rightarrow\quad \hat{\alpha}_1 - (\hat{\alpha}_1\cdot\hat{n})\,\hat{n} = 0 \quad\Rightarrow\quad \hat{e} = \mathrm{i}\,\frac{\hat{\alpha}_2 - (\hat{\alpha}_2\cdot\hat{n})\,\hat{n}}{\sqrt{|\hat{\alpha}_2|^2 - |\hat{\alpha}_2\cdot\hat{n}|^2}}\,.$$

d) It is sufficient to insert the (complex) unit vector $\hat{e}$ found above into the expression for the probability transition:

$$I(\hat{n}) \propto \frac{\omega_{\mathrm{f}i}^3}{h\,c^3}\left(|\vec{d}|^2 - |\vec{d}\cdot\hat{n}|^2\right), \qquad \vec{d} = -\mathrm{i}\,e\,\sqrt{\frac{\hbar}{2m\omega}}\,\hat{\alpha}\,, \quad \omega_{\mathrm{f}i} = \omega \quad\Rightarrow$$

$$I(\hat{n}) \propto \frac{e^2\omega^2}{4\pi\,mc^3}\left(1 - |\hat{\alpha}\cdot\hat{n}|^2\right)\,.$$

$I(\hat{n})$ is maximum for $\hat{\alpha}\cdot\hat{n} = 0$, namely along the direction $\hat{n} = \hat{k}\propto\hat{\alpha}_1\wedge\hat{\alpha}_2$ of propagation of the wave that induces the transition to the state $|\,E_1,\,\hat{\alpha}\,\rangle$, as well as in the opposite direction: in these directions the polarization is identical with that of the incident wave: $\hat{e} = \hat{\alpha}$.

14.5

a) The transition rate from the generic state $|\,E_1,\,\hat{\alpha}\,\rangle$ belonging to the first excited level to the ground state is given by (see Problem 14.4):

$$w_{0\leftarrow1} = \frac{e^2\omega^2}{4\pi\,mc^3}\int\left(1 - |\hat{\alpha}\cdot\hat{n}|^2\right)\mathrm{d}\Omega = \frac{e^2\omega^2}{4\pi\,mc^3}\left(4\pi - \alpha_i^*\,\alpha_j\int n_i\,n_j\,\mathrm{d}\Omega\right)\,.$$

Taking the z axis as the polar axis, one has:

$$n_x = \sin\theta\,\cos\phi\,, \quad n_y = \sin\theta\,\sin\phi\,, \quad n_z = \cos\theta\,.$$

For $i\neq j$ the integration over ϕ gives zero; for $i = j$:

$$\int n_x^2\,\mathrm{d}\Omega = \int n_y^2\,\mathrm{d}\Omega = \int n_z^2\,\mathrm{d}\Omega = \frac{4}{3}\,\pi \quad\Rightarrow\quad \int n_i\,n_j\,\mathrm{d}\Omega = \frac{4}{3}\,\pi\,\delta_{ij}$$

(alternatively: the integration over the angles of the tensor $n_i n_j$ must give rise to an isotropic – namely rotationally invariant – tensor, i.e. proportional to δ_{ij}; as the trace of the tensor $n_i n_j$ is a scalar, the proportionality constant is $(4/3)\pi$). Then:

$$w_{0 \leftarrow 1} = \frac{e^2 \omega^2}{4\pi\, mc^3}\left(4\pi - \frac{4}{3}\pi\, |\hat{a}|^2\right) = \frac{2e^2\omega^2}{3mc^3}\,.$$

b) $w_{0 \leftarrow 1} = 5.8 \times 10^7\,\mathrm{s}^{-1} \quad \Rightarrow \quad \tau = w_{0 \leftarrow 1}^{-1} = \dfrac{3mc^3}{2e^2\omega^2} = 1.7 \times 10^{-8}\,\mathrm{s}\,.$

c) As $\langle n{-}1 \mid x \mid n \rangle = \mathrm{i}\sqrt{\dfrac{\hbar}{2m\omega}}\,\sqrt{n}$, one has $w_{n-1\leftarrow n} = n\,\dfrac{2e^2\omega^2}{3mc^3}$. Therefore:

$$\tau(n \to n/e) = \tau(n \to n-1) + \tau(n-1 \to n-2) + \cdots$$
$$= \frac{3mc^3}{2e^2\omega^2}\sum_{n/e}^{n}\frac{1}{k} \simeq \frac{3mc^3}{2e^2\omega^2}\int_{n/e}^{n}\frac{1}{x}\,\mathrm{d}x = \frac{3mc^3}{2e^2\omega^2}\,.$$

14.6

a) Let $W_{b \leftarrow a} = W_{a \leftarrow b} \equiv W_{ab}$ be the probability for the induced transition between the two states and w_{ab} the spontaneous emission probability (integrated over the angles). The condition of thermal equilibrium among radiation and matter (the atoms we are considering) entails that:

$$N_a W_{ab} = N_b (W_{ab} + w_{ab})\,, \qquad \frac{N_a}{N_b} = e^{(E_b - E_a)/k_B T}$$

so the condition $W_{ab} = w_{ab}$ requires $N_a/N_b = 2 \Rightarrow (E_b - E_a)/k_B \overline{T} = \log 2 \Rightarrow \overline{T} = 1.67 \times 10^4\,\mathrm{K}\,.$

b) The intensity is proportional to the energy density:

$$u(\omega) = \frac{\hbar\,\omega^3}{\pi^2 c^3 (e^{\hbar\omega/k_B T} - 1)} \qquad \Rightarrow$$

$$\frac{I(\omega_{ba}, \overline{T})}{I(\omega_{ba}, T_L)} = \frac{e^{\hbar\omega_{ba}/k_B T_L} - 1}{e^{\hbar\omega_{ba}/k_B \overline{T}} - 1} = \frac{2^{\overline{T}/T_L} - 1}{2 - 1} = 46 \simeq \left(\frac{N_a}{N_b}\right)_{T_L}.$$

Therefore, since $w_{ab}/W_{ab} = N_a/N_b - 1$, in 'normal' conditions ($T_L \simeq 10^3\,\mathrm{K}$) the spontaneous emissions prevails on the induced one.

14.7

a) At the temperature and pressure given in the text the sodium vapours behave as an ideal gas, so the number of atoms is given by:

$$N = N_A\,\frac{p\,V}{R\,T} = 6 \times 10^{23} \cdot 2 \times 10^{-10} = 1.2 \times 10^{14}\,.$$

The ratio between the numbers of atoms in the first excited level to those in the lowest energy level is $N_1/N_0 = 3\,e^{-\Delta E/k_B T} = 0.9 \times 10^{-3}$, therefore $N_1 \simeq 0.9 \times 10^{-3} N = 1.1 \times 10^{11}$.

b) Since $N_0/N_1 \gg 1$, as seen in the Problem 14.6, it is possible to neglect the induced emission. The number of transitions to the lowest energy level in the unit time, due to spontaneous emission, is N_1/τ; so the emitted power is

$$W = \Delta E \times \frac{N_1}{\tau} = 2.3 \, \text{watt} \, .$$

14.8

a) Let us take the z axis along the direction of $\vec{\mathcal{E}}$. In the first case ($\hat{e} \parallel \vec{\mathcal{E}}$), owing to the selection rules for the operator D_z, two lines are observed, corresponding to the transitions to the states $|\, 2,0,0\,\rangle \pm |\, 2,1,0\,\rangle$ (energies $E_2^{(\pm)}$). In the second case ($\hat{e} \perp \vec{\mathcal{E}}$) only one absorption line is observed (to the level E_2): taking the x axis along the direction of $\hat{e}$, owing to the selection rules for D_x the final state is $|\, n = 2, l = 1, m_x = 0\,\rangle$, namely that combination of the states $|\, 2,1,1\,\rangle$, $|\, 2,1,-1\,\rangle$ whose wavefunction in the Schrödinger representation is $(x/r)\, R_{2,1}(r)$. Finally, if the radiation is circularly polarized, the complex unit vector $\hat{e}$ has both the z and the x components, so all the three lines are observed.

b) Only the line corresponding to the transition to the level E_2 is observed, independently of whether the radiation be polarized or not.

c) If $\Delta\nu/\nu \ll \Delta E/E_2$, by means of radiation polarized in the z direction it is possible to induce the transition to either one of the states $|\, 2,0,0\,\rangle \pm |\, 2,1,0\,\rangle$. In order to induce the transition to the generic stationary state with wavefunction $(\alpha\, x/r + \beta\, y/r)\, R_{2,1}(r)$ it is necessary that the polarization vector of the radiation be (see Problem 14.4) $\hat{e} \propto (\alpha, \beta, 0)$, that in general corresponds to an elliptic polarization in the x-y plane (α and β are complex numbers); therefore in this case the radiation must propagate in the z direction.

d) From Problem 12.13 we know that the central level (energy E_3) has degeneracy three and all the states belonging to this level have parity $+1$: one is the state $B \,|\, 3,0,0\,\rangle - A^* \,|\, 3,2,0\,\rangle$, the other two are states with $m = \pm 2$, therefore with $l = 2$; due to the selection rule on the space inversion, the transition from the ground state to this level is forbidden in the dipole approximation. All the other levels are simultaneously accessible since the corresponding states have no definite parity and, as seen in a), it is possible to induce transitions with $\Delta m = 0$ and with $\Delta m = \pm 1$.

14.9

a) The states that generate the eigenspace $\mathcal{H}^{(3)}$ belonging to E_3 are:

$$|\, 3,0,0\,\rangle, \qquad |\, 3,2,0\,\rangle, \qquad \frac{1}{\sqrt{2}}\left(|\, 3,2,2\,\rangle + |\, 3,2,-2\,\rangle\right).$$

The linear combinations of (notations as in Problem 12.15):

$$\frac{1}{\sqrt{2}}\left(\,|\,\mathcal{P}_1\,\rangle \pm |\,\mathcal{D}_{2,3}\,\rangle\right), \quad \frac{1}{\sqrt{2}}\left(\,|\,\mathcal{P}_2\,\rangle \pm |\,\mathcal{D}_{3,1}\,\rangle\right), \quad \frac{1}{\sqrt{2}}\left(\,|\,\mathcal{P}_3\,\rangle \pm |\,\mathcal{D}_{1,2}\,\rangle\right)$$

respectively belong to $E_3 \pm \Delta E$.

b) As the states of the level E_3 all have parity $+1$, only the transitions to the levels $E_3 \pm \Delta E$ – that do not have a definite parity – are possible.

c) Since $\Delta m = 0$ the transition probability to the generic state:

$$\frac{\alpha}{\sqrt{2}}\left(\,|\,\mathcal{P}_1\,\rangle + |\,\mathcal{D}_{2,3}\,\rangle\right) + \frac{\beta}{\sqrt{2}}\left(\,|\,\mathcal{P}_2\,\rangle + |\,\mathcal{D}_{3,1}\,\rangle\right) + \frac{\gamma}{\sqrt{2}}\left(\,|\,\mathcal{P}_3\,\rangle + |\,\mathcal{D}_{1,2}\,\rangle\right)$$

is proportional to $|\gamma|^2$, therefore the final state is $\frac{1}{\sqrt{2}}\left(\,|\,\mathcal{P}_3\,\rangle + |\,\mathcal{D}_{1,2}\,\rangle\right)$.

14.10   The atoms in the state $|\,2,0,0\,\rangle$ do not decay radiatively to the ground state since the transitions $l = 0 \to l = 0$ are forbidden, to the first order in perturbation theory, at any order of the multipole expansion.

a) The intensity of the radiation emitted in the transition $|\,2,1,0\,\rangle \to |\,1,0,0\,\rangle$ is vanishing along the z axis, so the fraction $r_+ \equiv \varrho_+/(\varrho_+ + \varrho_-)$ of the observed photons is in the circular polarization state $|\,e_{\sigma_+}\,\rangle$ whereas the fraction $r_- \equiv \varrho_-/(\varrho_+ + \varrho_-)$ is in the polarization state $|\,e_{\sigma_-}\,\rangle$; therefore the density matrix is

$$\varrho^{\text{ph}} = r_+\,|\,e_{\sigma_+}\,\rangle\langle\,e_{\sigma_+}\,| + r_-\,|\,e_{\sigma_-}\,\rangle\langle\,e_{\sigma_-}\,|\,.$$

So the radiation is partially circularly polarized and the degree of polarization is $|r_+ - r_-|$.

b) The probability of observing photons polarized along the x axis (i.e. in the state $|\,e_1\,\rangle$) is

$$P_1 = r_+|\langle\,e_1\,|\,e_{\sigma_+}\,\rangle|^2 + r_-|\langle\,e_1\,|\,e_{\sigma_-}\,\rangle|^2 = \frac{1}{2}\,(r_+ + r_-) = \text{Tr}\left[\,\varrho^{\text{ph}}\,|\,e_1\,\rangle\langle\,e_1\,|\,\right]$$

$$\Rightarrow \quad I_1^{(z)} = \frac{I^{(z)}}{2}\,(r_+ + r_-)\,.$$

c) The matrix ϱ^{ph} is invariant under rotations around the z axis (as it must be, given that such is ϱ^{at}) so the intensity is independent of α.

d) The radiation emitted in the transitions with $\Delta m = 0$ has an angular distribution proportional to $\sin^2 \theta$, whereas that emitted in a transition with either $\Delta m = +1$ or $\Delta m = -1$ has an angular distribution proportional to $\frac{1}{2}(1 + \cos^2 \theta)$. So:

$$I(\theta, \phi) \propto \left(\frac{\varrho_3}{\varrho_+ + \varrho_- + \varrho_3}\,\sin^2 \theta + \frac{\varrho_+ + \varrho_-}{\varrho_+ + \varrho_- + \varrho_3}\,\frac{1 + \cos^2 \theta}{2}\right).$$

e) The photons emitted in the transitions with $\Delta m = 0$ are revealed in the direction $\hat{n}$ with probability proportional to $\sin^2 \theta$ and are in the polarization state $|\,e_{n1}\,\rangle$; those emitted in the transitions with $\Delta m = \pm 1$ are revealed with probability proportional to $\frac{1}{2}(1 + \cos^2 \theta)$ and are in the elliptic polarization states:

$$|e_{n\pm}\rangle = \frac{\cos\theta\,|e_{n1}\rangle \pm i\,|e_{n2}\rangle}{(1+\cos^2\theta)^{1/2}}\,.$$

The statistical matrix relative to the polarization state of the photons is therefore:

$$\varrho^{\mathrm{ph}}(\theta,\phi) \propto \varrho_3 \sin^2\theta\,|e_{n1}\rangle\langle e_{n1}|$$
$$+ \tfrac{1}{2}(1+\cos^2\theta)\big(\varrho_+\,|e_{n+}\rangle\langle e_{n+}| + \varrho_-\,|e_{n-}\rangle\langle e_{n-}|\big)$$

(the proportionality factor can be determined by requiring $\mathrm{Tr}\,\varrho^{\mathrm{ph}} = 1$). One has:

$$I_{1,2} \propto \mathrm{Tr}\left(\varrho^{\mathrm{ph}}\,|e_{n1,2}\rangle\langle e_{n1,2}|\right);\qquad \frac{I_1}{I_2} = \frac{2\varrho_3\sin^2\theta + (\varrho_+ + \varrho_-)\cos^2\theta}{\varrho_+ + \varrho_-}\,.$$

14.11

a) $\hbar\omega \simeq \dfrac{e^2}{a_{\mathrm{B}}} = \dfrac{m_e c^2 e^4}{\hbar^2 c^2} = m_e c^2 \alpha^2\ (\simeq 27\,\mathrm{eV})\,.$

$$\tau_{\mathrm{rad}}^{-1} = \frac{4\,\omega^3}{3\hbar c^3}\,|\langle f\,|\,\vec{D}\,|\,i\rangle|^2 \simeq \omega\left(\frac{m_e c^2\alpha^2}{\hbar}\right)^2 \frac{e^2 a_{\mathrm{B}}^2}{\hbar c^3} = \omega\,\alpha^3.$$

In the present case $\omega = (E_2 - E_0)/\hbar = 2\pi c/\lambda_2 = 2.2\times 10^{15}\,\mathrm{s}^{-1}$ and $\tau_{\mathrm{rad}} \simeq 1.2\,\mathrm{ns}$ (indeed, putting $\hbar\omega = e^2/a_{\mathrm{B}}$ we have overestimated $\hbar\omega$ by a factor $\simeq 20$).

b) $n = N_{\mathrm{A}}\,p/RT = 2.4\times 10^{10}\,\mathrm{cm}^{-3},\quad \sigma \simeq 10^{-16}\,\mathrm{cm}^2 \Rightarrow \bar\lambda \simeq 4\times 10^5\,\mathrm{cm}\,;$

$$\bar v = \sqrt{3\,k_{\mathrm{B}}\,T/M_{\mathrm{Cs}}} \simeq 2.3\times 10^4\,\mathrm{cm/s} \Rightarrow \tau_{\mathrm{coll}} \simeq \bar\lambda/\bar v \simeq 17\,\mathrm{s}\,;$$

$$\tau_{\mathrm{walls}} \simeq \frac{1}{\bar v} = 4.3\times 10^{-5}\,\mathrm{s} > 10^3\,\tau_{\mathrm{rad}}^{\mathrm{exp}}\,.$$

c) The effect of collisions is inducing changes in the state of the atoms, as in the processes of thermalization; as $\tau_{\mathrm{rad}}^{\mathrm{exp}} \ll \tau_{\mathrm{walls}}$, most of the atoms decay before colliding with the walls, so only the transition $E_2 \to E_0$ is observed.

d) The mean velocity of helium atoms is $(133/4)^{1/2} = 5.8$ times that of cesium atoms and the mean free path of cesium atoms is practically the same as that of helium atoms, so $\tau_{\mathrm{coll}} \simeq 17/(5.8\cdot 10^8) = 2.9\times 10^{-8}\,\mathrm{s}$. Since $\tau_{\mathrm{coll}} \simeq \tau_{\mathrm{rad}}^{\mathrm{exp}}$, a fraction of the cesium atoms in the energy level E_2 undergoes a collision that allows them to pass to the level E_1 before decaying to the ground state: in the latter case two lines are observed.

14.12

a) The transition probability for spontaneous emission is proportional to $\omega^3\,|\langle E_{\mathrm{f}}\,|\,\vec{D}\,|\,E_{\mathrm{i}}\rangle|^2$ so, as $\omega \propto Z^2$, $\langle E_{\mathrm{f}}\,|\,\vec{D}\,|\,E_{\mathrm{i}}\rangle \propto Z^{-1}$, $\tau \propto Z^{-4}$:

$$\tau_{\mathrm{He^+}} = 10^{-10}\,\mathrm{s}\,;\qquad \tau_{\mathrm{Li^{++}}} = 2\times 10^{-11}\,\mathrm{s}\,;\qquad \tau_{\mathrm{CVI}} = 1.2\times 10^{-12}\,\mathrm{s}\,.$$

14.13

a) From Problem 14.4 one has:

$$\Delta M = 0: \quad w_{\text{fi}} = \frac{\omega_{\text{fi}}^3 |\vec{d}|^2}{2\pi\,\hbar\,c^3} \int \sin^2\theta \, d\Omega = \frac{4\,\omega_{\text{fi}}^3 |\vec{d}|^2}{3\hbar\,c^3}$$

$$\Delta M = \pm 1: \quad w_{\text{fi}} = \frac{\omega_{\text{fi}}^3 |\vec{d}|^2}{2\pi\,\hbar\,c^3} \int \frac{1 + \cos^2\theta}{2} \, d\Omega = \frac{4\,\omega_{\text{fi}}^3 |\vec{d}|^2}{3\hbar\,c^3}$$

where $|\vec{d}|^2 = |\langle E_{J=1}, M \mid \vec{D} \mid E_{J=0}\rangle|^2$, independent of M (see Problem 14.5).

b) As now J_z is not measured, the probability transition is the sum of the probability transitions either to the single states with $M = \pm 1, 0$ calculated above, or to any other orthonormal basis belonging to the level $E_{J=1}$:

$$w_{E_{J=1}\leftarrow E_{J=0}} = \frac{4\,\omega_{\text{fi}}^3 |\vec{d}|^2}{\hbar\,c^3}\,.$$

c) If $\hat{n}$ stands for the direction of observation,

$$I(\hat{n}) = \frac{\omega_{\text{fi}}^3 |\vec{d}|^2}{2\pi\,\hbar\,c^3}\left(\sin^2\theta + 2\frac{1 + \cos^2\theta}{2}\right) = \frac{\omega_{\text{fi}}^3 |\vec{d}|^2}{\pi\,\hbar\,c^3}$$

therefore the emitted radiation is isotropic, as one must expect, since the initial state is spherically symmetric $(J = 0)$.

d) The state of the system atom+photon is well determined, but it is not so for the atom alone: if the atom were in the pure state:

$$|\hat{a}\rangle = \alpha_1 \mid M = 1\rangle + \alpha_0 \mid M = 0\rangle + \alpha_{-1} \mid M = -1\rangle$$

the emitted radiation would not be isotropic but, as we have seen in Problem 14.4, $I(\hat{n}) \propto (|\vec{d}|^2 - |\vec{d}\cdot\hat{n}|^2)$. As the transition probability integrated over the angles is independent of the final state, the statistical matrix, relative to the atoms that have effected the transition to the level $E_{J=1}$, is $\frac{1}{3} \times$ the projector onto the states of the level $E_{J=1}$.

e) Let us take the z axis coinciding with the direction $\hat{n}$ of observation. Then the 50% of the photons are circularly right polarized, the 50% circularly left polarized (transitions with $\Delta M = \pm 1$). So the statistical matrix is $\frac{1}{2} \times$ the identity matrix.

15

Composite Systems and Identical Particles

Rotational energy levels of polyatomic molecules; entangled states and density matrices; singlet and triplet states; composition of angular momenta; quantum fluctuations; EPR paradox; quantum teleportation.

15.1 Given a system of $N > 1$ particles, the total angular momentum $\vec{L} \equiv \sum_\alpha \vec{q}^{\,\alpha} \wedge \vec{p}^{\,\alpha}$ can be decomposed as the sum of the angular momentum $\vec{L}^{(\mathrm{cm})}$ with respect to the center of mass and the angular momentum $\vec{Q} \wedge \vec{P}$ of the center of mass, where $M \equiv \sum_\alpha m_\alpha$, $\vec{Q} \equiv \sum_\alpha m_\alpha \vec{q}^{\,\alpha}/M$ and $\vec{P} \equiv \sum_\alpha \vec{p}^{\,\alpha}$: $\vec{L} = \vec{L}^{(\mathrm{cm})} + \vec{Q} \wedge \vec{P}$.

a) Verify that:

$$\vec{L}^{(\mathrm{cm})} \equiv \vec{L} - \vec{Q} \wedge \vec{P} = \sum_\alpha (\vec{q}^{\,\alpha} - \vec{Q}) \wedge \vec{p}^{\,\alpha} = \sum_\alpha \vec{q}^{\,\alpha} \wedge \left(\vec{p}^{\,\alpha} - \frac{m_\alpha}{M}\vec{P} \right).$$

b) Verify that the components of $\vec{L}^{(\mathrm{cm})}$ commute with $\vec{Q}$ and $\vec{P}$ (and therefore with the components of $\vec{Q} \wedge \vec{P}$).

c) Verify that $[L_i^{(\mathrm{cm})}, L_j^{(\mathrm{cm})}] = \mathrm{i}\,\hbar\,\epsilon_{ijk} L_k^{(\mathrm{cm})}$.

d) Which, among the following operators, have the commutation relations of the vectors with $\vec{L}^{(\mathrm{cm})}$: $[L_i^{(\mathrm{cm})}, V_j] = \mathrm{i}\,\hbar\,\epsilon_{ijk} V_k$?

$$\vec{q}^{\,\alpha}, \quad \vec{p}^{\,\alpha}, \quad \vec{q}^{\,\alpha} - \vec{Q}, \quad \vec{p}^{\,\alpha} - \vec{P}, \quad \vec{q}^{\,\alpha} - \vec{q}^{\,\beta}, \quad \vec{p}^{\,\alpha} - \vec{p}^{\,\beta}.$$

15.2 In order to determine the rotational energy levels of a polyatomic molecule, it is legitimate, in a first approximation, to consider the molecule as a rigid body (see Problem 11.13).

Let $\vec{L}$ be the angular momentum of the molecule in the center-of-mass frame (the one denoted as $\vec{L}^{(\mathrm{cm})}$ in the previous problem) and L_i ($i = 1, 2, 3$), the components of $\vec{L}$ with respect to the fixed axes x, y, z. Let $\hat{\xi}, \hat{\eta}, \hat{\zeta}$ the unit vectors of the principal axes of inertia of the molecule (the frame comoving with the molecule or "mobile frame") and $L_\xi \equiv \vec{L}\cdot\hat{\xi}$, $L_\eta \equiv \vec{L}\cdot\hat{\eta}$, $L_\zeta \equiv \vec{L}\cdot\hat{\zeta}$ the components of $\vec{L}$ with respect to the mobile frame.

a) Consider a molecule having the form of an isosceles triangle, consisting of three atoms that, for the sake of simplicity, we shall assume identical and

© Springer International Publishing AG 2017   301
E. d'Emilio and L.E. Picasso, *Problems in Quantum Mechanics*,
UNITEXT for Physics, DOI 10.1007/978-3-319-53267-7_15

pointlike. Express the unit vectors $\hat{\xi}$, $\hat{\eta}$, $\hat{\zeta}$ of the mobile frame in terms of the positions $\vec{q}_1$, $\vec{q}_2$, $\vec{q}_3$ of the three atoms.

b) Derive the commutation rules $[L_i, L_\xi]$, $[L_i, L_\eta]$, $[L_i, L_\zeta]$ and the commutation rules among the components of $\vec{L}$ with respect to the mobile frame: $[L_\xi, L_\eta]$, etc. Verify the following identity:

$$\vec{L}^2 \equiv L_x^2 + L_y^2 + L_z^2 = L_\xi^2 + L_\eta^2 + L_\zeta^2 .$$

How many are the linearly independent states corresponding to a given eigenvalue $\hbar^2 l(l+1)$ of $\vec{L}^2$?

The energy of the molecule in the center-of-mass frame is

$$H = \frac{L_\xi^2}{2I_\xi} + \frac{L_\eta^2}{2I_\eta} + \frac{L_\zeta^2}{2I_\zeta}$$

where I_ξ, I_η, I_ζ are the principal moments of inertia.

c) In the case the principal moments of inertia are all different from one another (asymmetric case), say which of the following observables commute with H: L_i, L_ξ, L_η, L_ζ, $\vec{L}^2$. Give the quantum numbers by which the energy levels of the molecule can be classified and the relative degeneracies.

d) Classify and determine the energy levels of the molecule and give the relative degeneracies in the case $I_\xi = I_\eta = I_\zeta \equiv I$ (spherical case).

e) Classify and determine the energy levels of the molecule and give the relative degeneracies in the case $I_\xi = I_\eta \equiv I \neq I_\zeta$ (symmetric case).

f) What can one say about the spectrum of H when $I_\zeta \to 0$ (linear molecule)?

15.3   Consider a molecule in the approximation of rigid body (see the Problem 15.2) and let $I_\xi \neq I_\eta \neq I_\zeta$ (asymmetric case).

a) Find the energy levels of the molecule in the center-of-mass frame corresponding to the values $l = 0$ and $l = 1$ of the angular momentum.

Assume now the molecule is in a laboratory rotating with constant angular velocity $\vec{\omega}_0$. Thanks to the presence of external forces that, however, do not influence the rigid structure of the molecule, its total momentum is $\vec{P} = 0$.

b) Say how the energy levels found in a) are modified.

15.4   Consider the system of two particles 1 and 2. Let $|a_1\rangle$, $|b_1\rangle$ be two orthonormal states for particle 1, let $|c_2\rangle$, $|d_2\rangle$ be two orthonormal states for particle 2; let finally:

$$|X\rangle = \alpha\,|a_1, c_2\rangle + \beta\,|b_1, d_2\rangle, \qquad \alpha \neq 0, \quad \beta \neq 0 \qquad |\alpha|^2 + |\beta|^2 = 1$$

be a state of the composite system.

a) Calculate the mean value in $|X\rangle$ of a generic observable $\xi_1 = f(q_1, p_1)$ relative to particle 1. Does a state $|x_1\rangle$ of particle 1 exist such that the mean value of any observable ξ_1 in the state $|x_1\rangle$ equals the mean value in the state $|X\rangle$?

b) Show that the density matrix, defined by $\langle\!\langle \xi \rangle\!\rangle = \mathrm{Tr}\,(\varrho\,\xi)$ (see Problem 4.8), is unique:

$$\mathrm{Tr}\,(\varrho'\,\xi) = \mathrm{Tr}\,(\varrho''\,\xi) \quad \forall\,\xi \quad \Rightarrow \quad \varrho' = \varrho''.$$

Use the result and find the density matrix ϱ_1 describing the state of particle 1. Does the density matrix ϱ_1 correspond to a pure state?

c) Write the density matrix $\varrho_{1,2}$ relative to the composite system in the state $|X\rangle$ and show that $\varrho_1 = \mathrm{Tr}_2\,\varrho_{1,2} \equiv \sum_{n_2}\langle n_2\,|\,\varrho_{1,2}\,|\,n_2\rangle$, where $|n_2\rangle$ is any orthonormal basis consisting of states relative to particle 2.

d) How do the answers to the preceding questions change if $\beta = 0$?

15.5 Consider a system of two particles of spin $\frac{1}{2}$. Denote by $|\pm_x\rangle$, $|\pm_y\rangle$, $|\pm_z\rangle$ respectively the eigenstates of σ_x, σ_y, σ_z for each particle.

a) Write the singlet state $|S = 0\rangle$ of the two particles both in terms of the eigenstates of $\sigma_x^{(1)}$, $\sigma_x^{(2)}$, and in terms of the eigenstates of $\sigma_y^{(1)}$, $\sigma_y^{(2)}$.

b) Show that $\langle S = 0\,|\,\sigma_i^{(1)}\sigma_j^{(2)}\,|\,S = 0\rangle = c\,\delta_{ij}$ and calculate the proportionality coefficient c.

c) Calculate the mean value of $(\vec{\sigma}^{(1)} \cdot \hat{a})(\vec{\sigma}^{(2)} \cdot \hat{b})$ $i)$ in the state $|S = 0\rangle$; $ii)$ in the statistical mixture $\{\,|+_z, -_z\rangle, \frac{1}{2};\ |-_z, +_z\rangle, \frac{1}{2}\,\}$; $iii)$ in the triplet state $|S = 1, S_z = 0\rangle$.

15.6 Consider a system of two distinguishable particles of spin $\frac{1}{2}$.

a) Write, in terms of the matrices $\vec{\sigma}^{(1)}$, $\vec{\sigma}^{(2)}$, the projection operators $\mathcal{P}_{S=0}$, $\mathcal{P}_{S=1}$ onto the singlet state and onto the triplet states.

Consider the isotropic mixture in which the two particles have opposite spins $|+\hat{n}, -\hat{n}\rangle: \sigma_n^{(1)}\,|+\hat{n}, -\hat{n}\rangle = |+\hat{n}, -\hat{n}\rangle$, $\sigma_n^{(2)}\,|+\hat{n}, -\hat{n}\rangle = -|+\hat{n}, -\hat{n}\rangle$, with the unit vector $\hat{n} \equiv \hat{n}(\theta, \phi)$ uniformly distributed on the unit sphere.

b) Calculate the mean value of $(\vec{\sigma}^{(1)} \cdot \hat{a})\,(\vec{\sigma}^{(2)} \cdot \hat{b})$ in the mixture.

c) Write, in terms of the matrices $\vec{\sigma}^{(1)}$, $\vec{\sigma}^{(2)}$, the density matrix:

$$\varrho = \frac{1}{4\pi}\int E_{+\hat{n}, -\hat{n}}\,\mathrm{d}\Omega \equiv \frac{1}{4\pi}\int |(+\hat{n})_1, (-\hat{n})_2\rangle\langle(-\hat{n})_2, (+\hat{n})_1|\,\mathrm{d}\Omega.$$

15.7 Particles (or atoms) endowed with a magnetic moment are produced either in a coherent superposition (pure state) or in a incoherent superposition (statistical mixture) of states with $J = 0$ and $J = 1$, $J_z = 0$.

a) Show that it is possible to select the particles with $J = 0$ by means of a Stern–Gerlach apparatus with the gradient of the magnetic field in a suitable direction.

Couples of particles (or atoms) of spin $\frac{1}{2}$ and endowed with a magnetic moment, are produced in a spin state with $S_z \equiv s_{1z} + s_{2z} = 0$.

b) Show it is possible to establish whether the particles are produced either in the singlet state, or in the statistical mixture $\{\,|+,-\rangle, \frac{1}{2}; \ |-,+\rangle, \frac{1}{2}\}$, or in the triplet state.

15.8   Consider two systems with angular momenta $j_1 = 1$ and $j_2 = 1$.

a) Write the states $|\,J, M\,\rangle$ that are simultaneous eigenstates of the total angular momentum and of its z component in terms of the states $|\,m_1, m_2\,\rangle \equiv |\,j_1, m_1; j_2, m_2\,\rangle$. Choose the phases of the states $|\,j, m\,\rangle$ as in equations (1) in the solution of Problem 8.2.

Consider now two systems with angular momenta $j_1 = 1$ and $j_2 = \frac{1}{2}$.

b) Write the states $|\,J, M\,\rangle$ in terms of the states $|\,m_1, m_2\,\rangle$.

Consider finally three particles of spin $\frac{1}{2}$.

c) Which are the possible values S of the total spin? How many independent states are there for each value of S?

d) Write a basis of eigenvectors $|\,S, M_S\,\rangle$ of the total spin and of its z component.

15.9   A system of two non-identical particles is in the state described by the normalized wavefunction:

$$\Psi(\vec{r}_1, \vec{r}_2) = N(\vec{r}_1 - \vec{r}_2)^2\, e^{-(r_1^2 + r_2^2)}, \qquad N = \sqrt{32/15\pi^3}$$

($\vec{r}_1$, $\vec{r}_2$ are measured in suitable units).

a) Say which are the possible results of measurements of $\vec{L}^2$ ($\vec{L} = \vec{L}_1 + \vec{L}_2$).

b) Say which are the possible results of measurements of $\vec{L}_1^2$ and the (not necessarily normalized) wavefunction of the system after each measurement.

c) Knowing that:

$$\int e^{-a r^2} dV = \left(\frac{\pi}{a}\right)^{3/2}, \qquad \int r^2 e^{-a r^2} dV = \frac{3}{2a}\left(\frac{\pi}{a}\right)^{3/2},$$

$$\int r^4 e^{-a r^2} dV = \frac{15}{4a^2}\left(\frac{\pi}{a}\right)^{3/2}$$

calculate the probability of the possible results of measurements of $\vec{L}_1^2$.

After each measurement of $\vec{L}_1^2$, L_{1z} is measured.

d) For each of the possible results obtained for $\vec{L}_1^2$, say which are the possible results of measurements of L_{1z} and the wavefunction after each measurement.

15.10   Let $\vec{L}_1$ and $\vec{L}_2$ be the angular momentum operators of two particles and $\vec{L} = \vec{L}_1 + \vec{L}_2$. In the following we will consider only the space of the states with $l_1 = 1$, $l_2 = 1$.

a) Write the most general wavefunction of the system (with $l_1 = 1$, $l_2 = 1$).

b) Show that the most general wavefunction with $L = 0$ has the form:

$$\Psi_0(\vec{r}_1, \vec{r}_2) = \vec{r}_1 \cdot \vec{r}_2 \, \Phi(r_1, r_2) .$$

c) Find the most general wavefunction of the system with $L = 1$.

d) Find the most general wavefunction of the system with $L = 2$.

15.11   A system of two particles is in a state $|A\rangle \xrightarrow{\text{SR}} \Psi_A(x_1, x_2)$. Let Δ_x be the interval $x - \frac{1}{2}\Delta$, $x + \frac{1}{2}\Delta$ and assume Ψ_A is normalized.

a) Write the expression of the probability $P(\Delta_x)$ of finding at least one particle in the interval Δ_x.

In the case of only one particle, the probability of finding it in Δ_x may be expressed as the mean value of the operator E_{Δ_x} that, in the Schrödinger representation, is the characteristic function of the interval Δ_x.

b) Express the probability $P(\Delta_x)$ determined in a) as the mean value of a suitable operator.

c) Show that, if Δ_1 and Δ_2 are two disjoint intervals, $P(\Delta_1 \cup \Delta_2) \neq P(\Delta_1) + P(\Delta_2)$ (therefore it is not possible to define a probability density $\rho(x)$ associated to $P(\Delta)$).

d) Write the expression for the average number $\bar{n}_{\Delta_x}$ of particles in the interval Δ_x.

15.12   Consider a system of two particles and the states:

$$|A_\pm\rangle \xrightarrow{\text{SR}} \frac{1}{\sqrt{2}}\big(\phi(x_1)\chi(x_2) \pm \chi(x_1)\phi(x_2)\big), \qquad |A_0\rangle \xrightarrow{\text{SR}} \phi(x_1)\chi(x_2)$$

with $\phi(x)$ and $\chi(x)$ normalized.

a) Calculate the mean value of the symmetric operator $f(q_1, q_2) = f(q_2, q_1)$ in the states $|A_+\rangle$, $|A_-\rangle$, $|A_0\rangle$.

b) Calculate the mean value of the operator $f(q_1, q_2) = f(q_2, q_1)$ in the states $|A_+\rangle$, $|A_-\rangle$, $|A_0\rangle$ in the case $\phi(x)$ and $\chi(x)$ have disjoint supports: $\phi(x)\chi(x) = 0$.

15.13   A particle in the state $|A_0\rangle$ impinges on a semi-transparent mirror. Let $|A\rangle$ be the state transmitted by the mirror and $|B\rangle$ the reflected state:

$$|A_0\rangle \to \tfrac{1}{\sqrt{2}}(|A\rangle + |B\rangle)$$

(the reflection and transmission coefficients are equal).

a) Let $|B_0\rangle$ (see figure) be the state of a particle that impinges on the mirror from the side opposite to that on which the particle in state $|A_0\rangle$ impinges – the incidence angle being the same in the two cases. Find the state of the particle beyond the mirror (the states $|A_0\rangle$ and $|B_0\rangle$, as well as the states $|A\rangle$ and $|B\rangle$, are mutually orthogonal).

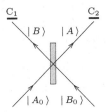

Assume now that two particles impinge on the mirror, one in the state $|A_0\rangle$, the other in the state $|B_0\rangle$. The two particles are then revealed by the counters C_1 and C_2.

b) Calculate the probability that the two counters click both (probability of having a coincidence) in the three cases in which   i) the two particles are distinguishable;   ii) the two particles are identical bosons;   iii) the two particles are identical fermions.

15.14   One of the consequences of the corpuscular nature of radiation (photons) is given by the existence of fluctuations not compatible with the classical theory of radiation [Hanbury Brown and Twiss effect, 1956]. Consider n identical photons, that impinge on a semi-transparent mirror with equal reflection and transmission coefficients, and are detected by the counters C_1 and C_2 (see figure).

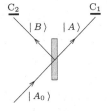

a) Calculate the probability P_{n_1,n_2} that n_1 photons are revealed by the counter C_1 and $n_2 = n - n_1$ by the counter C_2, in both cases when the n photons are sent to the mirror one at a time and when they are sent together.

b) Calculate the mean value $\overline{n_1 \times n_2}$ of the product of the countings of the two counters and the normalized correlation function:

$$g^{(2)} \equiv \frac{\overline{n_1 \times n_2}}{\overline{n}_1 \times \overline{n}_2}\,.$$

c) Compare the obtained result with the result that would be obtained by a classical treatment of the radiation, according to which the intensity of the transmitted wave and that of the reflected wave do not undergo quantum fluctuations (n_1 and n_2 are 'sure numbers').

15.15 The deuteron, the only bound state of the neutron–proton system (particles of spin $\frac{1}{2}$), has angular momentum $L = 0$ and spin $S = 1$ where $\vec{L} = \vec{r} \wedge \vec{p}$ is the relative angular momentum (i.e. $\vec{r}$ and $\vec{p}$ are the relative variables) and S is the total spin. The binding energy is $E_d = 2.2\,\mathrm{MeV}$. The neutron–proton interaction is schematized as:

$$H' = [-V_0 - a\,(\vec{\sigma}_p \cdot \vec{\sigma}_n - 1)]\,\chi(r)\,, \quad \chi(r) = \begin{cases} 1 & r \le r_0 = 2 \times 10^{-13}\,\mathrm{cm} \\ 0 & r > r_0 \end{cases}$$

where $\vec{r} \equiv \vec{r}_p - \vec{r}_n$. The masses are $m_p \simeq m_n \equiv m = 1.7 \times 10^{-24}\,\mathrm{g}$.

a) Knowing that the singlet state with minimum energy is not a bound state, but has energy $E_{S=0} \simeq 0$, find V_0 and a in terms of E_d and r_0 (see Problems 11.2 and 6.9: it may be necessary to solve numerically the equations that determine V_0; or in alternative, knowing a priori that the deuteron is a weakly bound system: $E_d \ll V_0$, approximate in a suitable way the above mentioned equations).

b) Say which are the possible values (L, S) for a couple of identical particles of spin $\frac{1}{2}$ where, as above, $\vec{L} = \vec{r} \wedge \vec{p}$ is the relative angular momentum and S is the total spin.

The neutron–neutron interaction is identical to the neutron–proton interaction.

c) Explain why no bound state consisting of two neutrons exists.

15.16 Two distinguishable particles of spin $\frac{1}{2}$ move apart as free particles, starting at time $t = 0$, the first toward the right in the orbital state $|\,r,t\,\rangle_1$ the second toward the left in the state $|\,l,t\,\rangle_2$. The two particles are in a state of total spin $S = 0$ and do not interact.

a) Which is the state (pure or statistic mixture) of particle 1 for $t > 0$?

At time $\bar{t}$ the component $\hat{n}$ of the spin of particle 2 is measured.

b) Show that, for $t > \bar{t}$, the result of a possible measurement of the component $\hat{n}$ of the spin of particle 1 is a priori determined.

c) In view of what has been found, and given that the particles do not interact, is it legitimate (according to the postulates of quantum mechanics) to state that already before $\bar{t}$ the particle 1 was in an eigenstate of $\vec{\sigma}_1 \cdot \hat{n}$? In this case, what can be said about the spin state of particle 2 for $0 < t \le \bar{t}$?

15.17 Two particles of spin $\frac{1}{2}$ (particles 2 and 3) are produced in a singlet state and move apart in different directions. Another particle of spin $\frac{1}{2}$ (particle 1), in an

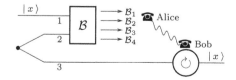

unknown spin state $|x\rangle$, travels along with particle 2. The spin state of the three particles is therefore $(\sigma_z|\pm\rangle = \pm|\pm\rangle)$:

$$|A\rangle = \tfrac{1}{\sqrt{2}}|x\rangle_1(|+\rangle_2|-\rangle_3 - |-\rangle_2|+\rangle_3), \qquad |x\rangle = \alpha|+\rangle + \beta|-\rangle.$$

On particles 1 and 2 Alice measures a nondegenerate observable $\mathcal{B}$, whose eigenstates ("Bell states") are:

$$|\mathcal{B}_1\rangle = \tfrac{1}{\sqrt{2}}(|+\rangle_1|+\rangle_2 + |-\rangle_1|-\rangle_2),$$

$$|\mathcal{B}_2\rangle = \tfrac{1}{\sqrt{2}}(|+\rangle_1|+\rangle_2 - |-\rangle_1|-\rangle_2),$$

$$|\mathcal{B}_3\rangle = \tfrac{1}{\sqrt{2}}(|+\rangle_1|-\rangle_2 + |-\rangle_1|+\rangle_2),$$

$$|\mathcal{B}_4\rangle = \tfrac{1}{\sqrt{2}}(|+\rangle_1|-\rangle_2 - |-\rangle_1|+\rangle_2).$$

a) Find the probability of the four possible results and for each of the obtained results find the state of particle 3 after the measurement.

The result of the measurement is communicated to Bob who, far from Alice, receives particle 3.

b) For each of the possible results of the measurement on particles 1 and 2, which rotation must Bob perform on the spin state of particle 3 in order that this is transformed back in the state $|x\rangle$ in which particle 1 was initially? (*Quantum teleportation* of an unknown state $|x\rangle$.)

Assume now that the spin state of the three particles is

$$|B\rangle = |x\rangle_1|+\rangle_2|-\rangle_3 .$$

c) Calculate the probabilities of the four possible results of the observable $\mathcal{B}$ and find the state of particle 3 after the measurement. Is it possible, in this case, to transform the state of particle 3 in the state $|x\rangle$ only knowing the result of the measurement of $\mathcal{B}$ on particles 1 and 2?

Solutions

15.1

a) The verification is immediate. The meaning of the last expression is given by the observation that $\vec{p}^{\alpha} - (m_\alpha/M)\vec{P} = m_\alpha(\vec{v}^{\alpha} - \vec{V}^{(cm)})$, i.e. the momentum of the α-th particle in the center-of-mass frame.

b) Since $\vec{Q}$ and $\vec{P}$ are vector operators both for $\vec{L}$ and for $\vec{Q} \wedge \vec{P}$ (therefore have the same commutation relations with both angular momenta), they commute with $\vec{L}^{(cm)} \equiv \vec{L} - \vec{Q} \wedge \vec{P}$.

c) Since $\vec{L}^{(cm)}$ is a vector operator for $\vec{L}$ and commutes with $\vec{Q} \wedge \vec{P}$,

$$[L_i^{(cm)}, L_j^{(cm)}] = [(\vec{L} - \vec{Q} \wedge \vec{P})_i, L_j^{(cm)}] = [L_i, L_j^{(cm)}] = i\,\hbar\,\epsilon_{ijk}\,L_k^{(cm)}.$$

d) Since $[(\vec{Q} \wedge \vec{P})_i, q_j^{\alpha}] = i\,\hbar\,\epsilon_{ijk}\,Q_k$ (indeed the angular momentum of the center of mass generates rotations of the coordinates of the center of mass) and since $[L_i, q_j^{\alpha}] = i\,\hbar\,\epsilon_{ijk}\,q_k^{\alpha}$, it follows that $[L_i^{(cm)}, q_j^{\alpha}] \neq i\,\hbar\,\epsilon_{ijk}\,q_k^{\alpha}$ (and similarly for $\vec{p}^{\alpha}$), but $[L_i^{(cm)}, (\vec{q}^{\alpha} - \vec{Q})_j] = i\,\hbar\,\epsilon_{ijk}\,(\vec{q}^{\alpha} - \vec{Q})_k$, likewise for $\vec{p}^{\alpha} - \vec{P}$. Therefore also $\vec{q}^{\alpha} - \vec{q}^{\beta}$, and $\vec{p}^{\alpha} - \vec{p}^{\beta}$ have with $\vec{L}^{(cm)}$ the commutation relations of the vectors.

15.2

a) The principal axes of inertia respectively have the direction that goes from the center of mass to atom 1, the direction orthogonal to it in the plane of the molecule and finally (not reported in the figure) the direction orthogonal to the plane of the molecule. So, if $\vec{Q} = \frac{1}{3}(\vec{q}_1 + \vec{q}_2 + \vec{q}_3)$ is the position operator of the center of mass, one has:

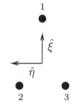

$$\hat{\xi} = \frac{\vec{q}_1 - \vec{Q}}{|\vec{q}_1 - \vec{Q}|}\,; \qquad \hat{\eta} = \frac{\vec{q}_2 - \vec{q}_3}{|\vec{q}_2 - \vec{q}_3|}\,; \qquad \hat{\zeta} = \hat{\xi} \wedge \hat{\eta}\,.$$

It is then evident that, even if the denominators are c–numbers (the molecule is rigid), $\hat{\xi}$, $\hat{\eta}$, $\hat{\zeta}$ are dynamical variables of the system and, therefore, operators.

b) L_ξ, L_η, L_ζ, contrary to L_x, L_y, L_z, are scalar operators because $\hat{\xi}$, $\hat{\eta}$, $\hat{\zeta}$ (contrary to $\hat{x}$, $\hat{y}$, $\hat{z}$) are vector operators. Therefore:

$$[L_i\,,\,L_\xi]=0\,,\qquad [L_i\,,\,L_\eta]=0\,,\qquad [L_i\,,\,L_\zeta]=0\,,\qquad i=1,\,2,\,3$$

and also $[\vec{L}^{\,2}\,,\,L_\xi]=[\vec{L}^{\,2}\,,\,L_\eta]=[\vec{L}^{\,2}\,,\,L_\zeta]=0$.

$$[L_\xi\,,\,L_\eta]=L_j\,[L_\xi\,,\,\eta_j]=L_j\,[L_i\,,\,\eta_j]\,\xi_i=i\,\hbar\,\epsilon_{ijk}\,L_j\,\eta_k\,\xi_i=-i\,\hbar\,L_j\,(\hat{\xi}\wedge\hat{\eta})_j$$

$$=-i\,\hbar\,L_\zeta$$

and likewise for the others. Note that the commutation rules we have found differ in sign from those among the components of $\vec{L}$ with respect to the frame at rest, but $-L_\xi,\,-L_\eta,\,-L_\zeta$ have the same commutation rules as $L_x,\,L_y,\,L_z$, and, as a consequence, the same properties.

The equality $L_x^2+L_y^2+L_z^2=L_\xi^2+L_\eta^2+L_\zeta^2$ is obvious; however, due to the occurrence of non commuting operators, by taking advantage of $[L_i\,\xi_i\,,\,L_j]=0$, one has:

$$L_\xi^2+L_\eta^2+L_\zeta^2=L_i\,\xi_i\,L_j\,\xi_j+\cdots=L_j\,L_i\,(\xi_i\,\xi_j+\eta_i\,\eta_j+\zeta_i\,\zeta_j)$$

$$=L_j\,L_i\,\delta_{ij}=\vec{L}^{\,2}\,.$$

The independent states for a given value of l are $(2l+1)^2$: since $\vec{L}^{\,2}$, L_z, L_ζ commute, they may be classified as $|\,l,m_z,m_\zeta\,\rangle$, and $-l\le m_z\le l$, $-l\le m_\zeta\le l$. They indeed are all independent: applying the operators $L_x\pm i\,L_y$ to the states $|\,l,m_z,m_\zeta\,\rangle$ m_z changes, m_ζ is instead unchanged, whereas applying the operators $L_\xi\pm i\,L_\eta$ m_ζ changes, m_z is unchanged.

c) When all the moments of inertia are different from one another, H commutes only with the L_i and, as a consequence, with $\vec{L}^{\,2}$. The (rotational) stationary states of the molecule are accordingly classified by the eigenvalue $\hbar^2 l(l+1)$ of $\vec{L}^{\,2}$ and by the eigenvalue $\hbar\,m$ of a component with respect to the frame at rest, e.g. L_z: $|\,\alpha;\,l,m_z\,\rangle$ where α, that stands for one or

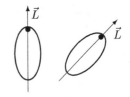

degeneracy with respect to m_z

more quantum numbers, is necessary: indeed, the degeneracy of each level $E_{\alpha,l}$ is $2l+1$: $-l\le m_z\le l$, whereas the number of states with fixed l is $(2l+1)^2$, therefore for any value of l there are, in general, $2l+1$ levels (see Problem 15.3).

Using the classical language, the degeneracy on m_z, stems from the fact that the orientation in space of the molecule is arbitrary. In this case the calculation of $E_{\alpha,l}$ is not possible for generic l.

d) In the present case $H = \vec{L}^2/2I$ commutes not only with all the L_i, but also with all the components of $\vec{L}$ with respect to the mobile frame. The stationary states are classified by l, m_z and by the eigenvalue of a component of $\vec{L}$ with respect to the moving frame, e.g. L_ζ: $|l, m_z; l, m_\zeta \rangle \equiv |l, m_z, m_\zeta \rangle$. The eigenvalue is

$$E_l = \frac{\hbar^2 l(l+1)}{2I}$$

and the degeneracy is $(2l+1)^2$: $-l \le m_z \le l$; $-l \le m_\zeta \le l$; the degeneracy on m_ζ stems from the fact that (still using the classical language) the molecule can be rotated, but the axis of rotation is kept fixed.

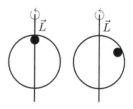

degeneracy with respect to m_ζ

In the latter case the system is invariant under both the group $SU(2)$ generated by L_i (see Problem 10.8), and the group $SU(2)$ generated by $-L_\xi$, $-L_\eta$, $-L_\zeta$: the invariance group is therefore $SU(2) \times SU(2)$ with the additional condition $L_x^2 + L_y^2 + L_z^2 = L_\xi^2 + L_\eta^2 + L_\zeta^2$.

e) One has:

$$H = \frac{L_\xi^2 + L_\eta^2}{2I} + \frac{L_\zeta^2}{2I_\zeta} = \frac{\vec{L}^2}{2I} + \left(\frac{1}{2I_\zeta} - \frac{1}{2I} \right) L_\zeta^2$$

so H commutes with all the L_i and with L_ζ. The states are once more classified as $|l, m_z, m_\zeta \rangle$. The eigenvalue:

$$E_{l,m_\zeta} = \frac{\hbar^2 l(l+1)}{2I} + \left(\frac{1}{2I_\zeta} - \frac{1}{2I} \right) \hbar^2 m_\zeta^2$$

is degenerate $2(2l+1)$ times if $m_\zeta \ne 0$, $2l+1$ if $m_\zeta = 0$.

f) As the principal moments of inertia fulfil the triangle inequality, $|I_\xi - I_\eta| \le I_\zeta$ and $I_\zeta \to 0$, we are in the symmetrical case. The limit $I_\zeta \to 0$ is possible only for the states with $m_\zeta = 0$, otherwise the energy diverges. Therefore:

$$E_l = \frac{\hbar^2 l(l+1)}{2I}$$

and the degeneracy is $2l+1$: this is the case, discussed in Problem 11.12, of diatomic molecules and, more generally, of linear molecules.

15.3

a) The state with $l = 0$ is also an eigenstate of L_ξ, L_η, L_ζ with vanishing eigenvalues, so $E_{l=0} = 0$.

Let us consider the states with $l = 1$: in any basis in which L_z is diagonal, the energy $H = L_\xi^2/2I_\xi + L_\eta^2/2I_\eta + L_\zeta^2/2I_\zeta$ is represented by a 9×9 matrix consisting of three identical 3×3 blocks, each relative to a given value of

m_z; having fixed m_z, for the operators L_ξ, L_η, L_ζ it is convenient to take the Cartesian representation found in Problem 8.2: one has:

$$L_\xi^2 = \begin{pmatrix} 0 & 0 & 0 \\ 0 & 1 & 0 \\ 0 & 0 & 1 \end{pmatrix}, \quad L_\eta^2 = \begin{pmatrix} 1 & 0 & 0 \\ 0 & 0 & 0 \\ 0 & 0 & 1 \end{pmatrix}, \quad L_\zeta^2 = \begin{pmatrix} 1 & 0 & 0 \\ 0 & 1 & 0 \\ 0 & 0 & 0 \end{pmatrix}$$

so H is diagonal and the eigenvalues in the space of the vectors with $l = 1$ are:

$$E_{\alpha_1} = \frac{\hbar^2}{2}\left(\frac{1}{I_\eta} + \frac{1}{I_\zeta}\right), \quad E_{\alpha_2} = \frac{\hbar^2}{2}\left(\frac{1}{I_\zeta} + \frac{1}{I_\xi}\right), \quad E_{\alpha_3} = \frac{\hbar^2}{2}\left(\frac{1}{I_\xi} + \frac{1}{I_\eta}\right)$$

each of them being three times degenerate.

b) As we have found in Problem 9.10, if H_I is the Hamiltonian in an inertial frame, the Hamiltonian in the laboratory is

$$H = \frac{\vec{P}^2}{2M} + \frac{L_\xi^2}{2I_\xi} + \frac{L_\eta^2}{2I_\eta} + \frac{L_\zeta^2}{2I_\zeta} - \omega_0 L_z^{tot}$$

where the z axis has been taken parallel to $\vec{\omega}_0$ and $\vec{L}^{tot}$ is the total angular momentum: $\vec{L}^{tot} = \vec{Q} \wedge \vec{P} + \vec{L}$.
Since $\vec{P} = 0$, the Hamiltonian reduces to:

$$H = \frac{L_\xi^2}{2I_\xi} + \frac{L_\eta^2}{2I_\eta} + \frac{L_\zeta^2}{2I_\zeta} - \omega_0 L_z .$$

The eigenstates of H still are $|\alpha_i; l = 1, m_z\rangle$, but the degeneracy on m_z is removed by the term $-\omega_0 L_z$: the eigenvalues now are $E_{\alpha_i} - m_z \hbar \omega_0$.

15.4

a) As ξ_1 acts as the identity on the states of particle 2 and $\langle c_2 \mid d_2 \rangle = 0$, one has:

$$\bar{\xi}_1 \equiv \langle X \mid \xi_1 \mid X \rangle = |\alpha|^2 \langle a_1 \mid \xi_1 \mid a_1 \rangle + |\beta|^2 \langle b_1 \mid \xi_1 \mid b_1 \rangle .$$

The state $|x_1\rangle$ does not exist: indeed, let $|x_1\rangle = \alpha'|a_1\rangle + \beta'|b_1\rangle + \gamma'|s_1\rangle$ with $\langle a_1 \mid s_1 \rangle = \langle b_1 \mid s_1 \rangle = 0$. If ξ_1 is the projector onto $|s_1\rangle$, one obtains $\gamma' = 0$; if $\xi_1 = |a_1\rangle\langle b_1| + |b_1\rangle\langle a_1|$ one obtains $\Re e(\alpha' \beta'^*) = 0$, whereas if $\xi_1 = i(|a_1\rangle\langle b_1| - |b_1\rangle\langle a_1|) \Rightarrow \Im m(\alpha' \beta'^*) = 0$ then either α' or β' vanishes. However in neither case the statistical averages $\bar{\xi}_1$ are reproduced.

b) Let $|n\rangle$ be whatever orthonormal basis. For any $|\bar{n}\rangle$, $|\bar{m}\rangle$, putting $\xi = |\bar{n}\rangle\langle \bar{m}| + |\bar{m}\rangle\langle \bar{n}|$, $\eta = i(|\bar{n}\rangle\langle \bar{m}| - |\bar{m}\rangle\langle \bar{n}|)$ and $\delta = \varrho' - \varrho''$, one has:

$$0 = \text{Tr}\,(\delta\,\xi) = \text{Tr}\,(\delta\,\eta) \Rightarrow \sum_n \langle n \mid \delta \mid \bar{n}\rangle\langle \bar{m} \mid n\rangle = \langle \bar{m} \mid \delta \mid \bar{n}\rangle = 0$$

$$\Rightarrow \varrho' = \varrho''.$$

For any ξ_1 it must be that $\mathrm{Tr}\,\varrho_1\,\xi_1 = |\alpha|^2\langle a_1 \mid \xi_1 \mid a_1\rangle + |\beta|^2\langle b_1 \mid \xi_1 \mid b_1\rangle$ whence $\varrho_1 = |\alpha|^2\,|a_1\rangle\langle a_1| + |\beta|^2\,|b_1\rangle\langle b_1|$ is, due to the above result, the only solution. As $\varrho_1^2 \neq \varrho_1$, ϱ_1 is not the statistical matrix corresponding to a pure state, in agreement with what we have found in a).

c) $\varrho_{1,2} = |X\rangle\langle X|$;

$$\mathrm{Tr}_2\,\varrho_{1,2} = |\alpha|^2\,|a_1\rangle\langle a_1|\Big(\sum_{n_2}|\langle n_2 \mid c_2\rangle|^2\Big)$$
$$+ |\beta|^2\,|b_1\rangle\langle b_1|\Big(\sum_{n_2}|\langle n_2 \mid d_2\rangle|^2\Big)$$
$$+ \cdots \Big(\sum_{n_2}\langle c_2 \mid n_2\rangle\langle n_2 \mid d_2\rangle\Big) + \cdots \Big(\sum_{n_2}\langle d_2 \mid n_2\rangle\langle n_2 \mid c_2\rangle\Big)$$
$$= |\alpha|^2\,|a_1\rangle\langle a_1| + |\beta|^2\,|b_1\rangle\langle b_1| = \varrho_1\,.$$

d) If $\beta = 0$ then $|x_1\rangle = |a_1\rangle$ and ϱ_1 corresponds to the pure state $|a_1\rangle$.

15.5

a) As the singlet state is invariant under rotations ($S = 0$), its expression is independent of the orientation of the Cartesian axes, so:
$$|S = 0\rangle = \tfrac{1}{\sqrt{2}}\big(|+_x,-_x\rangle - |-_x,+_x\rangle\big) = \tfrac{1}{\sqrt{2}}\big(|+_y,-_y\rangle - |-_y,+_y\rangle\big)\,.$$

b) Thanks to the rotational invariance of the singlet state $|0\rangle$, one has:
$$\langle 0 \mid \sigma_x^{(1)}\sigma_x^{(2)} \mid 0\rangle = \langle 0 \mid \sigma_y^{(1)}\sigma_y^{(2)} \mid 0\rangle = \langle 0 \mid \sigma_z^{(1)}\sigma_z^{(2)} \mid 0\rangle;$$
$$\langle 0 \mid \sigma_x^{(1)}\sigma_y^{(2)} \mid 0\rangle = 0$$

(the last equality follows, for example, from the fact that for a rotation by π around the x axis one has $\langle 0 \mid \sigma_x^{(1)}\sigma_y^{(2)} \mid 0\rangle = -\langle 0 \mid \sigma_x^{(1)}\sigma_y^{(2)} \mid 0\rangle$). The coefficient c obtains from:
$$\langle 0 \mid \vec{\sigma}^{\,(1)}\cdot\vec{\sigma}^{\,(2)} \mid 0\rangle = \sum_i \langle 0 \mid \sigma_i^{(1)}\sigma_i^{(2)} \mid 0\rangle = 3\,c$$
$$\vec{\sigma}^{\,(1)}\cdot\vec{\sigma}^{\,(2)} = \tfrac{1}{2}\Big(\big(\vec{\sigma}^{\,(1)} + \vec{\sigma}^{\,(2)}\big)^2 - \vec{\sigma}^{\,(1)}\cdot\vec{\sigma}^{\,(1)} - \vec{\sigma}^{\,(2)}\cdot\vec{\sigma}^{\,(2)}\Big) = 2S(S+1) - 3$$
whence $c = -1$.

c) i) $\langle 0 \mid (\vec{\sigma}^{\,(1)}\cdot\hat{a})(\vec{\sigma}^{\,(2)}\cdot\hat{b}) \mid 0\rangle = \langle 0 \mid \sigma_i^{(1)}\sigma_j^{(2)} \mid 0\rangle\,\hat{a}_i\,\hat{b}_j = -\hat{a}\cdot\hat{b}$.

ii) $\langle\!\langle\,(\vec{\sigma}^{\,(1)}\cdot\hat{a})(\vec{\sigma}^{\,(2)}\cdot\hat{b})\,\rangle\!\rangle = \tfrac{1}{2}\big(a_z\times(-b_z) + (-a_z)\times b_z\big) = -a_z b_z$.

iii) $\langle\,S = 1,\ S_z = 0 \mid (\vec{\sigma}^{\,(1)}\cdot\hat{a})(\vec{\sigma}^{\,(2)}\cdot\hat{b}) \mid S = 1,\ S_z = 0\,\rangle = a_x b_x + a_y b_y - a_z b_z$.

15.6

a) For $\mathcal{P}_{S=0}$ one can make the direct calculation; or, as $\mathcal{P}_{S=0}$ is rotationally invariant, one must have $\mathcal{P}_{S=0} = a\,\mathbb{1} + b\,\vec{\sigma}^{\,(1)}\cdot\vec{\sigma}^{\,(2)}$; as (see Problem 15.5) $\vec{\sigma}^{\,(1)}\cdot\vec{\sigma}^{\,(2)} = 2S(S+1) - 3$, the eigenvalues of $\vec{\sigma}^{\,(1)}\cdot\vec{\sigma}^{\,(2)}$ are -3 on the singlet state and $+1$ on the triplet states then, from $\mathcal{P}_{S=0}\mid S = 0\rangle = \mid S = 0\rangle$ and $\mathcal{P}_{S=0}\mid S = 1\rangle = 0$, it follows that $a - 3b = 1$, $a + b = 0$, therefore:
$$\mathcal{P}_{S=0} = \tfrac{1}{4}\big(\mathbb{1} - \vec{\sigma}^{\,(1)}\cdot\vec{\sigma}^{\,(2)}\big) \quad\Rightarrow\quad \mathcal{P}_{S=1} = \mathbb{1} - \mathcal{P}_{S=0} = \tfrac{3}{4}\,\mathbb{1} + \tfrac{1}{4}\,\vec{\sigma}^{\,(1)}\cdot\vec{\sigma}^{\,(2)}\,.$$

b) $\langle\!\langle (\vec{\sigma}^{\,(1)} \cdot \hat{a})(\vec{\sigma}^{\,(2)} \cdot \hat{b}) \rangle\!\rangle = \dfrac{1}{4\pi} \displaystyle\int \langle +\hat{n} \mid \vec{\sigma}^{\,(1)} \cdot \hat{a} \mid +\hat{n}\rangle\langle -\hat{n} \mid \vec{\sigma}^{\,(2)} \cdot \hat{b} \mid -\hat{n}\rangle \, d\Omega$

$\qquad = \dfrac{1}{4\pi} \displaystyle\int (-\hat{n} \cdot \hat{a})(\hat{n} \cdot \hat{b}) \, d\Omega = -\dfrac{1}{4\pi} \times \dfrac{4\pi}{3} \, \delta_{ij} \, a_i b_j = -\dfrac{1}{3} \hat{a} \cdot \hat{b} \, .$

c) $E_{+\hat{n},-\hat{n}} = \frac{1}{4}\big(\mathbb{1} + \vec{\sigma}^{\,(1)} \cdot \hat{n}\big)\big(\mathbb{1} - \vec{\sigma}^{\,(2)} \cdot \hat{n}\big) = \frac{1}{4}\Big(\mathbb{1} - \big(\vec{\sigma}^{\,(1)} \cdot \hat{n}\big)\big(\vec{\sigma}^{\,(2)} \cdot \hat{n}\big)\Big) + \cdots$

$\varrho = \dfrac{1}{16\pi} \displaystyle\int \Big(\mathbb{1} - \big(\vec{\sigma}^{\,(1)} \cdot \hat{n}\big)\big(\vec{\sigma}^{\,(2)} \cdot \hat{n}\big)\Big) \, d\Omega = \frac{1}{4}\Big(\mathbb{1} - \frac{1}{3}\vec{\sigma}^{\,(1)} \cdot \vec{\sigma}^{\,(2)}\Big)$

$\qquad = \dfrac{1}{2}\, \mathcal{P}_{S=0} + \dfrac{1}{6}\, \mathcal{P}_{S=1}$

(the omitted terms give zero after integration over the angles).

15.7

a) The states with angular momentum $J=0$ are eigenstates – with vanishing eigenvalue – of *all* the components of the angular momentum, whereas the states with angular momentum $J = 1$, $J_z = 0$ are superpositions of states with $J_n = \pm 1$, where $\hat{n}$ is whatever direction in the x-y plane. So, if the particles are sent in a Stern–Gerlach apparatus that measures – for example – J_x, the particles with $J = 0$ are not deflected, on the contrary those with $J = 1$, $J_z = 0$ are always deflected.

b) Let us assume that the particles travel together; in this case we send them in a Stern–Gerlach apparatus (otherwise we shall use two apparatuses with the magnets oriented in the same way). If the particles are produced in a singlet state, no matter how the Stern–Gerlach apparatus is oriented, the two particles belonging to the same pair are deflected in opposite directions. If instead the particles are produced in the statistical mixture and the Stern–Gerlach apparatus measures, for example, the x component of the angular momentum, in the 50% of the events the two particles are deflected in the same direction, i.e. with $S_x \equiv s_{1x} + s_{2x} = \pm 1$: indeed the states $\mid +, -\rangle$ and $\mid -, +\rangle$ are superpositions of states with $S_x = \pm 1, 0$. If, finally, the particles are produced in the triplet state, as:

$$\mid S = 1, \, S_z = 0\rangle = \tfrac{1}{\sqrt{2}}\big(\mid S = 1, \, S_x = 1\rangle + \mid S = 1, \, S_x = -1\rangle\big)$$

the two particles are always deflected in the same direction.

15.8

We shall write only the states with $M \geq 0$, because those with $M < 0$ are obtained by means of the substitution $m_1 \to -m_1$, $m_2 \to -m_2$.

a) $J = 0, 1, 2$. The states with $J = 2$ are symmetric under the exchange $1 \leftrightarrow 2$: indeed the state with $M = 2$ has $m_1 = m_2 = 1$, and the others ($M = 1, 0, -1, -2$) obtain by applying the operator $J_- = j_{1-} + j_{2-}$ that is symmetric. By taking advantage of (1) of Problem 8.2: $j_\pm \mid j, m\rangle = \hbar\sqrt{j(j+1) - m(m \pm 1)} \mid j, m \pm 1\rangle$. One has:

$|J = 2, M = 2\rangle = |1, 1\rangle$; $|J = 2, M = 1\rangle = \frac{1}{\sqrt{2}}(|1, 0\rangle + |0, 1\rangle)$;

$J_- |J = 2, M = 1\rangle = \hbar\sqrt{6}\,|J = 2, M = 0\rangle$

$$= \frac{1}{\sqrt{2}}(j_{1-} + j_{2-})(|1, 0\rangle + |0, 1\rangle) \Rightarrow$$

$|J = 2, M = 0\rangle = \frac{1}{\sqrt{6}}(|1, -1\rangle + 2|0, 0\rangle + |-1, 1\rangle)$.

$|J = 1, M = 1\rangle$ is orthogonal to $|J = 2, M = 1\rangle$, then:

$|J = 1, M = 1\rangle = \frac{1}{\sqrt{2}}(|1, 0\rangle - |0, 1\rangle)$

and, since it is antisymmetric, also $|J = 1, M = 0\rangle$ is antisymmetric:

$|J = 1, M = 0\rangle = \frac{1}{\sqrt{2}}(|1, -1\rangle - |-1, 1\rangle)$.

By orthogonality with $|J = 2, M = 0\rangle$ the state $J = 0$ (necessarily symmetric) obtains:

$|J = 0, M = 0\rangle = \frac{1}{\sqrt{3}}(|1, -1\rangle - |0, 0\rangle + |-1, 1\rangle)$.

b) $|J = \frac{3}{2}, M = \frac{3}{2}\rangle = |1, \frac{1}{2}\rangle$;

$J_- |J = \frac{3}{2}, M = \frac{3}{2}\rangle = \hbar\sqrt{3}\,|J = \frac{3}{2}, M = \frac{1}{2}\rangle = (j_{1-} + j_{2-})|1, \frac{1}{2}\rangle \Rightarrow$

$|J = \frac{3}{2}, M = \frac{1}{2}\rangle = \sqrt{\frac{2}{3}}|0, \frac{1}{2}\rangle + \sqrt{\frac{1}{3}}|1, -\frac{1}{2}\rangle$.

The vector $|J = \frac{1}{2}, M = \frac{1}{2}\rangle$ is obtained by requiring orthogonality to $|J = \frac{3}{2}, M = \frac{1}{2}\rangle$:

$|J = \frac{1}{2}, M = \frac{1}{2}\rangle = \sqrt{\frac{1}{3}}|0, \frac{1}{2}\rangle - \sqrt{\frac{2}{3}}|1, -\frac{1}{2}\rangle$.

c) The composition of the spin of two particles, e.g. 1 and 2, gives the values $S_{12} = 0, 1$ and the composition with the spin of the third particle gives once $S = \frac{3}{2}$ and twice $S = \frac{1}{2}$, so there are four states with $S = \frac{3}{2}$ and $2 + 2$ states with $S = \frac{1}{2}$.

d) The states with $S = \frac{3}{2}$ are symmetric ($|+, +, +\rangle$ is an abbreviation for the vector $|s_{1z} = \frac{1}{2}, s_{2z} = \frac{1}{2}, s_{3z} = \frac{1}{2}\rangle$, etc.):

$|S = \frac{3}{2}, M_S = \frac{3}{2}\rangle = |+, +, +\rangle$

$|S = \frac{3}{2}, M_S = \frac{1}{2}\rangle = \frac{1}{\sqrt{3}}(|+, +, -\rangle + |+, -, +\rangle + |-, +, +\rangle)$.

A pair of states with $S = \frac{1}{2}$ can be obtained, for example, by composing the spin $\frac{1}{2}$ of the third particle with the singlet state ($S_{12} = 0$) of particles 1 and 2:

$|S' = \frac{1}{2}, M_S = \frac{1}{2}\rangle = \frac{1}{\sqrt{2}}(|+, -\rangle - |-, +\rangle)|+\rangle$

$$= \frac{1}{\sqrt{2}}(|+, -, +\rangle - |-, +, +\rangle)$$.

The other pair of states with $S = \frac{1}{2}$ can be obtained either by composing the spin $\frac{1}{2}$ of the third particle with the triplet states ($S_{12} = 1$) of the first two, or by imposing the orthogonality with $|S = \frac{3}{2}, M_S = \frac{1}{2}\rangle$ and $|S' = \frac{1}{2}, M_S = \frac{1}{2}\rangle$:

$$| S'' = \tfrac{1}{2}, M_S = \tfrac{1}{2} \rangle = \sqrt{\tfrac{1}{6}} \, | +, -, + \rangle + \sqrt{\tfrac{1}{6}} \, | -, +, + \rangle - \sqrt{\tfrac{2}{3}} \, | +, +, - \rangle \, .$$

15.9

a) Ψ is invariant under rotations, so $L = 0$.

b) $\Psi(\vec{r}_1, \vec{r}_2) = N(r_1^2 + r_2^2) \, e^{-(r_1^2 + r_2^2)} - 2N(\vec{r}_1 \cdot \vec{r}_2) \, e^{-(r_1^2 + r_2^2)} \equiv \Psi_0 + \Psi_1 \, .$

   Ψ_0 has $l_1 = 0$ and Ψ_1 has $l_1 = 1$ (it is a linear combination of x_1, y_1, $z_1 \times$
   $e^{-(r_1^2 + r_2^2)}$), so the possible results are $l_1 = 0$ and 1.
   If the result is $l_1 = 0$, the wavefunction after the measurement is Ψ_0,
   whereas if $l_1 = 1$ the wavefunction after the measurement is Ψ_1.

c) $P_{l_1 = 0} = \dfrac{|\langle \Psi_0 \mid \Psi \rangle|^2}{\langle \Psi_0 \mid \Psi_0 \rangle} = \langle \Psi_0 \mid \Psi_0 \rangle$

$$= N^2 \int \left(r_1^4 + r_2^4 + 2 r_1^2 r_2^2 \right) e^{-2(r_1^2 + r_2^2)} dV_1 dV_2 = \frac{4}{5} \quad \Rightarrow \quad P_{l_1 = 1} = \frac{1}{5} \, .$$

d) If the result is $l_1 = 0$, obviously $m_1 = 0$ and the state remains unaltered
   (Ψ_0); if instead $l_1 = 1$, then the possible results are $m_1 = 0, \pm 1$ (for Ψ_1 is
   a linear combination of x_1, y_1, $z_1 \times$ etc.). In order to find the wavefunction
   after the measurement, it is convenient to express Ψ_1 as linear combination
   of eigenfunctions of L_{1z}:

$$\Psi_1 = 2N \left(z_1 z_2 + \tfrac{1}{2}(x_1 + i y_1)(x_2 - i y_2) + \tfrac{1}{2}(x_1 - i y_1)(x_2 + i y_2) \right) e^{-(r_1^2 + r_2^2)}.$$

   So, if $m_1 = 0$: $\Psi_1 \to z_1 z_2 \, e^{-(r_1^2 + r_2^2)}$;

   if $m_1 = +1$: $\Psi_1 \to (x_1 + i y_1)(x_2 - i y_2) \, e^{-(r_1^2 + r_2^2)}$;

   if $m_1 = -1$: $\Psi_1 \to (x_1 - i y_1)(x_2 + i y_2) \, e^{-(r_1^2 + r_2^2)}$.

15.10

a) The space of the states with $l_1 = 1$, $l_2 = 1$ is generated by the factorized
   states with wavefunctions $\sum_i \alpha_i x_{1i} \, \phi_1(r_1) \times \sum_j \beta_j x_{2j} \, \phi_2(r_2)$ and the most
   general superposition of this type is

$$\Psi(\vec{r}_1, \vec{r}_2) = \sum_{ij} C_{ij} \, x_{1i} \, x_{2j} \, \Phi(r_1, r_2) \, . \tag{1}$$

   Keeping the function $\Phi(r_1, r_2)$ fixed, one has $9 = 3 \times 3$ independent
   states.

b) The wavefunctions of the states with $L = 0$ have the form (1) and, in
   addition, they must be invariant under rotations, so $C_{ij} = c \, \delta_{ij}$.

c) As in Problem 8.1, from the commutation rules $[L_i, V_j] = i \, \hbar \, \epsilon_{ijk} V_k$ one
   has that the states with angular momentum $L = 1$ obtain by applying
   a vector operator V_i to states with $L = 0$. In this case, since $\Phi(r_1, r_2)$
   corresponds to $L = 0$, by applying $(\vec{r}_1 \wedge \vec{r}_2)_i$ to $\Phi(r_1, r_2)$, states of the
   form (1) with $L = 1$ are obtained, then:

$$\Psi_1(\vec{r}_1, \vec{r}_2) = \sum_i c_i \, (\vec{r}_1 \wedge \vec{r}_2)_i \, \Phi(r_1, r_2) \, .$$

d) From among the nine states having the form (1), the states linearly independent with respect to those with $L = 1$ are obtained from the symmetric combinations $\Psi_{\{ij\}}(\vec{r}_1, \vec{r}_2) \equiv (x_{1i} x_{2j} + x_{1j} x_{2i}) \, \Phi(r_1, r_2)$. However the latter are not orthogonal to those with $L = 0 : \sum_i \Psi_{\{ii\}} = \Psi_0$, therefore:

$$\Psi_2(\vec{r}_1, \vec{r}_2) = \sum_{ij} c_{ij} \left(x_{1i} x_{2j} + x_{1j} x_{2i} - \tfrac{2}{3} \left(\vec{r}_1 \cdot \vec{r}_2\right) \delta_{ij} \right) \Phi(r_1, r_2) \; .$$

To summarize, we have decomposed the matrix $x_{1i} x_{2j}$ in its trace $(L = 0)$, its antisymmetric part $(L = 1)$ and its symmetric, traceless part $(L = 2)$:

$$x_{1i} x_{2j} = \tfrac{1}{3}(\vec{r}_1 \cdot \vec{r}_2) \delta_{ij} + \tfrac{1}{2}\left(x_{1i} x_{2j} - x_{1j} x_{2i} \right)$$
$$+ \tfrac{1}{2}\left(x_{1i} x_{2j} + x_{1j} x_{2i} - \tfrac{2}{3}(\vec{r}_1 \cdot \vec{r}_2) \delta_{ij} \right) \; .$$

15.11

a) $P(\Delta_x)$ is the probability of finding particle 1 in and particle 2 out of Δ_x, $+\ 2$ in and 1 out, $+$ both in; or, also: 1 in and 2 anywhere, $+\ 2$ in and 1 anywhere, $-$ both in:

$$P(\Delta_x) = \int_{-\infty}^{+\infty} dx_2 \int_{x-\frac{1}{2}\Delta}^{x+\frac{1}{2}\Delta} |\Psi_A(x_1, x_2)|^2 \, dx_1 + \int_{-\infty}^{+\infty} dx_1 \int_{x-\frac{1}{2}\Delta}^{x+\frac{1}{2}\Delta} |\Psi_A(x_1, x_2)|^2 \, dx_2$$

$$- \int_{x-\frac{1}{2}\Delta}^{x+\frac{1}{2}\Delta} dx_1 \int_{x-\frac{1}{2}\Delta}^{x+\frac{1}{2}\Delta} |\Psi_A(x_1, x_2)|^2 \, dx_2 \; .$$

b) $P(\Delta_x) = \langle A \,|\, E_{\Delta_x}^1 \otimes \mathbb{1}^2 + \mathbb{1}^1 \otimes E_{\Delta_x}^2 - E_{\Delta_x}^1 \otimes E_{\Delta_x}^2 \,|\, A \rangle$.

c) Putting $E_\Delta^{1,2} \equiv E_\Delta^1 \otimes \mathbb{1}^2 + \mathbb{1}^1 \otimes E_\Delta^2 - E_\Delta^1 \otimes E_\Delta^2$, one has

$$E_{\Delta_1 \cup \Delta_2}^{1,2} = E_{\Delta_1}^{1,2} + E_{\Delta_2}^{1,2} - E_{\Delta_1}^1 \otimes E_{\Delta_2}^2 - E_{\Delta_2}^1 \otimes E_{\Delta_1}^2 \neq E_{\Delta_1}^{1,2} + E_{\Delta_2}^{1,2} \; .$$

d) $\overline{n}_{\Delta_x}$ is given by $1\times$ (the probability of finding particle 1 in Δ_x and particle 2 out $+\ 2$ in and 1 out) $+2\times$(both in):

$$\overline{n}_{\Delta_x} = \langle A \,|\, E_{\Delta_x}^1 \otimes \mathbb{1}^2 + \mathbb{1}^1 \otimes E_{\Delta_x}^2 \,|\, A \rangle = \int_{\Delta_x} \left(\rho_1(x') + \rho_2(x') \right) dx'$$

$$\rho_1(x) \equiv \int_{-\infty}^{+\infty} |\Psi_A(x, x_2)|^2 \, dx_2 \, , \qquad \rho_2(x) \equiv \int_{-\infty}^{+\infty} |\Psi_A(x_1, x)|^2 \, dx_1 \; .$$

So, in a probabilistic sense, $\rho_1(x)+\rho_2(x)$ represents the density of particles.

15.12

a) $\langle A_\pm \,|\, f(q_1, q_2) \,|\, A_\pm \rangle = \int |\phi(x_1) \chi(x_2)|^2 \, f(x_1, x_2) \, dx_1 dx_2$

$$\pm \int \phi^*(x_1) \chi(x_1) f(x_1, x_2) \chi^*(x_2) \phi(x_2) \, dx_1 dx_2$$

$\langle A_0 \,|\, f(q_1, q_2) \,|\, A_0 \rangle = \int |\phi(x_1) \chi(x_2)|^2 \, f(x_1, x_2) \, dx_1 dx_2 \; .$

The difference among the three mean values is given by the exchange integral.

b) The exchange integral is vanishing, so the three mean values are equal to one another. Even if the two particles are identical, in this case either the symmetrization or the antisymmetrization are not necessary.

15.13

a) As the initial states $|A_0\rangle$ and $|B_0\rangle$ are orthogonal to each other, such are also the final states; so, up to a phase factor:

$$|B_0\rangle \to \tfrac{1}{\sqrt{2}}(|B\rangle - |A\rangle) \,.$$

b) If the two particles are distinguishable, the initial state is $|A_0, B_0\rangle$, so:

$$|A_0, B_0\rangle \to \tfrac{1}{2}(|A\rangle + |B\rangle)(|A\rangle - |B\rangle)$$
$$= \tfrac{1}{2}(|A,A\rangle - |B,B\rangle - |A,B\rangle + |B,A\rangle)$$

therefore the probability for having a coincidence is $\tfrac14 + \tfrac14 = \tfrac12$.
If the two particles are identical particles:

$$\tfrac{1}{\sqrt{2}}(|A_0, B_0\rangle \pm |B_0, A_0\rangle) \to \begin{cases} \tfrac{1}{\sqrt{2}}(|A,A\rangle - |B,B\rangle) & \text{(bosons)} \\ \tfrac{1}{\sqrt{2}}(|A,B\rangle - |B,A\rangle) & \text{(fermions)} \end{cases}$$

so for bosons the probability for coincidences is 0, whereas for fermions it is 1: in fact two identical fermions cannot be in the same state.

15.14

a) Each photon has a probability $\tfrac12$ to be revealed by any of the counters; if the photons are sent one at a time, the probability P_{n_1,n_2} is given by the binomial distribution:

$$P_{n_1,n_2} = \frac{n!}{n_1!\,n_2!}\left(\frac{1}{2}\right)^n .$$

If the photons are sent together, each of them in the state $|A_0\rangle$, the initial state is $|A_0 \cdots A_0\rangle$ and $|A_0\rangle \to \tfrac{1}{\sqrt{2}}(|A\rangle + |B\rangle)$, so:

$$|\overbrace{A_0 \cdots A_0}^{n \text{ times}}\rangle \to \left(\tfrac{1}{\sqrt{2}}\right)^n (|A\rangle + |B\rangle) \cdots (|A\rangle + |B\rangle) \,.$$

In order to calculate the required probability we must take the scalar product between the vector to the right and the vector in which n_1 photons are in the state $|A\rangle$ and n_2 in the state $|B\rangle$, namely the normalized vector:

$$|n_1, n_2\rangle \equiv \left(\frac{n!}{n_1!\,n_2!}\right)^{-1/2} \mathcal{S}\,|\overbrace{A,\cdots A}^{n_1 \text{ times}}, \overbrace{B,\cdots B}^{n_2 \text{ times}}\rangle$$

where $\mathcal{S} \mid \overbrace{A, \cdots A}^{n_1 \text{ times}}, \overbrace{B, \cdots B}^{n_2 \text{ times}} \rangle$ is the symmetric combination of the $n!/n_1! n_2!$ vectors in which A appears n_1 times and B appears n_2 times:

$$P_{n_1,n_2} = \left| \left(\tfrac{1}{\sqrt{2}} \right)^n ((\langle A| + \langle B|) \cdots (\langle A| + \langle B|) \mid n_1, n_2 \rangle \right|^2$$

$$= \left(\tfrac{1}{2} \right)^n \left(\frac{n!}{n_1! n_2!} \right)^{-1} \times \left(\frac{n!}{n_1! n_2!} \right)^2 = \frac{n!}{n_1! n_2!} \left(\tfrac{1}{2} \right)^n$$

as in the case when the photons are independent.

b) One has $\bar{n}_1 = \bar{n}_2 = \tfrac{1}{2} n$.

$$\overline{n_1 \times n_2} = \sum_{n_1=1}^{n-1} n_1 n_2 \, P_{n_1,n_2} = \frac{1}{4} n(n-1) \sum_{n_1-1=0}^{n-2} \frac{(n-2)!}{(n_1-1)! \, (n_2-1)!} \left(\tfrac{1}{2} \right)^{n-2}$$

$$= \frac{1}{4} n(n-1) \quad \Rightarrow \quad g^{(2)} = 1 - \frac{1}{n} \,.$$

c) The absence of quantum fluctuations is equivalent to $n_1 \equiv \bar{n}_1$, $n_2 \equiv \bar{n}_2$, whence $g^{(2)} = 1$. The smaller the number of sent photons, the greater the difference between the classical and the quantum result (the fluctuations are more sizable).

Quantum fluctuations were given the first experimental evidence by Hanbury Brown and Twiss in 1956.

15.15

a) In the triplet state $H' \equiv V(r) = -V_0 \chi(r)$ (potential well of depth V_0). Taking $\hbar^2/mr_0^2 = 10 \, \text{MeV}$ as unit of energy (see Problem 6.11: the reduced mass of the n–p system is $\tfrac{1}{2} m$) and putting $v_0 \equiv V_0/(\hbar^2/mr_0^2)$, $\epsilon_d \equiv E_d/(\hbar^2/mr_0^2) = 0.22$, the equations that determine V_0 are:

$$\frac{\xi^2}{\tan^2 \xi} = \epsilon_d \quad (\xi \geq \pi/2); \qquad v_0 = \xi^2 + \frac{\xi^2}{\tan^2 \xi}$$

whose solution, found numerically, is $\xi = 1.82 \Rightarrow V_0 \simeq 36 \, \text{MeV}$. Alternatively, since the system is weakly bound, $\xi \simeq \pi/2$ whence $\xi^2/\tan^2 \xi \simeq (\pi/2)^2 (\xi - \pi/2)^2$ and in this way $\xi \simeq 1.87$ obtains.
In the singlet state $H' = [-V_0 + 4a] \chi(r)$ (well of depth $V_0 - 4a$) and since such a state has got energy $\simeq 0$, one has $\xi \simeq \pi/2 \Rightarrow V_0 - 4a \simeq (\pi/2)^2 \hbar^2/mr_0^2 \Rightarrow a \simeq 2.7 \, \text{MeV}$.

b) The singlet states are associated with states that are symmetric under the exchange of the orbital variables, and vice versa for the triplet states. If $\vec{q}, \vec{p}$ and $\vec{Q}, \vec{P}$ respectively are the relative and center-of-mass variables and Π_o the operator that exchanges the orbital variables, one has:

$$\Pi_o \, \vec{q} \, \Pi_o^{-1} = -\vec{q}, \; \Pi_o \, \vec{p} \, \Pi_o^{-1} = -\vec{p}; \quad \Pi_o \, \vec{Q} \, \Pi_o^{-1} = \vec{Q}, \; \Pi_o \, \vec{P} \, \Pi_o^{-1} = \vec{P} \,.$$

Therefore Π_o coincides with the space inversion of the relative variables. Recalling that for a particle the parity of the states of given L is $(-1)^L$, the states with even L are symmetric in the orbital variables and those with odd L are antisymmetric: so the possible pairs are (L even, $S = 0$), (L odd, $S = 1$). Of course also superpositions thereof are allowed.

c) As the unique bound state for the two particles interacting through H' is a state with $L = 0$ and $S = 1$, it cannot be the state relative to two identical fermions.

15.16

a) The state of the system is the pure state ($|+\rangle$ and $|-\rangle$ eigenvectors of any component, say $\hat{n}$, of $\vec{\sigma}$, see Problem 15.5):

$$|X, t\rangle = \tfrac{1}{\sqrt{2}}|r, t\rangle_1 |l, t\rangle_2 \Big(|+\rangle_1 |-\rangle_2 - |-\rangle_1 |+\rangle_2\Big)$$

whereas (see Problem 15.4) the state of particle 1 is the statistical mixture $\{|r, t\rangle|+\rangle, \tfrac{1}{2};\ |r, t\rangle|-\rangle, \tfrac{1}{2}\}$.

b) If the measurement of $\vec{\sigma}_2 \cdot \hat{n}$ yields the result $+1$, the system immediately after the measurement is in the state $|r, t\rangle_1 |l, t\rangle_2 |-\rangle_1 |+\rangle_2$, therefore particle 1 is in the spin state $|-\rangle_1$ and a measurement of $\vec{\sigma}_1 \cdot \hat{n}$ will yield the result -1; if instead the measurement of $\vec{\sigma}_2 \cdot \hat{n}$ has given the result -1, the measurement of $\vec{\sigma}_1 \cdot \hat{n}$ will give the result $+1$.

c) If for $t < \bar{t}$ particle 1 were in a pure spin state, e.g. $|-\hat{n}\rangle_1$, then particle 2 should be in the state $|+\hat{n}\rangle_2$, both by symmetry (no measurement has yet been made) and because the component S_n of the total spin is vanishing. So 50% of the couples should be in the spin state $|-\hat{n}\rangle_1 |+\hat{n}\rangle_2$ and the other 50%, in the state $|+\hat{n}\rangle_1 |-\hat{n}\rangle_2$, i.e. the system would be (for $t < \bar{t}$) in the statistical mixture $\{|-\hat{n}\rangle_1 |+\hat{n}\rangle_2, \tfrac{1}{2};\ |+\hat{n}\rangle_1 |-\hat{n}\rangle_2, \tfrac{1}{2}\}$, whereas by assumption the system has been prepared in the singlet spin state. In addition, since different $\hat{n}$'s correspond to different mixtures, the state (mixture) of the system for $t < \bar{t}$ would depend on which component of the spin one will decide to measure at time $\bar{t}$.

Therefore, according to quantum mechanics, it is wrong to assume that, in the absence of interactions between the two particles, the measurement effected on particle 2 cannot produce effects on the state of particle 1 (Einstein–Podolsky–Rosen paradox).

15.17

a) One has:

$$|A\rangle = \tfrac{1}{2}\Big(\alpha(|B_1\rangle + |B_2\rangle)|-\rangle_3 - \alpha(|B_3\rangle + |B_4\rangle)|+\rangle_3$$
$$+ \beta(|B_3\rangle - |B_4\rangle)|-\rangle_3 - \beta(|B_1\rangle - |B_2\rangle)|+\rangle_3\Big)$$

$$= \tfrac{1}{2} \,|\,\mathcal{B}_1\,\rangle \big(\alpha\,|-\rangle_3 - \beta\,|+\rangle_3 \big) + \tfrac{1}{2}\,|\,\mathcal{B}_2\,\rangle \big(\alpha\,|-\rangle_3 + \beta\,|+\rangle_3 \big)$$
$$+ \tfrac{1}{2}\,|\,\mathcal{B}_3\,\rangle \big(-\alpha\,|+\rangle_3 + \beta\,|-\rangle_3 \big) + \tfrac{1}{2}\,|\,\mathcal{B}_4\,\rangle \big(-\alpha\,|+\rangle_3 - \beta\,|-\rangle_3 \big)$$

so the four results all have probability $1/4$ and, for each of them, the state of the third particle after the measurement is

$$|\,\mathcal{B}_1\,\rangle : \quad |\,b_1\,\rangle_3 \equiv \alpha\,|-\rangle_3 - \beta\,|+\rangle_3 \,; \qquad |\,\mathcal{B}_2\,\rangle : \quad |\,b_2\,\rangle_3 \equiv \alpha\,|-\rangle_3 + \beta\,|+\rangle_3 \,;$$
$$|\,\mathcal{B}_3\,\rangle : \quad |\,b_3\,\rangle_3 \equiv \alpha\,|+\rangle_3 - \beta\,|-\rangle_3 \,; \qquad |\,\mathcal{B}_4\,\rangle : \quad |\,b_4\,\rangle_3 \equiv \alpha\,|+\rangle_3 + \beta\,|-\rangle_3 \,.$$

b) In the basis $|+\rangle$, $|-\rangle$ (the index 3 is hereafter omitted) one has:

$$|\,b_1\,\rangle \to \begin{pmatrix} -\beta \\ \alpha \end{pmatrix}; \quad \begin{pmatrix} 0 & 1 \\ -1 & 0 \end{pmatrix} \begin{pmatrix} -\beta \\ \alpha \end{pmatrix} = \begin{pmatrix} \alpha \\ \beta \end{pmatrix} \quad \Rightarrow \quad |\,x\,\rangle = -\mathrm{i}\,\sigma_y\,|\,b_1\,\rangle \,.$$

Likewise $|\,x\,\rangle = \sigma_x\,|\,b_2\,\rangle$, $|\,x\,\rangle = \sigma_z\,|\,b_3\,\rangle$, $|\,x\,\rangle = |\,b_4\,\rangle$.
The operator that performs the rotation of angle ϕ around the direction $\hat{n}$ on the spin states of a particle of spin $\tfrac{1}{2}$ is (see Problem 13.2):

$$U(\hat{n},\,\phi) = e^{\frac{1}{2}\mathrm{i}\phi\,\vec{\sigma}\cdot\hat{n}} = \cos(\phi/2) + \mathrm{i}\,\vec{\sigma}\cdot\hat{n}\,\sin(\phi/2)$$

so, up to phase factors, σ_x, σ_y, σ_z implement rotations by $180°$ respectively around the x, y, z axes. If the result is $\mathcal{B}_4$ no rotation must be performed.

c) $\quad |\,x\,\rangle_1\,|+\rangle_2\,|-\rangle_3 = \dfrac{1}{\sqrt{2}} \Big(\alpha \big(\,|\,\mathcal{B}_1\,\rangle + |\,\mathcal{B}_2\,\rangle\,\big)\,|-\rangle_3 + \beta \big(\,|\,\mathcal{B}_3\,\rangle - |\,\mathcal{B}_4\,\rangle\,\big)\,|-\rangle_3 \Big)$

so the probabilities respectively are $\tfrac{1}{2}|\alpha|^2$, $\tfrac{1}{2}|\alpha|^2$, $\tfrac{1}{2}|\beta|^2$, $\tfrac{1}{2}|\beta|^2$.
Particle 3 always is in the state $|-\rangle$ that does not contain information about the state $|\,x\,\rangle$ so, in this case, teleportation of the state $|\,x\,\rangle$ is not possible.

16

Applications to Atomic Physics

Perturbations on the fine structure energy levels of the hydrogen atom; electronic configurations and spectral terms; fine structure; Stark and Zeeman effects; intercombination lines.

Note. *Throughout the chapter angular momenta are measured in units of $\hbar$ and are, therefore, dimensionless.*

16.1 Due to relativistic effects and to the quantization of the electromagnetic field (Lamb shift), the hydrogen energy level with $n = 2$ splits into three levels, respectively with $l = 1$, $j = \frac{1}{2}$ $(2p_{\frac{1}{2}})$; $l = 0$, $j = \frac{1}{2}$ $(2s_{\frac{1}{2}})$; $l = 1$, $j = \frac{3}{2}$ $(2p_{\frac{3}{2}})$. The separations, expressed in wave numbers (see Problem 11.13) are $E_2 - E_1 = 0.0353\,\text{cm}^{-1}$, $E_3 - E_2 = 0.3652\,\text{cm}^{-1}$ (see figure).

a) Relying exclusively on what is reported above and summarized in the figure above, say, for any of the following observables, which certainly is not a constant of motion and which may be such:

$$L_x\,,\ L_y\,,\ L_z\,;\quad \vec{L}^{\,2}\,;\quad j_x\,,\ j_y\,,\ j_z\,;\quad \vec{j}^{\,2}\,;\quad I \text{ (space inversion)};\quad I_z\sigma_z$$

$(I_z: x \to x\,,\ y \to y\,,\ z \to -z$ and likewise for $p_x\,,\ p_y\,,\ p_z\,)$.

Owing to effects due to weak interactions, it may be necessary to add a term as the following to the Hamiltonian (whose levels are those described above):

$$H' = \tfrac{1}{2}\Big(g(r)\,(\vec{\sigma}\cdot\vec{p}) + (\vec{\sigma}\cdot\vec{p})\,g(r)\Big)\,.$$

b) Which, among the possible constants of motion found in a), certainly are no longer such?

We want now to study the effects of H' on the levels with $n = 2$.

c) Denote by $a, b \cdots$ the nonvanishing matrix elements of H' among the states with $n = 2$ (being so careful as to indicate equal matrix elements with equal letters, etc.). Restricting to the subspace of the states with $n = 2$, find the eigenvalues of the complete Hamiltonian $H_0 + H'$ in terms of the above parameters.

© Springer International Publishing AG 2017
E. d'Emilio and L.E. Picasso, *Problems in Quantum Mechanics*,
UNITEXT for Physics, DOI 10.1007/978-3-319-53267-7_16

d) How many lines are observed in the transitions between the energy levels with $n = 2$ and $n = 1$?

16.2   A hydrogen atom is subject to a constant uniform magnetic field $\vec{B}$. Neglecting the terms quadratic in B (the one in parenthesis is the most important), the Hamiltonian is

$$H = H_0 + \frac{e\,\hbar}{2mc}(\vec{L} + 2\vec{s}) \cdot \vec{B} \; + \left[\frac{e^2}{8mc^2}(\vec{r} \wedge \vec{B})^2\right] \tag{1}$$

where H_0 is the Hamiltonian of the atom in absence of magnetic field, whose energy levels with $n = 2$ are those described in Problem 16.1. In a first approximation, the radial wavefunctions of the states $2p_{\frac{1}{2}}$ and $2p_{\frac{3}{2}}$ can be taken equal.

a) Say which, among the eigenstates of H_0 with $n = 2$, are also eigenstates of H and find the corresponding eigenvalues.

Let $E_{2,i}(B)$, $i = 1 \cdots ?$, be the other eigenvalues of H restricted to the states with $n = 2$.

b) Taking in consideration only the matrix elements of H among the eigenstates of H_0 with $n = 2$, find the eigenvalues $E_{2,i}(B)$ and say for which values of B is the *weak field approximation* legitimate, and for which values, instead, is the *strong field approximation* more suitable. Since in (1) the terms in B^2 have been neglected, is it meaningful to keep the quadratic terms in the expression of $E_{2,i}(B)$?

c) How many lines are observed in the transitions to the ground state $1s_{\frac{1}{2}}$?

16.3   A hydrogen atom, whose levels with $n = 2$ are those described in Problem 16.1, is subject to a constant uniform electric field $\vec{\mathcal{E}}$.

a) Say which is the number of levels with $n = 2$ in the presence of the electric field and which is the behaviour (i.e. linear, quadratic, ...) of such energy levels as a function of $\mathcal{E}$ for $\mathcal{E} \to 0$.

b) Restricting to the first nonvanishing order of perturbation theory and ignoring the contributions from the levels with $n \neq 2$, calculate the effect of an electric field $\mathcal{E} = 600 \, \mathrm{V/cm}$ on the level $2p_{\frac{3}{2}}$. In order to calculate the matrix element of the electric dipole operator, use the value that is obtained from the nonrelativistic theory:
$\langle\, n = 2, l = 0, m = 0 \mid z \mid n = 2, l = 1, m = 0\,\rangle = 3\,a_{\mathrm{B}}$
and the results of Problem 15.8b.

Since $E_3 - E_2 \simeq 10\,(E_2 - E_1)$, for suitable values of $\mathcal{E}$ it is possible to calculate its effect of the on the levels $2p_{\frac{1}{2}}$ and $2s_{\frac{1}{2}}$ neglecting the matrix elements among the states of the mentioned levels and those of the level $2p_{\frac{3}{2}}$.

c) In the above mentioned approximation (and a fortiori ignoring the contributions from the levels with $n \neq 2$) calculate the effect of the electric field on the levels $2p_{\frac{1}{2}}$, $2s_{\frac{1}{2}}$ and say to what extent for $\mathcal{E} = 600\,\mathrm{V/cm}$ the suggested approximation is applicable.

16.4   The He atom has the levels (notation: (configuration) $^{2S+1}L_J$) $1s2s\,^1S_0$ and $1s2p\,^1P_1$ separated by a fraction of eV. A method to determine the value 2δ of such a separation consists in making measurements of spectroscopic type on atoms in a constant, uniform electric field: if the atom is subject to an electric field, even an intense one, the latter will perturb the two levels, without however appreciably mixing them with other states of the atom. The calculation of the effect of the electric field may, therefore, be made limiting oneself to the subspace spanned by the states belonging to the two considered levels.

a) Assume δ and the matrix elements of the perturbation are known. Find the stationary states in presence of the field and the corresponding energies.

The π lines (i.e. corresponding to transitions with $\Delta M = 0$), spontaneously emitted in the transitions from the considered levels (in presence of the field) to the lowest energy level, are observed. The difference between the frequencies of the mentioned lines corresponds to $\Delta E = 0.75\,\mathrm{eV}$ and the ratio R between the intensities is $R \simeq 1/9$.

b) Find the separation 2δ between the unperturbed energy levels.

c) If the electric field is absent, how many lines are observed in the transitions from the considered levels to the lowest energy level? In what region of the electromagnetic spectrum does the transition line between the two excited levels $1s2s\,^1S_0$, $1s2p\,^1P_1$ fall? Explain why, in the absence of the electric field, it is difficult to measure 2δ.

16.5   Consider atoms with two electrons and nuclear charge $Z \geq 2$ (helium-like ions). The first ionization energies (namely the minimum energy necessary to extract only one electron) from the neutral helium (He^I) to the four times ionized carbon (C^V) are, in eV:

He^I	Li^{II}	Be^{III}	B^{IV}	C^V
24.58	75.62	153.85	259.3	392.0

Since we are interested in the ground state of the above atoms, we shall ignore the spin of the electrons.

a) Demonstrate that the Hamiltonian:

$$H_Z = \frac{\vec{p}_1^{\,2}}{2m_e} - \frac{Ze^2}{r_1} + \frac{\vec{p}_2^{\,2}}{2m_e} - \frac{Ze^2}{r_2} + \frac{e^2}{r_{12}}$$

is unitarily equivalent to:

$$\tilde{H}_Z = Z^2 \left(\frac{\vec{p}_1^2}{2m_e} - \frac{e^2}{r_1} + \frac{\vec{p}_2^2}{2m_e} - \frac{e^2}{r_2} + \frac{e^2/Z}{r_{12}} \right) = Z^2 \left(H^{(0)} + \frac{e^2/Z}{r_{12}} \right) \quad (1)$$

and, as a consequence, that the energy of the ground state is given by the series:

$$E_Z = Z^2 E^{(0)} + Z\, a + b + \frac{1}{Z}\, c + \cdots .$$

b) Find $E^{(0)}$ and a. Use:

$$\langle 1,0,0; 1,0,0 \,|\, \frac{e^2}{r_{12}} \,|\, 1,0,0; 1,0,0 \rangle = \frac{5}{8} \frac{e^2}{a_B}$$

where $|1,0,0\rangle$ is the ground state of the hydrogen atom ($Z=1$) and, neglecting the terms $b + c/Z + \cdots$, find the first ionization energies for $Z = 2, \cdots, 6$. Compare the results with the experimental data given above.

c) Show that the correction to the first ionization energy due to the terms $-p_i^4/8m_e^3 c^2$ (see Problem 12.20) is of the order of $Z^4 \alpha^2 \times (e^2/a_B)$ and use the results of Problem 12.20 to calculate this relativistic correction from helium to carbon.

A simple way to obtain, with a good approximation, the energies of the lowest energy levels of the two-electron atoms consists in replacing the Coulombic repulsion between the electrons with a smaller nuclear charge (screening effect): $Z \to Z' = Z - \sigma$, namely:

$$H_Z \to H_{Z'}^{(0)} \equiv \frac{\vec{p}_1^2}{2m_e} - \frac{Z'e^2}{r_1} + \frac{\vec{p}_2^2}{2m_e} - \frac{Z'e^2}{r_2} .$$

d) The energy of the ground state of $H_{Z'}^{(0)}$ has the form $E_{Z'}^{(0)} = Z^2 E^{(0)} + Z\, a' + b'$: find σ in such a way that $a' = a$; calculate the first ionization energies for $Z = 2, \cdots, 6$ and compare the results with the experimental data.

16.6   Atoms with two electrons and nuclear charge $Z \geq 2$ ((helium-like ions) are subject to a constant, uniform magnetic field $\vec{B}$.

a) Calculate, in the approximation of independent electrons (i.e. ignoring the repulsion between them), the effect of the magnetic field on the lowest energy level to the first order in perturbation theory: exploit the identity given in Problem 12.14, suitably adapted to the case $Z \neq 1$.

b) Calculate, in the case of helium, the first order effect of the magnetic field, assuming as Hamiltonian in the absence of field the Hamiltonian $H_{Z'}^{(0)}$ defined in Problem 16.5.

16.7 The first excited configuration of the helium atom is $1s\,2s$.

a) Write the spectral terms ^{2S+1}L deriving from this configuration and their energies to the first order in the Coulombic repulsion, in terms of both the "direct integral" Δ_0 and the *exchange integral* Δ_1:

$$\Delta_0 = \langle\, (2s)_2 , (1s)_1 \,|\, \frac{e^2}{r_{12}} \,|\, (1s)_1 , (2s)_2 \,\rangle,$$

$$\Delta_1 = \langle\, (1s)_2 , (2s)_1 \,|\, \frac{e^2}{r_{12}} \,|\, (1s)_1 , (2s)_2 \,\rangle .$$

b) Show that the calculations of both Δ_0 and Δ_1 can be brought back to the calculation of the electrostatic energy of a charge distribution ρ_1 in the field of a charge distribution ρ_2 (possibly $\rho_1 = \rho_2$). Use this result and determine the sign of the exchange integral.

For the calculations of Δ_0 and Δ_1 it is also possible to exploit the fact that:

$$I \equiv \int \frac{d\Omega_1\, d\Omega_2}{r_{12}} \equiv \int \frac{d\Omega_1\, d\Omega_2}{\sqrt{r_1^2 + r_2^2 - 2r_1 r_2 \cos\theta_{12}}}$$

$$= \int d\Omega_1 \int \frac{d\Omega_{12}}{\sqrt{r_1^2 + r_2^2 - 2r_1 r_2 \cos\theta_{12}}}$$

and, after performing the angular integrations, to distinguish the integration regions $r_1 > r_2$ and $r_2 > r_1$.

c) Exploit the second method and calculate Δ_0. Use the methods of electrostatics (Gauss' theorem) and calculate Δ_1. Find the energies of the levels $1s\,2s$ and compare the results with the experimental data: the energies, referred to those of the ion He^+, are $E' = -4.76\,eV$, $E'' = -3.96\,eV$. Say which is the sign of the contribution of the second order to the computed energies.

16.8 The normal configuration of the carbon atom is $(1s)^2 (2s)^2\, 2p^2$.

a) Find the spectral terms ^{2S+1}L associated with this configuration.

In the *Russell–Saunders scheme*, in which the energy levels are classified by means of the quantum numbers L and S, the spin-orbit interaction is – to the first order – equivalent to $A_{LS}\, \vec{L} \cdot \vec{S}$, A_{LS} being a numerical coefficient. The levels E_{LSJ}, with the same L, S and J ranging from $|L - S|$ to $L + S$, give rise to a *fine structure multiplet*.

b) Show that the *Landé interval rule* $E_{LS,J} - E_{LS,J-1} = A_{LS}\, J$ holds.

The lowest energy levels of carbon are the fine structure multiplet $^3P_{0,1,2}$ (notation $^{2S+1}L_J$) with energies (expressed in wave numbers as in Problem 11.13) $E_0 = 0$, $E_1 = 16.4\,cm^{-1}$, $E_2 = 43.5\,cm^{-1}$.

c) Calculate the ratio $(E_2 - E_0)/(E_1 - E_0)$ and verify whether the result is in agreement with the Landé interval rule.

16.9 Consider the configuration: (filled shells) p^n, $n \le 6$.

a) Show that the number of independent (spin and orbital) states corresponding to this configuration equals that of the configuration (filled shells) p^{6-n}.

All the independent states corresponding to the configuration p^n may be obtained by calculating the determinants (*Slater determinants*) of the $n \times n$ minors of the $n \times 6$ matrix obtained by arbitrarily selecting n columns (we denote by $| m, s \rangle_i$ the state of the i-th electron: e.g. $| 1, - \rangle \equiv | m = 1, s = -\frac{1}{2} \rangle$):

$$\begin{pmatrix} | 1, + \rangle_1 & | 1, - \rangle_1 & | 0, + \rangle_1 & | 0, - \rangle_1 & | -1, + \rangle_1 & | -1, - \rangle_1 \\ \vdots & & & & & \\ | 1, + \rangle_n & | 1, - \rangle_n & | 0, + \rangle_n & | 0, - \rangle_n & | -1, + \rangle_n & | -1, - \rangle_n \end{pmatrix}.$$

b) Show that, for every state $| M_L, M_S \rangle$ belonging to the configuration p^n, there exists a state with the same quantum numbers belonging to the configuration p^{6-n}. Therefore the two configurations p^n and p^{6-n} give rise to the same spectral terms.

The result may be expressed by saying that the configuration p^{6-n} is equivalent to the configuration (filled shell p^6) $\bar{p}^n$, where $\bar{p}^n$ is the configuration of n "missing electrons", namely of n "holes" (this holds also for configurations consisting of d, f, $\cdots$ electrons): it is indeed possible to demonstrate that, to the first order in the Coulombic repulsion and in the spin-orbit interaction $H_{so} = \sum_i \xi(r_i)\, \vec{l}_i \cdot \vec{s}_i$, the holes behave, as far as energy is concerned, as particles of charge opposite to that of the electron.

c) The fluorine atom has atomic number $Z = 9$ and, in the periodic table, it precedes neon (noble gas). Its normal configuration is $(1s)^2(2s)^2\, 2p^5$. Find the corresponding spectral term and the number of independent states.

d) Find the spectral terms associated to the normal configuration of the oxygen atom $(1s)^2(2s)^2\, 2p^4$.

The lowest energy levels of the oxygen give rise to a fine structure triplet with energies $E_0 = 0$, $E_1 = 158.5\,\text{cm}^{-1}$, $E_2 = 226.5\,\text{cm}^{-1}$.

e) Say which are the quantum numbers L, S, J of the above mentioned three levels and with which approximation is the Landé interval rule verified (see Problem 16.8).

16.10   The normal configuration of nitrogen is $(1s)^2(2s)^2\,2p^3$.

a) Say whether this configuration can give rise to states with orbital angular momentum $L = 3$ (F states).

b) The states of the lowest energy level have total spin $S = 3/2$. What is the orbital angular momentum L?

c) Write the orbital part of the states of the lowest energy level in terms of the sates $|\,m_1, m_2, m_3\,\rangle$, the eigenstates of l_{1z}, l_{2z}, l_{3z}.

The configuration $(1s)^2(2s)^2\,2p^3$ gives also rise to P ($L = 1$) and D ($L = 2$) terms.

d) Say what is the total spin of such spectral terms and to how many fine structure energy levels does each of them give rise to.

16.11   Consider the excited configuration $(1s)^2(2s)^2\,2p\,3p$ of the carbon atom and let $R_2(r)$ and $R_3(r)$ be the radial wavefunctions of the p electrons, respectively with $n = 2$ and $n = 3$.

a) Find the possible values of the total orbital angular momentum L and, for each value of L, determine the values the total spin S may assume: note that the $2p$ and $3p$ electrons are not "equivalent electrons" ($R_2(r) \neq R_3(r)$).

b) Using Cartesian coordinates as in Problem 15.10, write the orbital wavefunctions of the two p electrons relative to the different spectral terms ^{2S+1}L found above.

For each value of L, the energy difference between the singlet and the triplet states – to the first order in the Coulombic repulsion – is determined by the behaviour of the wavefunction of the two p electrons for $\vec{r}_1 \approx \vec{r}_2$, namely in the region where e^2/r_{12} is 'large'.

c) For each fixed value of L, establish for which value of S does the spectral term have the lowest energy. Is *Hund's rule* – according to which the lowest energy belongs to the term with the highest spin – always fulfilled? Which, among all the considered levels, has the lowest energy?

16.12   Spectroscopic analysis has established that a certain atom, whose identity is unknown, has the following energy levels in eV (E_0 is the lowest energy level):

$$E_0 = 0, \qquad E_1 = 1.7 \times 10^{-2}, \qquad E_2 = 4 \times 10^{-2}, \qquad E_3 = 7 \times 10^{-2}$$

separated by about $0.3\,\text{eV}$ from the next higher level. It is therefore reasonable to assume that the levels $E_0, \cdots, E_3$ constitute a fine structure multiplet. Even if the Landé interval rule ($E_{LS,J} - E_{LS,J-1} = A_{LS}\,J$) is not expected to hold with good accuracy, some pieces of information may however be obtained from it, both in qualitative and in quantitative character.

a) Establish whether one is dealing with a direct multiplet ($A_{LS} > 0$) or, rather, with an inverted one ($A_{LS} < 0$). Say which is the value of the total spin S and the minimum value the orbital angular momentum L may have. Are the values of the total angular momentum J integers or half-integers?

b) Having established whether J is integer or half-integer, determine the value J_0 of the total angular momentum of the lowest energy level that gives rise to the best approximation to the Landé rule. Find L.

c) Taking into account all the configurations p^{n_1} and d^{n_2} for the electrons external to the filled shells, say which is the only configuration compatible with the found results. Knowing that the orbitals have to be filled according to the following order: $1s$, $2s$, $2p$, $3s$, $3p$, $4s$, $3d$, $4p$, $5s$, $\cdots$, say which is the first atom exhibiting the found configuration.

16.13 An excited configuration of the barium atom ($Z = 56$) is

$$(1s)^2 \cdots (5s)^2(5p)^6 \, 6s \, 6p \, .$$

In the central field approximation, the Hamiltonian of the outer electrons is

$$H = \sum_1^2 \left(\frac{\vec{p}_i^{\,2}}{2m} + V(r_i) + \xi(r_i) \, \vec{l}_i \cdot \vec{s}_i \right) + \frac{e^2}{r_{12}} \, .$$

In this case, due to the high value of Z, the intensity of the spin-orbit interaction $H_{so} = \sum_i \xi(r_i) \, \vec{l}_i \cdot \vec{s}_i$ of the two external electrons may be considered comparable with the intensity of the Coulombic repulsion.

a) Classify the energy levels of the configuration $6s \, 6p$ in the Russell–Saunders approximation, in which the spin-orbit interaction is considered as a first order perturbation (see Problem 16.8).

b) Classify the energy levels of the configuration $6s \, 6p$ in terms of the quantum numbers j_1, j_2 ($\vec{j} = \vec{l} + \vec{s}$) of the single electrons, in the approximation in which the Coulombic repulsion is totally neglected and find the relative degeneracies.

c) Find the possible values of the total angular momentum J for each of the levels found in b). Which is the effect of the Coulombic repulsion on these levels?

16.14 Consider the barium atom in the configuration: (filled shells) $6s \, 6p$, as in Problem 16.13. The energies in cm^{-1} of the four levels generated by this configuration, referred to the lowest energy level, are:

$$E_1 = 12266 \, , \; E_2 = 12637 \, , \; E_3 = 13515 ; \quad E_4 = 18060 \, .$$

In the central field approximation the Hamiltonian of the electrons $6s \, 6p$ is that given in the previous problem: however, in order to make quantitative predictions and since we are interested only in the states $6s\,6p$ of the two electrons, the Hamiltonian can be approximated as:

$$H = \sum_1^2 \left(\frac{\vec{p_i}^2}{2m} + V(r_i) + A\,\vec{l_i} \cdot \vec{s_i} \right) + \frac{e^2}{r_{12}}$$

where A is a numerical positive coefficient given by $\langle 6p \,|\, \xi(r) \,|\, 6p \rangle$.

a) Exploit the identity $\vec{l_1} \cdot \vec{s_1} + \vec{l_2} \cdot \vec{s_2} = \frac{1}{2}(\vec{L} \cdot \vec{S} + (\vec{l_1} - \vec{l_2}) \cdot (\vec{s_1} - \vec{s_2}))$ and calculate, in terms of A and in the Russell–Saunders approximation, the fine structure separations of the levels belonging to the $6s\,6p$ configuration, determined in Problem 16.13 a). Compare the value of the ratio $(E_3 - E_1)/(E_2 - E_1)$ with the theoretical prediction.

Take for known the value of the exchange integral:

$$\Delta_1 = \langle (6s)_2, (6p)_1 \,|\, \frac{e^2}{r_{12}} \,|\, (6s)_1, (6p)_2 \rangle \,.$$

b) Write the matrix representing $H' = H_{so} + e^2/r_{12}$ in the basis $^{2S+1}L_J$ of the four spectral terms deriving from the $6s\,6p$ configuration: in order to find the matrix elements that would require explicit calculation, it is convenient to exploit the fact that, in absence of Coulombic repulsion, one must find again the results of Problem 16.13 b). Once the rows and the columns of the matrix have been suitably ordered, find its eigenvalues.

c) Use the expressions of the eigenvalues in terms of A and Δ_1 found in b) and find two linear equations that allow one to determine the values of A and Δ_1 from the experimental data. Calculate numerically the value of the ratio A/Δ_1 and estimate the error one makes in treating the spin-orbit interaction in the Russell–Saunders scheme.

d) From the results of the preceding point, find a theoretical prediction for $E_4 - E_2$ and compare it with the experimental datum.

16.15   The first excited energy levels of mercury ($Z = 80$), generated by the configuration: $(1s)^2 \cdots (5s)^2 (5p)^6 (4f)^{14} (5d)^{10}\, 6s\,6p$, referred to the normal configuration: (filled shells) $(6s)^2$, expressed in eV, are the following:

$$E_1 = 4.667\,, \quad E_2 = 4.886\,, \quad E_3 = 5.461\,; \quad E_4 = 6.703$$

respectively classified – in the present case not appropriately, but according to the tradition – as $^3P_0$, $^3P_1$, $^3P_2$; $^1P_1$.

a) Show that the perturbative (first order) treatment of the spin-orbit interaction, particularly the Landé interval rule, cannot reproduce the fine structure of the ^3P levels with an approximation better than about 20%. If the classification of the energy levels in terms of L and S were correct, how many lines – in the transitions from the above mentioned levels to the lowest energy level – should one observe?

The correct answer to the above question, concerning the number of lines, actually disagrees with the observed number. In the light of this situation, the classification by means of the total spin S is inadequate, and we want here to find the correct expression for the stationary states. Since the other energy levels of mercury are not too close, one is allowed to take the spin-orbit interaction H_{so} into consideration in a nonperturbative way, limiting the calculation to the only states generated by the configuration: (filled shells) $6s\,6p$. In the present case, contrary to what we have done in the case of the barium atom (see Problem 16.14), due to the higher value of Z, we shall not try to establish relations among the matrix elements of H_{so} by considering the two external electrons as independent.

b) Assume that both the separation δ between the levels ^1P and ^3P in absence of H_{so} and the nonvanishing matrix elements of H_{so}, are known. Find the stationary states and the corresponding energies (up to a common additive constant E).

c) Find the value of δ and of the nonvanishing matrix elements of H_{so} in terms of the experimental values of the energy levels (better: in terms of their differences). Use the result and write the normalized stationary states with numerical coefficients.

At this point, some theoretical predictions can be made.

d) Calculate the Landé factors:

$$g_J \equiv 1 + \left(J(J+1)\right)^{-1}\langle E_J, J, M_J \mid \vec{S} \cdot \vec{J} \mid E_J, J, M_J \rangle, \qquad J \neq 0$$

of the above energy levels with $J > 0$ and compare the results with those observed in the Zeeman effect, that are, respectively: 1.479, 1.503, 1.019.

e) Calculate the ratio between the lifetimes (reciprocal of the transition probabilities per unit time) of the levels E_2 and E_4; the observed values are $\tau_2 \simeq 1.08 \times 10^{-7}$s, $\quad \tau_4 \simeq 1.30 \times 10^{-9}$s.

(The transition from the level E_2, improperly classified as $^3P_1$, is called an *intercombination line* (or *forbidden line*) inasmuch as it is – apparently – a transition between states with different spin.)

Solutions

16.1

a) By just looking at the figure one can deduce that $\vec{j}^{\,2}$ and $\vec{L}^{\,2}$ may be (and indeed are) constants of motion because the states are classified by means of these observables; certainly the components of $\vec{j}$ are constants of motion because the levels are degenerate on j_z (on the other hand, the atom is isolated), the space inversion I because its eigenvalues are $w = (-1)^l$. Certainly the components of $\vec{L}$ are not constants of motion, otherwise the states $p_{\frac{1}{2}}$ and $p_{\frac{3}{2}}$ should have the same energy. $I_z\sigma_z$ may be a constant of motion – and indeed it is – because its matrix elements between states belonging to different energy levels are vanishing: indeed it commutes with $\vec{L}^{\,2}$ and with $\vec{j}^{\,2} = \vec{L}^{\,2} + \vec{s}^{\,2} + 2\vec{L}\cdot\vec{s}$ $\;\left(I_z\sigma_z\,(L_{x,y})\,I_z\sigma_z = -L_{x,y},\; I_z\sigma_z\,(L_z)\,I_z\sigma_z = L_z\;$ and likewise for the spin$\right)$; or else: $I_z\sigma_z$ is the product of I times a rotation by π around the z axis, both for the orbital and the spin variables.

b) H' does not commute with I and, since it instead commutes with rotations, it does not commute with $I_z\sigma_z$; it does not commute either with $\vec{L}^{\,2}$ ($\vec{p}$ has nonvanishing matrix elements between states with $\Delta L = \pm 1$).

c) H' commutes with $\vec{j}$ (therefore with $\vec{j}^{\,2}$), whereas it is odd under space inversions: so its only nonvanishing matrix elements are between the states $2s_{\frac{1}{2}}$ and $2p_{\frac{1}{2}}$ with the same j_z and, in addition, those between the states with $j_z = +\frac{1}{2}$ equal those between the states with $j_z = -\frac{1}{2}$ (invariance under rotations). Putting:

$$a = \langle\, 2s_{\frac{1}{2}}, j_z = \pm\tfrac{1}{2} \mid H' \mid 2p_{\frac{1}{2}}, j_z = \pm\tfrac{1}{2}\,\rangle$$

(a could be taken real), one must diagonalize the matrix:

$$\begin{pmatrix} E_1 & a \\ a^* & E_2 \end{pmatrix} \quad \Rightarrow \quad E_\pm = \frac{1}{2}\left(E_1 + E_2 \pm \sqrt{(E_2 - E_1)^2 + 4\,|a|^2}\,\right).$$

So the level $2p_{\frac{3}{2}}$ is not perturbed, whereas states with the same j_z from the levels $2p_{\frac{1}{2}}$ and $2s_{\frac{1}{2}}$ mix and the corresponding eigenvalues move apart.

d) The ground state $1s_{\frac{1}{2}}$ is not perturbed. In the absence of H' one has only two lines between the levels $n = 2$ and the level $n = 1$, thanks to the selection rule on the space inversion ($2p_{\frac{3}{2}} \to 1s_{\frac{1}{2}}$, $2p_{\frac{1}{2}} \to 1s_{\frac{1}{2}}$); now instead, owing to the mixing between the states from the levels $2p_{\frac{1}{2}}$ and $2s_{\frac{1}{2}}$ there occur transitions from all the three levels with $n = 2$.

16.2

a) Take the z axis along the direction of $\vec{B}$. The eigenvectors of H_0 that are also eigenvectors of L_z and s_z are eigenvectors of H: such are the states $|2p_{\frac{3}{2}}, j_z = \pm\frac{3}{2}\rangle$ and $|2s_{\frac{1}{2}}, s_z = \pm\frac{1}{2}\rangle$. The corresponding eigenvalues are:

$$E_{2p_{3/2}, \pm 3/2} = E_3 \pm 2\mu_B B; \qquad E_{2s_{1/2}, \pm 1/2} = E_2 \pm \mu_B B.$$

b) The magnetic interaction has nonvanishing matrix elements among the states $2p_{\frac{1}{2}}$ and $2p_{\frac{3}{2}}$ with the same value of $j_z = \pm\frac{1}{2}$. For $j_z = \frac{1}{2}$ one has (see Problem 15.8b):

$$\langle 2p_{\frac{1}{2}}, \tfrac{1}{2} \mid j_z + s_z \mid 2p_{\frac{1}{2}}, \tfrac{1}{2} \rangle = \frac{1}{2} - \frac{1}{6} = \frac{1}{3}$$

$$\langle 2p_{\frac{3}{2}}, \tfrac{1}{2} \mid j_z + s_z \mid 2p_{\frac{3}{2}}, \tfrac{1}{2} \rangle = \frac{2}{3}$$

$$\langle 2p_{\frac{1}{2}}, \tfrac{1}{2} \mid j_z + s_z \mid 2p_{\frac{3}{2}}, \tfrac{1}{2} \rangle = \langle 2p_{\frac{1}{2}}, \tfrac{1}{2} \mid s_z \mid 2p_{\frac{3}{2}}, \tfrac{1}{2} \rangle = \frac{\sqrt{2}}{3}$$

so one is left with finding the eigenvalues of the matrix:

$$\begin{pmatrix} E_1 + \frac{1}{3}\mu_B B & \frac{\sqrt{2}}{3}\mu_B B \\ \frac{\sqrt{2}}{3}\mu_B B & E_3 + \frac{2}{3}\mu_B B \end{pmatrix} \Rightarrow$$

$$E_{\pm, j_z = +\frac{1}{2}} = \frac{1}{2}\Big(E_3 + E_1 + \mu_B B $$
$$\pm \sqrt{(E_3 - E_1)^2 + \tfrac{2}{3}(E_3 - E_1)\mu_B B + (\mu_B B)^2} \Big).$$

For $j_z = -\frac{1}{2}$, always from Problem 15.8b, it appears that changing $\mu_B B$ into $-\mu_B B$ is sufficient.

The weak field and strong field approximations respectively require $\mu_B B \ll E_3 - E_1$ and $\mu_B B \gg E_3 - E_1$ (the states $2s_{\frac{1}{2}}$ are exact eigenstates), whence respectively $B \ll 8.5 \times 10^3\,\mathrm{G}$ and $B \gg 8.5 \times 10^3\,\mathrm{G}$.

The quadratic term in the Hamiltonian (1) brings a contribution of the order $e^2 B^2 a_B^2 / 8m_e c^2$ ($a_B = \hbar^2/m_e e^2$ is the Bohr radius), whereas the terms

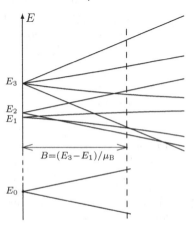

proportional to B^2 in the expression of $E_{2,i}(B)$ are of the order $(\mu_B B)^2/(E_3 - E_1)$ whence it appears that the ratio of the first to the second is of the same order as the ratio of $E_3 - E_1 \simeq 4 \times 10^{-5}\,\mathrm{eV}$ to $e^2/a_B \simeq 27\,\mathrm{eV}$.

c) Only transitions from the energy levels with $l = 1$ are possible: ten lines are observed, four of them being π ($\Delta j_z = 0$) and the other six being σ ($\Delta j_z = \pm 1$).

16.3

a) Taking the z axis parallel to $\vec{\mathcal{E}}$, the Hamiltonian writes $H = H_0 + e\mathcal{E}z$. Since H commutes with the reflections with respect to each plane that contains the z axis, one has the degeneracy $j_z \to -j_z$, so one has four levels (two from $2p_{\frac{3}{2}}$, one from $2p_{\frac{1}{2}}$ and one from $2s_{\frac{1}{2}}$). Since the eigenstates of H_0 have definite parity under space inversion, there is no first order effect. There is effect only from the second order on, so the behaviour of the levels as functions of $\mathcal{E}$ for $\mathcal{E} \to 0$ is quadratic. In addition, as H commutes with j_z, the energy of the states $2p_{\frac{3}{2}}$ with $j_z = \pm\frac{3}{2}$ stays unaltered (at least as long as $e\mathcal{E}a_B \ll E_{n=3} - E_{n=2} \simeq (\frac{1}{4} - \frac{1}{9})\times 13.6\,\mathrm{eV} \Rightarrow \mathcal{E} \ll 10^8\,\mathrm{V/cm}$).

b) In the second order one has (the states have the same $j_z = \pm\frac{1}{2}$ and for z the selection rule $\Delta l = \pm 1$ applies):

$$\Delta E_{3,\,j_z=\pm 1/2} \simeq -e^2\mathcal{E}^2 \frac{|\langle 2p_{\frac{3}{2}} \mid z \mid 2s_{\frac{1}{2}} \rangle|^2}{E_2 - E_3} = \frac{6(e\mathcal{E}a_B)^2}{E_3 - E_2} = 0.012\,\mathrm{cm}^{-1}.$$

As suggested, the contribution of the levels with $n \neq 2$ has been neglected.

c) Ignoring the level $2p_{\frac{3}{2}}$, the effect of a field of intensity $600\,\mathrm{V/cm}$ on the levels $2p_{\frac{1}{2}}$ and $2s_{\frac{1}{2}}$ is of the order of $(e\mathcal{E}a_B)^2/(E_2 - E_1) \simeq 0.026\,\mathrm{cm}^{-1}$, whereas the effect of the level $2p_{\frac{3}{2}}$ on the level $2s_{\frac{1}{2}}$ at the second order is equal to $-\Delta E_{3,\,j_z=\pm 1/2} = -0.012\,\mathrm{cm}^{-1}$. As a consequence, the approximation suggested in the text is barely legitimate. Anyway, let us determine approximately the effect of the electric field on the levels $2p_{\frac{1}{2}}$, $2s_{\frac{1}{2}}$: to this end it suffices to diagonalize the restriction of H to the states of the levels $2p_{\frac{1}{2}}$ and $2s_{\frac{1}{2}}$. One has (see Problem 15.8b):

$$\begin{pmatrix} E_1 & \sqrt{3}\,e\mathcal{E}\,a_B \\ \sqrt{3}\,e\mathcal{E}\,a_B & E_2 \end{pmatrix} \Rightarrow E_\pm = \frac{1}{2}\left(E_1 + E_2 \pm \sqrt{(E_2 - E_1)^2 + 12(e\mathcal{E}a_B)^2}\right)$$

therefore:

$$E_- = E_1 - \frac{1}{2}\left(E_1 - E_2 + \sqrt{(E_2 - E_1)^2 + 12(e\mathcal{E}a_B)^2}\right) = E_1 - 0.030\ \mathrm{cm}^{-1}$$

$$E_+ = E_2 + \frac{1}{2}\left(E_1 - E_2 + \sqrt{(E_2 - E_1)^2 + 12(e\mathcal{E}a_B)^2}\right) = E_2 + 0.030\ \mathrm{cm}^{-1}.$$

Note that the energy levels repel each other and that the calculation to the order $\mathcal{E}^2$ would provide level shifts given by $\mp 3(e\mathcal{E}a_B)^2/(E_2^0 - E_1^0) = \mp 0.056\,\mathrm{cm}^{-1}$ respectively.

If the presence of the level $2p_{\frac{3}{2}}$ is taken into account and the problem of finding the eigenvalues of the restriction of H to the states of the levels $2p_{\frac{3}{2}}$, $2p_{\frac{1}{2}}$ and $2s_{\frac{1}{2}}$ is solved numerically, one finds (in cm^{-1}):

$$\Delta E_{3,\,j_z=\pm 1/2} \simeq 0.011\,;\quad E_- = E_1 - 0.037\,;\quad E_+ = E_2 + 0.026$$

therefore the previous results for E_+ and E_- present, as expected, an error of about 20%.

16.4

a) Taking the z axis in the direction of the electric field, the term of interaction with the field is $H' = e\mathcal{E}(z_1 + z_2)$. In the basis $\big($notation $|\,L,M\,\rangle\big)$:

$$|\,0,0\,\rangle,\quad |\,1,0\,\rangle,\quad |\,1,1\,\rangle,\quad |\,1,-1\,\rangle$$

thanks to the selection rule on M $(\Delta M = 0)$ and on the parity (the states $|\,0,0\,\rangle$ and $|\,1,M\,\rangle$ have definite parities), one must diagonalize the 2×2 matrix (the states $|\,1,1\,\rangle$, $|\,1,-1\,\rangle$ are not perturbed):

$$\begin{pmatrix} E_0 & a \\ a & E_1 \end{pmatrix} = (E_0 + \delta)\times \mathbb{1} + \begin{pmatrix} -\delta & a \\ a & \delta \end{pmatrix}$$

where $a \equiv \langle\,0,0\,|\,H'\,|\,1,0\,\rangle$ can be taken real; the eigenvalues are:

$$E_\mp = E_0 + \delta \mp \sqrt{\delta^2 + a^2}$$

and the corresponding eigenvectors (not normalized to 1, but equally normalized) are:

$$|\,E_-\,\rangle = a\,|\,0,0\,\rangle + (\delta - \sqrt{\delta^2 + a^2})\,|\,1,0\,\rangle$$

$$|\,E_+\,\rangle = (\delta - \sqrt{\delta^2 + a^2})\,|\,0,0\,\rangle - a\,|\,1,0\,\rangle$$

(the eigenvectors, not being normalized, can be written in different ways: only the ratios of the coefficients is relevant). In addition, as already noted, the states $|\,1,1\,\rangle$ and $|\,1,-1\,\rangle$ both have the unperturbed energy E_1.

b) $\Delta E = 2\sqrt{\delta^2 + a^2}$.

The probability transitions between the states $|\,E_\pm\,\rangle$ and the ground state $(1s)^2\,{}^1S_0$ receive contribution only from the state $|\,1,0\,\rangle$, so their ratio (ignoring the term $(\nu_-/\nu_+)^3$ because $0.75\,eV \ll 20\,eV$) is

$$R = \frac{(\sqrt{\delta^2 + a^2} - \delta)^2}{a^2} = \frac{\Delta E - 2\delta}{\Delta E + 2\delta} \quad\Rightarrow\quad 2\delta = \Delta E\,\frac{1-R}{1+R} = 0.6\,eV\,.$$

c) In the absence of the electric field it is not possible to determine 2δ from the difference of the frequencies in the transitions between the lowest energy level to the two levels under consideration, because the transition $L=0 \to L=0$ is forbidden (therefore only one line is observed). The direct transition between the two levels falls in the far infrared region of the electromagnetic spectrum, so it is not easy to measure.

16.5

a) $\tilde{H}_Z$ is obtained by means of the canonical transformation $\tilde{p}_i = Z\,p_i$; $\tilde{q}_i = Z^{-1}q_i$. Perturbation theory applied to $Z^{-2}\tilde{H}_Z = H^{(0)} + (e^2/Z)/r_{12}$, in which $(e^2/Z)/r_{12}$ is considered as the perturbation, gives rise to E_Z in the form $Z^2\times$ (power series in Z^{-1}).

b) $E^{(0)} = -2\times\dfrac{e^2}{2a_{\mathrm{B}}} = -27.2\,\mathrm{eV};\qquad a = \dfrac{5}{8}\dfrac{e^2}{a_{\mathrm{B}}} = 17\,\mathrm{eV}.$

The first ionization energies are given, to the first order, by:

$$E_Z^{\mathrm{I}} = \left| E_Z - Z^2\frac{E^{(0)}}{2}\right| = \left(\frac{1}{2}Z^2 - \frac{5}{8}Z\right)\frac{e^2}{a_{\mathrm{B}}}.$$

One obtains (the energies are expressed in eV; in the third line the difference from the experimental value is reported):

He^{I}	$\mathrm{Li}^{\mathrm{II}}$	$\mathrm{Be}^{\mathrm{III}}$	B^{IV}	C^{V}
20.40	71.40	149.6	255.0	387.6
4.18	4.22	4.25	4.3	4.4

c) If $H_Z \to H_Z - \sum_i p_i^4/8m_e^3c^2$ then $Z^{-2}\tilde{H}_Z \to Z^{-2}\tilde{H}_Z - Z^2\sum_i p_i^4/8m_e^3c^2$, whence the first order correction to the first ionization energies (the term $(e^2/Z)/r_{12}$ is neglected) is proportional to Z^4 and, owing to the results of Problem 12.20, is given by $(5/4)\,Z^4\alpha^2\times(e^2/2a_{\mathrm{B}})$:

He^{I}	$\mathrm{Li}^{\mathrm{II}}$	$\mathrm{Be}^{\mathrm{III}}$	B^{IV}	C^{V}
0.014	0.07	0.23	0.57	1.17 .

As a consequence, trying to improve the agreement between the first order results found in b) and the experimental data, e.g. by calculating the c/Z term (see text), is not very meaningful – particularly for $Z\gtrsim 5$.

d) The energy of the ground state of $H_{Z'}^{(0)}$ is

$$E_{Z'}^{(0)} = -2\times\frac{Z'^2e^2}{2a_{\mathrm{B}}} = -(Z^2 - 2\sigma\,Z + \sigma^2)\frac{e^2}{a_{\mathrm{B}}} \;\Rightarrow\; 2\sigma\,\frac{e^2}{a_{\mathrm{B}}} = a \;\Rightarrow\; \sigma = \frac{5}{16}.$$

The result $E_{Z'}^{(0)} = -\left(Z - \frac{5}{16}\right)^2 e^2/a_{\mathrm{B}}$ coincides with the one obtained by making use of the variational method with the trial functions:

$$\frac{1}{4\pi}\left(\frac{Z'}{a_{\mathrm{B}}}\right)^3 4\,e^{-Z'\,r_1/a_{\mathrm{B}}}\,e^{-Z'\,r_2/a_{\mathrm{B}}}$$

where Z' is the variational parameter.

$$E_Z^{\mathrm{I}} = \left(\frac{1}{2}Z^2 - \frac{5}{8}Z + \frac{25}{256}\right)\frac{e^2}{a_{\mathrm{B}}} \;\Rightarrow$$

He^{I}	$\mathrm{Li}^{\mathrm{II}}$	$\mathrm{Be}^{\mathrm{III}}$	B^{IV}	C^{V}
23.1	74.1	152.3	257.7	390.3 .
1.5	1.6	1.6	1.6	1.7

16.6

a) Let us take the z axis parallel to $\vec{B}$ and $\vec{A}(\vec{r}) = \frac{1}{2}\vec{B}\wedge\vec{r}$. The Hamiltonian of the atom in the presence of the magnetic field is (see Problems 13.8 and 16.2 where $\vec{s}$ is replaced by the total spin $\vec{S}$):

$$H = H_0 + \frac{eB}{2m_e c}(L_z + 2S_z) + \frac{e^2 B^2}{8m_e c^2}(x_1^2 + y_1^2 + x_2^2 + y_2^2)$$

$$\equiv H_0 + H' + H''$$

where H_0 is the Hamiltonian of the atom in the absence of magnetic field. In the approximation of independent electrons the ground state $^1S_0$ is an eigenvector of $H_0 + H'$: $H' \, | \, L = 0, \, S = 0 \rangle = 0$, therefore to any order the perturbation to the lowest energy level is given only by H'' and one has (notation $| \, n, l, m; \, Z \, \rangle$):

$$\delta E_0 = \frac{e^2 B^2}{8m_e c^2} 2 \times \langle 1,0,0; \, Z \, | \, x^2 + y^2 \, | \, 1,0,0; \, Z \, \rangle$$

$$= \frac{e^2 B^2}{8m_e c^2} 2 \times \frac{2}{3} \langle 1,0,0; \, Z \, | \, r^2 \, | \, 1,0,0; \, Z \, \rangle = \frac{e^2 B^2}{8m_e c^2} 2 \times \frac{2}{3} \frac{1}{Z^2} 3a_B^2 \Rightarrow$$

$$\delta E_0 [\text{eV}] \simeq \frac{1}{Z^2} 2.4 \times 10^{-18} \times (B[\text{G}])^2.$$

As a consequence helium is diamagnetic and this is true for all the atoms whose ground state is $^1S_0$ (noble gases and alkaline earth atoms).

b) $\delta' E_0 = \delta E_0 \times (Z/Z')^2 = 1.4 \times \delta E_0$.

16.7

a) $L = 0$; $S = 0, 1$. To the zeroth order in the Coulomb repulsion the terms $^3S, \, ^1S$ have the same energy E_0 (exchange degeneracy); to the first order the energy of each term is determined by the orbital part of the state:

$$| \, ^3S \, \rangle \to \frac{1}{\sqrt{2}} \big(| \, 1s, 2s \, \rangle - | \, 2s, 1s \, \rangle \big), \quad | \, ^1S \, \rangle \to \frac{1}{\sqrt{2}} \big(| \, 1s, 2s \, \rangle + | \, 2s, 1s \, \rangle \big) \Rightarrow$$

$$E(^3S) = E_0 + \langle \, ^3S | \, \frac{e^2}{r_{12}} \, | \, ^3S \, \rangle = E_0 + \Delta_0 - \Delta_1,$$

$$E(^1S) = E_0 + \langle \, ^1S | \, \frac{e^2}{r_{12}} \, | \, ^1S \, \rangle = E_0 + \Delta_0 + \Delta_1;$$

$$E_0 = -Z^2 \frac{e^2}{2a_B} - \frac{Z^2}{4} \frac{e^2}{2a_B}.$$

b) $\displaystyle \Delta_0 = \int \psi_{1s}^*(r_1) \, \psi_{2s}^*(r_2) \frac{e^2}{r_{12}} \psi_{1s}(r_1) \, \psi_{2s}(r_2) \, dV_1 \, dV_2$

$$= \int \rho_2(r_2) \frac{1}{r_{12}} \rho_1(r_1) \, dV_1 \, dV_2$$

$$\rho_1(r) = e \, |\psi_{1s}(r)|^2, \qquad \rho_2(r) = e \, |\psi_{2s}(r)|^2.$$

$$\Delta_1 = \int \rho^*(r_2) \frac{1}{r_{12}} \rho(r_1) \, dV_1 \, dV_2, \qquad \rho(r) = e \, \psi_{1s}(r) \, \psi_{2s}^*(r).$$

$\Delta_0 > 0$ ($\rho_1 > 0$, $\rho_2 > 0$); $\Delta_1 > 0$ for it is twice the electrostatic energy of the (complex) charge distribution $\rho(r)$.

c) Calculation of the direct integral Δ_0: the angular integration suggested in the text gives:

$$I = \int d\Omega_1 \, 2\pi \, \frac{1}{2r_1 r_2} \, 2 \sqrt{r_1^2 + r_2^2 - 2r_1 r_2 x} \, \Big|_{x=+1}^{x=-1}$$

$$= (4\pi)^2 \, \frac{r_1 + r_2 - |r_1 - r_2|}{2r_1 r_2} = (4\pi)^2 \left(\frac{1}{r_1} \theta(r_1 - r_2) + \frac{1}{r_2} \theta(r_2 - r_1) \right)$$

where $\theta(x) = 1$ for $x > 0$ and $\theta(x) = 0$ otherwise. Whence:

$$\Delta_0 = \int_0^\infty dr_1 \, r_1^2 \int_0^{r_1} dr_2 \, r_2^2 \, \frac{e^2}{r_1} \, R_{10}^2(r_1) \, R_{20}^2(r_2)$$

$$+ \int_0^\infty dr_1 \, r_1^2 \int_{r_1}^\infty dr_2 \, r_2^2 \, \frac{e^2}{r_2} \, R_{10}^2(r_1) \, R_{20}^2(r_2)$$

$$= \frac{17}{81} \frac{Ze^2}{a_B} = 11.42 \, \text{eV} \quad (Z = 2) \, .$$

Calculation of the exchange integral Δ_1:

$$\rho(r) = \frac{e}{4\pi} \left(\frac{Z}{a_B} \right)^3 \frac{2}{2^{3/2}} \left(2 - \frac{Zr}{a_B} \right) e^{-3Zr/2a_B}$$

$$Q(r) = 4\pi \int_0^r \rho(r') \, r'^2 \, dr' = e \frac{\sqrt{2}}{3} \left(\frac{Z\,r}{a_B} \right)^3 e^{-3Zr/2a_B} \, ; \qquad E(r) = \frac{Q(r)}{r^2} \; \Rightarrow$$

$$\Delta_1 = \frac{2}{8\pi} \int E^2(r) \, dV = \frac{2}{9} \frac{Ze^2}{a_B} \int_0^\infty x^4 \, e^{-3x} \, dx$$

$$= \frac{16}{729} \frac{Ze^2}{a_B} = 1.2 \, \text{eV} \qquad (Z = 2) \, .$$

$$E(^3S) - Z^2 \frac{e^2}{2a_B} = -\frac{Z^2}{4} \frac{e^2}{2a_B} + \Delta_0 - \Delta_1 = -3.38 \, \text{eV} \, ,$$

$$E(^1S) = E(^3S) + 2\Delta_1 = -0.98 \, \text{eV}$$

where $-Z^2 e^2 / 2a_B$ is the energy of the ion He^+.

The result is rather disappointing. The contribution of the second order is negative in both cases, so it makes the disagreement with the observed data smaller: indeed the level ^3S is the lowest energy level of orthohelium and the contribution of the second order on the ground state always is negative; as for the level $1s \, 2s \, {}^1S$, the contribution of the level $(1s)^2$ is positive, but since it is as far as $20 \, \text{eV}$, it is overwhelmed by the negative contributions of the levels $1s \, 3s \, \cdots$.

16.8

a) The filled shells $(1s)^2 (2s)^2$ do not contribute either to the orbital angular momentum L or to the total spin S. The two p electrons give rise to states with $L = 0, 1, 2$ and spin $S = 0, 1$. As the two p electrons are equivalent ($n_1 = n_2$, $l_1 = l_2$) and the states with $L = 2$ and $L = 0$ are symmetric in the orbital variables (see Problem 15.8a), owing to Pauli principle they are singlet states ($S = 0$); the states with $L = 1$ are anti-symmetric, so they are triplet states ($S = 1$). One then has the following spectroscopic terms (ordered with increasing energy): ^3P, ^1D, ^1S.

b) $\vec{L} \cdot \vec{S} = \frac{1}{2}(\vec{J}^2 - \vec{L}^2 - \vec{S}^2)$ $\Rightarrow$

$$E_{LSJ} = E_{LS} + \langle E_{LS} \mid A_{LS} \vec{L} \cdot \vec{S} \mid E_{LS} \rangle$$
$$= E_{LS} + \frac{1}{2} A_{LS}(J(J+1) - L(L+1) - S(S+1))$$
$$\Rightarrow E_{LS,J} - E_{LS,J-1} = A_{LS} J .$$

c) $\dfrac{E_2 - E_0}{E_1 - E_0} = \dfrac{43.5}{16.4} = 2.65 .$

In the Russell–Saunders scheme, from the Landé interval rule:

$$\frac{E_{J=2} - E_{J=0}}{E_{J=1} - E_{J=0}} = 3$$

therefore the difference with the experimental data is approximately 10%.

16.9

a) The Pauli principle requires that each state can be occupied at most by one only electron. The number of ways one can choose n states out of the six available states is

$$\binom{6}{n} \equiv \frac{6!}{n! \, (6-n)!} = \binom{6}{6-n} .$$

b) The determinant of each $n \times n$ minor has a well defined value of both $M_L = \sum m_i$ and $M_S = \sum s_i$ and, since for any state $\mid m, s \rangle$ there exists $\mid -m, -s \rangle$, also the minor with $-M_L$ and $-M_S$ exists. Once a $n \times n$ minor has been chosen, the determinant of the $(6-n) \times (6-n)$ minor formed with the remaining $6-n$ states has $M'_L = -M_L$ and $M'_S = -M_S$: indeed $M_L + M'_L = \sum_1^n m_i + \sum_n^6 m_i = \sum_1^6 m_i = 0$ and likewise for spin.

c) The equivalent configuration is $(1s)^2(2s)^2 \, 2p$ (the same as for boron), therefore the spectral term is ^2P and the number of independent states is $(2S+1) \times (2L+1) = 6 .$

d) As $4 = 6 - 2$, the spectral terms are the same as for carbon: ^3P, ^1D, ^1S. The order in energy of such terms is determined by the Coulomb repulsion between the holes; the latter is positive as in the case of the electrons, so also the order in energy is the same as for carbon.

e) The multiplet of fine structure is "inverted": $^3P_{2,1,0}$ ($A_{LS} < 0$: the distance between adjacent levels decreases), because it is determined mainly by the interactions of the holes with the nucleus: the latter interaction has the sign opposite to that of the electrons. The Landé interval rule is verified to about 10% of accuracy:

$$\frac{E_0 - E_2}{E_1 - E_2} = \frac{226.5}{68} = 3.3 .$$

16.10

a) No: the states with $L = 3$ are completely symmetric (among them there is the state with $m_1 = m_2 = m_3 = 1$), and among the spin states of three electrons there are no completely antisymmetric states (see Problem 15.8).

b) The states with total spin $S = 3/2$ are symmetric, so the orbital part must be completely antisymmetric and this is possible only if the electrons occupy different orbital states: in the present case $m = 1$, $m = 0$, $m = -1$ so $L_z = 0$, no matter how the z axis is chosen, therefore also $L = 0$. In conclusion, the lowest energy level is ^4S.

c) The orbital state is obtained by means of a Slater determinant: denoting by $|m\rangle_i$ the state of the i-th electron, one has:

$$|L = 0\rangle = \frac{1}{\sqrt{3!}} \det \begin{pmatrix} |1\rangle_1 & |0\rangle_1 & |-1\rangle_1 \\ |1\rangle_2 & |0\rangle_2 & |-1\rangle_2 \\ |1\rangle_3 & |0\rangle_3 & |-1\rangle_3 \end{pmatrix}$$

$$= \frac{1}{\sqrt{6}} \Big(|1,0,-1\rangle + |0,-1,1\rangle + |-1,1,0\rangle$$

$$- |1,-1,0\rangle - |0,1,-1\rangle - |-1,0,1\rangle \Big) .$$

d) The only possible values of the total spin are $\frac{1}{2}$ and $\frac{3}{2}$; as seen above $S = \frac{3}{2} \Rightarrow L = 0$, so the total spin of the spectral terms P and D is $\frac{1}{2}$: ^2P, ^2D. Note that the total number of states of the terms ^4S, ^2P, ^2D is $4 + 6 + 10 = 20$, as it must be. Each of the terms ^2P, ^2D should give rise to a fine structure doublet: $^2P_{\frac{1}{2},\frac{3}{2}}$, $^2D_{\frac{3}{2},\frac{5}{2}}$. However, as the p orbital is exactly half filled, the effect of the spin-orbit interaction is vanishing to the first order (see Problem 16.9): the experimental data are consistent with this result.

16.11

a) The filled shells $(1s)^2(2s)^2$ do not contribute either to the orbital angular momentum L or to the total spin S. The two p electrons give rise to states with $L = 0, 1, 2$ and spin $S = 0, 1$. As the $2p$ and $3p$ electrons are not equivalent, for each value of L both the singlet and the triplet spin states are possible. So one has the following spectral terms: ^1S, ^3S; ^1P, ^3P; ^1D, ^3D.

b) The singlet states have symmetric orbital wavefunctions, the triplet states have instead antisymmetric orbital wavefunctions. Therefore:

$$^{1,3}S: \quad \hat{r}_1 \cdot \hat{r}_2 \left(R_2(r_1) R_3(r_2) \pm R_3(r_1) R_2(r_2) \right)$$

$$^{1,3}P: \quad (\hat{r}_1 \wedge \hat{r}_2)_i \left(R_2(r_1) R_3(r_2) \mp R_3(r_1) R_2(r_2) \right)$$

$$^{1,3}D: \quad (\hat{r}_{1i} \hat{r}_{2j} + \hat{r}_{1j} \hat{r}_{2i} - \tfrac{2}{3}(\hat{r}_1 \cdot \hat{r}_2) \delta_{ij}) \left(R_2(r_1) R_3(r_2) \pm R_3(r_1) R_2(r_2) \right).$$

Note the $\mp$ sign in the radial part of the wavefunction of the P states: it is due to the antisymmetry of the angular part $\hat{r}_1 \wedge \hat{r}_2$.

c) $E(^3S) < E(^1S)$ because $R_2(r_1) R_3(r_2) - R_3(r_1) R_2(r_2)$ vanishes for $\vec{r}_1 = \vec{r}_2$ so that the effect of the Coulomb repulsion is smaller. For the same reason $E(^3D) < E(^1D)$. Also in the case of the P terms, the one with wavefunction $R_2(r_1) R_3(r_2) - R_3(r_1) R_2(r_2)$ has a smaller energy, but – owing to the antisymmetry of the angular part – the wavefunction is symmetric, so in this case it is the singlet term that has the lowest energy. In the first two cases Hund's rule holds, whereas for the P states it is violated, at least in the form given in the text.

Thanks to the 'double antisymmetry' of the ^1P states, one may expect that the latter have energy smaller than the others. Indeed, the observed energies (in eV) of the levels we have considered are, neglecting fine structure and taking the lowest energy level as reference point:

^1P	^3D	^3S	^3P	^1D	^1S
8.53	8.64	8.77	8.85	9.00	9.04 .

16.12

a) As the distance between adjacent levels increases, one is dealing with a "direct multiplet". The number of levels of a fine structure multiplet is the least between $2S+1$ and $2L+1$ ($|L-S| \le J \le L+S$): as the levels are 4 and $2L+1$ is odd, it follows that $4 = 2S+1$, so $S = 3/2$, $L > S \Rightarrow L \ge 2$. The values of J are half-integers.

b) One may proceed in several ways. It is convenient to eliminate A_{LS} by taking ratios:

$$\frac{E_3 - E_2}{E_1 - E_0} = \frac{J_0 + 3}{J_0 + 1} = \frac{3}{1.7}, \qquad \frac{E_2 - E_1}{E_1 - E_0} = \frac{J_0 + 2}{J_0 + 1} = \frac{2.3}{1.7} \Rightarrow$$

$$1.3\, J_0 - 2.1 = 0 \Rightarrow J_0 = 1.6; \qquad 0.6\, J_0 - 1.1 = 0 \Rightarrow J_0 = 1.8$$

whence it follows (but the method of the least squares could be used as well) that the half-integer that best solves both equations is $J_0 = 3/2$, then $L - S = 3/2 \Rightarrow L = 3$ (spectral term ^4F).

c) As $S = 3/2$, the number of electrons must be odd and ≥ 3; in addition, the multiplet being direct, the outer orbital must be filled for less than its half (p^3 with $L = 3$ and $S = 3/2$ is excluded also because it is a completely symmetric state; d^5 because Hund's rule would require $S = 5/2$). There remains the configuration d^3. The first atom with such a configuration has $Z = 23$: $(1s)^2(2s)^2(2p)^6(3s)^2(3p)^6(4s)^2(3d)^3$, so it is vanadium.

16.13

a) The two outer electrons give rise to the spectral terms ^3P and ^1P. As the exchange integral is positive (see Problem 16.7), $E(^3P) < E(^1P)$. The term ^3P gives rise to states with total angular momentum $J = 0, 1, 2$ that are split in energy by the spin-orbit interaction. One then has, in order of increasing energy, the four levels $E(^3P_0)$, $E(^3P_1)$, $E(^3P_2)$; $E(^1P_1)$.

b) If the Coulomb repulsion is neglected, the Hamiltonian of the two outer electrons is separable, so the states can be classified by the quantum numbers n_i, l_i, j_i, j_{iz} of the single electrons: while for j_1 the only possible value is $\frac{1}{2}$, one has $j_2 = \frac{1}{2}, \frac{3}{2}$, whence the two energy levels: $E_{j_1=\frac{1}{2}, j_2=\frac{1}{2}}$, $E_{j_1=\frac{1}{2}, j_2=\frac{3}{2}}$, with degeneracies $(2j_1+1)(2j_2+1)$, i.e. 4 and 8: as the two electrons are not equivalent electrons, the antisymmetrization demanded by Pauli principle does not reduce the number of states (as instead it happens for the normal configuration $(6s)^2$, that gives rise to the only state: $j_1 = \frac{1}{2}, j_2 = \frac{1}{2} \to J = 0$, $\cancel{J=1}$).

c) The states belonging to the level $E_{j_1=\frac{1}{2}, j_2=\frac{1}{2}}$ have $J = 0, 1$; those belonging to $E_{j_1=\frac{1}{2}, j_2=\frac{3}{2}}$ have $J = 1, 2$. The Coulomb repulsion commutes with $\vec{J} = \vec{j}_1 + \vec{j}_2$, but does not commute either with $\vec{j}_1$ or with $\vec{j}_2$, so it removes the degeneracies on J and gives rise to two doublets of levels: $E_{j_1=\frac{1}{2}, j_2=\frac{1}{2}; J=0}$, $E_{j_1=\frac{1}{2}, j_2=\frac{1}{2}; J=1}$; $E_{j_1=\frac{1}{2}, j_2=\frac{3}{2}; J=1}$, $E_{j_1=\frac{1}{2}, j_2=\frac{3}{2}; J=2}$.

16.14

a) The term $(\vec{l}_1 - \vec{l}_2) \cdot (\vec{s}_1 - \vec{s}_2)$ is antisymmetric both in the orbital and in the spin variables, so its matrix elements among ^3P states vanish. As $\vec{L} \cdot \vec{S} = \frac{1}{2}(\vec{J}^2 - \vec{L}^2 - \vec{S}^2)$, calling $E(^3P)$ the energy in the absence of spin-orbit interaction, one has:

$$E(^3P_0) = E(^3P) - 2A, \quad E(^3P_1) = E(^3P) - A, \quad E(^3P_2) = E(^3P) + A \Rightarrow$$

$$\frac{E(^3P_2) - E(^3P_0)}{E(^3P_1) - E(^3P_0)} = 3, \quad \frac{E_3 - E_1}{E_2 - E_1} \simeq 3.37 .$$

b) Let E_0 be the value of the energy common to the ^1P and ^3P terms, inclusive of the "direct interaction" A_0, but in the absence of the "exchange interaction" that splits such terms by $2A_1$ (see Problems 16.7 and 16.13): $E(^3P) = E_0 - A_1$, $E(^1P) = E_0 + A_1$. The nonvanishing matrix elements of $\vec{L} \cdot \vec{S}$ are those calculated in a); the residual term in H_{so} has the nonvanishing matrix elements only between the $^3P_1$ and $^1P_1$ levels, we denote it by (real) B: ordering the basis according to $^3P_0$, $^3P_2$, $^3P_1$, $^1P_1$, one has:

$$H' \to \begin{pmatrix} E_0 - A_1 - 2A & 0 & 0 & 0 \\ 0 & E_0 - A_1 + A & 0 & 0 \\ 0 & 0 & E_0 - A_1 - A & B \\ 0 & 0 & B & E_0 + A_1 \end{pmatrix}.$$

In order to calculate B one can observe that for $\Delta_1 = 0$ the matrix must have only two distinct eigenvalues, so:

$$\begin{pmatrix} -A & B \\ B & 0 \end{pmatrix}$$

must have the eigenvalues $-2A$ and A. This implies that $B = \sqrt{2}\,A$. Therefore one has:

$$E(^3P_0) = E_0 - \Delta_1 - 2A$$

$$E(^3P_1) = E_0 - \frac{A}{2} - \sqrt{\left(\Delta_1 + \frac{A}{2}\right)^2 + 2A^2}$$

$$E(^3P_2) = E_0 - \Delta_1 + A$$

$$E(^1P_1) = E_0 - \frac{A}{2} + \sqrt{\left(\Delta_1 + \frac{A}{2}\right)^2 + 2A^2}$$

where not appropriately, but according to the use of spectroscopists, we have kept on calling $^1P_1$ and $^3P_1$ those states that are such only to the first order in A.

c) One has (in order to identify the levels, recall that $\Delta_1 > 0$, $A > 0$):

$$\begin{cases} E(^3P_2) - E(^3P_0) = 3A = E_3 - E_1 \\ E(^3P_1) + E(^1P_1) - \left(E(^3P_2) + E(^3P_0)\right) = 2\Delta_1 = E_4 + E_2 - (E_1 + E_3) \end{cases} \Rightarrow$$

$$\Delta_1 = 2458\,\text{cm}^{-1}, \quad A = 416\,\text{cm}^{-1}; \quad A/\Delta_1 \simeq 0.17\,.$$

In the Russell–Saunders approximation the terms quadratic in A (due to the matrix elements between $^1P_1$ and $^3P_1$) are neglected, so the error is of the order of $(A/\Delta_1)^2 \simeq 3\%$ (see the figure above: the dashed lines correspond to the Russell–Saunders approximation).

d) $E(^1P_1) - E(^3P_1) = 5461\,\text{cm}^{-1}$ that differs from $E_4 - E_2 = 5423\,\text{cm}^{-1}$ by less than 1%: an agreement with such a degree of accuracy should be, to some extent, considered fortuitous.

16.15

a) A first order perturbative treatment of the spin-orbit interaction gives:

$$\frac{E(^3P_2) - E(^3P_0)}{E(^3P_1) - E(^3P_0)} = 3$$

(Landé interval rule: see Problem 16.8), whereas the observed value is 3.63. If the classification of the energy levels in terms of the total spin S were correct, owing to the selection rule $\Delta S = 0$, only the transition from the level $^1P_1$ to the lowest energy level $^1S_0$ should be observed.

b) H_{so} has $\frac{1}{2}A_{LS}\left[J(J+1) - L(L+1) - S(S+1)\right]$ as diagonal matrix elements (in particular, those among the $^1P_1$ states are vanishing), whereas it has nondiagonal matrix elements only between the $^3P_1$ and $^1P_1$ states, all equal to one another (because independent of M_J). We shall denote them by B (that we will take real), and shall put $A \equiv A_{L=1,S=1}$. Calling E the (unknown) energy common to the ^3P and ^1P levels in the absence of both the Coulomb repulsion and H_{so}, we have to diagonalize the matrix (the basis is $^3P_1$, $^1P_1$):

$$E \times \mathbb{1} + \begin{pmatrix} -A & B \\ B & \delta \end{pmatrix}$$

whose eigenvalues and (nonnormalized) eigenvectors are:

$$E_{2,4} = E + \tfrac{1}{2}\big(\delta - A \mp \sqrt{(\delta + A)^2 + 4B^2}\,\big)$$

$$|\,E_2\,\rangle = B\,|\,^3P_1\,\rangle + \tfrac{1}{2}\big(\delta + A - \sqrt{(\delta + A)^2 + 4B^2}\,\big)\,|\,^1P_1\,\rangle$$

$$|\,E_4\,\rangle = B\,|\,^1P_1\,\rangle - \tfrac{1}{2}\big(\delta + A - \sqrt{(\delta + A)^2 + 4B^2}\,\big)\,|\,^3P_1\,\rangle .$$

In addition, the spin-orbit interaction being diagonal on the states $^3P_0$ and $^3P_2$, one has:

$$E_1 = E - 2A, \quad E_3 = E + A; \qquad |\,E_1\,\rangle = |\,^3P_0\,\rangle, \quad |\,E_3\,\rangle = |\,^3P_2\,\rangle .$$

c) One has:

$$\begin{cases} E_3 - E_1 = 3A & \Rightarrow \quad A = 0.265\,\text{eV} \\ E_4 + E_2 - (E_1 + E_3) = \delta = 1.461\,\text{eV} \\ E_4 - E_2 = \sqrt{(\delta + A)^2 + 4B^2} & \Rightarrow \quad B = 0.284\,\text{eV} \end{cases}$$

whence, after normalizing:

$$|\,E_2\,\rangle = 0.987\,|\,^3P_1\,\rangle - 0.159\,|\,^1P_1\,\rangle, \qquad |\,E_4\,\rangle = 0.987\,|\,^1P_1\,\rangle + 0.159\,|\,^3P_1\,\rangle .$$

d) $g_J = 1 + \dfrac{1}{2J(J+1)}\langle\,E_J\,|\,\vec{J}^2 + \vec{S}^2 - \vec{L}^2\,|\,E_J\,\rangle \quad \Rightarrow$

$$|\,E_2\,\rangle: \quad g(E_2)^{\text{th}} = (0.987)^2 g_{J=1,L=1,S=1} + (0.159)^2 g_{J=1,L=1,S=0} = 1.487$$

$$|\,E_3\,\rangle: \quad g(E_3)^{\text{th}} = g_{J=2,L=1,S=1} = 1.5$$

$$|\,E_4\,\rangle: \quad g(E_4)^{\text{th}} = (0.987)^2 g_{J=1,L=1,S=0} + (0.159)^2 g_{J=1,L=1,S=1} = 1.013 .$$

e) The transition probabilities between the $|\,E_2\,\rangle$ and $|\,E_4\,\rangle$ states and the ground state $^1S_0$ receive contribution only from the state $|\,L = 1, S = 0\,\rangle$ ($^1P_1$), therefore their ratio is (see Problem 14.12):

$$\frac{w_4}{w_2} = \left(\frac{E_4}{E_2}\right)^3 \left(\frac{0.987}{0.159}\right)^2 = \left(\frac{\tau_2}{\tau_4}\right)^{\text{th}} \simeq 99; \qquad \left(\frac{\tau_2}{\tau_4}\right)^{\text{exp}} \simeq 83 .$$

So the forbidden line is about 80 times less intense than the allowed line $|\,E_4\,\rangle \to {}^1S_0$.

The "mixing of spin" (i.e. the violation of the Russell–Saunders approximation) explains the intercombination lines present in many atoms, e.g. the alkaline earth atoms that have the $s\,p$ excited configurations.

17

Elementary Potential Scattering

One-dimensional systems; time-delay; optical theorem; hard sphere; spherical barrier; spherical potential well; spherical Dirac delta–shell; resonances at low energies; bound states and virtual levels as poles of the S-matrix; Breit-Wigner formula; Jost functions; Levinson theorem; Ramsauer-Townsend effect; Yukawa potential; Bragg reflection; identical particles.

17.1 Consider the one-dimensional problem of a particle of mass m and energy E that, coming from the left (see figure), encounters the potential well:

$$V(x) = \begin{cases} 0 & x \le 0 \\ -V_0 & 0 < x < a \\ 0 & x \ge a. \end{cases} \qquad V_0 > 0$$

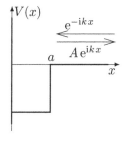

a) Determine both the transmission and the reflection coefficients $T \equiv |C|^2$ and $R \equiv |A|^2$.

b) Determine the energies for which $T = 1$ and compare the result with that of Problem 6.13.

17.2 A particle of mass m is confined in the (one-dimensional) region $x \ge 0$ and is subject to the potential

$$V(x) = \begin{cases} -V_0 & 0 < x < a \\ 0 & x \ge a. \end{cases} \qquad V_0 > 0$$

The particle is sent from the right with energy E, then is reflected back (see figure).

a) Determine the amplitude A of the reflected wave and the phase difference $2\,\delta(k)$ with respect to the wave reflected in the absence of the potential (the phase $\delta(k)$ will be given by an implicit equation). (The reason for writing 2δ instead of δ will be clarified in the solution.)

© Springer International Publishing AG 2017
E. d'Emilio and L.E. Picasso, *Problems in Quantum Mechanics*,
UNITEXT for Physics, DOI 10.1007/978-3-319-53267-7_17

Now consider the (more realistic) situation in which the incoming wave is a wave-packet narrowly concentrated about k_0 and significantly different from zero in a region far from the potential:

$$\psi(x, t \ll 0) = \int_{k \simeq k_0} \alpha(k)\, e^{-i\,(kx + Et/\hbar)}\, dk\,, \qquad E = \frac{\hbar^2 k^2}{2m}\,.$$

After the interaction with the potential the outgoing wave will be (within suitable approximations)

$$\psi(x, t \gg 0) = -\int_{k \simeq k_0} \alpha(k)\, e^{i\,(kx + 2\delta(k) - Et/\hbar)}\, dk\,.$$

b) Approximate $\delta(k)$ by $\delta(k_0) + (k - k_0)\,\delta'(k_0)$ and show that the meaning of $2\,\delta'(k_0)$ is that of a (backward) shift in the position of the center of the wave-packet (with respect to the case without potential) and, therefore, $2\hbar\, d\,\delta/dE$ is the *time-delay* produced by the scattering potential.

17.3   As in the previous problem, a particle of mass m is confined in the (one-dimensional) region $x \geq 0$. The particle with energy E is sent from the right against a rectangular barrier of height $V_0 > 0$:

$$V(x) = \begin{cases} V_0 & 0 < x < a \\ 0 & x \geq a\,. \end{cases} \qquad V_0 > 0$$

a) With $E < V_0$, determine the amplitude A of the reflected wave and the phase difference $2\,\delta(k)$ with respect to the wave reflected in the absence of the potential (as in the previous problem the phase $\delta(k)$ will be given by an implicit equation).

b) Calculate the limit of $\delta(k)$ for $V_0 \to \infty$ with E fixed.

17.4   Consider, from the classical point of view, the scattering of a beam of point particles off a target of (fixed) hard spheres of radius a. The *luminosity* of the beam is $\mathcal{N}$ (i.e. $\mathcal{N}$ particles cross the unit area in the unit time).

a) Calculate the total cross section σ^{tot} (the total cross section is given by the number Δn of scattered particles per unit time by a single scatterer divided by $\mathcal{N}$: $\sigma^{\text{tot}} = \Delta n/\mathcal{N}$.

b) Calculate the differential cross section $\sigma(\theta, \phi)$ (i.e. the fraction of particles scattered in the solid angle $\Delta\Omega$, divided by $\Delta\Omega$).

17.5   The *optical theorem*, which relates the total cross section to the imaginary part of the forward scattering amplitude:

$$\sigma^{\text{tot}}(k) = \frac{4\pi}{k}\,\Im m\, f_{\vec{k}}(\theta = 0)$$

(the polar axis is parallel to $\vec{k}$), is a general theorem of wave propagation whose validity does not require special assumptions about the potential responsible for the scattering. The proof is quite simple in the case of potentials with spherical symmetry. In this case the scattering amplitude can be written as

$$f_k(\theta) = \frac{1}{k} \sum_{l=0}^{\infty} \sqrt{4\pi(2l+1)} \, e^{i\,\delta_l(k)} \sin \delta_l(k) \, Y_{l,0}(\theta)$$

(partial wave expansion).

a) Making use of the relation $Y_{l,0}(0) = \sqrt{(2l+1)/4\pi}$ and of the orthonormality condition $\int Y_{l'm'}^*(\theta,\phi) Y_{lm}(\theta,\phi) \, d\Omega = \delta_{ll'} \, \delta_{mm'}$, prove the optical theorem in the case of potentials with spherical symmetry.

b) To what extent are the phases $\delta_l(k)$ determined? ($\delta_l(k) \to \delta_l(k)+?$)

17.6 In Problem 17.4 we considered the scattering of point particles off a hard sphere from the classical point of view. We will face the same problem in the framework of quantum mechanics and we will see that the result exhibits a striking disagreement with the classical one.

Let us begin with the case of a 'soft sphere', i.e. with the scattering from the potential

$$V(r) = \begin{cases} V_0 & \text{for } r < a \qquad V_0 > 0 \\ 0 & \text{for } r \geq a. \end{cases}$$

Assume that the energy E of the particles is sufficiently low that only the component of the wavefunction with angular momentum $l = 0$ is scattered ('s-wave scattering').

a) Which condition guarantees that only the s-wave gives a significant contribution to the scattering cross section?

b) Determine the phase shift $\delta_{l=0}(k)$ and the (s-wave) total cross section (as in Problems 17.2 and 17.3, $\delta_{l=0}(k)$ will be given by an implicit equation).

c) Now, with E fixed, take the limit $V_0 \to \infty$ (hard sphere) and calculate the s-wave cross section. Calculate also directly the phase shift $\delta_{l=0}(k)$ when $V_0 = \infty$ from the very beginning, then the cross section ('low energy scattering from a hard sphere').

d) Calculate, in the limit $V_0 = \infty$, the 'time-delay' (see Problem 17.2) of the scattered wave.

17.7 Consider the scattering of a particle of mass m from the potential

$$V(r) = \begin{cases} -V_0 & \text{for } r < a \qquad V_0 > 0 \\ 0 & \text{for } r \geq a. \end{cases}$$

Assume that the energy E of the particles is sufficiently low so that only the component of the wavefunction with angular momentum $l = 0$ is scattered.

a) Determine the equation for the phase shift $\delta_0(k)$.

With $k_0 \equiv \sqrt{2mV_0/\hbar^2}$, assume $k_0 a \neq (2n+1)\pi/2$.

b) Determine the total cross section $\sigma^{\text{tot}}(E)$ for $E = 0$.

c) Calculate the *scattering length* $a_s \equiv -\lim_{k\to 0} \delta_0(k)/k$ for $m = m_e$ (the electron mass), $a = 1.1\,\text{Å}$ in the two cases: $V_0 = 7\,\text{eV}$ and $V_0 = 12\,\text{eV}$.

d) In the two cases $V_0 = 7\,\text{eV}$ and $V_0 = 12\,\text{eV}$ determine the number of bound states of the system with angular momentum $l = 0$.

17.8 A particle of mass m is subject to a the potential well $V(r) = -V_0$ ($V_0 > 0$) for $r < a$, $V(r) = 0$ for $r \geq a$.

a) Determine the lowest value V_0^{min} of V_0 such that the s-wave scattering is resonant for $E = 0$ (the scattering with angular momentum l is said to be resonant when the *unitarity bound* $\sigma_l^{\text{tot}}(k) = 4\pi\,(2l+1)/k^2$ is reached, even if, because of the k^2 in the denominator, this does not exactly correspond to a maximum of the cross section. Moreover we will see in Problem 17.14 that an additional condition is opportune for the definition of resonance).

b) Calculate V_0^{min} if $m = m_e$ (the electron mass) and $a = 1.1\,\text{Å}$.

17.9 Consider the $l = 0$ scattering of a particle of mass m from a potential well $V(r) = -V_0$ for $r < a$, $V(r) = 0$ for $r \geq a$. The presence of a bound state of energy E_B close to zero (i.e. $|E_B| \ll V_0$, $|E_B| \equiv \hbar^2\kappa^2/2m$) produces an enhancement of the cross section at low energies. We will assume that $|E_B|$ can be neglected with respect to V_0 and ka with respect to $\delta_0(k)$ that, as we will see, when $k \approx \kappa$, is close to $-\pi/2$.

a) For which values of V_0 does a bound state with angular momentum $l = 0$ and energy E_B close to zero exist?

b) Under the stated approximations write the equation determining $\tan\delta_0(k)$.

c) Neglecting, as suggested above, $|E_B|$ with respect to V_0, write the equation determining $|E_B|$, then show that $\tan\delta_0(k) = -k/\kappa$ $(\kappa = \sqrt{2m|E_B|/\hbar^2})$.

d) Write the s-wave scattering amplitude

$$f_0(k) = \frac{1}{k}\,e^{i\,\delta_0(k)} \sin\delta_0(k) = \frac{1}{k}\,\frac{1}{\cot\delta_0(k) - i}$$

and the total cross section $\sigma_0^{\text{tot}}(k)$. Is the optical theorem satisfied? Compare the result with that of Problem 17.7.

e) Use the equation $\tan k_0 a/k_0 = -1/\kappa$, found while solving point c), to determine E_B.

17.10   Consider the $l = 0$ scattering of a particle of mass m from a potential well of depth V_0 for $r < a$. Suppose that $k_0 a = (2n + 1)\pi/2 - \epsilon$, $n \geq 0$, $0 < \epsilon \ll 1$, where $k_0 a \equiv \sqrt{2m V_0 a^2/\hbar^2}$. As in the previous problem we will assume that ka can be neglected with respect to $\delta_0(k)$ that, as we will see, in this case can be close to $+\pi/2$.

a) How many bound states does the potential well admit?

b) Under the stated approximation determine $\tan \delta_0(k)$.

c) Write the s-wave scattering amplitude

$$f_0(k) = \frac{1}{k}\, e^{i\,\delta_0(k)} \sin \delta_0(k) = \frac{1}{k}\, \frac{1}{\cot \delta_0(k) - i}$$

and the total cross section $\sigma_0^{\rm tot}(k)$. Compare the results, both for the scattering amplitude and the total cross section, with those of Problem 17.9.

17.11   Consider the $l = 0$ scattering of a particle of mass m from the Dirac delta–shell potential $V(r) = -\lambda\,\delta(r - a)$, $\lambda > 0$. From Problem 6.18 we know that the δ-potential gives rise to a discontinuity in the first derivative of the reduced radial function $u_0(r)$ at the point $r = a$ given by

$$\Delta u_0'(a) \equiv u_0'(a^+) - u_0'(a^-) = -\frac{2m\lambda}{\hbar^2}\, u_0(a) = -g\, u_0(a)\,, \quad g \equiv \frac{2m\lambda}{\hbar^2} > 0\,.$$

a) For which values of the energy $E = \hbar^2 k^2/2m$ does the cross section vanish?

b) Imposing the suitable conditions on the function $u_0(r)$ at $r = a$, write the equation determining the phase shift $\delta_{l=0}(k)$.

From Problem 6.23 we know that, for suitable values of λ, a bound state exists (the odd-parity one of Problem 6.23). Moreover it is a general fact that, as we have seen in particular in Problem 17.9, the cross section at zero energy is resonant if there is a bound state with zero energy.

c) Consider the low energy scattering and exploit the above information to establish for what values of λ does the bound state exist and compare the result with that of Problem 6.23.

17.12   In the case of central potentials the scattering operator S is defined as the operator represented, in the angular momentum representation, by the matrix whose non vanishing elements are the diagonal ones $S_l(k) \equiv e^{2i\,\delta_l(k)}$. One can show in general that $S_l(k)$ admits an analytic continuation in a suitable region of the complex k-plane and the poles on the upper imaginary axis $\Re e\, k = 0$, $\Im m\, k > 0$ correspond to the bound states. For simplicity we will show this for the potential of the previous problem $V(r) = -\lambda\,\delta(r - a)$ in the case of zero angular momentum.

a) What is the relation between the S-matrix and the scattering amplitude?

b) Consider the expression of $\tan \delta_0(k)$ found in the previous problem that we rewrite in the following form:

$$\tan \delta_0(k) = ga\, \frac{1 - \cos 2ka}{2ka - ga\, \sin 2ka}\,.$$

Show that the analytic continuation to the imaginary axis of the complex plane $z = x + iy$ of the sin and cos functions are:

$$\sin(i\,y) = i \sinh y\,; \qquad \cos(i\,y) = \cosh y$$

and write the expression of $\tan \delta_0(i\,\kappa)$ on the imaginary axis of the $k + i\,\kappa$ plane.

c) Exploit the identity

$$S_0(k) = e^{2i\,\delta_0(k)} = \frac{1 + i\,\tan \delta_0(k)}{1 - i\,\tan \delta_0(k)}$$

and find the equation determining the position of the (unique) pole of $S_0(k)$ on the positive imaginary axis ($ga > 1$; if $1 > ga > 0$ the pole, no longer corresponding to a bound state, is on the negative imaginary axis: see the next question).

d) If $ga = 1 + \epsilon$, $0 < \epsilon \ll 1$, the pole (i.e. the bound state) is close to zero. Determine its position, i.e. κ. To this purpose expand the l.h.s. of the equation $e^{-2\kappa a} = 1 - 2\kappa a/ga$ found in Problem 6.23 (and also in the answer to the previous question) to 2nd order in κa. Then show that if $ga = 1 - \epsilon$ (no bound state) the pole occurs on the negative imaginary axis: this is the *virtual level* mentioned in Problem 17.10.

17.13   After having learned how to find the poles of the S-matrix corresponding to the bound states, the reader can try his hand for the case of the potential well $V(x) = -V_0$ for $r \le a$, $V(r) = 0$ for $r > a$.

a) Starting from the equation (see Problems 17.7–17.10)

$$\tan\left(ka + \delta_0(k)\right) = (k/k_1)\, \tan k_1 a, \qquad k_1 = \sqrt{2m(V_0 + E)/\hbar^2} \equiv \sqrt{k_0^2 + k^2}\,,$$

show that the $l = 0$ component of the S-matrix can be written as

$$S_0(k) = e^{-2i\,ka}\, \frac{1 + i\,(k/k_1)\, \tan k_1 a}{1 - i\,(k/k_1)\, \tan k_1 a}\,.$$

b) Pay due attention to the analytic continuation of $\tan k_1 a$ to the imaginary axis, then find the equations determining the positions of the poles of $S_0(k)$ on the positive imaginary axis. Compare the result with that of Problem 6.11.

c) Exploit the above expression for the S-matrix and the results of Problems 17.9 and 17.10 to give a low energy expression ($ka \ll 1$) of the S-matrix both in the case of a bound state and of a 'virtual level' with energies close to zero.

17.14   In the line of Problems 17.12, 17.13, we continue to exemplify the structure of the analytic continuation of the S-matrix to the complex $k + i\kappa$ plane. It is a general feature of such a continuation that, apart from the poles on the positive imaginary axis ($k = 0$, $\kappa > 0$), the bound states, the other poles can only occur in the lower half complex plane.

Consider the scattering of a particle from a potential $V(r)$. Suppose that, for a given angular momentum l, a pole occurs at $k + i\kappa = k_r - ib$ ($b > 0$, $k_r = \sqrt{2mE_r/\hbar^2}$).

a) Exploit the fact that the S-matrix is unitary to show that S_l can be written as

$$S_l(k) = \frac{k - k_r - ib}{k - k_r + ib}\, e^{i\gamma(k)}, \quad \gamma(k) = \gamma^*(k) .$$

Assume that the pole in $k_r - ib$ is sufficiently far from other poles and that for $k \approx k_r$, $e^{i\gamma(k)} = 1$.

b) Use the above expression of the S-matrix (with $e^{i\gamma(k)} = 1$) to determine the scattering amplitude $f_l(k,\theta)$ and the total cross section $\sigma_l^{tot}(k)$ for $k \approx k_r$ and verify that the l-wave is resonant for $k = k_r$.

c) Calculate the time-delay Δt at the resonance. What is the behaviour of the phase $\delta_l(k)$ near the resonance? In particular, what is the difference between $\delta_l(k)$ for $k \gg k_r$ and $\delta_l(k)$ for $k \ll k_r$?

d) Write the total cross section $\sigma_l^{tot}(E)$ (near the resonance) in terms of the energy E of the particle and $\Gamma/2 \equiv 2\hbar/\Delta t$.

17.15   Levinson theorem asserts that the number of bound states equals $1/\pi \times$ the difference between the phase at zero energy and at infinity:

$$\delta(0) - \delta(\infty) = n\,\pi \quad (n \text{ is the number of bound states}).$$

The aim of this problem is to provide the means to verify Levinson theorem in the simple $l = 0$ case of the attractive spherical shell potential $-\lambda\,\delta(r - a)$, the same we dealt with in Problems 17.11 and 17.12. This will also give us the opportunity to introduce some concepts that are fundamental in the theory of scattering.

The Jost function $\mathcal{J}_0(k)$ is defined as the value at $r = 0$ of the solution $\mathcal{J}_0(k,r)$ of the reduced Schrödinger equation ($l = 0$):

$$-\frac{d^2 \mathcal{J}_0(k,r)}{dr^2} + U(r)\,\mathcal{J}_0(k,r) = k^2 \mathcal{J}_0(k,r), \quad U(r) = \frac{2m}{\hbar^2} V(r)$$

with the asymptotic condition $\mathcal{J}_0(k,r) \to e^{-ikr}$ for $r \to \infty$.

a) Determine the Jost functions $\mathcal{J}_0(k)$ when $V(r) = -\lambda\,\delta(r - a)$.

b) Verify that the ($l = 0$ component of the) S-matrix can be written as

$$S_0(k) \equiv e^{2i\delta_0(k)} = \frac{\mathcal{J}_0(k)}{\mathcal{J}_0(-k)}$$

(to this end it is sufficient to verify that $|\mathcal{J}_0(k)/\mathcal{J}_0(-k)| = 1$ and that the phase $\gamma(k)$ of $\mathcal{J}_0(k)$ coincides with the phase $\delta_0(k)$ found in Problem 17.11).

c) After observing that $\mathcal{J}_0(k) = |\mathcal{J}_0(k)| e^{i\delta_0(k)}$ is defined also for negative values of k with $|\mathcal{J}_0(k)|$ an even function of k and $\delta_0(k)$ an odd one, show that

$$\int_{-\infty}^{+\infty} \frac{\mathcal{J}_0'(-k)}{\mathcal{J}_0(-k)} \, dk = 2i \left(\delta_0(0) - \delta_0(\infty) \right) .$$

d) Exploiting the fact that $\mathcal{J}_0(k) \to 1$ for $|k| \to \infty$ and that (as it can be shown) $\mathcal{J}_0'(k)$ is $O(|k|^{-2})$ at infinity in the upper half complex k–plane, show the validity of Levinson theorem.

17.16   When electrons are scattered by the atoms of a noble gas it happens that the total cross section exhibits very small values at electron energies of about $1\,\mathrm{eV}$. In these cases the mean free path of the electrons in the gas is very long and the gas appears as almost transparent to the electrons. This is the Ramsauer-Townsend effect that is analogous to the perfect transmission found at particular energies in one-dimensional scattering from a square well (see Problem 17.1).

We will consider the case of the electron scattering from Xenon atoms. Xenon atoms have atomic number $Z = 54$ and a radius $a = 1.4\,\text{Å}$. To make the problem viable, we schematize the Xenon atom as a spherical potential well of depth V_0 and radius a.

a) If the energy of the electrons is $E = 0.7\,\mathrm{eV}$, calculate ka, where $k = \sqrt{2mE}/\hbar$.

b) Write the equation that determines the energies for which the cross section vanishes.

c) Knowing that the first value of the energy such that $\sigma_{\mathrm{tot}}(E) = 0$ is $E = 0.7\,\mathrm{eV}$, determine the depth V_0: to this end, either solve the equation for V_0 using one of the many equation-solver software or, as an alternative, verify that $V_0 \simeq 39\,\mathrm{eV}$ is the solution.

17.17   The Yukawa potential

$$Y(r) = -Y_0 \frac{e^{-r/r_0}}{r/r_0} , \qquad Y_0 > 0$$

was introduced by H. Yukawa in the thirties of the past century to account for the interaction between nucleons. In the Born approximation it gives rise to the scattering amplitude $(q = 2k \sin(\theta/2))$

$$f_Y(k, \theta) = -\frac{2m}{\hbar^2 q} \int \sin(qr') Y(r') \, r' dr' = \frac{2mY_0}{\hbar^2} \frac{r_0^3}{1 + (qr_0)^2} .$$

Since it is not possible to exactly calculate the scattering amplitude, it might be useful to approximate the Yukawa potential with the spherical potential well $V(r) = -V_0$ for $r < a$, $V(r) = 0$ for $r \geq a$.

a) Determine the scattering amplitude $f_V(k, \theta)$ for the spherical square potential in the Born approximation.

b) With $r_0 = a$ determine V_0 so that the *volume-integrals* (i.e. the integral over the all space) of the two potentials are the same.

In both cases (Yukawa and spherical potential well) it is possible to calculate the total cross section by taking $x = 2ka\sin(\theta/2)$ as the variable of integration. At high energies, the total cross section for the scattering off the spherical square potential is:

$$\sigma_V^{tot}(k) = \frac{9}{8} \frac{\sigma_V^{tot}(0)}{(ka)^2}, \qquad ka \gg 1.$$

c) Calculate the total cross section at high energies for the Yukawa potential and determine V_0 and a so that the total cross sections coincide both at zero energy and in the limit of high energies.

17.18   The elastic scattering of a particle off a molecule consisting of two identical atoms a distance a apart can be, as a first approximation (i.e. neglecting the distortion induced by the particle), considered as the scattering from the potential $V_{mol}(\vec{r}) = V_{at}(\vec{r}) + V_{at}(\vec{r} - \vec{a})$. Suppose that the scattering amplitude f_{at} from a single atom in the Born approximation is known:

$$f_{at}(\vec{q}) = -\frac{1}{4\pi} \frac{2m}{\hbar^2} \int e^{-i\vec{k}_f \cdot \vec{r}'} V_{at}(\vec{r}') e^{i\vec{k}_i \cdot \vec{r}'} d\vec{r}', \qquad \vec{q} = \vec{k}_f - \vec{k}_i.$$

a) Calculate, in the Born approximation, the scattering amplitude $f_{mol}(\vec{q})$ and the cross section $\sigma_{mol}(\vec{q})$ for the scattering off the molecule.

b) Calculate the scattering amplitude and the cross section if $\lambda \equiv 2\pi/k \gg |\vec{a}|$.

In real situations the particles are scattered by a gas of molecules. As usual, we assume that the molecules are sufficiently far apart so that any collision process involves only one of them. Furthermore, assume that the single atom potential V_{at} is a central potential: $V_{at}(r)$.

c) Taking into account that the molecules in the gas are randomly oriented, calculate the cross section: to this end which between the amplitude and the cross section should be averaged over all directions of the molecules?

17.19   Consider the scattering of a beam of particles by a cubic crystal with lattice spacing a: $\vec{r}_n = n_1 a \, \hat{x} + n_2 a \, \hat{y} + n_3 a \, \hat{z}$ are the positions of the atoms (e.g silicon) forming the crystal, $\hat{x}, \hat{y}, \hat{z}$ are the unit vectors of the Cartesian axes and $0 \leq n_i \leq N^{1/3}$, $N \gg 1$ (N is the number of atoms involved in a single scattering process: it is of the order of the dimension of the wave packet divided by a^3: according to the estimate in Problem 7.9 it can be quite large, say $N \simeq 10^{12}$). Suppose that the differential cross section $\sigma_{at}(\vec{q})$ ($\vec{q} \equiv \vec{k}_f - \vec{k}_i$) in the Born approximation for the scattering from the single silicon atom is known and is sufficiently regular.

a) Write, in the Born approximation, the differential cross section for the scattering by the crystal.

Suppose that the particles are sent against the crystal in a direction orthogonal to one of the planes, say with momentum $\hbar \vec{k}$ parallel to the z-axis.

b) Determine for which values of k backward scattering is possible (more precisely: not negligible). Exploit the result to show the validity of the Bragg formula ($2a \sin\theta = n\lambda$, see Problem 2.14) for the reflection by the crystal.

c) Still with incidence of the beam orthogonal to the crystal plane $z = 0$, determine the angular width of the forward diffraction peak.

At high temperature the atoms of the crystal undergo nonnegligible vibrations about their equilibrium positions: a simple way to take into account the effect on the scattering cross section is to consider the relative positions of the atoms $\vec{r}_n - \vec{r}_m$ as randomly distributed.

d) Under the above hypothesis determine the scattering cross section.

17.20   Neutrons and protons differ in the charge and their masses are equal within 1.3‰. Neutron–neutron and neutron–proton potentials are the same and for simplicity we assume that they are central and spin-independent.

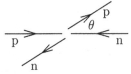

Let, for a given energy E, $f_E^{pn}(\theta)$ be the p–n scattering amplitude in the center-of-mass frame, i.e. the scattering amplitude for the observation of one of the two particles, e.g. the proton. The scattering angle θ is taken with respect to the incident direction.

a) Write the differential cross section $\sigma_E^{nn}(\theta)$ in the center-of-mass frame for the n–n scattering when the two colliding neutrons are in the same spin state.

b) Write cross section $\sigma_E^{nn}(\theta)$ when the two colliding neutrons are in orthogonal spin states.

c) Determine the differential cross section in the more realistic situation in which the two colliding beams are unpolarized, i.e. they are a uniform statistical mixture of the neutron spin states.

Now consider the proton–neutron scattering. The detector is not supposed to recognize either the charge of the nucleons or their mass difference.

d) Determine the neutron–proton differential cross section $\sigma(\theta)$ in the center-of-mass frame.

17.21 Consider the neutron–proton scattering in the center-of-mass frame. Assume that the counters only detect the protons and let $\sigma_E^p(\theta_p)$ be the differential cross section for the observation of the protons at the angle θ_p.

a) Determine the neutron differential cross section $\sigma_E^n(\theta)$ in terms of that of the protons (that are the ones that are revealed).

In the 30's of the last century the 'Via Panisperna boys' led by E. Fermi undertook experiments where slow neutrons were scattered by different substances among which paraffin that, being rich with hydrogen atoms, allows for the study of neutron–proton scattering.

b) If the neutron–proton scattering amplitude $f_E^{np}(\theta)$ in the center-of-mass frame is known, determine both the differential and the total cross section in the laboratory system, i.e. the one where the protons are initially at rest (the mass difference between protons and neutrons is neglected). To this end, first determine the relation between the scattering angles θ, ϕ in the center-of-mass and θ_0, ϕ_0 in the laboratory frame, then exploit the fact that the same number of particles are scattered in the corresponding solid angles $d\Omega$ and $d\Omega_0$.

Solutions

17.1

a) With $k = \sqrt{2mE}/\hbar$ and $k_1 = \sqrt{2m(E+V_0)}/\hbar$ one has

$$\psi_E(x) = \begin{cases} e^{ikx} + A\,e^{-ikx} & x \le 0 \\ B\,e^{ik_1 x} + B'\,e^{-ik_1 x} & 0 \le x \le a \\ C\,e^{ik(x-a)} & x \ge a\,. \end{cases}$$

The coefficients A, B, B', C are determined by the continuity conditions of ψ_E and ψ'_E in $x = 0$ and $x = a$:

$$\begin{cases} 1 + A = B + B' \\ (k/k_1)\,(1-A) = B - B' \end{cases}; \quad \begin{cases} C = B\,e^{ik_1 a} + B'\,e^{-ik_1 a} \\ C = (k_1/k)\,(B\,e^{ik_1 a} - B'\,e^{-ik_1 a}) \end{cases} \Rightarrow$$

$$\begin{cases} C = (B + B')\,\cos k_1 a + i\,(B - B')\,\sin k_1 a \\ C = (k_1/k)\,[(B - B')\,\cos k_1 a + i\,(B + B')\,\sin k_1 a] \end{cases} \Rightarrow$$

$$\begin{cases} C = (1 + A)\,\cos k_1 a + i\,(k/k_1)\,(1 - A)\,\sin k_1 a \\ C = (1 - A)\,\cos k_1 a + i\,(k_1/k)\,(1 + A)\,\sin k_1 a \end{cases} \Rightarrow$$

$$(1 + A)\,\cos k_1 a + i\,(k/k_1)\,(1 - A)\,\sin k_1 a$$
$$= (1 - A)\,\cos k_1 a + i\,(k_1/k)\,(1 + A)\,\sin k_1 a \qquad \Rightarrow$$

$$A = i\,\frac{[(k_1/k) - (k/k_1)]\,\sin k_1 a}{2\cos k_1 a - i\,[(k_1/k) + (k/k_1)]\,\sin k_1 a} \qquad \Rightarrow$$

$$\begin{cases} |A|^2 = \dfrac{V_0^2\,\sin^2 k_1 a}{4E\,(E + V_0) + V_0^2\,\sin^2 k_1 a} \\[3mm] |C|^2 = 1 - |A|^2 = \dfrac{4E\,(E + V_0)}{4E\,(E + V_0) + V_0^2\,\sin^2 k_1 a} \end{cases}$$

where use has been made of $T^2 + R^2 = 1$ (the number of the transmitted plus that of the reflected particles equals the number of the incident ones).

Solutions

b) $T = 1$ when $A = 0$, i.e. $\sin k_1 a = 0$: as in Problem 6.13 this occurs when between $x = 0$ and $x = a$ (the region where the potential is different from zero) an integer number of half wave lengths is contained. The corresponding energies are given by

$$E = \frac{n^2 h^2}{8ma^2} - V_0 , \quad n \geq \sqrt{8mV_0 a^2/h^2} \quad (E \geq 0) .$$

17.2

a) First consider the case $V_0 = 0$: here the only condition is the vanishing of the wavefunction at $x = 0$, then $\psi_E(x)$ is proportional to $\sin kx$, therefore $A = -1$. If $V_0 \neq 0$, with $k = \sqrt{2mE}/\hbar$ and $k_1 = \sqrt{2m(V_0 + E)}/\hbar$ one has

$$\psi_E(x) = \begin{cases} B \sin k_1 x & 0 \leq x \leq a \\ e^{-ikx} + A e^{ikx} & x \geq a . \end{cases}$$

The continuity conditions at $x = 0$ and $x = a$ read

$$\begin{cases} B \sin k_1 a & = e^{-ika} + A e^{ika} \\ k_1 B \cos k_1 a = -ik\left(e^{-ika} - A e^{ika}\right) \end{cases} \Rightarrow$$

$$k_1 \left(e^{-ika} + A e^{ika}\right) \cos k_1 a = -ik\left(e^{-ika} - A e^{ika}\right) \sin k_1 a \quad \Rightarrow$$

$$A = -e^{-2ika} \frac{\cos k_1 a + i\,(k/k_1) \sin k_1 a}{\cos k_1 a - i\,(k/k_1) \sin k_1 a} .$$

Clearly $|A| = 1$, therefore $A = -e^{2i\,\delta(k)}$ with $2\,\delta(k) = -2ka + 2\alpha$ where $\alpha = \tan^{-1}\left[(k/k_1) \tan k_1 a\right]$, thus

$$2\,\delta(k) = -2\,ka + 2\,\tan^{-1}\left[(k/k_1) \tan k_1 a\right]$$

or, equivalently

$$\tan\left(ka + \delta(k)\right) = (k/k_1) \tan k_1 a .$$

The reason for writing $2\,\delta$ instead of δ is the following: in the absence of the potential, $\psi_E(x)$ is proportional to $\sin kx$ therefore, when the potential is present, it is commonplace to write, for $x \geq a$, $\psi_E(x)$ as $\sin\left(kx + \delta(k)\right)$, but then ($\propto =$ 'proportional to'):

$$\sin\left(kx + \delta(k)\right) \propto e^{i\,\delta(k)} e^{ikx} - e^{-i\,\delta(k)} e^{-ikx} \propto e^{-ikx} - e^{2i\,\delta(k)} e^{ikx} .$$

b) $kx + 2\delta(k) \rightarrow kx + 2\delta(k_0) + 2(k - k_0)\,\delta'(k_0)$

$$= k\left(x + 2\,\delta'(k_0)\right) + \text{constant terms}$$

therefore the center of the packet is shifted by $-2\,\delta'(k_0)$ and, since the (mean) velocity of the packet is $\hbar k_0/m$, the time-delay is

$$\Delta t = 2\frac{m\,\delta'(k_0)}{\hbar k_0} = \frac{2\hbar\,\delta'(k_0)}{d\,E/d\,k} = 2\hbar\,\frac{d\,\delta(E)}{d\,E}\bigg|_{E=E_0} , \quad E_0 = \frac{\hbar^2 k_0^2}{2m} .$$

17.3

a) With $k = \sqrt{2mE}/\hbar$ and $\kappa = \sqrt{2m(V_0 - E)}/\hbar$ one has

$$\psi_E(x) = \begin{cases} B \sinh \kappa x & 0 \leq x \leq a \\ e^{-ikx} + A e^{ikx} & x \geq a . \end{cases}$$

The continuity conditions at $x = 0$ and $x = a$ read

$$\begin{cases} B \sinh \kappa a = e^{-ika} + A e^{ika} \\ \kappa B \cosh \kappa a = -ik \left(e^{-ika} - A e^{ika} \right) \end{cases} \Rightarrow$$

$$\kappa \left(e^{-ika} + A e^{ika} \right) \cosh \kappa a = -ik \left(e^{-ika} - A e^{ika} \right) \sinh \kappa a \qquad \Rightarrow$$

$$A = -e^{-2ika} \frac{\cosh \kappa a + i \left(k/\kappa \right) \sinh \kappa a}{\cosh \kappa a - i \left(k/\kappa \right) \sinh \kappa a} .$$

Clearly $|A| = 1$, therefore $A = -e^{2i\,\delta(k)}$ with $2\,\delta(k) = -2ka + 2\alpha$ where $\alpha = \tan^{-1}\left[(k/\kappa) \tanh \kappa a \right]$, thus

$$2\,\delta(k) = -2\,ka + 2 \tan^{-1}\left[(k/\kappa) \tanh \kappa a \right]$$

or, equivalently

$$\tan \left(ka + \delta(k) \right) = (k/\kappa) \tanh \kappa a .$$

b) For $V_0 \to \infty$ $(k/\kappa) \tanh \kappa a \to 0$, then $\delta(k) \to -ka$.

Actually, if $V_0 = \infty$, the wavefunction $\psi_E(x) = 0$ for $0 \leq x \leq a$ and $\psi_E(x) = \sin k(x - a)$ for $x \geq a$, then $\delta(k) = -ka$.

17.4

a) $\sigma^{\text{tot}} = \pi a^2$. Indeed, in the unit time $\mathcal{N} \times \pi a^2$ particles hit the single scatterer (the sphere) and each of them is scattered.

b) The distance b of the trajectory of a particle from the parallel through the center of the sphere is called *impact parameter*. It is related to the scattering angle θ in the following way: since (see figure) $2\,\alpha = \pi - \theta$, we have

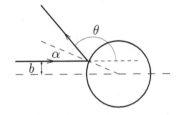

$$b = a \sin(\pi/2 - \theta/2) = a \cos(\theta/2) .$$

The particles with impact parameters between b and $b+db$, i.e. within the area $2\pi\, b\, db$, due to the symmetry around the (polar) axis parallel to the beam and passing through the center of the sphere, are scattered within the solid angle $2\pi \sin \theta \, |d\theta|$, then

$$\sigma(\theta) = \frac{\mathcal{N} \times 2\pi\, b\, db}{\mathcal{N} \times 2\pi \sin \theta \, |d\theta|} = \frac{a \cos(\theta/2)}{\sin \theta} \left| \frac{db}{d\theta} \right| = \frac{a^2}{4} .$$

therefore the particles are scattered isotropically and we find again
$\sigma^{\text{tot}} = 4\pi\,\sigma(\theta) = \pi\,a^2$.

17.5

a) $\sigma^{\text{tot}}(k) = \displaystyle\int |f_k(\theta)|^2 d\Omega$, and thanks to the orthogonality of the spherical harmonics

$$\sigma^{\text{tot}}(k) = \sum_{l=0}^{\infty} \sigma_l(k) = \sum_{l=0}^{\infty} \frac{4\pi}{k^2}(2l+1)\sin^2\delta_l(k)\,.$$

Moreover

$$
\begin{aligned}
f_k(\theta=0) &= \frac{1}{k}\sum_l \sqrt{4\pi(2l+1)}\;e^{i\,\delta_l(k)}\sin\delta_l(k)\,Y_{l,0}(0)\\
&= \frac{1}{k}\sum_l (2l+1)\,e^{i\,\delta_l(k)}\sin\delta_l(k)\,,
\end{aligned}
$$

$$\Im m\, f_k(\theta=0) = \frac{1}{k}\sum_l (2l+1)\sin^2\delta_l(k) = \frac{k}{4\pi}\sigma^{\text{tot}}(k)\,.$$

b) $e^{i\,\delta_l(k)}\sin\delta_l(k) = e^{i\,(\delta_l(k)+\pi)}\sin\big(\delta_l(k)+\pi\big)$, therefore all the phases are defined mod π (not mod 2π).

17.6

a) By dimensional analysis, since the relevant parameters in the problem are the range a of the potential and the wave-number k of the particles, it can be inferred that the condition should read $ka \ll 1$. This conclusion is strengthened by the following semi-classical argument: since only particles with impact parameter (see Problem 17.4) less than a are scattered, their maximum angular momentum is $\hbar\,ka$, therefore the condition is $\hbar\,ka \ll \hbar$.

b) The incoming wavefunction is

$$e^{i\,\vec{k}\cdot\vec{r}} = \sum_{l=0}^{\infty} i^l\sqrt{4\pi(2l+1)}\;j_l(kr)\,Y_{l,0}(\theta)$$

where the radial functions $j_l(kr)$ are the Bessel functions and $j_l(kr)\,Y_{l,0}(\theta)$ is the free solution with energy $E = \hbar^2 k^2/2m$ and angular momentum l. Because of the assumption of s-wave scattering, we consider only the $l=0$ component. The scattering wavefunction ψ_E is then of the form

$$\psi_E(r) = j_0(kr) + \chi_0(kr) = \frac{\sin kr}{kr} + \chi_0(kr)$$

where $\chi_0(kr)$ is the scattered wave that for $r \ge a$ must be an outgoing spherical wave. The *reduced radial function* $u_0(r) \equiv r\,\psi_E(r)$ satisfies the equation

$$-\frac{d^2 u_0(r)}{dr^2} + U(r)\,u_0(r) = k^2 u_0(r)\,, \qquad u_0(0) = 0\,, \quad U(r) = \frac{2m}{\hbar^2}V(r)$$

therefore, for $r \geq a$, $u_0(r)$ is of the form $u_0(r) = A \sin\left(kr + \delta_0(k)\right)$ and the constant A has to be determined in such a way that the difference, for $r \geq a$, between $\psi_E(r)$ and the free solution $j_0(kr)$:

$$\frac{A}{2i}\left(e^{+i\delta_0(k)}\,\frac{e^{ikr}}{kr} - e^{-i\delta_0(k)}\,\frac{e^{-ikr}}{kr}\right) - \frac{1}{2i}\left(\frac{e^{ikr}}{kr} - \frac{e^{-ikr}}{kr}\right)$$

is an outgoing wave (the scattered wave). Then

$$A = e^{i\delta_0(k)}, \quad \chi_0(kr) \equiv \psi_E(r) - j_0(kr)$$

$$= \frac{1}{2ik}\left(e^{2i\delta_0(k)} - 1\right)\frac{e^{ikr}}{r} \equiv f_0\,\frac{e^{ikr}}{r}\;.$$

Therefore the scattering amplitude $f_0(k)$ and the total cross section $\sigma_0^{\rm tot}(k)$ are:

$$f_0 = \frac{1}{k}\,e^{i\delta_0(k)}\sin\delta_0(k), \qquad \sigma_0^{\rm tot}(k) = \frac{4\pi}{k^2}\sin^2\delta_0(k)\;.$$

It remains to determine the phase shift $\delta_0(k)$. This is done by imposing the continuity conditions for $u_0(r)$ at $r = a$: with $\kappa = \sqrt{2m(V_0 - E)}/\hbar$

$$u_0(r) = \begin{cases} B\sinh\kappa r & 0 \leq r \leq a \\ e^{i\delta_0(k)}\sin\left(kr + \delta_0(k)\right) & r \geq a \end{cases}$$

then:

$$\begin{cases} B\sinh\kappa a = e^{i\delta_0(k)}\sin\left(ka + \delta_0(k)\right) \\ B\cosh\kappa a = (k/\kappa)\,e^{i\delta_0(k)}\cos\left(ka + \delta_0(k)\right) \end{cases} \Rightarrow$$

$$\tan\left(ka + \delta_0(k)\right) = (k/\kappa)\tanh\kappa a$$

i.e. the same result we found in Problem 17.3: indeed, the reduced radial function $u_0(r)$ satisfies the same equation as $\psi(x)$ in Problem 17.3.

c) As in Problem 17.3 for $V_0 \to \infty$ $\delta_0(k) \to -ka$ then

$$\sigma_0^{\rm tot}(k) = \frac{4\pi}{k^2}\sin^2 ka \simeq 4\pi\,a^2 \qquad (ka \ll 1)$$

four times the classical outcome!

The same result could have been obtained more easily by taking from the beginning $V_0 = \infty$: indeed, in this case

$$u_0(r) = \begin{cases} 0 & 0 \leq r \leq a \\ \sin k(r - a) & r \geq a \end{cases} \Rightarrow \delta_0(k) = -ka\;.$$

d) $\Delta t = 2\hbar\,\mathrm{d}\delta_0/\mathrm{d}E = 2m\,\delta_0'(k)/\hbar\,k = -2ma/\hbar\,k$, i.e. $\Delta t = -2a/v$ therefore the time-delay of the scattered wave is negative, i.e. the wave is 'anticipated', in agreement with intuition: indeed, $2a/v$ is precisely the time necessary to go from $r = a$ to $r = 0$ and back to $r = a$.

17.7

a) The reduced radial function $u_0(r)$ is of the form

$$u_0(r) = \begin{cases} B \sin k_1 r & \text{for } r \le a \\ A \sin \left(kr + \delta_0(k)\right) & \text{for } r \ge a \end{cases}$$

and the continuity conditions at $r = a$:

$$\begin{cases} B \sin k_1 a = A \sin \left(ka + \delta_0(k)\right), & k_1 = \sqrt{2m(V_0 + E)/\hbar^2} \equiv \sqrt{k_0^2 + k^2} \\ k_1 B \cos k_1 a = kA \cos \left(ka + \delta_0(k)\right) \end{cases}$$

imply $\tan \left(ka + \delta_0(k)\right) = (k/k_1) \tan k_1 a$.

b) Provided $k_1 a \ne (2n + 1)\pi/2$, which is guaranteed by the assumption $k_0 a \ne (2n+1)\pi/2$ and the requirement $E \to 0$, $ka + \delta_0(k)$ is $O(k)$, then

$$\tan \left(ka + \delta_0(k)\right) \simeq ka + \delta_0(k) \quad \Rightarrow \quad \delta_0(k) \simeq -ka + \frac{k}{k_1} \tan k_1 a .$$

Moreover for $E \to 0$ $k_1 \to k_0$, therefore the cross section is

$$\sigma_0^{\text{tot}}(E = 0) = \lim_{k \to 0} \frac{4\pi}{k^2} \sin^2 \delta_0(k) \simeq \lim_{k \to 0} \frac{4\pi \, \delta_0^2(k)}{k^2}$$

$$= 4\pi \, a^2 \lim_{k \to 0} \left(\frac{\tan k_1 a}{k_1 a} - 1\right)^2 = 4\pi \, a^2 \left(\frac{\tan k_0 a}{k_0 a} - 1\right)^2 .$$

c) $\lim_{k \to 0} \dfrac{\delta_0(k)}{k} = a \left(\dfrac{\tan k_0 a}{k_0 a} - 1\right)$.

In order to calculate expressions as $k_0 a = \sqrt{2m_e V_0 a^2/\hbar^2}$ it is always a good rule to express it in terms of fundamental quantities. So, for instance, in the present case it is useful to recall that $e^2/2a_B = 13.6\,\text{eV}$ where $a_B = \hbar^2/m_e e^2 = 0.53\,\text{Å}$ is the Bohr radius, so:

$$\frac{2m_e V_0 a^2}{\hbar^2} = V_0 \times \left(\frac{a}{a_B}\right)^2 \frac{2m_e a_B^2}{\hbar^2} = V_0 \times \left(\frac{a}{a_B}\right)^2 \frac{2 a_B}{e^2} = \left(\frac{a}{a_B}\right)^2 \times \frac{V_0}{13.6} .$$

If $V_0 = 7\,\text{eV}$:

$$\frac{2m_e V_0 a^2}{\hbar^2} \simeq 4.31 \times \frac{7}{13.6} \simeq 2.22 ; \qquad k_0 a = \sqrt{2.22} \simeq 1.49$$

$$a_s = -a \left(\frac{\tan 1.49}{1.49} - 1\right) \simeq -7.3\,a \simeq -8\,\text{Å} .$$

If $V_0 = 12\,\text{eV}$:

$$\frac{2m_e V_0 a^2}{\hbar^2} \simeq 3.8 ; \qquad k_0 a = \sqrt{3.8} \simeq 1.95$$

$$a_s = -a \left(\frac{\tan 1.95}{1.95} - 1\right) \simeq +2.3\,a \simeq 2.5\,\text{Å} .$$

The reason for the different sign in the two cases is due to the change of sign of $\tan k_0 a$ (from positive to negative if $k_0 a$ increases) when $k_0 a$ crosses an odd multiple of $\pi/2$ (in the present case $1.49 < \pi/2 < 1.95$).

d) With $V_0 = 7\,\text{eV}$ $k_0 a = 1.49 < \pi/2 = 1.57$ therefore, as discussed in Problems 12.2 and 6.9, there is no bound state. If $V_0 = 12\,\text{eV}$ $k_0 a = 1.95$ and since $\pi/2 < k_0 a < 3\pi/2$ there is just one bound state. Therefore the change of sign of a_s occurs exactly when a new bound state appears: we will get back to this point in the forthcoming problems.

17.8

a) The unitary bound is reached when $\sin^2 \delta_l(k) = 1$, i.e. when $\delta_l(k) = \pi/2 \pmod{\pi}$. In the equation $\tan\left(ka + \delta_0(k)\right) = (k/k_1)\tan k_1 a$ (see the previous problem) the l.h.s. – with $\delta_0(k) = \pi/2 \pmod{\pi}$ – diverges for $k \to 0$, while the r.h.s goes to 0 unless $k_0 a \equiv \sqrt{2mV_0 a^2/\hbar^2} = (2n+1)\pi/2$. Then

$$V_0^{\min} = \left(\frac{\pi}{2}\right)^2 \frac{\hbar^2}{2ma^2}\,.$$

In this case ($E = 0$) – however – the cross section diverges because of the k^2 in the denominator (waves not perfectly monochromatic, i.e. wave packets, should be considered).

Viceversa, if $V_0 = V_0^{\min}$ and $E = 0$, $\delta_0(k) = \pi/2 \pmod{\pi}$.

b) With $m = m_e$ and $a = 1.1\,\text{Å}$:

$$V_0^{\min} = \left(\frac{\pi}{2}\right)^2 \left(\frac{a_B}{a}\right)^2 \times 13.6 \simeq 7.8\,\text{eV}\,.$$

17.9

a) As discussed in Problems 12.2 and 6.9, the number of bound states due to a potential well of depth V_0 is given by the number of odd multiples of $\pi/2$ contained in $k_0 a \equiv \sqrt{2mV_0 a^2/\hbar^2}$: every time $k_0 a$ goes across an odd multiple of $\pi/2$ a new bound state appears with energy equal to zero. Therefore the condition for the existence of a bound state of energy E_B close to zero is that
$$k_0 a = (2n+1)\pi/2 + \epsilon, \quad n \geq 0, \quad 0 < \epsilon \ll 1.$$

b) From the equation $\tan\left(ka + \delta_0(k)\right) = (k/k_1)\tan k_1 a$, $k_1 = \sqrt{k_0^2 + k^2}$, neglecting ka with respect to $\delta_0(k)$ and, 'a fortiori' putting $k_1 a = k_0 a$, one gets:

$$\frac{\tan \delta_0(k)}{k} = \frac{\tan k_0 a}{k_0}\,.$$

c) The reduced radial function $u_0(r)$ for $r \leq a$ is

$$u_0(r) = \sin\left(\sqrt{2m(V_0 - |E_B|)/\hbar^2}\; r\right) \simeq \sin\left(\sqrt{2mV_0/\hbar^2}\; r\right) = \sin k_0 r$$

while for $r \geq a$ $u(r) = A\,e^{-\kappa r}$, therefore the continuity conditions read:

$$\begin{cases} \sin k_0 a = A\,e^{-\kappa a} \\ k_0 \cos k_0 a = -\kappa A\,e^{-\kappa a} \end{cases} \Rightarrow \quad \frac{\tan k_0 a}{k_0} = -\frac{1}{\kappa}$$

then from $\tan\delta_0(k)/k = \tan k_0 a/k_0$:

$$\tan\delta_0(k) = -\frac{k}{\kappa} = -\sqrt{\frac{E}{|E_B|}} \; .$$

d) $f_0(k) = \dfrac{i}{k - i\kappa}$, $\sigma_0^{\text{tot}}(k) = \dfrac{4\pi}{\kappa^2 + k^2}$;

$$\frac{4\pi}{k}\,\Im m f_0(k) = \frac{4\pi}{k}\frac{k}{\kappa^2 + k^2} = \sigma_0^{\text{tot}}(k) \; .$$

To compare the result with that of Problem 17.7, that however holds only for $k = 0$, we write the cross section as

$$\sigma_0^{\text{tot}}(k) = \frac{4\pi a^2}{(\kappa a)^2 + (ka)^2} = \frac{4\pi a^2}{(k_0 a)^2 \cot^2 k_0 a + (ka)^2}$$

that for $k \to 0$ becomes

$$\sigma_0^{\text{tot}}(0) = \frac{4\pi a^2}{(\kappa a)^2} = 4\pi a^2 \left(\frac{\tan k_0 a}{k_0 a}\right)^2 \; .$$

The missing 1 is due to the fact that, since $k_0 a \simeq (2n+1)\pi/2$, $\tan k_0 a$ is very large with respect to 1.

e) $\tan k_0 a = \tan\big[(2n+1)\pi/2 + \epsilon)\big] = -\cot\epsilon \simeq -1/\epsilon \;\Rightarrow\; \kappa = \epsilon\,k_0$, therefore:

$$E_B = -\frac{\hbar^2}{2m}\,(k_0\epsilon)^2 \; .$$

Thanks to the above result ($\kappa \propto \epsilon$) and because of $ka \ll 1$, the cross section can be considerably larger than the 'geometrical' one, i.e. $4\pi a^2$. The same for the scattering length.

17.10

a) There are $n-1$ bound states.

b) From the equation $\tan\big(ka + \delta_0(k)\big) = (k/k_1)\tan k_1 a$, $k_1 = \sqrt{k_0^2 + k^2}$, neglecting ka with respect to $\delta_0(k)$ and, up to terms $O(k^2/k_0^2)$, putting $k_1 a = k_0 a$, one gets:

$$\frac{\tan\delta_0(k)}{k} = \frac{\tan k_0 a}{k_0} = \frac{\cot\epsilon}{k_0} \simeq \frac{1}{k_0\epsilon} ; \qquad \tan\delta_0(k) = \frac{k}{k_0\epsilon} \; .$$

c) $f_0(k) = \dfrac{1}{k_0\epsilon - ik}$, $\sigma_0^{\text{tot}} = \dfrac{4\pi}{(k_0\epsilon)^2 + k^2}$.

According to the result found at the end of Problem 17.9, $(k_0\epsilon)^2 = \kappa^2 = 2m|E_B|/\hbar^2$ where E_B is the n-th bound state (the one close to zero) when the depth of the potential well is such that $k_0 a = (2n+1)\pi/2 + \epsilon$, $\epsilon > 0$. Therefore the scattering amplitude f_0 is obtained from that of Problem 17.9 with $\kappa \to -\kappa$ (or $\epsilon \to -\epsilon$). The cross sections are the same in the two cases ($ka \ll 1$).

Even if with $k_0 a = (2n+1)\pi/2 - \epsilon$ there is no bound state with energy close to zero, one says that $|E_B|$ is a 'virtual level' since it sufficient to increase 'infinitesimally' the depth V_0 of the potential, or equivalently make the infinitesimal transition from $k_0 a = (2n+1)\pi/2 - \epsilon$ to $k_0 a = (2n+1)\pi/2 + \epsilon$, to transform it into the real bound state E_B (see also Problem 17.12).

17.11

a) For $E > 0$, $u_0(r)$ for $r \le a$ is proportional to $\sin kr$. The cross section vanishes when the wave function is the same as the one without the potential: this happens when $u_0(a) = 0$ since in this case there is no discontinuity, as if the δ-potential were absent. Therefore $ka = n\pi$. Clearly, in this case there is no time-delay.

b) The conditions at $r = a$ are

$$
\begin{cases}
\sin ka = A \sin \left(ka + \delta_0(k)\right) \\
k \cos ka - g \sin ka = kA \cos \left(ka + \delta_0(k)\right)
\end{cases}
\Rightarrow
$$

$$
k \cot ka - g = k \cot \left(ka + \delta_0(k)\right) .
$$

c) One way to determine the low energy scattering amplitude is to solve the above equation with respect to $\tan \delta_0(k)$ and then to extract the low energy behaviour:

$$
\frac{ka}{\tan ka} - ga = ka \, \frac{1 - \tan ka \tan \delta_0(k)}{\tan ka + \tan \delta_0(k)} \quad \Rightarrow
$$

$$
\tan \delta_0(k) = \frac{ga \sin^2 ka}{ka - ga \sin ka \cos ka}
$$

then, to order $O(ka)$:

$$
\tan \delta_0(k) \simeq \frac{ga\,ka}{1 - ga}
$$

and the resonance, i.e. $\delta_0 = \pi/2$, occurs for $ga = 1$. Therefore the existence of the (unique) bound state requires $ga \ge 1$, i.e. $\lambda \ge \hbar^2/2ma$ in accordance with the result of Problem 6.23. However, from the above equation one could be induced to conclude that $\delta_0 = \pi/2$ also for any $ka > 0$ (still $ka \ll 1$). To avoid this erroneous conclusion it is sufficient to go one step further in approximating the denominator appearing in the expression of $\tan \delta_0(k)$, then:

$$
\tan \delta_0(k) \simeq \frac{ga\,ka}{1 - ga\left(1 - 2(ka)^2/3\right)}
$$

and now is evident that $\delta_0 = \pi/2$ only for $E = 0$.

17.12

a) $S_l(k) - 1 = e^{2i\,\delta_l(k)} - 1 = 2i\,e^{i\,\delta_l(k)}\sin\delta_l(k)$

$$f_l(k,\theta) \equiv \frac{\sqrt{4\pi\,(2l+1)}}{k}\,e^{i\,\delta_l(k)}\sin\delta_l(k)\,Y_{l,0}(\theta)$$

$$= \frac{\sqrt{4\pi\,(2l+1)}}{2i\,k}\,(S_l(k)-1)\,Y_{l,0}(\theta)\ .$$

b) $\sin(x + iy) = (e^{i(x+iy)} - e^{i(x-iy)})/2i \ \Rightarrow \ \sin(iy) = i\sinh y$ and likewise $\cos(iy) = \cosh y$. Therefore

$$\tan\delta_0(i\,\kappa) = ga\,\frac{1 - \cosh 2\kappa a}{i(2\kappa a - ga\,\sinh 2\kappa a)}\ .$$

c) The position of the pole is given by the equation $1 = i\,\tan\delta_0(i\,\kappa)$, i.e.

$$2\kappa a - ga\,\sinh 2\kappa a = ga - ga\,\cosh 2\kappa a \quad \Rightarrow$$

$$2\kappa a - ga = \frac{ga}{2}\left(e^{2\kappa a} - e^{-2\kappa a} - e^{2\kappa a} - e^{-2\kappa a}\right)$$

$$= -ga\,e^{-2\kappa a}$$

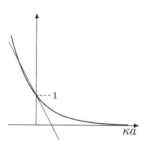

therefore $e^{-2\kappa a} = 1 - 2\kappa a/ga$, in agreement with the result of Problem 6.23, which also implies that the pole on the positive imaginary axis exists only if $ga \geq 1$.

From the figure it is clear that if $0 < ga \leq 1$ the pole is on the negative imaginary axis and if $ga \leq 0$ there is no pole.

d) $1 - 2\kappa a + 2\,(\kappa a)^2 = 1 - 2\kappa a/ga \ \Rightarrow \ \kappa a = (ga - 1)/ga \simeq \epsilon$.
Then $E_B = -\hbar^2\epsilon^2/2ma^2$. If instead $ga = 1-\epsilon$, $\kappa a \simeq -\epsilon$: the pole is on the negative imaginary axis: 'virtual level' also referred to as 'antibound state'.

17.13

a) $\dfrac{1 + i\,(k/k_1)\tan k_1 a}{1 - i\,(k/k_1)\tan k_1 a} = \dfrac{1 + i\tan\left(ka + \delta_0(k)\right)}{1 - i\tan\left(ka + \delta_0(k)\right)} = e^{2i(ka+\delta_0(k))} = e^{2i\,ka}\,S_0(k).$

b) If $k \to i\kappa$, $k_1 \to \sqrt{k_0^2 - \kappa^2}$, $\tan k_1 = \tan\sqrt{k_0^2 + k^2} \to \tan\sqrt{k_0^2 - \kappa^2}$.

The positions of the poles of $S_0(k)$ on the positive imaginary axis are given by the equation

$$1 = i\,\frac{i\kappa}{\sqrt{k_0^2 - \kappa^2}}\,\tan\sqrt{(k_0^2 - \kappa^2)a^2}\ .$$

With the notations of Problem 6.11: $\xi = \sqrt{(k_0^2 - \kappa^2)a^2}$, $\eta = \kappa a$ the above equation reads $\eta = -\xi/\tan\xi$, $\xi^2 + \eta^2 = k_0^2 a^2 = 2mV_0a^2/\hbar^2$, in accordance with the result of Problem 6.11.

c) When one of the poles is close to zero, i.e. $\kappa \ll k_0$, we find again, as in Problem 17.9, $\tan k_0 a = -k_0/\kappa$. In this case from the expression of $S_0(k)$ given in the text and ignoring the factor e^{-2ika} ($ka \ll 1$),

$$S_0(k) = \frac{1 + i(k/k_0)\tan k_0}{1 - i(k/k_0)\tan k_0} = \frac{1 - ik/\kappa}{1 + ik/\kappa} = -\frac{k + i\kappa}{k - i\kappa} \qquad \kappa \ll k_0, \qquad ka \ll 1$$

that is consistent with the expression of the scattering amplitude $f_0(k)$ found in Problem 17.9:

$$f_0(k) = \frac{S_0(k) - 1}{2ik} = \frac{i}{k - i\kappa}.$$

If the pole corresponds to a 'virtual level' (as in Problems 17.10, 17.12), it is sufficient to change κ into $-\kappa$:

$$S_0(k) = -\frac{k - i\kappa}{k + i\kappa} \qquad (\kappa > 0).$$

17.14

a) $S_l(k) = \dfrac{g(k)}{k - k_r + ib}$; $\quad S_l^*(k) = S_l^{-1}(k) \Rightarrow g(k) = h(k) \times (k - k_r - ib)$

with $|h(k)| = 1$, then $h(k) = e^{i\gamma(k)}$.

b) From Problem 17.12 and $e^{i\gamma(k)} = 1$:

$$f_l(k, \theta) = \frac{\sqrt{4\pi(2l + 1)}}{2ik}(S_l(k) - 1)Y_{l,0}(\theta), \qquad S_l(k) - 1 = \frac{-2ib}{k - k_r + ib}$$

therefore

$$f_l(k, \theta) = -\frac{\sqrt{4\pi(2l + 1)}}{k}\frac{b}{k - k_r + ib}Y_{l,0}(\theta)$$

$$\sigma_l^{tot}(k) = \frac{4\pi(2l + 1)}{k^2}\frac{b^2}{(k - k_r)^2 + b^2}.$$

$\sigma_l^{tot}(k_r) = 4\pi(2l + 1)/k_r^2$, therefore the unitarity bound (see Problem 17.8) is reached, i.e. the l-wave is resonant. This is a consequence of $e^{i\gamma(k_r)} = 1$.

c) Since the l-wave is resonant, $\delta_l(k_r) = \pi/2$. Differentiating

$$S_l(k) = e^{2i\delta_l(k)} = \frac{k - k_r - ib}{k - k_r + ib}$$

with respect to k and then putting $k = k_r$, we get $\delta_l' \equiv d\delta_l(k_r)/dk = 1/b$ therefore the time-delay $\Delta t = 2m\,\delta_l'(k_r)/\hbar k_r$ is positive. The phase $\delta_l(k)$ crosses $\pi/2$ from below ($\delta_l'(k_r) > 0$). Actually, from the above expression of $S_l(k)$ it is clear that going from left to right $\delta_l(k)$ undergoes an increase of π (just like the classical forced harmonic oscillator).

d) Since

$$(k - k_r)^2 \simeq \frac{m^2}{\hbar^4 k_r^2} (E - E_r)^2, \quad b = \frac{1}{\delta_l'(k_r)} = \frac{2m}{\hbar\, k_r\, \Delta t}$$

$$\sigma_l^{tot}(E) \simeq \frac{4\pi\, (2l+1)}{k^2} \frac{(2\hbar/\Delta t)^2}{(E - E_r)^2 + (2\hbar/\Delta t)^2}.$$

It is customary to put $\Gamma/2 = 2\hbar/\Delta t$:

$$\sigma_l^{tot}(E) \simeq \frac{4\pi\, (2l+1)}{k^2} \frac{(\Gamma/2)^2}{(E - E_r)^2 + (\Gamma/2)^2}$$

and in this form it is known as the *Breit-Wigner* resonance formula.

Two observations are in order: first, Γ – that has the dimensions of an energy – is called the width of the resonance; second, Γ is inversely proportional to the time-delay: the sharper is the resonance, the longer is the time spent by the particle in the region where the potential is effective.

It should be noted that the saturation of the unitarity bound not always corresponds to a pole of the S-matrix: in these cases the phase $\delta_l(k)$ crosses $\pi/2$ from above ($\delta_l'(k_r) < 0$). While a pole of the S-matrix can be arbitrarily close to the real axis, and therefore the resonance peak can be arbitrarily narrow, when $\delta_l(k_r) = \pi/2$ but $\delta_l'(k_r) < 0$ the peak in the cross section is "large" since it can be shown (E.P. Wigner 1955) that $\delta_l'(k)$ is bounded below by a quantity of the order of $-a$ (a is the range of the potential). Actually, some Authors reserve the name of "resonance" only to those with $\delta_l'(k_r)$ positive and large.

17.15

a) Let $\mathcal{J}_0(k, r) = A\, e^{ikr} + B\, e^{-ikr}$ for $r \le a$. The conditions to be satisfied, analogous to those in Problem 17.11, are:

$$\begin{cases} A\, e^{ika} + B\, e^{-ika} = e^{-ika} \\ A\, e^{ika} - B\, e^{-ika} = -e^{-ika} - i\dfrac{g}{k}\, e^{-ika} \end{cases} \qquad \left(g \equiv \frac{2m\lambda}{\hbar^2} > 0\right) \qquad \Rightarrow$$

$$A = -i\frac{g}{2k}\, e^{-2ika}, \qquad B = 1 + i\frac{g}{2k}$$

$$\mathcal{J}_0(k) = A + B = 1 + i\frac{g}{2k}\left(1 - e^{-2ika}\right).$$

b) $\mathcal{J}_0(-k) = \mathcal{J}_0^*(k)$.

$$\mathcal{J}_0(k) = 1 - \frac{g}{k}\sin ka \cos ka + i\frac{g}{k}\sin^2 ka$$

$$\mathcal{J}_0(k) = |\mathcal{J}_0(k)|\, e^{i\gamma(k)}, \qquad \tan\gamma(k) = \frac{g\sin^2 ka}{k - g\sin ka \cos ka} = \tan\delta_0(k).$$

c) The modulus of $\mathcal{J}_0(k)$ is obviously an even function. The phase is odd since the imaginary part of $\mathcal{J}_0(k)$ is odd while the real part is even. Let us put $G(k) = |\mathcal{J}_0(k)|$, then $\mathcal{J}_0(-k) = G(k)\, e^{-i\delta_0(k)}$. Then

$$\frac{\mathcal{J}_0'(-k)}{\mathcal{J}_0(-k)} = \frac{G'(k)}{G(k)} - i\,\delta_0'(k)\,.$$

The first term $G'(k)/G(k)$, being an odd function, does not contribute to the integral, therefore:

$$\int_{-\infty}^{+\infty} \frac{\mathcal{J}_0'(-k)}{\mathcal{J}_0(-k)}\,dk = -2i\int_0^{+\infty} \delta_0'(k)\,dk = 2i\left(\delta_0(0) - \delta_0(\infty)\right)\,.$$

d) The zeroes of $\mathcal{J}_0(-k)$ are the poles of $S_0(k)$; on the upper half complex plane the only poles are those on the positive imaginary axis corresponding to the bound states (see Problem 17.12). Therefore, if the integral of $\mathcal{J}_0'(-k)/\mathcal{J}_0(k)$ from $-\infty$ to $+\infty$ is closed with the semi–circle at infinity in the upper half complex plane, by the Cauchy theorem we get

$$\int_{-\infty}^{+\infty} \frac{\mathcal{J}_0'(-k)}{\mathcal{J}_0(-k)}\,dk = 2i\,\pi\,n\,.$$

In the present case $n = 1$ or 0 according to whether $g \gtrless 1$.

17.16

a) $\dfrac{2mE\,a^2}{\hbar^2} = E\times\left(\dfrac{a}{a_B}\right)^2 \dfrac{2m a_B^2}{\hbar^2} \simeq 7\times\dfrac{E}{13.6} \simeq 0.36$; $ka = \sqrt{0.36} = 0.6$.

b) For any energy $E > 0$, as in Problem 17.2:

$$r\,\psi_E(x) = \begin{cases} B\sin k_1 r & 0 \le r \le a \\ \sin\left(kr + \delta_0(k)\right) & r \ge a\,. \end{cases}$$

The cross section vanishes when $\delta_0(k) = 0,\ \pi,\ \cdots$, and in this case the continuity conditions read $\left(k = \sqrt{2mE}/\hbar,\quad k_1 = \sqrt{2m(E + V_0)}/\hbar\right)$:

$$\begin{cases} B\sin k_1 a = \pm\sin ka \\ k_1 B\cos k_1 a = \pm k\cos ka \end{cases} \qquad\Rightarrow\qquad k_1 a\,\tan ka = ka\,\tan k_1 a\,.$$

c) The equation to be solved is

$$\frac{k_1 a}{\tan k_1 a} = \frac{ka}{\tan ka} = \frac{0.6}{\tan(0.6)} = 0.88\,.$$

We will content ourselves with the verification that $V_0 \simeq 39\,\mathrm{eV}$ is a solution. Indeed:

$$k_1 a = \sqrt{\frac{2mV_0 a^2}{\hbar^2} + (ka)^2} = \sqrt{\left(\frac{a}{a_B}\right)^2 \frac{V_0}{13.6} + (0.6)^2} = 4.52$$

and $4.52/\tan(4.52) = 0.88$.

17.17

a) $f_V(k,\,\theta) = \dfrac{2mV_0}{\hbar^2 q}\displaystyle\int_0^a \sin(q\,r')\,r'\,dr' = \dfrac{2mV_0}{\hbar^2}\,\dfrac{\sin qa - qa\cos qa}{q^3}$.

b) The volume–integral coincides, up to the prefactor $-2m/4\pi\hbar^2$, with the scattering amplitude at zero momentum transfer $\vec{k}_f = \vec{k}_i$ ($\vec{k}_i \equiv \vec{k}$):

$$f_{\vec{k}}(\theta,\phi) = -\frac{1}{4\pi}\frac{2m}{\hbar^2}\int e^{-i\vec{k}_f\cdot\vec{r}'}V(\vec{r}')e^{i\vec{k}_i\cdot\vec{r}'}\,d\vec{r}'$$

and $\vec{k}_f = \vec{k}_i \Rightarrow q \equiv 2k\sin\theta/2 = 0$. Since

$$\lim_{q\to0}\frac{\sin qa - qa\cos qa}{(qa)^3} = \frac{1}{3}$$

and $r_0 = a$, then $V_0 = 3Y_0$.

c) The integral to be calculated, with $x = 2kr_0\sin(\theta/2)$, is

$$\frac{1}{(2kr_0)^2}\int d\phi \int_0^{2kr_0}\frac{1}{(1+x^2)^2}4x\,dx = \frac{4\pi}{1+4(kr_0)^2}$$

therefore

$$\sigma_Y^{tot}(k) = \frac{4m^2 Y_0^2}{\hbar^4}\frac{4\pi\,r_0^6}{1+4(kr_0)^2} = \frac{\sigma_Y^{tot}(0)}{1+4(kr_0)^2} \simeq \frac{\sigma_Y^{tot}(0)}{4(kr_0)^2}\ .$$

Apart from common factors, the two conditions to be satisfied respectively for the low and high energies agreement are:

$$Y_0 r_0^3 = \frac{1}{3}V_0 a^3 \quad\text{and}\quad Y_0 r_0^2 = \frac{1}{\sqrt{2}}V_0 a^2$$

then

$$V_0 = \frac{2\sqrt{2}}{9}Y_0\ ,\qquad a = \frac{3\sqrt{2}}{2}r_0\ .$$

To sum up, the potential $V(r) = -(2\sqrt{2}/9)Y_0$ for $r \le (3\sqrt{2}/2)\,r_0$ is a good approximation to the Yukawa potential given in the text insofar it gives rise, in the Born approximation, to the same zero energy and high energy cross sections (see figure).

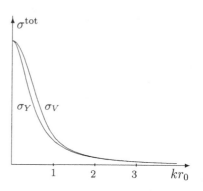

17.18

a) $f_{mol}(\vec{q}) = -\frac{1}{4\pi}\frac{2m}{\hbar^2}\left[\int e^{-i\vec{k}_f\cdot\vec{r}'}\left(V_{at}(\vec{r}') + V_{at}(\vec{r}'-\vec{a})\right)e^{i\vec{k}_i\cdot\vec{r}'}\,d\vec{r}'\right]$

$= f_{at}(\vec{q})\left(1 + e^{-i\vec{q}\cdot\vec{a}}\right)$

$\sigma_{mol}(\vec{q}) = \sigma_{at}(\vec{q})\left|1 + e^{-i\vec{q}\cdot\vec{a}}\right|^2 = 2\,\sigma_{at}(\vec{q})\left(1 + \cos(\vec{q}\cdot\vec{a})\right)\ .$

The term proportional to $\cos(\vec{q}\cdot\vec{a})$ is due to the interference between the waves scattered by the two atoms.

b) $|\vec{q}\cdot\vec{a}| \le qa = 2ka\sin\theta/2 \le 2ka \ll 1 \quad\Rightarrow$

$$f_{\text{mol}}(\vec{q}) = 2 f_{\text{at}}(\vec{q}), \qquad \sigma_{\text{mol}}(\vec{q}) = 4 \sigma_{\text{at}}(\vec{q})$$

therefore when $\lambda \gg a$ the interference is completely constructive, i.e. the two atoms are 'seen' as a single atom and the potential is $2 V_{\text{at}}(\vec{r})$.

c) Each particle reaching the detector is, by assumption, scattered by a single molecule, therefore all the scattering processes are independent and there is no interference among them, therefore the cross section must be averaged, not the amplitude (while instead, as we have seen, there is interference in the scattering by the two atoms of any single molecule).
Then, with the polar axis in the direction of $\vec{q}$,

$$\overline{\cos(\vec{q} \cdot \vec{a})} = \frac{1}{4\pi} \int \cos(qa \cos\theta')\, d\Omega' = \frac{\sin qa}{qa}$$

therefore, since the single atom potential is central,

$$\overline{\sigma_{\text{mol}}(q)} = 2\,\sigma_{\text{at}}(q)\left(1 + \frac{\sin qa}{qa}\right).$$

Therefore the interference due to the two atoms is relevant only when $\sin qa/qa \simeq 1$, i.e. near the forward direction: $\theta \lesssim 1/ka$ or, as we have already seen, when $ka \ll 1$.

17.19

a) If $V_{\text{at}}(\vec{r})$ is the potential of the single atom of the crystal, the scattering amplitude and the cross section for the scattering by the crystal in the Born approximation are $\quad (n \equiv (n_1, n_2, n_3))$:

$$f_{\text{crys}}(\vec{q}) = -\frac{1}{4\pi} \frac{2m}{\hbar^2} \sum_n \int e^{-i\vec{k}_f \cdot \vec{r}'} V_{\text{at}}(\vec{r}' - \vec{r}_n)\, e^{i\vec{k}_i \cdot \vec{r}'}\, d\vec{r}'$$

$$= f_{\text{at}}(\vec{q}) \sum_n e^{-i\vec{q}\cdot\vec{r}_n}, \quad \sigma_{\text{crys}} = \sigma_{\text{at}}(\vec{q})\left|\sum_n e^{-i\vec{q}\cdot\vec{r}_n}\right|^2; \quad \vec{q} = \vec{k}_f - \vec{k}_i.$$

Now:

$$\sum_{n_1} e^{-i n_1 q_1 a} \times \sum_{n_2} e^{-i n_2 q_2 a} \times \sum_{n_3} e^{-i n_3 q_3 a} = \prod_{j=1}^{3} \frac{1 - e^{-i n_j q_j a\, N^{1/3}}}{1 - e^{-i n_j q_j a}}$$

therefore:

$$\sigma_{\text{crys}}(\vec{q}) = \sigma_{\text{at}}(\vec{q}) \prod_{j=1}^{3} \frac{\sin^2(q_j a\, N^{1/3}/2)}{\sin^2(q_j a/2)}.$$

b) With $\vec{k}_f = -\vec{k}_i$, $\vec{q} = \vec{k}_f - \vec{k}_i = (0, 0, -2k_i) \equiv (0, 0, -2k)$

$$\sigma_{\text{crys}} = \sigma_{\text{at}}(\vec{q}) \frac{\sin^2(ka\, N^{1/3})}{\sin^2(ka)}.$$

The factor $\sin^2(ka\, N^{1/3})/\sin^2(ka)$, as shown in the figure, presents pronounced peaks of height $N^{2/3}$ and width $\Delta k = \pi/aN^{1/3}$ when the denominator vanishes, i.e when $ka = n\pi$, i.e. $2a = n\lambda$ in agreement with the Bragg formula $2a \sin \theta = n\lambda$ for $\theta = \pi/2$ (backward scattering). If the particle are sent toward the crystal at an angle θ with respect to the $z = 0$ plane ('glancing angle') and the 'Bragg reflected' scattered particles are observed (see Problem 2.14), $\vec{k}_f - \vec{k}_i = (0, 0, -2k \sin \theta)$ then, as before, $ka \sin \theta = n\pi$ and the Bragg formula follows.

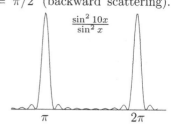

c) With $\vec{k}_i = (0, 0, k)$ and, taken $\vec{k}_f$ in the y-z plane, for small angles $\vec{k}_f = (0, k\theta, k)$, then $\vec{q} = (0, k\theta, 0)$. Therefore the width of the forward diffraction peak is determined by $ka\theta N^{1/3}/2 \lesssim \pi$, then $\theta^{max} \simeq \lambda/(N^{1/3}a)$. Its height is proportional to N^2.

d) The scattering cross section contains the factor

$$\left| \sum_n e^{-i\vec{q}\cdot\vec{r}_n} \right|^2 = \sum_n \left| e^{-i\vec{q}\cdot\vec{r}_n} \right|^2 + \sum_{n \neq m} e^{-i\vec{q}\cdot(\vec{r}_n - \vec{r}_m)} .$$

The first term on the r.h.s. is N while the second term, due to the hypothesis of randomness of the phases $\vec{q}\cdot(\vec{r}_n - \vec{r}_m)$ and the large number of them, is taken to be zero. Therefore, in the considered case, $\sigma_{crys}(\vec{q}) = N \sigma_{at}(\vec{q})$.

17.20

a) Since the potential is central the scattering amplitude only depends on the angle θ between the direction of the two beams after and before scattering. Given that the two neutrons are in a spin–symmetric state, the n–n scattering amplitude is ($f_E(\theta) \equiv f_E^{pn}(\theta)$):

$$f_E^{\uparrow\uparrow}(\theta) = f_E(\theta) - f_E(\pi - \theta)$$

and the cross section

$$\sigma_E^{\uparrow\uparrow}(\theta) = \left| f_E(\theta) \right|^2 + \left| f_E(\pi - \theta) \right|^2 - 2 \Re\left[f_E^*(\theta) f_E(\pi - \theta) \right] .$$

The cross section is symmetrical about $\theta = \pi/2$ (in the center-of-mass frame).

b) The problem can be tackled in two different ways: first, the two neutrons are distinguishable since they are (and will be) in orthogonal spin states, but the detector does not detect the spin so the measured cross section is

$$\sigma_E^{\uparrow\downarrow}(\theta) = \left| f_E(\theta) \right|^2 + \left| f_E(\pi - \theta) \right|^2 .$$

Indeed, the detector at the angles θ, ϕ would respond both when the neutron with spin up is scattered in the direction θ, ϕ, and also when that same neutron is scattered in the opposite direction $\pi - \theta$, $\phi + \pi$, since in that case it is the neutron with spin down that hits the detector. Alternatively, the (antisymmetric) scattered wavefunction for the relative motion is:

$$\chi(\vec{r}_1 - \vec{r}_2) \,|+-\rangle - \chi(\vec{r}_2 - \vec{r}_1) \,|-+\rangle \propto f_E(\theta) \,|+-\rangle - f_E(\pi-\theta) \,|-+\rangle \,.$$

The two amplitudes $f_E(\theta)$ and $f_E(\pi - \theta)$ cannot interfere because are associated with orthogonal spin states.

c) On the average, every four collisions two are in the spin states $|++\rangle$, $|--\rangle$ and two in the spin states $|+-\rangle$, $|-+\rangle$. The interference term $-2\Re[f_E^*(\theta)f_E(\pi - \theta)]$ is present in the first case (parallel spins), not in the second, therefore

$$\sigma_E^{\text{unpol}}(\theta) = |f_E(\theta)|^2 + |f_E(\pi - \theta)|^2 - \frac{2}{4} \times 2\,\Re[f_E^*(\theta)f_E(\pi - \theta)] \,.$$

Equivalently, every four collisions three are in the triplet state and one in the singlet state and the interference term must be taken 3 times with the $-$ sign and only once with the $+$ sign: therefore, obviously, the result is the same as before.

d) The problem is identical with that of the collision between two neutrons in orthogonal spin states: the two particles are distinguishable, but identical for the detectors. Then:

$$\sigma(\theta) = |f_E(\theta)|^2 + |f_E(\pi - \theta)|^2 \,.$$

17.21

a) When a neutron is scattered at the angle θ, the proton is revealed by the counter at the angle $\pi - \theta$, i.e. in the opposite direction, therefore $\sigma_E^n(\theta) = \sigma_E^p(\pi - \theta)$.

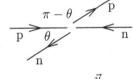

b) In order to pass from the center-of-mass to the laboratory frame the velocity $\vec{v}$ of the neutrons in the center-of-mass frame must be added to all velocities: the results for the velocities of the scattered particles in the laboratory system are

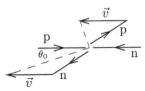

drawn in the figure as dashed lines. From simple geometry it follows that $\theta_0 = \theta/2$ and the angle of the recoil proton with respect with the direction of the incoming neutrons is $(\pi - \theta)/2$. Obviously $\phi_0 = \phi$. Then both the neutrons and the protons are scattered in the forward direction in the laboratory frame, i.e. at angles less then $\pi/2$. From

$$\sigma_0(\theta_0,\,\phi_0)\sin\theta_0\,d\theta_0\,d\phi_0 = \sigma(\theta,\,\phi)\sin\theta\,d\theta\,d\phi \qquad \Rightarrow$$

$$\sigma_0(\theta_0,\,\phi_0) = 4\cos\theta_0\,\sigma(2\theta_0,\,\phi_0) \,.$$

The total cross section is the same in both frames.

Appendix A

Physical Constants

Electronvolt	eV	$1.6 \times 10^{-12}\,\mathrm{erg}$
Speed of light	c	$3 \times 10^{10}\,\mathrm{cm/s}$
Elementary charge	e	$4.8 \times 10^{-10}\,\mathrm{esu} = 1.6 \times 10^{-19}\,\mathrm{C}$
Electron mass	m_e	$0.91 \times 10^{-27}\,\mathrm{g} = 0.51\,\mathrm{MeV}/c^2$
Hydrogen mass	m_H	$1.7 \times 10^{-24}\,\mathrm{g} = 939\,\mathrm{MeV}/c^2$
Planck constant	h	$6.6 \times 10^{-27}\,\mathrm{erg\ s} = 4.1 \times 10^{-15}\,\mathrm{eV\ s}$
Reduced Planck constant	$\hbar = \dfrac{h}{2\pi}$	$1.05 \times 10^{-27}\,\mathrm{erg\ s} = 0.66 \times 10^{-15}\,\mathrm{eV\ s}$
Boltzmann constant	k_B	$1.38 \times 10^{-16}\,\mathrm{erg/K} \simeq \dfrac{1}{12000}\,\mathrm{eV/K}$
Avogadro constant	N_A	$6.03 \times 10^{23}\,\mathrm{mol}^{-1}$
Fine structure constant	$\alpha = \dfrac{e^2}{\hbar c}$	$7.3 \times 10^{-3} \simeq \dfrac{1}{137}$
Bohr radius	$a_\mathrm{B} = \dfrac{\hbar^2}{m_\mathrm{e} e^2}$	$0.53\,\text{Å} = 0.53 \times 10^{-8}\,\mathrm{cm}$
Bohr magneton	$\mu_\mathrm{B} = \dfrac{e\hbar}{2m_\mathrm{e} c}$	$0.93 \times 10^{-20}\,\mathrm{erg/G} = 5.8 \times 10^{-9}\,\mathrm{eV/G}$
Rydberg constant	$R_\infty = \dfrac{e^2}{2a_\mathrm{B} hc}$	$109737\,\mathrm{cm}^{-1}$
Compton wavelength	$\lambda_\mathrm{c} = \dfrac{h}{m_\mathrm{e} c}$	$0.024\,\text{Å}$
Classical electron radius	$r_\mathrm{e} = \dfrac{e^2}{m_\mathrm{e} c^2}$	$2.8 \times 10^{-13}\,\mathrm{cm}$
Atomic unit of energy	$\dfrac{e^2}{a_\mathrm{B}} = \alpha^2 m_\mathrm{e} c^2$	$27.2\,\mathrm{eV}$
A useful mnemonic rule	$h\,c$	$12400\,\mathrm{eV\ Å}$
A useful conversion rule		$E[\mathrm{eV}] = hc \times E[\mathrm{cm}^{-1}] = 1.24 \times 10^{-4}\,E[\mathrm{cm}^{-1}]$
		$E[\mathrm{cm}^{-1}] = 0.8 \times 10^4\,E[\mathrm{eV}]$

© Springer International Publishing AG 2017
E. d'Emilio and L.E. Picasso, *Problems in Quantum Mechanics*,
UNITEXT for Physics, DOI 10.1007/978-3-319-53267-7

Appendix B

Useful Formulae

Normalized Gaussian wavefunctions:

$$|A\rangle \xrightarrow{\text{SR}} \psi_A(x) = \left(\pi a^2\right)^{-1/4} e^{-x^2/2a^2};$$

$$|A\rangle \xrightarrow{\text{MR}} \varphi_A(p) = \left(\pi \hbar^2/a^2\right)^{-1/4} e^{-p^2 a^2/2\hbar^2}$$

$$\overline{x^2} = \tfrac{1}{2}a^2, \qquad \overline{x^4} = \tfrac{3}{4}a^4; \qquad\qquad \overline{p^2} = \hbar^2/2a^2, \qquad \overline{p^4} = 3\hbar^2/4a^4.$$

Normalized eigenfunctions of the harmonic oscillator:

$$\psi_n(x) = \frac{1}{\sqrt{2^n\, n!}} \left(\frac{m\,\omega}{\pi\,\hbar}\right)^{1/4} H_n(\sqrt{m\,\omega/\hbar}\,x)\, e^{-(m\,\omega/2\,\hbar)\,x^2}$$

$$H_0(\xi) = 1, \quad H_1(\xi) = 2\xi, \quad H_2(\xi) = 4\xi^2 - 2.$$

Spherical harmonics: $\qquad \displaystyle\int \big|Y_{l,m}(\theta,\,\phi)\big|^2 \, d\Omega = 1, \qquad d\Omega = \operatorname{sen}\theta \, d\theta \, d\phi$

$$Y_{0,0}(\theta,\phi) \quad = \quad \sqrt{\frac{1}{4\pi}}$$

$$Y_{1,\pm1}(\theta,\phi) \quad = \quad \sqrt{\frac{3}{8\pi}}\,\operatorname{sen}\theta\, e^{\pm i\phi} \quad = \quad \sqrt{\frac{3}{8\pi}}\,\frac{x\pm i\,y}{r}$$

$$Y_{1,0}(\theta,\phi) \quad = \quad \sqrt{\frac{3}{4\pi}}\,\cos\theta \quad = \quad \sqrt{\frac{3}{4\pi}}\,\frac{z}{r}$$

$$Y_{2,\pm2}(\theta,\phi) \quad = \quad \sqrt{\frac{15}{32\pi}}\,\operatorname{sen}^2\theta\, e^{\pm2i\phi} \quad = \quad \sqrt{\frac{15}{32\pi}}\,\frac{(x\pm i\,y)^2}{r^2}$$

$$Y_{2,\pm1}(\theta,\phi) \quad = \quad \sqrt{\frac{15}{8\pi}}\,\operatorname{sen}\theta \cos\theta\, e^{\pm i\phi} \quad = \quad \sqrt{\frac{15}{8\pi}}\,\frac{z\,(x\pm i\,y)}{r^2}$$

$$Y_{2,0}(\theta,\phi) \quad = \quad \sqrt{\frac{5}{16\pi}}\,(1 - 3\cos^2\theta) \quad = \quad \sqrt{\frac{5}{16\pi}}\,\frac{x^2 + y^2 - 2z^2}{r^2}.$$

© Springer International Publishing AG 2017
E. d'Emilio and L.E. Picasso, *Problems in Quantum Mechanics*,
UNITEXT for Physics, DOI 10.1007/978-3-319-53267-7

Energy levels of hydrogen-like ions: (infinite nuclear mass)

$$E_n = -Z^2 \frac{m_e\,e^4}{2\hbar^2 n^2} = -Z^2 \frac{e^2}{2n^2 a_B} = -Z^2 \frac{\alpha^2 m_e\,c^2}{2n^2} = -Z^2 \frac{R_\infty\,h\,c}{n^2} = -Z^2 \frac{13.6}{n^2}\ \text{eV}\,.$$

Radial functions for hydrogen-like ions:

$$\int_0^\infty \left| R_{n,l}(\rho)\right|^2 \rho^2 \mathrm{d}\rho = 1\,, \qquad \rho = Z\,r/a_B$$

$$R_{1,0}(\rho) = 2\,e^{-\rho}$$

$$R_{2,0}(\rho) = \frac{1}{\sqrt{2}}\left(1 - \frac{1}{2}\rho\right) e^{-\rho/2}$$

$$R_{2,1}(\rho) = \frac{1}{2\sqrt{6}}\,\rho\,e^{-\rho/2}$$

$$R_{3,0}(\rho) = \frac{2}{3\sqrt{3}}\left(1 - \frac{2}{3}\rho + \frac{2}{27}\rho^2\right) e^{-\rho/3}$$

$$R_{3,1}(\rho) = \frac{8}{27\sqrt{6}}\,\rho\left(1 - \frac{1}{6}\rho\right) e^{-\rho/3}$$

$$R_{3,2}(\rho) = \frac{4}{81\sqrt{30}}\,\rho^2 e^{-\rho/3}$$

Note: $\displaystyle\int_0^\infty \left|\left(\frac{Z}{a_B}\right)^{3/2} R_{n,l}(Zr/a_B)\right|^2 r^2\mathrm{d}r = 1\,.$

Pauli matrices:

$$\sigma_1 = \begin{pmatrix} 0 & 1 \\ 1 & 0 \end{pmatrix}, \qquad \sigma_2 = \begin{pmatrix} 0 & -i \\ i & 0 \end{pmatrix}, \qquad \sigma_3 = \begin{pmatrix} 1 & 0 \\ 0 & -1 \end{pmatrix}.$$

Index

© Springer International Publishing AG 2017
E. d'Emilio and L.E. Picasso, *Problems in Quantum Mechanics*,
UNITEXT for Physics, DOI 10.1007/978-3-319-53267-7

Printed in the United States
By Bookmasters